Toxicology of Herbal Products

Olavi Pelkonen • Pierre Duez
Pia Maarit Vuorela • Heikki Vuorela
Editors

Toxicology of Herbal Products

Springer

Editors
Olavi Pelkonen
Department of Pharmacology & Toxicology
University of Oulu
Oulu, Finland

Pia Maarit Vuorela
Faculty of Pharmacy
University of Helsinki
Helsinki, Finland

Pierre Duez
Service de Chimie Thérapeutique et de Pharmacognosie
University of Mons
Mons, Belgium

Heikki Vuorela
Faculty of Pharmacy
University of Helsinki
Helsinki, Finland

ISBN 978-3-319-43804-7 ISBN 978-3-319-43806-1 (eBook)
DOI 10.1007/978-3-319-43806-1

Library of Congress Control Number: 2017933879

Printed on acid-free paper

This Springer imprint is published by Springer Nature
The registered company is Springer International Publishing AG
The registered company address is: Gewerbestrasse 11, 6330 Cham, Switzerland

Preface

The origins of this book were based on my experiences as a co-opted expert in toxicology on the Committee of Herbal Medicinal Products (HMPC) of the European Medicines Agency (EMA) in London. Even though herbal medicines have been reviewed through the EU Medicines Directive since 2001, their evaluation and acceptance are based on a much more "relaxed" set of guidelines than that required for conventional medicines (Fig. 1, by Pierre Duez). Admittedly, the database for herbal medicines' efficacy and toxicity was, and still is, far inferior to that for conventional medicines; furthermore, traditional wisdom regards these products as harmless because they are "natural." Therefore, the EU did not deem it necessary to require a dossier similar to that for conventional drugs.

Despite indications from long-standing use, it should be remembered, however, that the toxicities of many herbs have been discovered only by serendipity or unfortunate clinical findings, and mid-term and long-term toxic effects certainly appear as the most important factors in the field of predictive toxicology. Given the widespread use of herbal medicines, the study of herb-drug interactions is an important part of the evaluation of their safety.

From the start, publishing a book on the toxicity of herbal products was quite a laborious undertaking. First, there were no clear examples available in the field of herbal medicinal products. Second, and perhaps more importantly, herbal products are complex, plant-derived mixtures, and most scientific techniques, methods, and approaches are designed for single chemical substances or, at the most, mixtures of just a few. In many aspects, this field is more akin to environmental toxicology than to conventional drug toxicology. Third, at the time of writing, international standards for herbal products were either completely lacking or under development, and it was difficult to see whether such harmonization was at all possible or desirable. Finally, there is a huge variety of traditions and practices regarding the local, regional, and national uses of herbal products, and it is simply impossible to embark on the writing of any comprehensive treatise.

As a result, this book does not attempt to be an exhaustive presentation of the toxicities of herbal products. At the outset it was clear that methodological aspects should claim a sizable share of the contents; for this reason, chapters on individual

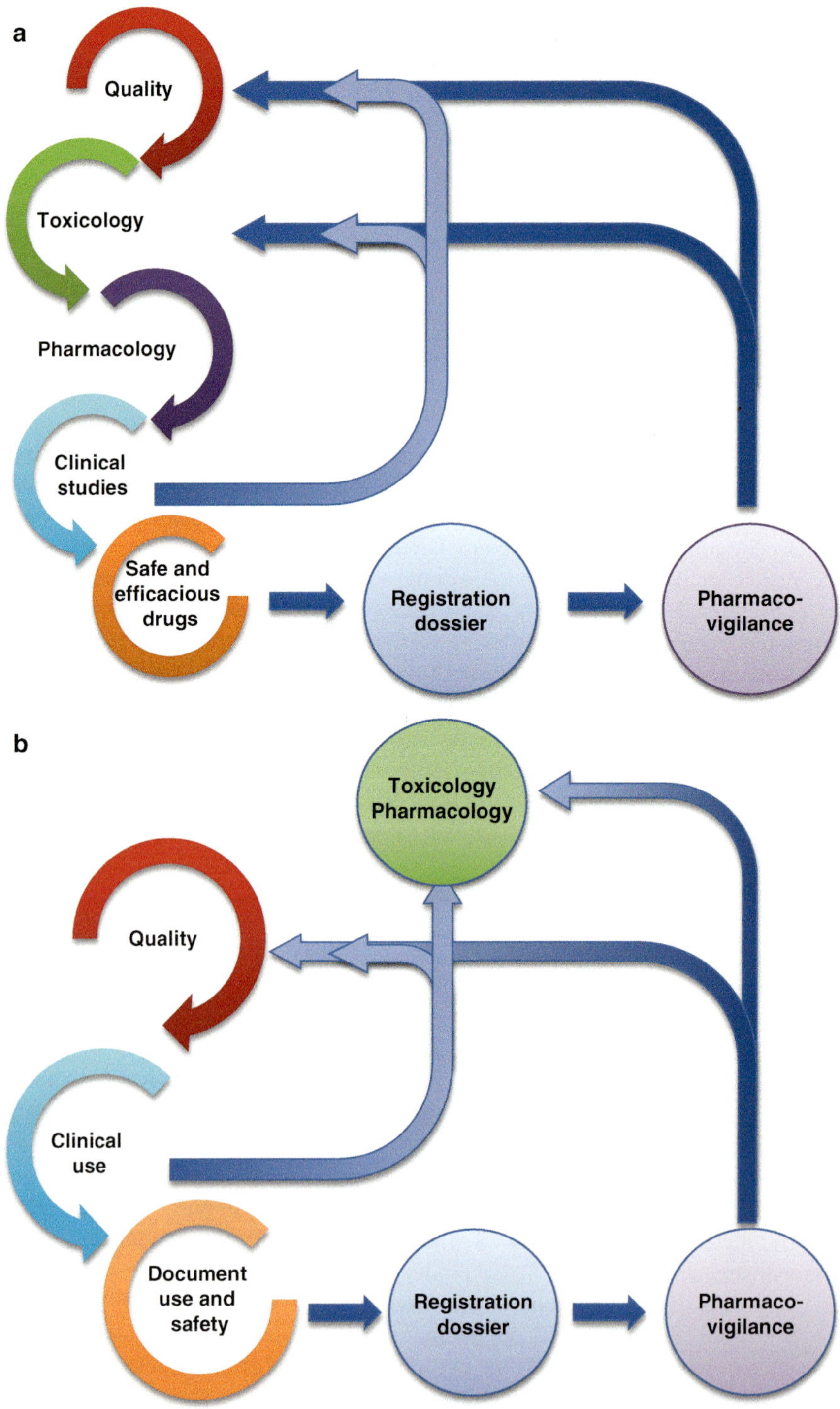

Fig. 1 Pharmacology and toxicology and their relationships with other aspects of herbal studies. Classical (**a**) and Traditional (**b**) drug registration procedures

toxicities serve simply as examples, and the particular products were chosen for examination based on the availability of significant numbers of quality studies and on the significance of adverse outcomes. It was also important to present approaches that were more suitable for the study of complex mixtures – i.e., “'omics” methods and systems toxicology. The final chapter provides a look to the future, and as such, proposes avenues of inquiry to pursue – which may prove to be not fully realizable.

This book is intended to serve as a starting point or template for others who will produce subsequent treatises. I am waiting – hoping – for this to happen before too long.

I would like to thank my fellow co-editors for their ideas, help, and laborious reading in the midst of their arduous, stressful academic lives (I myself am a free-wheeling retiree!), as well as all the writers for their excellent contributions. I am also thankful to all the scientists and regulators at the HMPC, some of whom even contributed to the book. I learned much from all of them and, as a novice in the field, as I certainly was in the beginning, I greatly appreciate their friendship, help, and views on various aspects of herbal medicinal products. Now I must confess, after leaving HMPC, that I miss them dearly.

Oulu, Finland Olavi Pelkonen
Brussels Airport Lounge, June 23, 2016

About the Editors

Olavi Pelkonen

Professor of Pharmacology (Emeritus) and Former Head of the Department
Department of Pharmacology and Toxicology, University of Oulu, Oulu, Finland.

Olavi Pelkonen received M.D. and Ph.D. degrees from the University of Oulu, Finland. From 1978 until 2010 he was Associate and Full Professor of Pharmacology as well as Head of the Department in Oulu, Finland. He has published more than 300 original and review articles in international scientific journals, mainly on various aspects of drug and carcinogen metabolism, especially by cytochrome P450 enzymes, and their regulation by genetic and environmental factors. For the last 10 years, he was a co-opted toxicology expert on the Committee of Herbal Medicinal Products of the European Medicines Agency.

Pierre Duez

Professor of Pharmacognosy, Pharmacology, and Nutrition, and Head of the Unit
Unit of Therapeutic Chemistry and Pharmacognosy, University of Mons (UMONS), Mons, Belgium

Pierre Duez is a Pharmacist with a Ph.D. from the Université Libre de Bruxelles (ULB), Belgium; his thesis was on DNA photo-oxidative damage. From 2002 until 2013 he was Professor of Pharmacognosy and Bromatology at ULB. Prof. Duez has published 140 original and review articles in international scientific journals, mainly on various aspects of DNA damage and repair, analytical chemistry, and pharmacognosy. His major research interests concern the modulation of DNA repair mechanisms by natural products, with the development of analytical methods for probing various mechanisms of DNA repair. For the last eight years, he has been a member of the "Traditional Chinese Medicines" working group of the European Directorate for the Quality of Medicines and Healthcare (European Pharmacopoeia) of the Council of Europe.

Pia M. Vuorela

Professor of Pharmaceutical Biology and Vice-Dean of the Faculty of Pharmacy
University of Helsinki, Helsinki, Finland

Pia M. Vuorela has a Ph.D. (Pharm.) from the University of Helsinki, Finland. From 2005 until 2013, she was Full Professor of Pharmaceutical Chemistry and Acting Head of Pharmacy at Åbo Akademi University in Turku, Finland; since 2013 she has been Full Professor of Pharmaceutical Biology at the University of Helsinki, as well as the Vice-Dean. She has also been Adjunct Professor in Pharmacognosy in Helsinki since 1994 , and in Pharmaceutical Chemistry in Turku since 2003. Prof. Vuurela has published almost 200 original and review articles in international scientific journals, mainly on bioassays, pharmaceutically relevant microbial targets (biofilms, intracellular pathogens), and, more recently, on the development of novel anti-infectives for pharmaceutically relevant formulations (medical devices).

Heikki Vuorela

Professor of Pharmacognosy, University of Helsinki, Former Vice-Dean for Research at the Faculty of Pharmacy, University of Helsinki.

Previously, Prof. Vuorela worked as a researcher at Novartis AG (Sandoz AG), Basel, and ETH-Zurich, Switzerland. Prof. Vuorela has authored over 130 peer-reviewed scientific articles and three patent families in pharmacy, mostly in the field of drug discovery and chromatography. He has been principal supervisor of 16 doctoral dissertations and has been the official opponent of 10 Ph.D. and Ph.D. (Pharm.) exams in Finland and Switzerland. For the last few years, Prof. Vuorela has focused his scientific interest on natural product drug discovery.

Contents

Contributors

Jorma Ahokas School of Medical Sciences, RMIT-University, Bundoora, Australia

Nachtergael Amandine Unit of Therapeutic Chemistry and Pharmacognosy, Faculty of Medicine and Pharmacy, Research Institute for Health Sciences and Technology, University of Mons – UMONS, Mons, Belgium

Thomas Baudoux, MD Nephrology Department, Erasme Hospital and Laboratory of Experimental Nephrology, Faculty of Medicine, Université Libre de Bruxelles, Brussels, Belgium

Alessandro Buriani Center Maria Paola Belloni Regazzo for Pharmacogenomics and Personalized Medicine, Gruppo Data Medica, Padova, Italy

Department of Pharmacological and Pharmaceutical Sciences, University of Padova, Padova, Italy

Maria Carrara Department of Pharmacological and Pharmaceutical Sciences, University of Padova, Padova, Italy

Thomas Efferth Department of Pharmaceutical Biology, Institute of Pharmacy and Biochemistry, Johannes Gutenberg University, Mainz, Germany

Hani El-Nezami School of Biological Sciences, Faculty of Science, Kadoorie Biological Sciences Building, The University of Hong Kong, Pokfulam, Hong Kong

Stefano Fortinguerra Center Maria Paola Belloni Regazzo for Pharmacogenomics and Personalized Medicine, Gruppo Data Medica, Padova, Italy

Department of Pharmacological and Pharmaceutical Sciences, University of Padova, Padova, Italy

Kati Hanhineva Institute of Public Health and Clinical Nutrition, University of Eastern Finland, Kuopio, Finland

Werner Knöss Federal Institute for Drugs and Medical Devices (BfArM), Bonn, Germany

Gert Laekeman Clinical Pharmacology and Pharmacotherapy, Faculty of Pharmaceutical Sciences, KU Leuven, Leuven, Belgium

Poivre Mélanie Unit of Therapeutic Chemistry and Pharmacognosy, Faculty of Medicine and Pharmacy, Research Institute for Health Sciences and Technology, University of Mons – UMONS, Mons, Belgium

Okusa Philippe Ndjolo Unit of Therapeutic Chemistry and Pharmacognosy, Faculty of Medicine and Pharmacy, Research Institute for Health Sciences and Technology, University of Mons – UMONS, Mons, Belgium

Joëlle L. Nortier, MD, PhD Nephrology Department, Erasme Hospital and Laboratory of Experimental Nephrology, Faculty of Medicine, Université Libre de Bruxelles, Brussels, Belgium

Markku Pasanen School of Pharmacy, University of Eastern Finland, Kuopio, Finland

Olavi Pelkonen Department of Pharmacology and Toxicology, Institute of Biomedicine, Aapistie 5 B FIN-90014, University of Oulu, Oulu, Finland

Duez Pierre Unit of Therapeutic Chemistry and Pharmacognosy, Faculty of Medicine and Pharmacy, Research Institute for Health Sciences and Technology, University of Mons – UMONS, Mons, Belgium

Jandirk Sendker University of Münster, Institute of Pharmaceutical Biology and Phytochemistry, Münster, Germany

Ean-Jeong Seo Department of Pharmaceutical Biology, Institute of Pharmacy and Biochemistry, Johannes Gutenberg University, Mainz, Germany

Danielle Sevior Department of Pharmacology and Toxicology, University of Oulu, Oulu, Finland

School of Medical Sciences, RMIT-University, Bundoora, VIC, Australia

Helen Sheridan Trinity College Dublin, School of Pharmacy and Pharmaceutical Sciences, Dublin 2, Ireland

Bunel Valérian Unit of Therapeutic Chemistry and Pharmacognosy, Faculty of Medicine and Pharmacy, Research Institute for Health Sciences and Technology, University of Mons – UMONS, Mons, Belgium

Laboratory of Pharmacognosy, Bromatology and Human Nutrition, Faculty of Pharmacy, Université Libre de Bruxelles – ULB, Brussels, Belgium

Pieter van der Bijl Department of Internal Medicine, Faculty of Medicine and Health Sciences, Stellenbosch University, Cape Town, South Africa

Jacqueline Wiesner Federal Institute for Drugs and Medical Devices (BfArM), Bonn, Germany

Elizabeth M. Williamson School of Pharmacy, University of Reading, Reading, UK

Chit Shing Jackson Woo School of Biological Sciences, Faculty of Science, Kadoorie Biological Sciences Building, The University of Hong Kong, Pokfulam, Hong Kong

Chapter 1
Toxicity of Herbal Medicines: From Past to Present to Future

Werner Knöss

Abstract Herbal medicines were among the first remedies used by human beings for healing. Without any doubt, this use was also accompanied by the poisonous or toxic effects of plants. In order to improve chances of survival, in parallel, mankind increased its knowledge of the therapeutic uses and toxicity of herbal medicines. Individual experiences were extended to the systematic collection of all data related to the effects of plants and herbal medicines. Progress and development were initiated by spreading herbal medicines along the trade routes and the scientific demand for insight into toxic principles and their modes of action. Moreover, methodologies applied to the life sciences provided more and more efficient tools for researching the toxicology of herbal medicines. Toxicology has developed into a multidisciplinary field with numerous sub-disciplines. Nevertheless, because herbal medicines are multi-component mixtures, it is still a major challenge to investigate their toxicity. Usually, available data never meet all requirements for perfect assessment; therefore, the evaluation of herbal medicines' toxicity in regulatory frameworks is a matrix-based assessment that should ideally conclude with adequate recommendations for risk management.

Keywords Herbal medicines • Toxicity • Toxicology • Medicinal plants • Assessment

---

Disclaimer The views presented in this chapter represent the personal views of the author and should not be cited or understood as the position of the Federal Institute for Drugs and Medical Devices (BfArM) or the European Medicines Agency (EMA).

W. Knöss
Federal Institute for Drugs and Medical Devices (BfArM),
Kurt-Georg-Kiesinger-Allee 3, 53175 Bonn, Germany
e-mail: werner.knoess@bfarm.de

O. Pelkonen et al. (eds.), *Toxicology of Herbal Products*,
DOI 10.1007/978-3-319-43806-1_1

Introduction

Herbal medicines currently contribute to health care in most parts of the world. Depending on the degree of development of a particular country, herbal medicines may be an important part of primary health care where the availability of therapeutic alternatives is still limited. In highly developed countries, herbal medicines are often accepted as an option that complements standard therapies (Heinrich 2001). Generally, there is a globally growing interest in the integration of traditional medicines that often includes herbal medicines as one of their therapeutic options besides, for example, nutrition, physical treatment, or therapeutic movement. Whenever medicinal plants are applied, there is a basic expectation of quality, efficacy, and safety. The toxicity of herbal medicines has an intrinsic potential because of their multi-component nature. Such mixtures of multiple natural constituents can differently synergize, antagonize, and/or interact with the human body and metabolism by numerous mechanisms; some of them have therapeutic value, while others may be closer to what is regarded as a "toxic effect." Over the centuries, the toxicity of herbal medicines evolved from a recognition-based experience to a well-investigated field of research. Moreover, potential toxicity has become a major topic in drug development and herbal medicine regulation.

Herbal Medicines and Human Beings

Throughout human history, people have used plants in their daily lives. Plants offer a diverse spectrum of specific and fascinating characteristics and have accordingly been used for nutrition, as spices, for clothing, for construction, for coloring and decoration, or as energy resources (Lewington 2003). When observing animal behavior, using plants for nutrition, or perhaps accidentally (by trial and error), people discovered medicinal properties in plants. Such properties have covered a broad range, from healing to poisoning – and even life- threatening. Primarily, the medicinal use of plants had developed on a regional scale and was dependent on regionally available plants. Based on the social and cultural backgrounds of communities, specific plants were selected and even cultivated, when possible. The application of plants was steadily improved and the knowledge of healing and poisoning properties was conserved. In some regions, such as Asia and Europe, the transfer of information had been based on written documentation since ancient times. In other regions, such as Africa or South America, information was spread to succeeding generations by word of mouth (Heinrich 2001).

A first trend towards the globalization of herbal medicines coincided with the development of major trade routes during the Middle Ages, especially through the distant Arabian empire (Abbassides Dynasty, 750–1258). Together with commercial products, herbal medicines and spices were exchanged between continents. Well-known examples of herbal medicines that have been used in European phytotherapy for hundreds of years include fruits of the *Syzygium aromaticum* (L.) Merr. & L.M. Perry from Madagascar, the roots of the *Ipecacuanha* species from South America, or the barks of the *Cinnamomum* species from Asia. From a global perspective, the use of herbal medicines was not restricted to a particular group of persons; in some regions, applications were in the hands of doctors, and in others, they were left to individual persons or specifically trained healers. Especially during the Middle Ages, groups of people conserved and spread information on herbal medicines all over the world.

Definitions

"The Toxicity of Herbal Medicines" is a simple title, and at first glance it seems pretty appropriate to create a distinct topic for the reader. Surprisingly, it turned out to have been quite a challenge to reduce the title to common definitions of the included topics. "Herbal medicines" is a common term for lay people as well as for scientists. However, there is great variability in regulatory frameworks on herbal medicines, and existing definitions are diverse. The World Health Organization (WHO)'s definition of a herbal medicine (WHO 2015) is "Herbal medicines include herbs, herbal materials, herbal preparations and finished herbal products, that contain as active ingredients parts of plants, or other plant materials, or combinations."

In the European Union (European Parliament 2001) there are legal definitions for herbal medicinal products, herbal substances (a synonym for herbal drugs), and herbal preparations, which are the principal basis of the regulation of these products.

Herbal medicinal product – Any medicinal product, exclusively containing active ingredients of one or more herbal substances or one or more herbal preparations, or one or more such herbal substances in combination with one or more such herbal preparations.

Herbal substances – All mainly whole, fragmented, or cut plants, plant parts, algae, fungi, lichen in an unprocessed, usually dried, form, but sometimes fresh. Certain exudates that have not been subjected to a specific treatment are also considered to be herbal substances. Herbal substances are precisely defined by the plant part used and the botanical name according to the binomial system (genus, species, variety, and author).

Herbal preparations – Preparations obtained by subjecting herbal substances to treatments such as extraction, distillation, expression, fractionation, purification, concentration, or fermentation. These include comminuted or powdered herbal substances, tinctures, extracts, essential oils, expressed juices, and processed exudates.

In other parts of the world, similar products are called "natural health products" (Canada), or are distributed in a different context, e.g., as food supplements.

According to Webster's *New Encyclopedic Dictionary*, "toxic" means "poisonous" [late Latin *toxicus*, from the Latin *toxicum* ("poison"), and from the Greek *toxikon* ("arrow poison"), derived from *toxon* "bow, arrow"], whereas "toxicology" is defined as "a science that deals with poisonous materials and their effect and with the problems involved in their use and control." Toxicology is a multidisciplinary field of science that may focus on various areas such as pharmacy, medicine, or the environment. A key reference in toxicology is a statement by Theophrastus Bombastus von Hohenheim, who was better known as Paracelsus (1493–1541). He was the first to claim that toxicity is dependent on posology and concentration, thus creating a fine line between therapeutic activity and poisoning: "All substances are poisons; there is none which is not a poison. The right dose differentiates a poison and a remedy." (Doull and Bruce 1986). A good illustrative example of this adage is foxglove (*Digitalis purpurea* L.); its leaves contain cardiac glycosides that can have a therapeutic benefit at a very low dosage, but are life-threatening at slightly higher dosages because they trigger arrhythmias. Although modern toxicology is gathering all available data, evaluating mechanisms, investigating molecular properties of compounds, assessing risks, and finally drawing conclusions and giving recommendations to risk management (Eaton and Gilbert 2013), it is still relatively difficult to properly evaluate complex mixtures.

Nowadays, with special reference to medicinal products, toxicology is included at different stages:

- Drug development. Screening programs include assays in order to obtain insight into potentially toxic activities of compounds. Positive results may lead to exclusion from further developments or, from the opposite point of view, could point to a therapeutic useful activity on a distinct target (e.g., anti-cancer activity).
- Drug evaluation. A defined set of preclinical data is required for the authorization of medicinal products. Case by case, the need for data or for options to waive data is evaluated. Overall, the available data contribute to the safety evaluation of medicinal products.
- Environmental toxicity. In developed regulatory frameworks, medicinal products are also evaluated for their potential risk to environments. Mostly, this is of limited relevance for herbal medicines. The active constituents are usually derived from nature and are highly biodegradable, but only in rare cases excipients may pose a potential risk to the environment.
- Pharmacovigilance and clinical toxicology. If signs of poisoning are found, it is necessary to consider the option of poisoning by herbal medicines.

Time Periods in the Toxicity of Herbal Medicines

Early experiences in ancient times. The first observations on the toxicity of herbal medicines were based on the association of severe effects – for example, on heart or lungs, with herb consumption; such observations are eased whenever remarkable toxic effects manifest relatively quickly after administration. Poisonous medicinal plants were hardly used afterwards in medicine, but were sometimes applied as poisons in hunting, defense, ordeals ("judgment of the gods"), or death sentences – the most famous example of which is the death of Socrates by an extract of *Conium maculatum* L. The first documents on herbal medicines can be traced to Asia and date back more than 2,000 years. A well-known example from China is a collection of descriptions and information on 365 drugs, which was named "*Shen nong ben cao jing*." Similarly, the Egyptian (the "Eber papyrus") and Greek schools from the Mediterranean area documented the therapeutic use of medicinal plants, e.g., the "*Materia medica*" of Dioscorides, reporting on 600 medicinal plants and their basic use in phytotherapy (Doull and Bruce 1986; Heinrich 2001; Gallo 2013).

Systematic collection in the Middle Ages. The Middle Ages had important impacts on herbal medicines. As already noted, there was an increasing exchange following major trade routes connecting Europe with Asia, notably through the Arabian empire and the crusades. In the early Middle Ages in Europe, health care was in the hands of monks from various sects. Consequently, the monks contributed to the documentation of information, but grew medicinal plants in the monasteries as well. The Roman Empire of Byzantium also significantly contributed to the preservation and refinement of antique knowledge. Secondly, progress in botany research led to a systematic collection of species and the description of morphological and anatomical characteristics and to other properties and effects.

Innovations during the Renaissance. Following the invention of book printing in Europe, several standard books were written that included information on the therapeutic benefits of plants, e.g., the *New Kreuterbook* by Leonhart Fuchs (1543). With the founding of universities, pharmacists acquired a more important role; their education included training and expertise in the preparation and use of herbal medicines, but also information on toxic and poisonous plants (Fuchs, reprint 2002; Meyer et al. 2005; Müller-Jahncke et al. 2005). The upheaval of new commercial routes to the Americas highlights new botanical species and original medicinal uses.

Research in natural product chemistry. In the seventeenth century, science provided the first tools for investigating the chemistry of plants in detail (Müller-Jahncke et al. 2005). In particular, natural product chemistry gave insight into specific molecules or groups of molecules that contributed to a therapeutic effect or resulted in poisoning. During subsequent centuries, information was collected about several hundred thousand compounds of plant origin. In some cases, the modes of action could be explained and provided a platform for strategies to develop even stronger-acting compounds or antidotes.

Macromolecules and targets. Since second half of the twentieth century, a major focus in science has been given to understanding macromolecules and their potential role as targets. Proteins, carbohydrates, and DNA interact in many ways with natural constituents and, in contrast to the early recognition of organ failures, it became obvious that plants may have medium- to long-term negative effects that are linked to genotoxic, nephrotoxic, hepatotoxic, or carcinogenic effects.

A new era founded on "omics" technologies. In the last few decades, new technologies have also been offered. Metabolomics, transcriptomics, and proteomics, etc., have been opening new dimensions in research (Pelkonen et al. 2012). Herbal medicines and their multi-component compounds may act on multiple targets in multiple ways. Understanding the complex networks of bioavailability, metabolism, and effecting of targets will lead to new insight about toxic processes in the human body.

Scientific Progress and the Development of Subdisciplines

The short overview on the various periods in the toxicity of herbal medicines shows that the development of toxicology has never followed a linear pathway. Progress in new scientific tools, knowledge about toxic principles, recognition of targets, and understanding toxic mechanisms and interactions were major drivers of the development. First classifications addressed toxicity towards specific organs, mostly associated with acute toxicity. Research on chronic toxicity is based on the observation of long-term effects that may sometimes be deduced from pharmacovigilance data or systematic pre-clinical studies. In the context of herbal medicines, liver toxicity, nephrotoxicity, genotoxicity, carcinogenicity, and reproductive toxicity are potential hazards that are hard to identify. Other subdisciplines with special relevance for herbal medicines are allergenicity, immunotoxicity, phototoxicity, cardiotoxicity, vascular toxicity, neurotoxicity, and interactions that have been described for specific plants (e.g., Johns Cupp 2000; Frohne and Pfänder 2004; Mills and Bone 2005; Williamson et al. 2013).

In comparison with chemical entities, the availability of adequate data is often a major obstacle in the assessment of herbal medicines' toxicity. Information on the components of the multi-component mixtures of herbal medicines, their pharmacokinetics, their pharmacodynamics, and their bioavailabilities is limited. Consequently, basic models and methods of calculation cannot always be applied.

Development of Regulation Related to the Toxicity of Herbal Medicines

Textbooks from ancient and medieval times were mainly scientific compilations that did not contain any regulatory context. Perhaps the development of pharmacopoeias, can be seen as a step towards the regulation of herbal medicines. Mostly, pharmacopoeias provided detailed descriptions of herbal drugs and a set of tests and assays to ensure pharmaceutical quality. If there is information on the potentially toxic components of an herbal drug, e.g., the modern European Pharmacopeia, which includes an assay and defines a threshold, must be complied with. Regulatory frameworks on herbal medicines as part of legislation have been developed only since the twentieth century. Worldwide, there are various concepts governing the regulation of herbal medicines (Fan et al. 2012; Wiesner and Knöss 2014). Basically, all these concepts share the approach to assure quality, efficacy, and safety.

Sometimes regulatory frameworks are based on notification of herbal medicines and a surveillance of the market. Other regulatory frameworks request an assessment of quality, safety, and efficacy prior to being put on the market; subsequently, pharmacovigilance and inspections contribute to supporting safe use during marketing. In the European Union, herbal medicines may be assigned to one of three categories (European Parliament 2001):

- New herbal medicinal products
- Herbal medicinal products with well-established use
- Traditional herbal medicinal products

For new herbal medicinal products with a new active substance (i.e., not previously authorized), a full application is necessary and the complete set of toxicological tests should be available: primary pharmacodynamics, secondary pharmacodynamics, safety pharmacology, and pharmacodynamic interactions.

Pharmacokinetics. Analytical methods and validation reports, absorption, distribution, metabolism, excretion, pharmacokinetic interactions (non-clinical), and other pharmacokinetic studies.

Toxicology. Single-dose toxicity, repeat-dose toxicity, genotoxicity, *in vitro* and *in vivo* (including supportive toxicokinetics evaluations), carcinogenicity, long-term studies, short- or medium-term studies, other studies, reproductive and developmental toxicity, fertility and early embryonic development, embryo-fetal development, prenatal and postnatal development, studies in which the offspring of juvenile animals are investigated and/or further evaluated, and local tolerance.

Other Toxicity Studies. Antigenicity, immuno-toxicity, mechanistic studies, dependence, metabolites, impurities.

For herbal medicinal products with well-established uses, the essential data should have been available in the scientific literature and been consistent for at least 10 years; based on documented use, the requirement of some data – such as carcinogenicity – may be waived. The safety of traditional herbal medicinal products is concluded after safe medicinal use for at least 30 years; only under specific circumstances may additional data be required.

The specific challenge of the safety assessment of herbal medicines is the limited availability of data. A careful and responsible evaluation of all existing data, the identification of gaps, and adequate decision-making can be seen as preconditions for the balanced handling of the toxicity of herbal medicines within regulatory frameworks.

Global Diversity and Future Challenges

Toxicology is a rapidly developing discipline. The scientific community is always discovering new technologies and methodologies that provide additional tools to investigate the toxicity of natural components. Accordingly, our knowledge of the toxicity of herbal medicines will continue to increase. It is likely that new inventions in pharmaceutical technology and the development of new galenic forms will allow the therapeutic use of compounds that were previously considered to be too toxic; also, some detoxification effects induced by multi-herb preparations may be clarified. Perhaps further testing will show the toxicity of compounds that have so far been considered safe. Also, it will be necessary to investigate the potential toxicity of herbal medicines within contexts that have been neglected so far, e.g., prolongation of the QT-interval (Schramm et al. 2011), interference with pharmacokinetics and metabolism, or modulation of DNA repair systems (indirect genotoxicity) (Nachtergael et al. 2015).

Expecting further globalization of herbal medicines, but facing the diverse regulatory systems and classifications for herbal medicines, a major challenge is to establish suitable communication platforms at the interface between toxicology as a science and toxicology as a part of regulatory requirements and safety assessment. For traditionally used herbal medicines, there are major gaps in available data and information, but incentives to generate additional information and to close the gaps are limited. Many fundamental questions have not been resolved, such as how can data from experiments with animals be applied to human beings, or what is the difference between temporary use and application over a lifetime? "Omics" technologies may help answer questions on synergisms of toxic constituents, but may also suggest models to explain the decrease in potential toxicity in multi-component mixtures. The future challenge is to apply new technologies to herbal medicines and to step into a new dimension of data handling. The new information should contribute to an adequate assessment of the toxicity of herbal medicines and to valuable use.

References

Doull J, Bruce MC (1986) Origin and scope of toxicology. In: Klaassen CD, Amdur MO, Doull J (eds) Casarett and Doull's toxicology – the basic science of poisons, 3rd edn. Macmillan publishing company, New York, pp 3–32

Eaton DL, Gilbert SG and Gallo (2013). Principles of toxicology. In: Klaassen CD (ed) Casarett and Doull's toxicology – the basic science of poisons, 8th edn. McGraw Hill Education, New York, pp 13–48

European Parliament and EU Council, Directive 2001/83/EC; available at http://ec.europa.eu/health/files/eudralex/vol

Fan TP, Deal G, Hoi-Lun K, Rees D, Sun H, Chen S, Dou JH, Makaraov VG, Poharitskaya ON, Shikov AN, Kim YS, Huang YT, Chang YS, Jia W, Dias A, Wong V, Chan K (2012) Future development of global regulations of Chinese herbal products. J Ethnopharmakol 140:568–586

Frohne D, Pfänder HJ (2004) Giftpflanzen, 5th edn. Wissenschaftliche Verlagsgesellschaft, Stuttgart

Fuchs L (2002, reprint of the Basel edition 1543) New Kreuterbuch, VMA-Verlag

Gallo MA (2013) History and scope of toxicology. In: Klaassen CD (ed) Casarett and Doull's toxicology – the basic science of poisons, 8th edn. McGraw Hill Education, New York, pp 13–48

Heinrich M (2001) Ethnopharmazie und Ethnobotanik. Wissenschaftliche Verlagsgesellschaft, Stuttgart

Johns Cupp M (ed) (2000) Toxicology and clinical pharmacology of herbal products. Humana Press, New Jersey

Lewington A (2003) Plants for people. Transworld Publishers, London

Meyer JG, Uehleke B, Saum K (2005) Das große Handbuch der Klosterheilkunde. Verlag Zabert Sandmann, München

Mills S, Bone K (eds) (2005) The essential guide to herbal safety. Elsevier, Sydney

Müller-Jahncke WD, Friedrich C, Meyer U (2005) Arzneimittelgeschichte, 2nd edn. Wissenschaftliche Verlagsgesellschaft, Stuttgart

Nachtergael A, Poivre M, Belayew A, Duez P (2015) In vitro genotoxicity tests point to an unexpected and harmful effect of a Magnolia and Aristolochia association. J Ethnopharmacol 174:178–186

Pelkonen O, Pasanen M, Lindon JC, Chan K, Zhao L, Deal G, Xu Q, Fan TP (2012) Omics and its potential impact on R&D and regulation of complex herbal products. J Ethnopharmacol 140:587–593

Schramm A, Baburin I, Hering S, Hamburger M (2011) hERG Channel inhibitors in extracts of *Coptidis rhizome*. Planta Med 77:692–697

WHO, Geneva (accessed 2015) http://www.who.int/medicines/areas/traditional/definitions/en/

Wiesner J, Knöss W (2014) Future visions for traditional and herbal medicinal products – a global practice for evaluation and regulation? J Ethnopharmacol 158:516–518

Williamson E, Driver S, Baxter K (eds) (2013) Stockley's herbal medicines interactions, 2nd edn. Pharmaceutical Press, London

Chapter 2
History and Current Status of Herbal Medicines

Jandirk Sendker and Helen Sheridan

Abstract Herbal medicine, or "phyto-medicine," refers to the practice of using plant material for medicinal purposes. Herbal medicine has a long tradition of use outside of conventional medicine, its earliest evidence of human use being recorded during excavations of Neanderthal sites, such as the Shanidar caves in northern Iraq. Many herbal medicines have evolved through traditional use within a specific cultural context. For some cultures, the traditional use is documented in written texts, and for others the traditional knowledge and its use have been passed down orally from one generation to the next. Several herbal drugs have yielded important modern therapeutic agents e.g., aspirin (*Salix spp* L.), taxol (*Taxus baccata* L.), and the Vinca alkaloids (*Catharanthus roseus* (L.) G.Don). Herbal medicines also play a significant and increasingly important role in global healthcare, where they are finding new and expanding markets as health foods and preventative medicines. The sources of the supply of medicinal plants are wild harvested and cultivated materials, and there are increasing demands for a sustainable supply of quality material. The worldwide annual market for herbal products approaches US $60 billion. The global "functional food" and dietary supplement markets are growing at a significant pace and have requirements for increasing quantities of high quality herbal materials.

Keywords Apothecaries • Chinese herbal medicine • Classification • Cultivation • Herbal medicine • Middle Ages • Neanderthal • Neolithic • Nutraceuticals Pharmaceuticals • Phytochemistry • Pre-Christian • Species identification • Traditional Chinese medicine • Traditional herbal medicine • Wild-harvesting

J. Sendker
University of Münster, Institute of Pharmaceutical Biology and Phytochemistry, 48149 Münster, Germany
e-mail: jandirk.sendker@uni-muenster.de

H. Sheridan (✉)
Trinity College Dublin, School of Pharmacy and Pharmaceutical Sciences, Dublin 2, Ireland
e-mail: hsheridn@tcd.ie

O. Pelkonen et al. (eds.), *Toxicology of Herbal Products*,
DOI 10.1007/978-3-319-43806-1_2

History and Current Status of Herbal Medicines and Herbal Medicinal Products

Early Evidence of Use of Herbal Medicine

Neanderthal Period

The terms "herbal medicine" (HM), "traditional herbal medicine" (THM), or "phytomedicine" refer to the use of plant material, such as whole plant, roots, bark, leaves, flowers, berries, or seeds, for medicinal purposes (WHO 2005). HM has a long tradition of use outside conventional medicine, with the discovery of archaeological evidence of the use of herbal medicine by humans, dating as far back as the Neanderthal period. Excavations at the Neanderthal sites (65,000 years old) at the Shanidar Caves in northern Iraq (Solecki 1971) have identified the presence of pollen of modern herbal medicines such as yarrow (*Achillea millefolium* L., Asteraceae), chamomile (*Matricaria recutita* L., Asteraceae), centaury (*Centaurium erythraea* Rafn., Gentianaceae), mallow species (Malvaceae), and ephedra species (Ephedraceae), suggesting their early ritual or medicinal use (Yaniv and Bachrach 2005) (Fig. 2.1). Traces of yarrow and chamomile were also found on the teeth of five Neanderthals at the El Sidrón site in northern Spain, dating back some 50,000 years (Hardy et al 2012). Another significant find was in Texas (Leach et al 1996), where again, plant pollen of yarrow and chamomile have been documented in addition to ephedra and coltsfoot (*Tussilago farfara* L., Asteraceae). In many countries, these species are still used in HM, and yarrow's role as a panacea for all sorts of aliments is well recognized (Jarić et al 2007; Applequist and Oerman 2011).

Fig. 2.1 Chamomile, mallow and yarrow, some of the earliest medicinal plants

Neolithic Period

Significant evidence of early use of HM from the Neolithic period has been documented in the analysis of the human remains and belongings of Otzi, the 5,300-year-old Ice Man, which were recovered from a thawed glacier in the Tyrolean Alps. It has been established that Otzi suffered from whipworm (*Trichuris trichiura* Linnaeus). Also found in his posession were some pieces of the fruiting bodies of a fungus, birch polypore (*Piptoporus betulinus* (Bull.ex Fr.) P. Karst), which we now know contains bio-active compounds that display ascaricidal, anti-inflammatory, and laxative activities (Capasso 1998). These discoveries support the belief that primitive humans understood and used herbal medicine. However, as pointed out by Edwards et al (2012), any hypothesis behind the medicinal use in these early times (of what we now know to be herbal medicine) can only be Conjecture as we have no understanding of the Neanderthal or Otzi's comprehension of illness or of any potential benefits relating to their herbal collections.

Pre-Christian Period

The extensive use of HM has been shown globally through the millennia. Important contemporary HM systems have their foundations in pre-Christian times in Mesopotamia, Asia, Africa, the Americas, and Australasia. Ayurvedic medicine is an example, where many of the important herbal practices date back over millennia to the Rig-Veda, the collection of sacred Hindi verses (Saboo et al 2014). Chinese Herbal Medicine (CHM) has written records indicating the use of herbal medicines such as *Shi Jing* or *Shan Hai Jing* dating from around 1000 to 500 B.C. (Pan et al 2014; Mohamed et al 2012). Many different medical papyri have survived from Ancient Egypt; they include, among others, the Edwin Smyth (Breasted 1930) and the Eber's Papyri from around 1500 B.C. The Eber's papyrus contains information on over 850 plant medicines, including remedies in which aloe vera was used for numerous indications, including the treatment of ulcers, skin diseases, burns, and allergies, and *Ocimum basilicum* L. (Basil), which was documented for its use to treat heart problems. It is believed that the entries in the papyrus may have been copied from earlier texts (Bryan 1930). Not all cultures have written records of their use of HMs, and for some, including many aspects of African HM, indigenous knowledge was just passed down orally from generation to generation (Waite 1992; Feierman and Janzen 1992).

First Century to the Middle Ages

In the Middle Ages in Europe, the knowledge of medicine that was fostered by the Greek, Roman, and Egyptian civilizations stagnated and did not change significantly until the seventeenth/eighteenth centuries (Hajar 2012). The study of herbal medicinal plants (HMPs) was associated with monks and their monasteries. Most monasteries had medicinal gardens in which they grew healing herbs from which they produced herbal cures, which remained a part of folk medicine (Duin and Sutcliffe 1992). Dioscorides' *Materia Medica*, written in 65 A.D., was the foundation of a significant portion of the herbal medicine practiced until 1500. There were over 600 HMs recorded in this text, which is a precursor to all modern pharmacopoeias, and is considered one of the most influential HM books in history. It remained in use until about 1600 A.D. (Riddle 1985).

In the mid-seventeenth century. herbal treatment reached its peak of popularity when the physician, botanist, and herbalist Nicholas Culpeper wrote a book called *The English Physician*, in which he linked HM with the signs of the zodiac. *The English Physician* was followed by his *Complete Herbal* in 1653 (McCarl 1996). Although the zodiac theory has been dismissed, Culpepper did identify many therapeutically active herbs, including wintergreen, which contains the painkiller salicin, which was of course also documented by Hippocrates ca 400 B.C. as "*willow leaf brew eases pain of childbirth*" (Aspirin Foundation 2015). Herbalism was also influenced by the discovery of the Americas. The arrival of tobacco and chocolate in Europe had a great influence on medicinal practice as they were viewed as medicines. A Spanish doctor, Nicholas Monardes, claimed that tobacco could cure 36 health problems (Frampton 1967). By the 1630s, Jesuit's bark (quinine) had arrived in Europe from Peru (Achan et al 2011).

In Asia, unlike in Europe, the practice of using HM did not decline in the first century A.D. In fact, in traditional Chinese medicine (TCM), the *Chinese Manual of Materia Medica* (*Shennong Ben Cao Jing*) was compiled in the first century A.D. during the Han Dynasty, describing some 252 HMs (Yang 2007). Another later work, written during the Ming Dynasty (1368–1644), *Compendium of Materia Medica* (*Ben Cao Gang mu*), is considered to be one of the most significant *Materia Medica* ever written. It was compiled over a period of 27 years by the physician Li Shizhen, who was also an herbalist and pharmacologist. In this compendium, 12,000 herbs and herbal prescriptions were recorded (Hoizey and Hoizey 1993; Gwei-djen 1976; Mu 2008). During the Ming Dynasty (1368–1644), medicine continued to flourish in China and medical exchanges took place between China, Korea, Japan, and even Europe (Hinrichs and Barnes 2013). TCM and CHM made further strides from the Song (221–207 B.C.) through to the Qing Dynasties (1644–1911).

The use of HM in Indian Ayurvedic medicine originated in the sub-continent in pre-historic times (Svoboda 2000). Ayurveda flourished throughout the Middle Ages in India and several important medical works, including those by Sushruta and Charaka, were compiled during this period. Acharya Charak produced the *Charak*(*a*) *Samhita*. This was in essence a pharmacopeia that remains fundamental to Ayurvedic medicine. It records the qualities and applications of over 100,000 plants and plant derivatives (Manojkumar 2013; IPF 2012). Botanical medicine increased in popularity in nineteenth-century America, when Thomsonianism was established by Samuel Thomson (1769–1843), who began practicing as an herbalist in New Hampshire in 1805. This popular medical approach was introduced to Britain by Albert Coffin, where it gained a great deal of support (Denham 2013).

Non-human Use of Herbal Medicine

As human use of HM has grown and developed throughout the ages, it is also worth noting that primates other than humans, such as chimpanzees (*Pan troglodytes*), self-medicate with medicinal plants to treat infections and illness (Huffman 2003). Birds have also been shown to harness the insecticidal activity of plants such as *Nicotiana* to reduce their ectoparasite load (Suárez-Rodríguez et al 2013).

Herbal Medicine to Pharmaceuticals

Apothecaries: The Foundation of Pharmaceutical Companies

In the middle of the nineteenth century, apothecaries (or early pharmacists) prepared traditional plant-based remedies by extraction and concentration, and were involved in the production of medicines such as morphine, quinine, and strychnine. It is in the small production environments of the apothecary that major pharmaceutical companies have their origins. Merck, for example, had its roots in the wholesale production of drugs in an apothecary shop in Darmstadt, Germany, in 1668 (Merck 2015). By 1887 Merck had a successful export business to the U.S., when they opened an office in New York. Similarly, Schering (Germany), Hoffmann-La Roche (Switzerland), and many other modern pharmaceutical companies started as apothecaries and medicine producers between the early 1830s and late 1890s (*Chem. and Engineering News* 2005). Within a short time, these small companies moved from natural products to synthetic chemistry. By the start of the twentieth century, most medicines were sold without a prescription and many were compounded by pharmacists. Physicians often administered directly to patients, and formulations were supplied to the doctors by small companies. The middle third of the twentieth century witnessed breakthroughs in the development of synthetic, semi-synthetic, and naturally occurring pharmaceuticals.

Plants as the Source of Early Pharmaceuticals

Many ancient and traditional herbal drugs, from which principal bioactive compounds have been isolated and characterised, have yielded important modern therapeutic agents. These include medicines such as aspirin, with its precursor, salicin, being derived from willow (*Salix* species) and first used as an analgesic over 2,000 years ago (Smith et al 2014). Aspirin was first synthesised 110 years ago, and in 1899, it was trademarked by Bayer under the Imperial Office of Berlin. In 2010, sales of aspirin generated € 916 million ($1.27 billion) for Bayer (Bloomberg 2014).

The modern anti-mitotic and anti-microtubule *vinca* alkaloids, such as vinblastine, vincristine, vindesine, and vinorelbine, are derived from the Madagascar periwinkle plant, *Catharanthus roseus* (L.) G. Don (Apocynaceae) (Moudi et al 2013). This plant has a long history as a folk medicine to alleviate diabetes in Asia, Africa, and Central America (Don 1999; Nammi et al 2003; Patel et al 2012; Ong et al 2011). Historically it has been used topically in Ayurvedic medicine to treat wasp stings and in CHM to treat diabetes and malaria (Nejat et al 2015). It was only in the 1950s that Robert Noble and Charles Beer, in their efforts to verify the antidiabetic properties of the *vinca* extracts, instead revealed the anticancer properties of this HM and isolated the active principles vinblastine and vincristine (Noble 1990). Taxol (Paclitaxel), a taxane, is another important anti-mitotic and anti-microtubule drug (Jordan and Wilson 2004); it is currently one of the most effective drugs against breast and ovarian cancers. It was isolated in the 1970s from a sample of the Pacific yew collected (1962) as part of the National Cancer Institute's search for novel anti-cancer drugs. For thousands of years, yew leaves, seeds, and bark have been used for suicide and as a chemical weapon during hunting and warfare, and it was even considered dangerous to sleep beneath the shade of a yew bush (Burrows and Tyrl 2013; Wilson et al 2001). Taxol is not the major toxic principle in this plant; the cardiotoxic taxine alkaloids are responsible. Sales of plant-derived anti-cancer drugs, which include the taxanes and the *vinca* alkaloids, have grown by about 16% in the past five years, and have been valued at US $4.8 billion in 2010 in the U.S., Japan, Germany, France, the U.K., Spain, and Italy (*IMS Health* 2011).

Two HMPs are the source of modern antimalarial drugs. Cinchona bark (*Cinchona officinalis* L., Rubiaceae) known as the "fever tree" in Peru, was first brought to Europe in the 1630s. The Jesuit Order advocated the bark, also known as the "Jesuit's bark" or powder (or quinine bark). The first instructions for the use of the bark had been published in the late 1640s, as the *Schedula Romana* (Jarcho 1993). Quinine bark was first sold in Britain in 1658, and was the first entry in the British Pharmacopoeia in 1677 (Chan et al 2011). In 1820, Pelletier and Bienaimé Caventou isolated quinine from cinchona bark (Song 2009). At this point, quinine became the standard treatment for malaria (Meshnick and Dobson 2001); quinine remained the mainstay of malaria treatment until the 1920s and continues to contribute to its management (Achan et al 2011). The more recent discovery of the anti-malarial compound artemisinin (*Artemesisia annua* L.) has its origin in 1967, when a malaria-targeted project was set up in China under the leadership of the Project 523 office. However, this medicine was also recorded in early CHM as *qinghao*, which first appeared in the mid-1500s in Ge Hong's *A Handbook of Prescriptions for Emergencies* (Tu 2011). The important position of naturally derived medicines and the role of artemisinin in the treatment of malaria were fully acknowledged in the awarding of the 2015 Nobel Prize for Medicine to Professor Youyou Tu, a pharmacologist at the China Academy of Chinese Medical Sciences in Beijing for her work on *A. annua* L. Tu and her team screened more than 2,000 Chinese herbal remedies to search for drugs with anti-malarial activity. Analysis of the aqueous extract prepared by immersing a handful of qinghao in 2 liters of water, and squeezing it out, was central to the discovery of artemisinin. Tu realized that the conventional phytochemical approach involving extraction with heat could destroy the active components. Artemisinin-based combination therapies (ACTs) are recommended by the WHO as the first-line treatment for uncomplicated *P. falciparum* malaria (WHO 2015).

Snowdrops (*Galanthus elwesii Hook f and Galanthus woronowii Losinsk. Amaryllidaceae*) are another interesting example of the discovery of a therapeutic compound from a traditional medicine. They produce galanthamine, a cholinesterase inhibitor that is currently used clinically to treat mild to moderate vascular dementia and Alzheimer's disease (Olin and Schneider 2002). Galanthamine is isolated from snowdrops (and related genera such as daffodils (*Narcissus*, *Leucojum*) and *Lycoris*). The lead bioactive compound was discovered serendipitously in the early 1950s by a Bulgarian pharmacologist travelling in a rural region of Bulgaria. He observed the practice of local inhabitants who massaged their foreheads with snowdrop bulbs; they explained that this was a traditional treatment for headaches (Marco and do Carmo Carreiras 2006; Heinrich and Teoh 2004). However, it is believed that the therapeutic properties of this herb were recorded in the eighth century B.C. by the Greek poet Homer in his epic poem *The Odyssey*, where he describes the herb *moly*, the flowers as white as milk.

There are numerous other plant-derived therapeutic molecules, including morphine (*Papaver somniferum* L.), ephedrine (*Ephedra* sp), capsaicin (*Capsicum* sp) – and too many others to present in this chapter – but there are many excellent articles that discuss them in detail (Gurib-Fakim 2006; Salim et al 2008; Pereira et al 2012; Atanasov et al 2015). Other therapeutic molecules from the natural world have also been identified as part of academic, NIH, or pharma-driven drug discovery programs (Kingston 2011).

Distribution of Medicinal Plant Species

Over the last few decades, HMs have been gaining increasing importance across the globe – as sources of active pharmaceutical compounds or synthetic intermediates, as components of health foods, and as preventative medicines in their own right. Plants with medicinal properties are scattered across the Plant Kingdom. Some families are rich in HMPs and others have just one or two medicinal species, while some species, distributed over a range of families, are rich in similar metabolites. Some species occur across cultures and geographical regions, while others are unique to specific ecosystems, geographical regions, and to specific cultural herbal medicine systems. In 2002, Schippmann et al estimated that there were 297,000 plant species, of which some 52,885, distributed globally, were used as herbal medicines. In 2004 the total estimated number of plant species was increased to 350,000 species of existing plants (including seed plants, bryophytes, and ferns) (Pan et al 2014). Of these, few species are used in any significant volume and only a very small percentage is used medicinally. For example, in TCM, 4,941 botanical materials are in use, of which only some 500 are commonly used. Some regions contain vast proportions of the earth's biodiversity and therefore the corresponding herbal medicinal systems contain unique species; for example, the sub-continent of India

Table 2.1 Global numbers of plant species and HMPs (Pan et al 2014; Kustantinah 2010; Schippmann et al 2002; Sharma et al 2008)

Country	Species	Medicinal species
China	26,082	4,941
India	45,000	3,500
Indonesia	22,500	180
USA	21,641	2,564
World	350,000	52,885

contains approximately 45,000 species, of which 3,500 species of plants have medicinal value and 500 are used in the Ayurvedic industry (Table 2.1). In other cases, some key medicinal herbal species have limited distribution; one example is *Echinacea* (Asteraceae), which is found growing wild in North America and was used frequently by Native Americans for hundreds of years but is now cultivated extensively in Europe as a modern herbal medicine (Kumar and Ramaiah 2011).

Source of Herbal Material

The sources of supply of medicinal plants are (1) wild harvested, and (2) cultivated material. A limited number of plant species are cultivated, while the majority of species are sourced by wild harvesting (Bodeker et al 1997).

Wild Harvest

The collection of plant material, either whole plants or plant parts, such as aerial parts (*herba*), roots (*radix*), flowers (*flos*), leaves (*folia*) or wood (*lignum*) from wild sources provides material for pharmaceutical and herbal products, ranging from *Taxus* spp. used to treat cancer, to lime flower for herbal teas (Table 2.2). As a result of wild harvesting, almost one in five plant species from all over the world is under threat of extinction. Selected species at high risk have been monitored, but there is no marked recovery (Traffic 2014). Although many common species can be harvested in a sustainable way, there are significant numbers of species with limited and sporadic distribution that cannot be harvested in a sustainable way (Earthscan 2011). The World Wildlife Fund (WWF) found that wild harvesting for local requirements was not detrimental to plant survival in several African countries due to the small quantities and common varieties involved (Kuipers 1997). The quantities of wild harvested material that are now traded commercially are, however, of growing concern. Over 90 % of HMPs used traditionally are wild, and gathered in Third World countries (Handa 2005). In India, which has more than 6.6 % of the world's medicinal plant species, only 82 medicinal plants have recommended agro practices (NMPB 2012; Kandavel and Sekar 2015) and in China, only 20 % of their herbal materials are harvested from cultivation. In the U.S., wild HMPs are being over-harvested, and more than 2,400 acres of natural habitat are eroded every day,

Table 2.2 Sources of some common HMPs

Product	Botanical name	Cult/wild	Origin
American ginseng root	*Panax quinquefolius*	Cultivated	USA (Wisconsin) Minnesota. Canada
Asian ginseng root	*Panax ginseng*	Cultivated	China
Buchu leaf	*Agathosma betulina*	Cultivated	South Africa
Goldenseal rhizome	*Hydrastis canadensis*	Cultivated	Wisconsin
Saffron style and stigma	*Crocus sativa*	Cultivated	Kashmir
Andrographis herb	*Andrographis paniculata*	Wild/cultivated	India
Narrow leaved coneflower root (Echinacea)	*Echinacea angustifolia*	Wild/cultivated	USA
Shatavari root	*Asparagus racemosus*	Wild/cultivated	India
American ginseng root	*Panax quinquefolius*	Wild	USA
Costus root	*Saussurea costus*	Wild	India
Goldenseal rhizome	*Hydrastis canadensis*	Wild	USA
Schisandra fruit	*Schisandra chinensis*	Wild	China (North)
Bilberry fruit	*Vaccinium myrtillus*	Wild	Bosnia Herzegovina, Croatia, Poland

Adapted from Lubbe and Verpoorte (2011)

and up to 29 % of plants historically used by native North Americans are threatened with extinction (Mathe and Mathe 2008). In Bulgaria, 80 % of exported HMPs are wild harvested from natural habitats (Alter Agro 2004).

Cultivation

The quality and biomass of HMPs can often be improved by medicinal plant cultivation. Efficient cultivation, harvesting, storage, and processing are the targets of HMP producers. However, market forces often drive and influence cultivation. Certified organic cultivation of medicinal herbs increases the acceptance of herbs by buyers and brings higher prices. Hungary, Poland, India (psyllium, periwinkle, and opium), China, Spain (liquorice), and Argentina (chamomile, psyllium) cultivate medicinal plants on a commercial scale (Kuipers 1997; Seed 2014).

Wild Harvest vs. Cultivation

The reason for the scale of wild harvesting relative to cultivation may be due in part to the fact that in many THM systems, wild harvested herbal material is considered to be superior. As a result, in China, for example, wild ginseng roots are significantly more valuable than artificially produced roots. The value of ginseng is associated with the appearance of the gnarled wild roots that symbolize the vitality and potency of the root (Kala et al 2006), and they command higher prices (Schippmann et al 2002). Similarly, traditional medicine practitioners in Botswana believe that cultivated material is not acceptable, as it does not have the "power of the material" harvested from the wild. This power is associated with the specific area in which the plants are grown and may well correlate with climatic and regional variations that have an impact on metabolomic composition of herbal plants and therefore therapeutic potency (Kandavel and Sekar 2015; Schippmann et al 2002).

The Current Position of Herbal Medicine and Traditional Herbal Medicine

Global Use

The current position is that HM and THM play significant and increasing roles in global healthcare in Asia, Africa, the Americas, Australasia, and Europe (WHO 2000a, 2002, 2003, 2005). The WHO estimates that 80 % of African and 70 % of Indian populations use some form of THM (WHO 2000b). Traditional medicine (TM), which can include HM and THM, is described by the WHO (2002, 2013) as the "sum total of the knowledge, skills, and practices based on the theories, beliefs, and experiences indigenous to different cultures, whether explicable or not, used in the maintenance of health as well as in the prevention, diagnosis, improvement or treatment of physical and mental illness." Within this context, THM incorporates plant, animal, and mineral-based medicines. In many cultures, TMs, such as traditional Indian (Ayurvedic) medicine, TCM, and traditional Arabian (Unani) medicine, have developed over thousands of years and are a part of medical systems (Patwardhan 2014). In China, traditional medicine, of which a major component relates to the use of THMs, represents approximately 40 % of all health care delivered. This is reflected in the fact that a significant portion (>90 %) of Chinese hospitals have traditional medicine units (WHO 2005). However, the use of HM and THM is not limited to the developing world. The use and interest in complementary and alternative medicines (CAM), including HM and TM, has increased greatly in industrialised countries. In the United States in 2002, it was estimated that 20 % of adults reported taking an herbal product (Barnes et al 2004). More recently, Ernst et al (2005) and Barnes et al (2008) showed that this has increased to 38 % of adults, and additionally 12 % of children who were using some form of TM or HM. These figures are also reflected in the growing value of the HM markets worldwide (*Nutraceuticals World* 2012).

New and Emerging Uses

HMs are finding new and expanding markets, and they can be incorporated into health foods and preventative medicines. They also have an increasing presence in functional foods, nutraceutical, and natural health products (Chauhan et al 2013; Wachtel-Galor and Benzie 2011). They can be processed and formulated in a great number of ways, including tablets, capsules, teas, tinctures, creams, oils, and liquids. The demand for crude medicinal plants – as well as their formulations – represents a large and growing worldwide demand.

Herbal Medicine Industries

HMPs are used by a large number of industries. These are rapidly growing multidisciplinary industries of global significance that can be classified into a number of groups (Fig. 2.2). The industry uses HMPs that are either cultivated or wild-crafted (collected from the wild) for their medicinal value. Of the 350,000 species of vascular plants globally, 52,885 of them are used medicinally (Table 2.1) (Schippmann et al 2002). Such plants can be used to derive single pharmaceutical compounds and precursors that are modified by the pharmaceutical industry; plant material for inclusion in herbal medicines or for extraction of essential oils or formulation of standardised extracts; or plant material or extracts that can be used as food and dietary supplements.

Pharmaceutical Companies

Medicinal plants have many applications in the pharmaceutical sector: (1) the isolation of purified metabolites, e.g., Taxol isolated from *Taxus baccata* (L.), *vinca* alkaloids, isolated from *Catharantheus roseus* (L.) g. Don, digitoxin extracted from Digitalis spp. (2) for the isolation of high-value extracts (Don 1999). In such circumstances, the extracts are standardized in terms of their main chemical components. In many cases, these are in mixtures with other ingredients, e.g., senna extract from *Cassia senna* (L). and (3) as key intermediates in API (active pharmaceutical ingredient) synthesis; or as precursors for the synthetic production of more complex pharmacologically active substances. For example, diosgenin, isolated from yams in the genus *Dioscorea*, is used in the commercial manufacture of progesterone (Djerassi 1992). There is a tremendous demand for plant species from this category. An estimated 25 % of prescription drugs in the U.S. consists of plant-derived principal components or crude extracts (*Interactive European Network for Industrial Crops and their Applications* (2000–2005)).

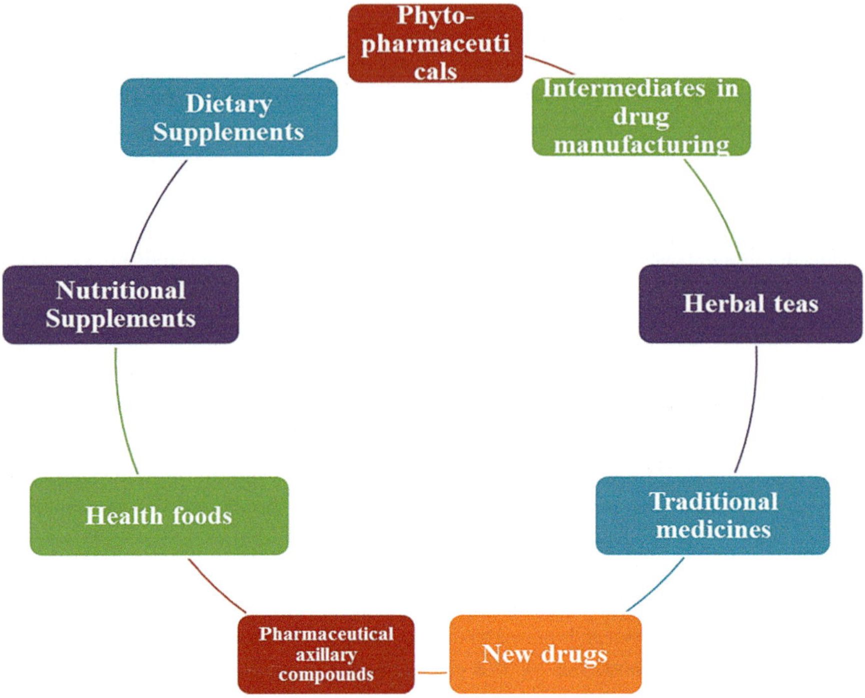

Fig. 2.2 Industrial use of HMPs

Phytopharmaceutical Companies

Standardised extracts of herbal plant material are used to make pharmaceutical and phyto-pharmaceutical products, and there is often little distinction between these product classes. Phyto-pharmaceutical companies use herbal material together with plant extracts to make their range of products, which may include tinctures, teas, or extracts that can be formulated in different ways. There are a variety of plant products that are in considerable demand for herbal companies e.g., garlic, ginseng, and psyllium seed husks. The classification as health products and phyto-pharmaceuticals is quite important since the former are marketed as such in order to avoid the need to license a product as a medicine (Directive 2004/24/EC).

Nutraceutical Companies

The term Nutraceutical was first used in 1989 by Stephen DeFelice; it comes from the combination of the terms "nutrition" and "pharmaceutical." The definition extends to a food or food part that has associated medical or health benefits. These may include their therapeutic benefits in addition to the basic nutritional value found in foodstuffs (Brower 2005). Nutraceutical ingredients are primarily used in (1) functional foods, and (2) dietary supplements; they include naturally occurring chemicals that have medicinal properties or can assist in health maintenance or disease prevention, and they can include herbal products. In Britain, the Ministry of Agriculture, Fisheries and Food has developed a definition of a functional food as "a food that has a component incorporated into it to give it a specific medical or physiological benefit, other than purely nutritional benefit" (Cockbill 1994; Kalra 2003). The term "dietary supplement" has been defined using several criteria as defined in the Dietary Supplement Health and Education Act of 1994 (DHEA 1994). The definition includes "… a herb or other botanicals, amino acids or a dietary substance for use by man to supplement the diet by increasing the total daily intake or a concentrate, metabolite, constituent, extract, or combinations of these ingredients." The DHEA established medicinal botanicals as dietary supplements. This may be due to the fact that a number of products used as herbal medicines, such as garlic (*Allium sativum* L.) and ginger (*Zingiber officinale* Roscoe) are also common food ingredients. The FDA does not regulate dietary supplements, and there is no onus on the manufacturers to prove safety or efficacy (although they must have a history of safety). The global "functional food" and dietary supplement markets are growing at a significant pace (Nutraceuticals World 2012) and have requirements for increasing quantities of the high quality of HMPs.

References

Achan J, Talisuna AO, Erhart A, Yeka A, James K, Tibenderana JK et al (2011) Quinine, an old anti-malarial drug in a modern world: role in the treatment of malaria. Malar J 10:144. doi:10.1186/1475-2875-10-144, http://www.malariajournal.com/content/pdf/1475-2875-10-144.pdf

Alter Agro (2004) Biological cultivation of herbs. EL/2004/B/F/PP-148258. http://www.euranek.com/alter-agro/pdf/unit7_en.pdf

Applequist W, Oerman D (2011) Yarrow (*Achillea millefolium* L.): a neglected panacea? A review of ethnobotany, bioactivity, and biomedical research. Econ Bot 65(2):209–225

Aspirin Foundation (2015) http://www.aspirin-foundation.com/what/100.html

Atanasov AG, Waltenberger B, Pferschy-Wenzig EM, Linder T, Wawrosch C, Uhrin P, Temml V, Wang L, Schwaiger S, Heiss EH et al (2015) Discovery and resupply of pharmacologically active plant-derived natural products: a review. Biotechnol Adv 33(2015):1582–1614

Barnes PM, Powell-Griner E, McFann K, Nahin RL (2004) Complementary and alternative medicine use among adults: United States, 2002. Advanced data from vital and health statistics; no 343. National Center for Health Statistics, Hyattsville

Barnes PM, Bloom B, Nahin R (2008) Complementary and alternative medicine use among adults and children. United States, http://www.cdc.gov/nchs/data/nhsr/nhsr012.pdf

Bloomberg (2014) http://www.bloomberg.com/news/articles/2014-05-05/limit-use-of-aspirin-to-prevent-heart-attack-fda-says

Bodeker G, Bhat KKS, Burley J, Vantomme P (eds) (1997) Medicinal plants for forest conservation and health care, UN Food and Agriculture Organisation (FAO) Non-Wood Forest Products Series, No.11. Food and Agriculture Organization of the United Nations, Rome

Breasted JH (1930) The Edwin Smith surgical papyrus. Oriental Institute Publications, vol III–IV. The University Press, Chicago

Brower V (2005) A nutraceutical a day may keep the doctor away. EMBO Rep 6(8):708–711, 10.1038

Bryan PW (1930) The Papyrus Ebers. Geoffrey Bles, London
Burrows E, Tyrl RJ (2013) Toxic plants of North America, 2nd edn. Wiley-Blackwell, Ames. ISBN 978-0-8138-2034-7
Capasso L (1998) 5300 years ago, the ice man used natural laxatives and antibiotics. Lancet 352(9143):1864. doi:10.1016/S0140-6736(05)79939-
Chan A, Talisuna AO, Erhart A, Yeka A, Tibenderana JK et al (2011) Quinine, an old anti-malarial drug in a modern world: role in the treatment of malaria. Malar J 10:144, http://www.malariajournal.com/content/pdf/1475-2875-10-144.pdf
Chauhan B, Kumar G, Kalam N, Andari SH (2013) Current concepts and prospects of herbal nutraceutical: A review. J Adv Pharm Technol Res. 4(1): 4–8. doi: 10.4103/2231-4040.107494 https://dx.doi.org/10.4103%2F2231-4040.107494" \t "pmc_ext
Cockbill CA (1994) Food law and functional foods. Br Food J 96:3–4
Denham AM (2013) Herbal medicine in nineteenth century England: the career of John Skelton. MA by Research University of York Department of History. http://etheses.whiterose.ac.uk/5003/1/Herbal%20Medicine%20in%20Nineteenth%20Century%20England,%20Career%20of%20John%20Skelton%20.pdf
DHEA (1994) The dietary supplement health and education act of 1994 http://www.health.gov/dietsupp/ch1.htm
Directive 2004/24/EC http://eur-lex.europa.eu/LexUriServ/LexUriServ.do?uri=OJ:L:2004:136:0085:0090:en:PDF
Djerassi C (1992) Steroid research at Syntex: "the pill" and cortisone. Steroids 57(12):631–641. doi:10.1016/0039-128X(92)90016-3.PMID1481227
Don G (1999) Catharanthus roseus. In: Ross IA (ed) Medicinal plants of the world. Human Press, Totowa, pp 109–111
Duin N, Sutcliffe J (1992) A history of medicine. From prehistory to the year 2020. Simon and Schuster, London
Earthscan (2011) The state of the world's land and water resources for food and agriculture. Managing Systems at Risk. Http://Www.Fao.Org/Docrep/017/I1688e/I1688e.Pdf
Edwards E, Da-Costa-Rocha I, Lawrence MJ, Cable C, Heinrich M (2012) Use and efficacy of herbal medicines: part 1 — historical and traditional use. Pharm J 288:685–686
Chem and Engineering News (2005) The pharmaceutical golden era. pp. 1930–1960
Ernst E, Schmidt K, Wider B (2005) CAM research in Britain: the last 10 years. Complement Ther Clin Pract 11:17–20
Feierman S, Janzen JM (eds) (1992) The social basis of health and healing in Africa (comparative studies of health systems and medical care, No 30). University of California Press, Berkeley
Frampton J (1967) Joyfull newes out of the newe founde worlde (trans: Nicholas Monardes). Published by AMS Press, New York
Gurib-Fakim A (2006) Medicinal plants: traditions of yesterday and drugs of tomorrow. Mol Asp Med 27(1):1–93. doi:10.1016/j.mam.2005.07.008
Gwei-djen L (1976) China's greatest naturalist: a brief biography of Li Shizhen. Am J Chin Med 4(3):209–218
Hajar R (2012) The air of history (part II) medicine in the Middle Ages. Heart Views 13(4):158–162
Handa SS (2005) Quality assurance of raw material of medicinal and aromatic plants. In: Proceedings of Expert Group Meeting on "Approaches for the sustainable exploitation of biodiversity of medicinal and aromatic plants in South East Asian countries, exploring the possibility of application of combinatorial chemistry" Bangkok, 29 June – 1 July. p 8.
Hardy K, Buckley S, Collins MJ, Estalrrich A, Brothwell D et al (2012) Neanderthal medics? Evidence for food, cooking and medicinal plants entrapped in dental calculus. Naturwissenschaften Sci Nat. doi:10.1007/s00114-012-0942-0
Heinrich M, Teoh HL(2004) Galanthamine from snowdrop—the development of a modern drug against Alzheimer's disease from local Caucasian knowledge. J Ethnopharmacol 92:147–162
Hinrichs TJ, Barnes LL (2013) Chinese medicine and healing: an illustrated history. Harvard University Press, Cambridge, MA

Hoizey D, Hoizey MJ (1993) A History of Chinese Medicine (trans: Paul Bailey). Edinburgh University Press Ltd

Huffman MA (2003) Animal self-medication and medicine: exploration and exploitation of the medicinal properties of plants. Proc Nutr Soc 62(2):371–381

IMS Health (2011) https://www.imshealth.com/files/web/IMSH%20Institute/Reports/The%20Global%20Use%20of%20Medicines%202015/Global_Use_of_Medicines_Report2015.pdf

Interactive European Network for Industrial Crops and Their Applications (2000 2005) Summary report for the European Union. QLK5-CT-2000-00111. [www.ienica.net/reports/ienicafinalsummaryreport2000-2005.pdfFreefull-text]

IPF (2012) International Pharmaceutical Federation. From making medicines to optimizing outcomes: the evolution of a profession 1912 to 2012. https://www.fip.org/centennial/files/static/UCL_BOOKLET_Web.pdf

Jarcho S (1993) Quinine's Predecessor: Francesco Torti and the Early History of Cinchona, vol 10. Johns Hopkins University Press, Baltimore, p 280

Jarić S, Popović Z, Mačukanović-Jocić M, Djurdjević L, Mijatović M et al (2007) An ethnobotanical study on the usage of wild medicinal herbs from Kopaonik Mountain (Central Serbia). J Ethnopharmacol 111(1):160–175

Jordan MA, Wilson L (2004) Microtubules as a target for anticancer drugs. Nat Rev Cancer 4:253–265

Kala CP, Dhyani PP, Sajwan BS (2006) Journal of ethnobiology and ethnomedicine. J Ethnobiol Ethnomed 2:32

Kalra EK (2003) Nutraceutical – definition and introduction. AAPS Pharm Sci 5:2–3

Kandavel D, Sekar S (2015) Impact of certain biotic and abiotic factors on phyllanthin and hypophyllanthin content of *Phyllanthus Amarus*. Schum & Thonn. From three different habitats. Int J Pharm Pharm Sci 7(5):258–265

Kingston DGI (2011) Modern natural products drug discovery and its relevance to biodiversity conservation. J Nat Prod 74(3):496–511. doi:10.1021/np100550t

Kuipers SE (1997) Trade in medicinal plants. In: FAO (ed) Medicinal plants for forest conservation and healthcare, Non-Wood Forest Products 11. Food and Agriculture Organization of the United Nations, Rome, p 45

Kumar KM, Ramaiah S (2011) Pharmacological importance of *Echinacea purpurea*. Int J Pharm Bio Sci 2(4):304–314

Kustantinah D (2010) Challenges and opportunities in regulating Herbal Medicines in Indonesia. National Agency of Drug and Food Control. The Republic of Indonesia. Http://Www.Who.Int/Medicines/Areas/Quality_Safety/Regulation_Legislation/Icdra/WB-1-A_Herbalmedicines_Indonesia.Pdf

Leach JD, Holloway RG, Almarez FA (1996) Prehistoric evidence for the use of *Chenopodium* (Goosefoot) from the Hueco Bolson of West Texas. Tex J Sci 2:163–165

Lubbe A, Verpoorte R (2011) Cultivation of medicinal and aromatic plants for specialty industrial materials. Ind Crop Prod 34:785–801. doi:10.1016/j.indcrop.2011.01.019

Manojkumar VC (2013) Virechana (purgation therapy) – a review. Int J Ayurvedic Herb Med 3(5):1315–1329

Marco L, do Carmo Carreiras M (2006) Galanthamine, a natural product for the treatment of Alzheimer's disease. Recent Pat CNS Drug Discov 1(1):105–111

Mathe A, Mathe I (2008) Quality assurance of cultivated and gathered medicinal plants. Acta Hort (ISHS) 765:67–76, http://www.actahort.org/books/765/765_8.htm

McCarl MR (1996) Publishing the works of Nicholas Culpeper, astrological herbalist and translator of Latin medical works in seventeenth-century London. Can Bull Med Hist 13(2):225–276

Merck (2015) http://www.merckgroup.com/en/company/merck_the_original/merck_the_original.html

Meshnick SR, Dobson MJ (2001) The history of antimalarial drugs. In: Rosenthal PJ (ed) Antimalarial chemotherapy: mechanisms of action, resistance, and new directions in drug discovery. Humana Press, Totowa, pp 15–25

Mohamed I, Shuid A, Borhanuddin B, Fozi N (2012) The application of phytomedicine in modern drug development. Internet J Herb Plant Med 1(2). https://ispub.com/IJHPM/1/2/14278

Moudi M, Go R, Yien CYS, Nazre M (2013) Vinca alkaloids. Int J Prev Med 4(11):1231–1235

Mu BCG (2008) Encyclopaedia of the history of science, technology, and medicine in non-western cultures 399–400

Nammi S, Boini MK, Lodagala SD, Behara RBS (2003) The juice of fresh leaves of *Catharanthus roseus* Linn. reduces blood glucose in normal and alloxan diabetic rabbits. BMC Complement Altern Med 3:4, http://www.ncbi.nlm.nih.gov/pmc/articles/PMC194756/

Nejat N, Valdiani A, Cahill D, Tan Y-H, Maziah M, Abiri R (2015) Ornamental exterior versus therapeutic interior of Madagascar Periwinkle (*Catharanthus roseus*): the two faces of a versatile herb. Volume 2015:19 pages. http://www.hindawi.com/journals/tswj/2015/982412/

NMPB (2012) Techniques of selected medicinal plants. Available from: http://www.nmpb-mpdb.nic.in

Noble RL (1990) The discovery of the Vinca alkaloids – chemotherapeutic agents against cancer. Biochem Cell Biol 68(12):1344–1351

Nutraceuticals World (2012) http://www.nutraceuticalsworld.com/issues/2012-11.

Olin J, Schneider L (2002) Galantamine for Alzheimer's disease. Cochrane Database Syst Rev 3, CD001747

Ong HC, Ahmad N, Milow P (2011) Traditional medicinal plants used by the Temuan villagers in Kampung Tering, Negeri Sembilan, Malaysia. Ethnol Med 5:169–173

Pan S-Y, Litscher G, Gao S-H, Zhou S-F, Yu Z-L et al (2014) Historical perspective of traditional indigenous medical practices: the current renaissance and conservation of herbal resources. Evid Based Complement Alternat Med 2014:525340, http://www.hindawi.com/journals/ecam/2014/525340/

Patel DK, Kumar R, Laloo D, Hemalatha S (2012) Natural medicines from plant source used for therapy of diabetes mellitus: an overview of its pharmacological aspects. Asian Pac J Trop Dis 2:239–250

Patwardhan B (2014) Bridging Ayurveda with evidence-based scientific approaches in medicine. EPMA J 5(1):19, http://doi.org/10.1186/1878-5085-5-19

Pereira DM, Valentao P, Correia-da-Silva G, Teixeira N, Andrade PB (2012) Plant secondary metabolites in cancer chemotherapy: where are we? Curr Pharm Biotechnol 13(5):632–650

Riddle JM (1985) Dioscorides on pharmacy and medicine. University of Texas Press, Austin

Saboo DSS, Chavan RW, Tapadiya GG, Khadabadi SS (2014) An important ethnomedicinal plant *Balanite Aegyptiaca*. Int J Phytopharmacy Rev Artic 4(3):75–78

Salim AA, Chin Y-W, Kinghorn AD (2008) Chapter 1, drug discovery from plants. In: Ramawat KG, Merillon JM (eds). Bioactive molecules and medicinal plants. Chapter 1. doi:10.1007/978-3-540-74603-4_1,c Springer

Schippmann U, Cunningham AB, Leaman DJ (2002) Impact of cultivation and gathering of medicinal plants on biodiversity: global trends and issues. Biodiversity and the ecosystem approach in agriculture, forestry and fisheries. Satellite event on the occasion of the ninth regular session of the commission on genetic resources for food and agriculture. FAO, Rome, 12–13 October

Seed 2014: https://www.seed.uno/awards/all/safi-organics-limited.html

Sharma A, Shanker C, Tyagi LK, Singh M, Rao C-V (2008) Herbal medicine for market potential in India: an overview. Acad J Plant Sci 1(2):26–36

Smith T, Elwood P, Keating C, Rothwell P, Detering E et al (2014) The Aspirin Foundation Scientific Conference: the history, the present state and the future of aspirin prophylaxis. Ecancer 8:388, http://www.aspirin-foundation.com/documents/can-8-388.pdf

Solecki R (1971) Shanidar: the first flower people. Alfred A. Knopf Incorporated, New York: pp 245–250

Song CE (2009) An overview of cinchona alkaloids in chemistry. In: Song CE (ed) Cinchonal alkaloids in synthesis and catalysis, ligands, immobilization and organocatalysis. Wiley-VCH, Weinheim

Suárez-Rodríguez M, López-Rull I, Macías Garcia C (2013) Incorporation of cigarette butts into nests reduces nest ectoparasite load in urban birds: new ingredients for an old recipe? Biol Lett 9(1):20120931, http://rsbl.royalsocietypublishing.org/content/roybiolett/9/1/20120931.full.pdf
Svoboda RE (2000) Ayurveda: life, health and longevity. Penguin, UK, p 58
Traffic (2014) http://www.traffic.org/medicinal-plants/
Tu Y (2011) The discovery of artemisinin (qinghaosu) and gifts from Chinese medicine. Nat Med 17(10):1217–1220, http://www.laskerfoundation.org/awards/pdf/2011_c_youyou.pdf
Wachtel-Galor S, Benzie IFF (eds) (2011) Herbal medicine: an introduction to its history, usage, regulation, current trends, and research needs. In: Herbal medicine: biomolecular and clinical aspects. 2nd edn. CRC Press, Boca Raton. Chapter 1. Available from: http://www.ncbi.nlm.nih.gov/books/NBK92773/
Waite GM (1992) A history of traditional medicine and health care in Pre-colonial East-Central Africa. E. Mellen Press, New York
WHO (2000a) General guidelines for methodologies on research and evaluation of traditional medicine, WHO/EDM/TRM/2000.1. World Health Organization, Geneva
WHO (2000b) Traditional medicine in the African region. An initial situation analysis (1998–1999). WHO Regional Office for Africa, Harare, 2000
WHO (2002) WHO traditional medicine strategy 2002–2005. http://www.wpro.who.int/health_technology/book_who_traditional_medicine_strategy_2002_2005.pdf
WHO (2003) Traditional medicine fact sheet N°134. Revised May 2003. http://www.who.int/mediacentre/factsheets/2003/fs134/en/
WHO (2005) National policy on traditional medicine and regulation of herbal medicines. Report of a WHO global survey. http://whqlibdoc.who.int/publications/2005/9241593237.pdf
WHO (2013) Traditional medicine strategy: 2014–2023. http://apps.who.int/iris/bitstream/10665/92455/1/9789241506090_eng.pdf
WHO (2015) Q&A on artemisinin resistance http://who.int/malaria/media/artemisinin_resistance_qa/en/
Wilson CR, Sauer J, Hooser SB (2001) Taxines: a review of the mechanism and toxicity of yew (Taxus spp.) alkaloids. Toxicon 39:175–185
Yang S-Z (2007) The Divine Farmer's Materia Medica: a Translation of the Shen Nong Ben Cao Jing. Blue Poppy Press, Boulder
Yaniv Z, Bachrach U (2005) Handbook of medicinal plants. Retrieved from http://books.google.com/books?id=QN78GmmPGBQC&pg=PA131&dq=Achillea+millefolium&hl=en&sa=X&ei=caUxUdHTKYaq8ATy4IHQDQ&ved=0CFgQ6AEwBw#v=onepage&q=Achillea%20millefolium&f=false

Chapter 3
Composition and Quality Control of Herbal Medicines

Jandirk Sendker and Helen Sheridan

Abstract Herbal medicine, or "phyto-medicine," refers to the practice of using plant materials for medicinal purposes. Across the globe, traditional herbal medicines play a significant role in healthcare, and the worldwide annual market for these products approaches US $60 billion. Several challenges face the increasing growth in the herbal medicine and herbal medicinal product markets. These challenges relate to the lack of harmonization of international standards; sustainable production of high-quality herbal material in the face of overharvesting of wild species; and determining and establishing the quality, safety, and efficacy of these materials, which can range from simple, one-herb formulas to complex, multi-component formulas as seen in some traditional Chinese herbal medicines. The increasing understanding of the minor components' importance and synergism in the bioactivity of herbal medicine poses numerous scientific questions that need to be addressed by targeted research programs. In this chapter we deal with the important topics of the composition of herbal medicines and the quality control of these medicinal products, with special emphasis on the role of concomitant compounds and co-effectors; we examine the production chain and the complex factors that impact the composition of the final herbal material; we discuss and evaluate the current methods and accepted gold standards for quality control of herbal materials and finished products; and we also examine emerging technologies and consider changes in international regulations and the impact they may have on this area.

Keywords Analytical characterization • Chemical fingerprint • Chemometrics • Chinese herbal medicine • Chromatography • Chromatographic fingerprinting • Classification • Co-effectors • Concomitant compounds • Cultivation • Exploratory data analysis • GACP • GMP • Herbal medicine • Metabolomics • Multi-component

J. Sendker (✉)
University of Münster, Institute of Pharmaceutical Biology and Phytochemistry, 48149 Münster, Germany
e-mail: jandirk.sendker@uni-muenster.de

H. Sheridan
Trinity College Dublin, School of Pharmacy and Pharmaceutical Sciences, Dublin 2, Ireland
e-mail: hsheridn@tcd.ie

O. Pelkonen et al. (eds.), *Toxicology of Herbal Products*,
DOI 10.1007/978-3-319-43806-1_3

mixture • Natural deep eutectic solvents (NADES) • Nutraceuticals pharmaceuticals • Phytochemistry • Quality control • Species identification • Spectroscopy • Solubilization • Stability • Synergy • Traditional Chinese medicine • Traditional herbal medicine • Wild harvesting

Introduction

The global increase in the demand for herbal medicinal products (HMPs) has led to a significant expansion of this market (He et al. 2015). The increasing demand for herbal products, the progressive integration of traditional medicinal systems and thus herbal medicines (HM) into national health care systems (WHO 2005), and the inclusion of expanding markets such as nutraceuticals into the HMP arena, have made the quality of herbal products a major concern for health authorities, pharmaceutical industries, and for consumers. HMPs are composed of thousands of chemical entities that in their entirety account for the desired and undesired properties of these materials, and, therefore, their quality, safety, and efficacy. Examples where HMP properties are explainable by the presence of one compound or compound class are few, and there is increasing interest in the role of trace compounds that influence HMP quality by subtle means. There is growing evidence that various production steps during the preparation of HMP material may affect its chemical composition, and therefore quality control becomes an ever more challenging field, as reflected by recent regulatory guidelines and increasing application of chemometric methods (Guo et al. 2014; Qi and Kelley 2014; Sheridan et al. 2015).

Guidelines on Production of Quality Herbal Material and Herbal Medicinal Products

Good Agricultural and Collection Practices (GACP)

Ultimately, the safety and efficacy of herbal products is governed by their quality (WHO 2003a, b, 2007). Quality aspects of HMP arise throughout the production chain. Factors such as misidentification of herbal material; intentional adulteration; contamination, including heavy metals, microorganisms, or pesticides; generation of artifacts caused by enzymatic activity; or fumigation of plant material can lead to serious and potentially fatal consequences (Nortier et al. 2000; Okem et al. 2015). It is widely accepted that quality in the cultivation, collection, and formulation of crude herbal material and finished HMP must be built in from the beginning of the production cycle. For HMP and finished products, the cultivation, production and processing (if any) directly impacts product quality. Good agricultural and collection practice for herbal starting materials (GACP) are required to ensure consistent

Fig. 3.1 Important factors to be considered for GACP

quality of HMP. Regulations relating to guidelines on GACP and GMP for medicinal plants have been published by the World Health Organization (WHO) (2003b, 2007). These guidelines contain addenda on GACP for Traditional Chinese, and also Japanese Medicinal Materials. GACP guidelines were reintroduced by the Committee on Herbal Medicinal Products (HMPC) of the European Medicines Agency (EMEA) in 2006. The GACP sub-committee of EUROPAM (European Herb Growers Association) have also produced guidelines for the Good Agricultural and Wild Collection Practice of Medicinal and Aromatic (Culinary) plants traded and used in the European Union (EUROPAM 2006).

GACP promotes in-built quality from the very beginning of the production chain. These guidelines provide details on the methodology required to cultivate and harvest HMP to acceptable international standards. They also direct the information that needs to be collated to ensure the good standard operating procedures (SOP) that are required for appropriate reproducible quality. These SOPs include records of plant/seed source, site conditions, chemical applications, water source, etc. Details relating to aspects of GACP practice are shown in Fig. 3.1. For local gatherers, collecting wild herbal material is in order to generate income. Due to low financial returns, commercial plant gatherers often overharvest the natural resources rather than manage them (Sher et al. 2012). Collection from the wild presents additional challenges that include the misidentification of plants and overharvesting in addition to errors that arise because collectors may lack specific knowledge in terms of plant identification, plant parts required, or favored environmental growth conditions that can have an impact on quality (Li et al. 2014). Between 2003 and 2010, the number of traditional Chinese medicine (TCM) herbs grown in agricultural sites complying with GCAP increased from 13 to 49 (Siow et al. 2005; Zhang et al. 2010).

Good Manufacturing Practices (GMP)

Good manufacturing practice (GMP) for all aspects of the manufacture and storage of active pharmaceutical ingredients (API) also applies to HMP and herbal substances (EMEA 2006; WHO 2007). Good manufacturing practice for HMP differs globally, and harmonization has not yet been attained (WHO 2007; He et al. 2015; Calixto 2000; EU 2008). GMP is the mechanism by which manufacturing standards and quality control measures comply with requirements; thus, GMP guarantees the quality both of the herbal materials and the finished products, such as capsules, powders, and tinctures. Essentially, the requirements for GMP of HMP are common to GMP for pharmaceutical products. The WHO has produced and upgraded guidelines relating to the GMP of HMP (WHO 2007; EU 2008). These reference guidelines are designed to assist national health authorities to develop their own GMP requirements, taking into account the regional and local socio-economic factors that impact the medicinal plant cultivation and production of finished HMP. Fundamentally, adequate quality assurance (QA) systems are required, especially as a consequence of the in-built complexity of medicinal plants and HM that can be further enhanced by the formulation of multi-component HM.

Pharmacopoeias and Herbal Medicinal Plants

Pharmacopoeias provide specifications that ensure the quality of specific pharmaceutical substances and medicinal products that include HMP. According to the WHO, 140 independent countries at present use 30 national as well as African, European, and International Pharmacopoeias (WHO 2001). The WHO has published *The International Pharmacopoeia* (Ph. Int. 2014) (Fourth Edition, including supplements 2006, 2008, 2013) in an aim to achieve broad global uniformity of quality specifications. In contrast with national and regional pharmacopoeias that are legally binding, *The International Pharmacopoeia* (Ph. Int.) is issued by the WHO as a recommendation with the aim of providing international standards – including less technically demanding alternatives where needed – for adoption by member states, and to help achieve global harmonization of quality specifications. Although there is a move to harmonization, strategies of individual pharmacopoeias differ for geographical and economic reasons as well as depending on the level of integration to respective regional medical systems.

In Europe, a regional approach is used. *The European Pharmacopoeia* (Ph. Eur.) is legally binding in the 38 ratifying parties of the *Convention on the Elaboration of a European Pharmacopoeia* (37 member states and the European Union) (WHO 2013; Bouin and Wierer 2014). In the *U.S. Pharmacopoeial Convention* (USP), standards for the identity, strength, quality, and purity of medicines, food ingredients, and dietary supplements are presented. These can be enforced in the U.S. by the Food and Drug Administration (FDA). Over 140 countries use these standards. The USP proposed the first 23 HMPs to be included in the new *Herbal Medicines Compendium* (HMC) (Biopharma 2013). The *Chinese Pharmacopoeia* 2010 (Volume 1, English Edition) contains 2165 national herbal monographs, of which 1019 are new admissions (439 for prepared slices) and 634 are revisions; the content includes Chinese Material Medica, Prepared Slices of Chinese Crude Drugs, Vegetable Oil, Fats and Extracts, and Patented Chinese Traditional Medicines together with Single Ingredients of Chinese Crude Drug Preparations. The Chinese Pharmacopoeia is legally binding, as are the *Korean* (1959) and *The Japanese Pharmacopoeia* (2001) (WHO 2005). In the rest of Asia, there is great variation in the approaches to the regulation of HMP. The WHO (2005) published data on a trans-national survey of the status of pharmacopoeias across the globe. Full detail is laid out in the *WHO National Policy on Traditional Medicine and Regulation of Herbal Medicines – Report* of a WHO Global Survey (WHO 2005).

Herbal Monographs

An herbal monograph is a document that defines a botanical drug and supplies information that allows for the herbal drug's proper quality assurance, or for evaluating its safety and efficacy. The *International Pharmacopoeia*, national pharmacopoeias, and other documents, contain monographs that are central to the authentication and quality determination of herbal materials. There are many sources of published monographs available, including the *WHO Monographs* (WHO 2008; WHO 2015; volumes 1–4) and *Monographs on Selected Medicinal Plants and on Medicinal Plants Commonly Used in the Newly Independent States (NIS)*, which aim "to provide scientific information on the safety, efficacy, and quality control of widely used medicinal plants; provide models to assist Member States in developing their own monographs or formularies for these and other herbal medicines; and facilitate information exchange among Member States."

In the EU, herbal monographs reflect the scientific opinion of HMPC on safety and efficacy data of an herbal substance and its preparations intended for medicinal use. Final EU monographs can be used in application reference material by a marketing-authorization applicant (well-established-use part) and by a traditional-use-registration applicant (traditional-use part) (EMEA 2015). Other important herbal monographs include the USP, the American Herbal Pharmacopoeia (AHP), and the Complete German Commission E Monographs, or European Scientific Cooperative on Phytotherapy (ESCOP) monographs (Awang 1997).

The need for harmonized monographic standards is evidenced particularly in Chinese herbal medicine (CHM), where diverse national and regional practices arise due to different monographic standards in many international pharmacopoeias (Chan et al. 2009). Nearly 5000 drugs (herbal, animal, and mineral) have been used in China (Xiao 1986). In Volume I of the Chinese Pharmacopoeia 2010 (English edition), there are 2165 monographs. A series of 60 CHM monographs have been implemented for the Ph. Eur. as a result of a program of work by expert groups. There are a number of related publications and other ongoing projects. Differences in detail between the ChP and Ph. Eur. Monographs mean that all new TCM herbal drugs for the Ph. Eur. had to be re-examined using available methods (Bauer and Franz 2010). Some of the accepted published herbs include *Bistortae rhizoma (Quan shen)* (Ph. Eur. 6.0), *Notoginseng radix (San qi)* (Ph. Eur. 6.0), *Sanguisorbae radix (Di yu)* (Ph. Eur. 6.1), *Schisandrae fructus* (*Wu wei zi*) (Ph. Eur. 6.3), *Carthami flos* (*Hong hua*) (Ph. Eur. 6.4), *Ephedrae herba (Ma huang)* (Ph. Eur. 6.7), *Stephaniae tetrandrae radix (Han Fangji)* (Ph. Eur. 7.0), *Astragali mongholici radix (Huang qi)* (Ph. Eur. 7.0), and *Scutellariae baicalensis radix (Huang qin)* (Ph. Eur. 7.1).

EU Directive 2004/24/EC

The introduction of the European Directive 2004/24/EC has increased the need for quality standards for herbal materials. This directive introduced a simplified registration procedure for traditional HMP and is central to the efforts toward harmonizing current legislative frameworks for European HMP. This directive impacts the registration of non-European HMP, of which only a limited number have been registered. The aim of the simplified registration procedure is to "safeguard public health, remove the differences and uncertainties about the status of traditional herbal medicinal products that existed in the past in the Member States and facilitate the free movement of such products by introducing harmonized rules in this area." A simplified registration is now in place in Europe for traditional HMPs that have a traditional or long-established use. A full license is required for traditional HMPs that do not fall into these categories. In most cases, efficacy has to be proven by clinical trials, while for those traditional HMPs with established or traditional use, a bibliographic dossier addressing quality, safety, and efficacy on the basis of well-established use is sufficient (European Commission 2004).

Quality of Herbal Material and Herbal Medicinal Products

Naming of Plants and Herbal Materials

Plant Species

A fundamental aspect of any HMP's quality is the consistent identity of the herbal ingredients. In the first instance, utilization of the correct organisms must be ensured. Typically, a particular herbal ingredient is required to originate from one particular plant species. Any plant species is taxonomically fixed by its Latin scientific binomial, followed by a naming authority, e.g., *Papaver somniferum* L., which is permanently attached to a particular holotype specimen (McNeill et al. 2012). As a consequence, any use of a compliant name refers to this particular specimen. This naming system is still confused by the circumstance that about 350,000 currently known plant species are associated with more than 1,000,000 Latin scientific names. These include the preferentially used accepted names, their respective synonyms, and unresolved names that are currently consolidated and evaluated from numerous data sets. This is collated in "The Plant List," an on-line data base of all known plant species initialized by a consortium of Royal Botanical Gardens, Kew and Missouri Botanical Garden (Kalwij 2012; The Plant List 2013). The importance of a consequent use of Latin scientific names including the naming authority is well illustrated by the occurrence of homonymous binomen: *Illicium anisatum* L. is the poisonous Japanese shikimi tree, whereas *I. anisatum* Lour. is used as a synonym for the accepted name of *I.verum* Hook f., the source of the medicinally used star anise fruits. However, even the consequent use of Latin scientific names, including naming authority, does not prevent confusion arising from different taxonomic concepts of species (Paton 2009). A continuous problem is that existing standards for botanical nomenclature are not sufficiently put into practice (Bennett and Balick 2014; Rivera et al. 2014).

Pharmacopoeias list the medical plant species or sometimes multiple species to be used for the production of a monographed herbal drug. As desired medicinal properties are of course not strictly correlated with the taxonomic concept of species, pharmacopoeia monographs sometimes also refer to infraspecific taxa (*Capsicum annuum* var. *minimum* (Mill.) Heizer), species complexes (*Valeriana officinalis* L. sensu lato) or – in the case of willow bark – allow harvesting from any species of the *Salix* genus, as long as the requirements regarding the plant part and content of active ingredients are met (Council of Europe 2015; Länger 2014).

Herbal Products

According to WHO definitions, HMPs are classified into herbs, herbal materials and (finished) herbal preparations (WHO 2000). Within the production chain of any HMP, herbs represent the first stable product derived from living plants and include entire, fragmented, or powdered plant parts such as leaves, flowers, bark, or other plant parts. The term herbal materials additionally embraces materials originating from some kind of processing or manufacturing process, such as gums, oils, resins or, for example, roasted plant parts. It is hence congruent with the term herbal drug as used by, for example, Ph. Eur. Some perceptions of herbal materials or herbal drugs include fresh plant materials, but the essential meaning of the term occurring in any definition refers to dried plants or plant parts (Júnior et al. 2014; Council of Europe 2015).

In contrast to plant species' names, no international standardized system for the naming of herbal products is in use. Some clarity is provided by modern pharmacopoeia or comparable official monographs that describe herbal materials along with their desired properties and give Latin names in addition to their vernacular names. Frequently, but not necessarily, these terms refer to the Latin scientific name of the medicinal plant species and the plant part used, e.g., *Uvae-ursi folium* (bearberry leaves, Ph. Eur.) or *Radix Astragali* (huang-chi, WHO Monographs). These Latin names are also frequently used in scientific literature, although the linkage to pharmacopoeia monographs often remains uncertain. Care must be taken, as similar or even identical Latin names are used by different pharmacopoeias to describe herbal materials with more or less deviant definitions.

Some important properties of herbal materials may not be seen from the Latin drug name. TCM knows so- called *daodi* qualities that are expected to show superior therapeutic quality. *Daodi* quality is defined as "medicinal material that is produced and assembled in specific geographic regions with designated natural conditions and ecological environment, with particular attention to cultivation technique, harvesting, and processing." (Zhao et al. 2012). Furthermore, herbal materials may or may not have been subjected to some kind of traditional processing, such as fermentation or heat treatment, sometimes involving adjuvant materials. In TCM, various "decoction pieces" can originate from the same raw herbal drug by subjecting it to different methods of *paozhi* processing (The term *paozhi* describes a group of TCM-specific methods of processing herbal materials. See subchapter "Specific Processing" for further details). Such decoction pieces, assumed to show distinct therapeutic properties, can be distinguished by their vernacular names but not by their Latin drug names. Because of these ambiguities, it was considered to be good practice in the context of CHM to use a minimum name standard for single herbal drugs that combines the Latin scientific species name including the naming authority with the plant part used, and (if relevant) the processing method, e.g., *Alisma orientale* (Sam.) Juz., tuber, salted (Chan et al. 2012).

(Finished) Herbal preparations, according to the WHO, may consist of one or more herbal materials or extractives and also excipients but not synthetic compounds or isolated constituents (WHO 2000; Aarland et al. 2014). Traditional HMPs are frequently prepared from mixtures of two or often more than 10 single herbal drugs (Yeong-Deug and Chang 2004). This applies for many traditional HMs from all over the world, as well as for HMs based on modern extraction technology. Such preparations have traditional or proprietary names that usually do not give information about their composition. Traditional complex HMs may also differ in their composition, depending on regional practices (Chan et al. 2012).

Factors Affecting the Chemical Composition of an Herbal Preparation

Plant Growth and Collection

As mentioned, herbs are gathered either by collecting wild plants or, increasingly, by cultivation. GAPC has the potential to minimize quality shortcomings or even health hazards arizing from contaminations (heavy metals, mycotoxins, microorganisms, pesticides), misidentification, cross-contamination, or undesired chemical alterations arising from inconsistent agricultural practices (Zhang et al. 2012). It was considered that optimizing cultivation practices towards an optimal biomass production of medicinal plants may result in herbs with a lower content of pharmacologically active secondary metabolites (Li et al. 2014). Notably, numerous studies have found great increases in various relevant secondary metabolites in drought-stressed plants, resulting from an oversupply of reduction equivalents that is channelled by the biosynthesis of highly reduced secondary metabolites (Selmar and Kleinwächter 2013). Furthermore, the chemical phenotype of a particular plant species is influenced by numerous spatial, temporal, and ecological factors (Tanko et al. 2005; Govindaraghavan 2008; Moore et al. 2014).

Drying

A first basic processing technique to conserve fresh plant material is drying. Simple air-drying techniques at an ambient temperature in the shade or in the sunlight reduce the residual water content to 5–10 %, related to the absolute dry mass, thereby halting enzymatic activity inside the plant material as well as microbial growth. Other drying techniques for plant material use higher temperatures in discontinuous (oven), continuous (e.g., belt dryer) processes (Müller and Heindl 2006), or freeze-drying (Abascal et al. 2005; Tanko et al. 2005). However, the stabilized dry herbs may significantly differ from the fresh product, which has been extensively studied for aromatic herbs, where the amount and composition (especially of volatile chemical components) depend on the drying protocol, but other components may be affected as well (Rocha et al. 2011; Bucar et al. 2013). Notably, also freeze-dried herbal material has repeatedly been shown to chemically differ from fresh material as well as from air-, oven- or vacuum-dried material (Abascal et al. 2005).

Depending on the chemical and physicochemical properties of the affected constituents, several factors may influence the chemical composition during drying, which were previously discussed in detail (Fig. 3.2) (Flück 1968; Wichtl 1970). Low drying temperatures may allow for the decomposition of plant constituents by the activity of herbal or microbial enzymes. Towards the end of a drying process, the structural collapse of the plant material may enable contact between otherwise separated enzymes and substrates, e.g., by merging protoplasm with vacuole content

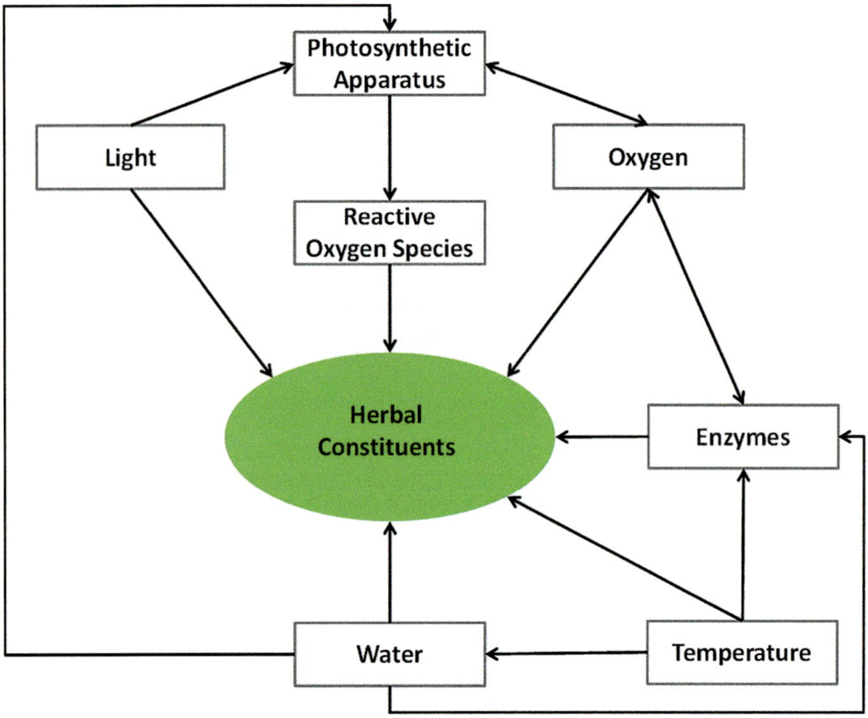

Fig. 3.2 Factors affecting herbal constituents during drying (Sendker 2006)

(Prothon et al. 2003; Kreis 2007). High drying temperatures may prevent enzymatically derived alterations, but instead cause the decomposition of thermolabile constituents or evaporation of volatile compounds. Furthermore, oxidative reactions may occur during drying, with or without direct involvement of plant enzymes such as peroxidases. Illumination during drying may intensify such changes (Sendker and Nahrstedt 2009; Bucar et al. 2013). The specified alterations will occur as long as the water content in the drying material allows such an action, and will stop when the residual water content of the drying material reaches about 10 %, related to the absolute dry mass.

In some cases, the chemical changes taking place during drying are essential for product quality. Well-known examples are the enforced fermentation of foodstuffs such as black tea, coffee beans, or vanilla pods. Significant examples involving medicinal plants are the formation of sennosides in drying leaves of *Cassia angustifolia* Vahl or of the typical aromatic components of *Valerianae radix* (Atzorn et al. 1981; Sticher 2007). Garlic oil that was hydro-distilled from air-dried, oven-dried, or freeze-dried plant material was distinguishable by the presence or absence of particular components in their GC profiles (Dziri et al. 2014). Finally, dried herbal material generally shows an increased porosity when compared to fresh plant material; the extent of the porosity depends on the drying method (Prothon et al. 2003) and may become relevant for extraction.

Stability During Storage

Numerous factors may affect the stability of HMP; these include pH, light, temperature, enzymatic activity, solvents, metal ions, or concomitant compounds (Gafner and Bergeron 2005). Unless some heat treatment has irreversibly denatured an herb's or herbal material's enzymatic activity, remoistening will reactivate enzymatic activity and thus continue to alter the chemical composition or allow microbial or fungal growth. A relative air-humidity of more than about 65 % is regarded as sufficient for reactivation of enzymes which in turn continue the actions already described for drying (Flück 1968). This is reflected in the WHO storage instruction for the label "protect from moisture," which allows "no more than 60 % relative humidity" (WHO 2003b).

A study on the stability of *Sennae folium* demonstrated that decomposition of sennosides into their agylcones occurred when storing the respective herbs at ICH-conform testing conditions (25–40 °C and 60–75 % relative air-humidity), with the degradation speed correlated to temperature. The decomposition was attributed to enzymatic activity, as no significant formation of aglycones was observed with a commercial dry extract that should be free of enzymatic activity (Goppel and Franz 2004). The same study demonstrated a strong influence of the packing material on the stability of the dry extract: while sennoside content was quite stable over one-year storage at 40 °C and 75 % relative air-humidity in PE-wide neck bottles or sealed aluminium bags, storage in PE bags led to a strong decline, with almost all sennosides being decomposed to unidentified products after one year. *Frangulae cortex* is an example of long-term storage being a prerequisite for the generation of the active anthraquinone glycosides by oxidation of the genuine anthrone glycosides that occur in the freshly harvested bark.

Chemical alterations also occur in herbal preparations. Residual enzymatic activity of glycosidases and polyphenol oxidase have been shown to affect the stability of herbal extracts even in 40 % ethanol or 50 % methanol (Nüsslein et al. 2000; Li et al. 2008). Besides, chemical reactions have been reported to occur between the alcoholic and other organic solvents and plant constituents (Li et al. 2008; Maltese et al. 2009; Ellendorff 2014).

One of the most interesting aspects is the stabilizing or destabilizing influence of concomitant compounds which thereby may indirectly influence an HM's activity without showing relevant pharmacological activity themselves. Herbal antioxidants have repeatedly been shown to stabilize other phytochemicals, such as carotenoids in black tea or carnosic acid in rosemary extracts (Gafner and Bergeron 2005). Organic acids, such as citric or malonic acid, have been shown to cause the rearrangement of flavonoids (Gafner and Bergeron 2005) or to increase the stability of phenolics in a glycerin extract of *Echinacea purpurea* L. (Bergeron et al. 2002). Similar effects were achieved in the same study by adding an extract from hibiscus flowers, which may indicate a possible reason for herbal mixtures as established in many traditional medicinal systems (q.v. subchapter "Stabilization" for stabilizing effects of concomitant compounds). An interesting observation during the storage of a Thai traditional herbal mixture was the generation of new fatty acid esters that did not occur in individually stored herbs (Nualkaew et al. 2004).

Specific Processing

According to the WHO, specific processing represents post-harvest measures beyond drying to improve the purity of the plant part being used; reducing drying time; preventing damage from mold, other microorganisms, and insects; detoxifying indigenous toxic ingredients; and enhancing therapeutic efficacy (WHO 2003a). A practice not falling under this definition is the remoistening or "softening" of dry herbs, usually applied to whole plant parts in order to make their texture more suitable to comminution and thus avoid fracturing so as to yield more or less uniform fragments. This is especially relevant for the production of Chinese decoction pieces, which are often cut into product-specific shapes. Studies on the chemical impact of remoistening are lacking, but we have observed a rapid and complete decline of prunasin after remoistening leaves of *Olinia ventosa* (L.) Cuf,, thereby continuing an oxidative degradation to amides that already occurred during drying at 30 °C (Sendker 2006; Sendker and Nahrstedt 2009).

Examples where specific processing long ago became an integral part of traditional medicinal systems are *samskara* techniques from Ayurveda, or *paozhi* techniques from TCM (Zhao et al. 2010; Chand et al. 2013). In particular, *paozhi* has received significant attention in the past few years, and the need for research and standardization has been recognized. Meanwhile, numerous studies indicate a significant impact of *paozhi* techniques on the activity and chemical composition of several CHMs (Zhao et al. 2010; Jiang et al. 2013). The *Chinese Pharmacopoeia* lists 24 different *paozhi* techniques, such as frying and calcining of steaming with or without the addition of adjuvant materials such as vinegar or honey (State Pharmacopoeia Committee 2010). The application of *paozhi* processing to raw herbs aims at instilling in them the desired properties for therapeutic application and thereby transforming them into decoction pieces. Many complex traditional Chinese formulas are formed by decoction and application of the resultant preparation. Prominent examples can be found in the detoxification of herbal materials from different highly toxic *Aconitum* species by *paozhi* and also by Ayurvedic samskara processing techniques (Singhuber et al. 2009; Jaiswal et al. 2013). Notably, different kinds of decoction pieces can be formed from the same raw material by applying different processing techniques with each product claimed to show distinct therapeutic profiles (Zhao et al. 2010). Comparing raw herbs with their processed decoction pieces has brought to light significant differences in their chemical profiles, e.g., allowing the analytical differentiation of five differently processed products from *Coptis chinensis* Franch. Also, several studies show that during *paozhi* processing, new compounds are formed that are not present in the untreated raw material (Jiang et al. 2013).

Extraction

Herbal medicines can be administered to patients in several forms, including the ingestion of comminuted herbal materials. However, besides rather rare preparations involving fresh plant material, a vast majority of finished herbal preparations are produced by extracting dry herbal materials. The resulting extracts may be ingested directly in the form of an aqueous infusion, or decoction, or an extract with, for example, aqueous ethanol, or may be further processed into dry or spissum extracts to form granules, single dosage forms, or other compliant phytomedicines. The chemical composition of any extract generally differs from that of the herbal material, in that only part of the herbal constituents will be extracted, depending on extraction methods, temperature, kind and amount of solvent used, and pressure (Gaedcke and Steinhoff 2000). Any of these factors can affect a herbal preparation's chemical profile and hence its therapeutic outcome.

Modern herbal preparations are often based on dried extracts prepared with rather lipophilic extraction solvents, thereby excluding large amounts of unspecific polar compounds such as sugars or amino acids that impose technological problems and are usually not considered therapeutically relevant. At the same time, an extensive extraction of less polar plant constituents, such as flavonoids, terpenes, or saponins is achieved – that is, including those constituents typically considered to be *constituents with known therapeutic activity*, *active markers*, *and analytical markers* and concomitant compounds or co-effectors with solubilizing or stabilizing properties (Gaedcke and Steinhoff 2000). In contrast, traditional HMs are usually extracted with hot or cold water and often prepared from herbal mixtures following specific and sometimes complex protocols, which is especially true for decoctions utilized in TCMs (Sheridan et al. 2012).

A particularly interesting aspect is that several studies found interactions between the various herbal materials within a joint decoction or infusion of two or more herbs (mixed extraction or co-extraction). Infusions prepared from leaves of *Thymus vulgaris* L., in combination with the roots of either *Primula veris* L. or *Glycyrrhiza glabra* L., contained a reduced amount of essential oil in the final product, which was mainly attributable to a loss of thymol. The effect was hypothesized to result from the presence of saponins, which represent the main constituents of both root drugs (Tschiggerl and Bucar 2011). An impact of saponins on the solubility of herbal constituents from co-extracted herbal materials is indicated by multiple studies (Güçlü-Üstündağ and Mazza 2007) and can be complex as shown by studies on the solubilization of saikosaponins from *Bupleuri radix* with ginseng saponins. Neutral ginseng saponins cannot solubilize saikosaponins on their own, but potentiate the solubilizing effect of acidic ginseng saponins (Watanabe et al. 1988; Zhou et al. 1991).

A series of studies on the traditional Chinese two-herb formula *of Danggui buxue tang* (DBT), composed of *Astragali radix* and *Angelicae sinensis radix*, shows that co-extracted decoctions contained particular compounds at higher concentrations when compared to the mixture of individually decocted herbs (Mak et al. 2006; Choi et al. 2011a). Also, the herb ratio was found to have an impact (Gao et al. 2006) on the time span of soaking DBT in cold water before decoction (Zhang et al. 2014a). A very relevant finding was that a factor from earlier in the production chain, namely the *paozhi* processing of *Angelicae sinensis radix*, increased the extraction yield of astragaloside IV and calcycosin from *Astragali radix* under conditions of co-extraction (Dong et al. 2006).

Finally, a further processing of liquid extracts to spissum or dry extracts allowing for the production of convenient pharmaceutical products such as granules or tablets may affect the chemical composition. The drying of such extracts involves evaporation with the eventual addition of water to the spissum extract in order to remove residual organic solvents, and then usually a final drying step with the addition of excipients to improve the technological properties of the final dry extract. In order to reduce the microbial count, a short steam treatment may be applied before or during the drying process. Chemical alterations to consider here are the loss of thermolabile or steam-volatile components such as essential oils, coumarins, or sennosides, or precipitation of compounds that become insoluble during the process (Gaedcke and Steinhoff 2000).

Relevance of Chemical Adulterations

Besides variations in the plant metabolome due to different genetic resources or growth conditions, the manufacturing chain of an herbal preparation involves numerous production steps, each of which may intentionally or unintentionally affect the composition of the finished herbal preparation that is consumed by a patient (Table 3.1). In cases where a therapeutic property of an herbal material can be satisfactorily attributed to a limited number of chemical entities, any influential factor can be optimized towards an optimum yield of these compounds. Examples of such cases are cayenne pepper, where the relieving effect on muscle pain is attributable to the capsaicinoid compounds, or a number of herbal materials whose purgative effect is attributable to anthraquinone glycosides. However, for most HMs, there is insufficient information for evaluating the therapeutic properties from their chemical composition. Hence, the meaning of chemical alterations for biological activity can neither be evaluated, regardless if the alteration is caused by growth conditions, drying, specific processing or artifact formation during storage in an alcoholic solution. The degradation of herbal components may be detrimental to product quality, but also beneficial when the degradation affects toxic compounds or gives rise to components with greater activity (Cordell 2014). *Frangulae cortex* is an excellent example of this, as the genuine toxic anthrone glycosides are degraded to the active anthraquinone glycosides by oxidation during storage for at least one year (Werner and Merz 2007).

Herbal Material and Products: Chemical Composition

Classification of Plant Constituents

Unspecific Matrix Constituents

Today about 200,000 different plant metabolites have been reported, and up to 15,000 have been estimated to occur in a single plant species (Wolfender et al. 2015). Proteins, reserve carbohydrates and ubiquitous primary metabolites such

Table 3.1 Selection of factors affecting the quality of HMP based on *Hypericum perforatum* L

Production step	Impact	Reference
Wild collection	Chemotype without rutin	Pavol et al. (2001)
	Subspecies without hypericine and hyperforin	Bruni and Sacchetti (2009)
	Adulteration with other *Hypericum* species (distinct chemical profiles, in particular lacking hyperforin)	Umek et al. (1999)
		Murch and Saxena (2006)
Growth	Nickel-exposed plants (25–50 mM) contained no hyperforin and only ~10 % of the control group's naphthoquinones	Murch et al. (2003)
	Insect herbivory increases contents of hypericin and hyperforin by 30-100 %	Murch and Saxena (2006)
	Water stress increases hyperforin three- to fourfold and decreases naphthodianthrone content	Bruni and Sacchetti (2009)
	Exposure to bright light increases hypericin content	
Harvest	Highest contents of naphthoquinones and flavonoids around early flowering period and of hyperforin during fruiting period	Seidler-Łożykowska (2003)
	Raised cutting height yields herbal material with increased concentration of hypericine	
Drying	Light exposure lowers protonaphthodianthrones/naphthodianthrone ratio	Bruni and Sacchetti (2009)
	80 % increase in hypericin and total flavonoids contents when drying younger plant material at low temperatures (10–30 °C)	
	Decrease of flavonol glycosides at drying temperatures of 40–80 °C	
	95 % decrease in hyperforin content after air-drying	
Stability during storage	Hyperforin and protohypericin disappear under 4 h illumination	Ang et al. (2004)
	Hyperforin and naphthodianthrones concentrations decline faster at low pH	
Extraction	Extract yield and composition depends on diminution of herbal material, extraction solvent, -time and -temperature	Gaedcke (2003)

as sugars, amino acids, organic acids, or photosynthetic pigments constitute a major mass fraction of an herbal materials soluble content as well as of their polar extracts. This is reflected in declining extract yields when increasing the amount of ethanol in a hydro-alcoholic extraction solvent (Gaedcke and Steinhoff 2000). As the therapeutic properties of HMPs are usually attributed to secondary metabolites of lower polarity, these ubiquitous substances are often regarded as dispensable matrix material that causes technological problems or increases the required dosage of dry extracts in single-dosage forms (Eder and Mehnert 2000). Therefore, efforts are usually made to reduce this matter in the industrial production of dry extracts.

Plant Secondary Metabolites

With such exceptions as the polysaccharides of *Echinacea* products that have been considered to partake at their immunomodulatory activity (Janeš 2010), constituents contributing to therapeutic properties of an herbal preparation usually belong to the plants' secondary metabolites. They represent a huge variability of substances of different chemical, physicochemical, and biological properties. Each individual plant species can be expected to show a distinct spectrum of secondary metabolites with individual quantitative distributions depending on genetic pre-conditions and ontogenesis (Moore et al. 2014). No uniform system exists to classify plant secondary metabolites. Frequently referred substance groups are defined according to their biosynthetic origin (e.g., terpenes, polyketides, phenylpropanes), functional groups (e.g., phenolics, phenolic acids, amides, amino acids), specific chemical constitution (e.g., flavonoids, coumarins, iridoids, lignans, glycosides), reactivity (e.g., cyanogenic glycosides), occurrence of nitrogen (alkaloids), physicochemical properties (compatible solutes, saponins, tannins, bulking agents), or biological properties (pungent or bitter compounds, cardiac glycosides) with significant overlap between the definitions.

Therapeutic Relevance

Apart from academic issues, it is not the presence or quantity of a particular chemical entity, but the therapeutic properties that are passed into a product through the herbal constituents that are of interest. With regard to the relevance of particular compounds, or groups of compounds, to a particular therapeutic property, plant constituents have been rated as *constituents with known therapeutic activity* when their content widely correlates with a product's biological activity. By contrast, compounds are rated as *active markers* when they cannot solely explain a product's efficacy to a satisfactory degree or are known to have an indirect impact as co-effectors on, for example, stability or solubility (EMEA 2011). In the context of quality control, assessment of these two groups of constituents allows for evaluating a product's therapeutic properties and hence its quality. Notably, a constituent with known therapeutic activity, an *active marker*, or a co-effector, is not only linked to a particular herbal product but also to its indication. Good examples are preparations of *Rheum palmatum* L. and *R. officinale* Baillon, which are used in Europe as laxatives for their content of anthraquinone glycosides and as styptic for their content of tannins (Knöss 2007). In the case where no sufficient link between activity and constituents is possible, *analytical markers* substitute the role of *constituents with known therapeutic activity* or *active markers* for quality control purposes. *Analytical markers* should be characteristic for the herbal material and may give indications on stability, or can be used for in-process controls. *Negative markers* (such as allergens) impair product quality and *accompanying constituents* represent a widely polar matrix, which is not assumed to be relevant to product quality (Busse 2000).

Actives Directly Related to Quality

For relatively few herbal preparations, quality can be assessed by the quantification of *constituents with known therapeutic activity*, as is the case for capsaicinoids in cayenne pepper used for the relief of muscle pain or total hydroxyanthracene glycosides in senna pods, and some other herbal drugs used against constipation. The therapeutic quality of such preparations can be straightforwardly controlled by rather simple and robust methods such as HPLC or photometric assays and adjusting the preparations to a defined concentration of *constituents with known therapeutic activity* (standardized extracts). For most herbal preparations, however, quality seems to depend more or less on multiple constituents; otherwise we have no reliable knowledge about active constituents at all. In such cases, quality control methods targeting single chemical entities cannot satisfactorily guarantee sufficient product quality.

For about the last 15 years, this situation has been increasingly attributed to the idea of synergy (Eder and Mehnert 2000; Williamson 2001; Wagner and Ulrich-Merzenich 2009; Ma et al. 2009; Gertsch 2011; Yang et al. 2014). Wagner and Ulrich-Merzenich (2009) have described mechanisms that may account for synergistic effects, including (1) pharmacological synergy in the sense of Berenbaum (1989), where the combination of two substances shows an over-additive effect on a particular target, and (2) synergistic multi-target effects based on either the polyvalency of single constituents, or multiple constituents with distinct targets – or both. In any case, synergy towards a therapeutic effect is achieved by addressing multiple targets. Examples of synergy within herbal preparations related to the multiple target mechanism are found in the spasmolytic activity of cannabis extracts, the antidiabetic activity of *Coptidis Rhizoma* extracts, and the antidepressant or anxiolytic activities of St. John's wort, and kava-kava extract, respectively (Wagner and Ulrich-Merzenich 2009; Ma et al. 2009; Yang et al. 2014). The constituents involved in these and other examples can be expected to contribute directly to an HM's quality.

Actives and Co-effectors Indirectly Related to Quality

In contrast, synergy may also arise from constituents that do not show pharmacologic activities related to the indication of an herbal preparation, but indirectly contribute by increasing the bioavailability of pharmacologically active constituents by improving their solubility or stability, or affecting their absorption, metabolism, or elimination. These mechanisms have been classified as physicochemical or pharmacokinetic synergy, respectively (Wagner and Ulrich-Merzenich 2009). Numerous observations of such interactions have been made, and physicochemical interactions are expected to generally occur in herbal preparations (Eder and Mehnert 2000; Yang et al. 2014). The responsible co-effectors or mechanisms often remain

Table 3.2 Possible antidepressive functions of substances from St. John's wort (Nahrstedt and Butterweck 2010)

Function	Substance
Monoamine oxidase inhibition	Hypericin
	Xanthones
	Flavonoids
GABA receptor agonism	Amentoflavon
Neurotransmitter reuptake inhibition	Hyperforin
Corticotropin-releasing-factor receptor 1 antagonism	Hypericin
Catechol-*O*-methyltransferase inhibition	Extract
Interleukin 6 suppression	Extract
Normalizing augmented hypothalamic-pituitary-adrenal axis activity	Flavonoids
Enhancing hypericin bioavailability	Flavonoids
	Proanthocyanidins

unknown, probably because they are not directly detectable by classical target-based approaches such as bioactivity-guided isolation.

The Example of St. John's Wort

Herbal medicines based on St. John's wort illustrate how multiple active markers jointly contribute to their antidepressant activity and are partly co-effectors in that they have an impact on the bioavailability of other compounds. The widely applied hydro-alcoholic extracts of St. John's wort are phytochemically characterized to an extent of 60–70 %. The naphthodianthrones hypericin and pseudohypericin, the unstable prenylated polyketide hyperforin, and several flavonoid glycosides show activities that influence the complex pathological situation of depression by various mechanisms (Table 3.2). Additionally, flavonoid glycosides and procyanidins have been shown to increase the bioavailability of hypericin, which, as a pure compound, is hardly soluble in water and poorly absorbed as determined in the Caco-2 model. Besides, other phenolics – especially dimeric and trimeric procyanidins – were shown to increase hypericin's water solubility up to 120-fold and decrease its octanol/water partition coefficient (physicochemical synergism), while flavonoid glycosides in particular could increase the basolateral transport of hypericine through Caco-2 cells and oral bioavailability in rats (pharmacokinetic synergism) (Butterweck et al. 1998; Nahrstedt and Butterweck 2010). Nöldner and Schötz (2002) found that a St. John's wort extract with a conspicuously low content of rutin was inactive in the antidepressant model of a forced swimming test, but achieved usual activity after replenishing the extract with rutin, which was inactive when administered alone. The *European Pharmacopoeia* accommodates this situation by demanding – for St. John's wort and St. John's wort dry extract – quantified minimum concentrations for total naphthodianthrones and flavonoids; however, it limits

Table 3.3 Selected examples of solubilization in water

Solubilizing substance	Solubilized substance	Factor	Reference
Frangulae cortex extract	Anthraquinone glycosides	n.s.	Eder and Mehnert (2000)
Digitalis purpurea folium extract	Digitoxin	n.s.	Eder and Mehnert (2000)
Saponin fractions of different herbal materials	Digitoxin	0.5–1.5	Walthelm et al. (2001)
	Rutin	0.8–2.8	
	Aesculin	0.7–1.3	
Ammi visnagae fructus extract	Khellin	4	Eder and Mehnert (2000)
Buckwheat extract	Rutin	5	Eder and Mehnert (2000)
Rhein-8-*O*-glucoside, potassium salt	Polymeric proanthocyanidin fraction of rhubarb	2.5–13	Tanaka et al. (1997)
Kava rhizome extracts	Kava pyrones	5–40	Eder and Mehnert (2000)
Digitalis lanata leaf extract	Digoxin	>10	Eder and Mehnert (2000)
Ginsenoside R_0	Saikosaponin a	29	Güçlü-Üstündağ and Mazza (2007)
Hemsleya macrosperma bisdesmosides		36–62	
Glycyrrhizin	Isoliquiritigenin	57	Eder and Mehnert (2000)
Steroidsaponins	Convallaria cardiac glycosides	100	Eder and Mehnert (2000)
Procyanidin B2	Hypericin	120	Nahrstedt and Butterweck (2010)
Procyanidin C1		80	

Factors apply for experimental conditions given in the references and the literature cited there
n.s. not specified

the concentration of hyperforin which, despite its apparent activity as a serotonin and dopamine reuptake inhibitor, has proven itself to cause numerous problematic interactions by CYP3A4 induction.

Phenolics as Solubilizers

Herbal constituents are frequently reported to show an increased solubility or dissolution rate in extracts when compared to the isolated compound, which is usually explained by the presence of concomitant substances (Eder and Mehnert 2000) (Table 3.3). The identity of these co-effectors, the mechanisms of interaction, and the therapeutic relevance for particular herbal preparations often remain unknown, yet numerous studies give an impression of the relevance and potential of such physicochemical interactions. An aspect of particular interest, especially with regard to mixed herbal preparations, are the interactions such as between naphtodianthrones and rather unspecific and widespread compounds such as procyanidin B2, rutin, hyperoside, or chlorogenic acid (Nahrstedt and Butterweck 2010). Comparable interactions affecting tannins' *n*-octanol-water-partition-coefficient, solubility, or adsorption on hide powder have been described to occur between tannins and paeoniflorin, amygdalin, glycyrrhizin potassium salt, aconitine trifluoroacetate, liquiritin apioside, and rhein-8-*O*-glucoside, respectively (Tanaka et al. 1997).

Saponins as Solubilizers

Another widespread substance class known for physicochemical interactions with other herbal constituents and other chemical drugs are saponins. There are numerous examples of the interactions of various saponins and diterpene glycosides with each other or with other herbal constituents, often resulting in an increased water-solubility of poorly soluble substances (Güçlü-Üstündağ and Mazza 2007; Liu 2009, 2010). Systematic studies on the impact of different saponins on the water-solubility of quercetin, rutin, digitoxin, and aesculin, however, found both increased and decreased solubility of these substances, depending on the kind and amount of saponin added. Notably, a positive impact on solubility was not exclusively attributable to micellar solubilization, as some saponins were effective below their critical micellar concentration, thus indicating an alternative mechanism of solubilization (Walthelm et al. 2001; Güçlü-Üstündağ and Mazza 2007).

Complexes and Colloidal Dispersions

The formation of molecular complexes involving saponins has been reported and might account for the described alterations in solubility, as has also been discussed for the solubilizing effects on tannins (Tanaka et al. 1997). Model studies on complexes involving *Hedera* saponins and other molecules indicate that such molecular complexes may have an impact on biological activity (Yakovishin et al. 2011, 2012). However, complexation may also lead to reduced solubility or the formation of sediments within an aqueous environment leading to lowered extraction yields of these compounds. A well-known example is the complexation of alkaloids with tannins or acidic constituents (Eder and Mehnert 2000).

Patel et al. (2012) examined a 2:1 mixture (molar ratio) of the alkaloid berberine and tannic acid that formed a sediment in an aqueous solution. If tannic acid was mixed with one part of glycyrrhizin before berberine was added, a stable colloidal dispersion was formed instead of sediment. Colloid-like aggregates have been also been detected in numerous tinctures (St. John's wort, among others), and aqueous herbal decoctions of single herbal materials and herbal mixtures as used in TCM (Zhuang et al. 2008; Chenery 2009). In a recent study, more than 95 % of the ephedrine and pseudoephedrine present in a decoction of three herbal components and gypsum was found to be associated with colloid-like aggregates. It was suggested that proteins, phospholipids, and primary and secondary metabolites – including bioactive compounds – might constitute the aggregates observed by dynamic light scattering or the Tyndall effect. The aggregates may affect bioavailability and activity of the preparations and the aggregate formation may be a molecular mechanism accounting for synergetic effects (Zhuang et al. 2008; Hu et al. 2009; Ma et al. 2009; Chenery 2009; Zhou et al. 2014). Evaluating the actual therapeutic importance of such aggregates is difficult, as compounds susceptible to aggregation tend to interfere with bioassays and may yield false positive or false negative results as well as bell-shaped concentration-activity relationships (Bisson et al. 2015).

Solubilization by Polar Matrix Constituents

The impact on the solubility of herbal constituents can be attributed to the polar matrix that is mainly composed of carbohydrates, organic acids, and amino acids, which represents a large percentage of the soluble matter in polar extracts as well as in herbal materials. These matrix compounds promote an amorphous state of other constituents in solid dispersions, facilitate wettability, and thus allow for increased dissolution rates and also for transiently increased solubility by formation of supersaturated solutions (Eder and Mehnert 2000).

A recent finding is the solvating capacity of natural deep eutectic solvents (NADES), which are formed of low molecular polar compounds such as sucrose, glucose, fructose, proline, choline, malic acid, and citric acid, etc., which also abundantly occur in plants. These compounds, while being solid in their pure state, may form viscous liquids in combination. NADES have been shown to dissolve poorly water-soluble plant constituents; there is an implication that they account for *in vivo* concentrations of flavonoids and anthocyanidins that exceed their solubility in water as pure compounds (Choi et al. 2011b). Notably, NADES can be prepared by the solvent-evaporation method, where (aqueous) solutions of the ingredients are combined and the water evaporates afterwards (Wikene et al. 2015), which is perfectly comparable to the situation of drying plant materials or extracts. Hence it seems reasonable that NADES play a role as dissolution enhancers in herbal preparations.

Stabilization

Little attention has been paid to the possible stabilizing or destabilizing effects of concomitant herbal substances. A rather prominent example is the protection of unsaturated fatty oils by concomitant antioxidants against becoming rancid.

A study on the stability of iridoids (valepotriates and baldrinales) found that these compounds are quickly decomposed in hydroalcoholic preparations from *Valeriana officinalis* root, but persist in hydroalcoholic preparations of *Centranthus ruber* roots. This difference was explained by the presence of large amounts of arginine in *V. officinalis* roots, which is thought to react with the iridoids (cited in Nahrstedt 1989). NADES have been described to enhance the stability of safflower colorants against various stressors compared to aqueous and hydro-alcoholic solutions, which was explained by intense H-bond interaction between the solute and NADES (Dai et al. 2014). Many of the compounds forming NADES, plus numerous comparable polar compounds such as trehalose or glycinbetaine, have been described as so-called "compatible solutes" before, protecting numerous organisms such as desiccation-tolerant plants against various kinds of abiotic stress, including oxidative stress and dehydration. A possible protective mechanism is the replacement of hydratizing water molecules from the surface of macromolecules and cellular structures by compatible solutes ultimately leading to the vitrification of these structures by forming an amorphous and highly viscous glassy state (Rascio and Rocca 2005). Although compatible solutes are only known for the stabilization of larger structures, their stabilizing effects via H-bond interactions strongly resembles NADES and thus evince relevance for the stabilization of secondary plant metabolites as well. However, both NADES and compatible solutes are reported to have an effect at very high concentrations of at least 100 mM; it has also been shown for NADES that water diminishes their stabilizing effect (Dai et al. 2014). Consequently, stabilization of these compounds should be restricted to dry herbal material and dry or spissum extracts.

Pharmacokinetic Synergy

It has been shown in many studies that herbal constituents may have an improved bioavailability when administered as part of an extract rather than as a pure compound (Table 3.4). Improved bioavailability of pharmacologically active substances may result from solubilization or stabilization by co-effectors as already mentioned, or from the effects on their absorption, metabolism, or elimination. A classic example is the Amazonian psychotropic product of *ayahuasca*, where the oral bioavailability of the psychotropic dimethyltryptamine from one herbal component depends on the presence of MAO-inhibiting harmane alkaloids from another herbal component (Gertsch 2011). Today, many examples of plant constituents have been shown to interact with intestinal transporters and metabolic enzymes (Gurley 2012; Yang et al. 2014). Besides the potential relevance of pharmacokinetic synergy for phytotherapy, there is also rising concern about undesired herb-drug interactions. In particular, St. John's wort extracts have been recognized as causing drastically reduced plasma concentrations of the immunosuppressant cyclosporine. However, most of the research is based on *in vitro* examinations with limited significance for therapeutic reality (Gurley 2012).

A number of herbal constituents, the most prominent of which is piperine, have been designated as "bioavailability enhancers" or "bioenhancers" for their pharmacokinetic interaction potential. The application of piperine as bioavailability enhancer

Table 3.4 Selected examples for pharmacokinetic interactions of herbal products or their constituents

Preparation/compound	Effect	Reference
Decoction of *Gardenia jasminoides*, *Fructus auranti immaturus* and *Cortex magnolia officinalis*	Increased AUC of geniposide after administration of complete decoction when compared to administration of *Gardenia jasmonoides* decoction	Sun et al. (2012)
Decoction of *Angelicae dahuricae radix* (ADR), *Scutellariae radix* (SR)	Essential oil and coumarins from ADR increase intestinal absorption of baicalin from SR (rats)	Yang et al. (2014)
Atropa belladonna, fresh leaf extract	Increased permeation through isolated rat small intestine due to the presence of flavonol glycosides	Eder and Mehnert (2000)
Digitalis purpurea extract	Increased absorption of cardiac glycoside gitoxin in the presence of concomitant compounds modeled by a *Digitalis purpurea* extract fraction that was inactivated by alkali treatment	Eder and Mehnert (2000)
	Retarded absorption of Digitoxin through mucilaginous concomitant compounds	
Piper methysticum extract	Increased bioavailability concentrations of kava-lactones when administered as extract (mice, dog)	Eder and Mehnert (2000)
Hypericum perforatum extract	Increased bioavailabilty of hypericine in the presence of phenolic compounds	Nahrstedt and Butterweck (2010)
Ammi visnaga extract	Faster absorption of Khellin when administered as extract	Wagner and Ulrich-Merzenich (2009)
Grapefruit juice	Impact on pharmacokinetics of numerous drugs by:	Bailey (2010)
	Inhibition of CYP34, CYP1A2 Inhibition of intestinal transporter OATP	Bailey et al. (2012)
6,7-Dimethylescaletin, Rhein, Geniposide	Oral administration of triple-combination increases AUC of each single compound compared to individual administration (rats)	Yang et al. (2014)
Piperine	Increases bioavailability of numerous chemical drugs and herbal constituents by inhibition of metabolic enzymes	Dudhatra et al. (2012) Ajazuddin et al. (2014)

results from the ancient Ayurvedic preparation of "Trikatu," composed of *Piper nigrum* L., *P. longum* L. and *Zingiber officinale* Roscoe. Piperine has been known to inhibit several drug metabolizing enzymes, increase blood plasma concentrations, and retard the elimination of several chemical drugs (Dudhatra et al. 2012; Ajazuddin et al. 2014). In a study involving eight human volunteers, piperine increased the AUC of curcumin by about 2000 % (Shoba et al. 1998). Notably, quercetin is also designated as a bioactivity enhancer for its interaction with CYP3A4 and P-glycoprotein. Quercetin is one of the most widespread flavonoids occuring in many plants in glycosidic forms. The quercetin glycosides of rutin and hyperoside are present in St. John's wort extract and have been shown to increase the transport of hypericine through Caco-2 cells, apparently by inhibiting p-glycoprotein (Nahrstedt and Butterweck 2010).

Chemometrics for Quality Control of Finished Herbal Medicines

As we have seen, because HMPs are frequently prepared from two or more herbs, they contain a large number of various secondary metabolites representing different chemical classes (Yuan and Lin 2000). Of these metabolites, some are active, some are inactive, and some can be toxic. Because of this complexity, the correlation between chemical composition and bioactivity of many HMPs is incomplete (Chau et al. 2009).

Apart from a few drugs where *constituents with known therapeutic activity* can widely explain an HM's therapeutic property, there is a general knowledge gap about the nature and function of minor bioactive constituents in HMPs. This lack of knowledge hinders the satisfactory quality control of many herbal preparations. Many of the quality control methods in use today are surrogates, and even for abundantly researched herbs such as hawthorn and valerian, no clear link could be made between plant constituents and therapeutic quality. But also multiplicity of information about active constituents can hamper the establishment of straightforward quality control methods, as is the case for chamomile flowers or eleutherococcus root.

Quality Control by Only One or a Few Constituents

When considering the many factors affecting the chemical composition of herbal preparations throughout the production chain and the indirect contributions of concomitant compounds to therapeutic properties, it seems likely that in such cases numerous constituents account for quality rather than a single, hitherto (un)known constituent. A pragmatic and reasonable approach accommodating this situation is given by the handling of quantified extracts and other extracts in *Ph. Eur.*: in order to meet their requirements, e.g., towards the minimum total flavonoid content of passion flower dry extract, the extracts may be adjusted only by mixing different batches. Thus the composition of concomitant compounds is less affected than by adjusting the content with excipients such as dextrin, as allowed for standardized extracts whose therapeutic quality depends solely on *constituents with known therapeutic activity*. As long as only one minimum concentration has to be met (typical for the category of other extracts), the mixing of extract batches is straightforward. If more than one requirement has to be met, adjustment becomes increasingly difficult. The monograph for ginkgo dry extract, refined and quantified, defines concentration ranges for each total flavonoid, bilobalides, and ginkgolides, plus a maximum concentration for the allergen ginkgolic acid. Three different quantitative assays are necessary to measure these contents. For St. John's wort dry extract, a concentration range for total hypericins, a minimum concentration for total flavonoids and a maximum concentration for hyperforin are required, and are tested using two different quantitative assays (Council of Europe 2015). These examples show how increasing knowledge about relevant constituents has already led to quality standards based on rather complex characteristics of constituent profiles.

Quality Control by Chemical Fingerprints

Today, there is increasing interest in evaluating herbs, herbal materials, and herbal preparations based on chemical fingerprinting – that is, analytical data incorporating the broadest possible range of herbal constituents. Provided that such chemical fingerprints are strictly comparable with regard to signal positions and detector response characteristics, they can be evaluated by methods of multivariate statistics in order to classify samples, predict their quality, or reflect on particular constituents' contributions to a specific property. A particular advantage of using such comprehensive fingerprints is that any detectable constituent, whether known or unknown, can be accounted for evaluation, weighted according to their meaning for, predicting an herbal preparation's biological activity or an herb's regional origin (Gad 2013).

Chromatographic Techniques

Traditionally, the chemical fingerprint, in its most simplistic form as the thin layer chromatographic (TLC) profile, has been used for the monographic identification of various herbal drugs by qualitative or semi-quantitative analysis. Quantitative measurements are accessible by densitometry or video-densitometry. Other fingerprinting techniques that have become widely accessible and acceptable are high-performance liquid chromatography (HPLC/UHPLC), and gas chromatography (GC). Qualitative and quantitative data on such fingerprints allow for capturing the ratios between their single constituents as well as the presence of adulterants. For fingerprint analysis, a similarity index (SI) is used to determine how much sample fingerprints deviate from the fingerprint of an authentic reference standard. This can be applied to HMPs from different varieties and hybrids, various geographical regions, from various producers, or to HMPs that have been processed in different ways. The SI provides a quantitative measure of this kind, which compares the intensities of such signals of two analyses that correspond to each other in signal position (retention time plus eventual *m/z* or chemical shift value). The chemical composition of two HMPs is considered almost identical to each other if the SI value is close to 1.0 (Yi et al. 2009; Chau et al. 2011). Several (dis)similarity approaches, such as (dis)similarity metrics or exploratory analysis approaches applied on chemical fingerprints, have been reviewed by Goodarzi et al. (2013).

Spectroscopic and Hyphenated Techniques

In recent years, chemical fingerprinting has become increasingly more sophisticated and now various automated and hyphenated chromatographic and spectroscopic techniques are used to characterize the complex metabolomic patterns in many HMPs (Govindaraghavan et al. 2011; Chen et al. 2013; Hu and Xu 2014). Analytical methods with the greatest ability to display the composition of the entire extract by fingerprint data are based on NMR and MS detection, either used on the samples directly or after separation (hyphenated techniques: LC-MS, GC-MS, LC-NMR). Each analytical method has its advantages and limits that determine their applicability. NMR-based methods have a very good reproducibility with regard to signal positions and intensity patterns but lack sensitivity. MS-based methods are highly sensitive and thus able to display minor compounds, but the appearance of mass spectra is strongly influenced by instrumental and experimental factors so that data obtained from different instruments or laboratories are hardly comparable. Similarly, LC separations from different sources vary in retention time, peak shapes, etc., that depend on instrumental details, column age and so forth, while GC separation will limit the fingerprints to volatile constituents (Verpoorte et al. 2008). Hyphenated instruments show greatly improved performances in terms of selectivity, chromatographic separation abilities, measurement, and precision (Sheridan et al. 2012; Chen et al. 2013), and they can generate vast quantities of data.

Chemometric Methods

Chemometrics is a useful tool for evaluating data relating to the quality of HMP. It can be used to optimize data from such hyphenated experiments, reducing or eliminating unwanted sources of variation (such as reducing noise levels to a minimum), and aligning peaks. These techniques include principal component analysis (PCA), linear discriminate analysis (LDA), spectral correlative chromatography (SCC), information theory (IT), local least square (LLS), heuristic-evolving latent projections (HELP), and orthogonal projection analysis (OPA) (Sheridan et al. 2012; Bansal et al. 2014). It is now widely accepted that a standardized chemical fingerprint for an herbal product is not always the best method of standardization of such complex mixtures, although such methodology is the basis of a number of herbal monographs (Tistaert et al. 2011). The reason for this is that activity can be due to multiple trace constituents and synergisms, which are not the dominant signals in a spectrum. Therefore, the optimal method of analysis should capture these relationships – if not for day-to-day quality control, then certainly to help understand the active components and their true mechanisms of action. This knowledge could then be used to optimize HMs and to design more reproducible modernized formulations that still retain the activity and potency of their ancestors: Wang et al. (2006) used a computational approach to predict the bioactivity of Qi-Xue-Bing-Zhi-Fang, a CHM used to decrease cholesterol. They examined the proportion of two active components of the CHM, which was optimized based on its metabolomic-bioactivity model to obtain a new formulation for this botanical drug. Currently there is a large quantity of research being carried out to determine optimal methods to assess the quality of HM from a combined metabolomic-bioactivity profile perspective (Fan et al. 2006; Jiang et al. 2013).

Examples for the Chemometric Analysis of Chemical Fingerprints

A chemometric approach has been used by Zhu et al. (2014) to study the efficacy of *Radix Lateralis Preparata* (*Fuzi*), a TCM first listed in *Shennong Bencao Jing* in 22–250 A.D. Five different samples of fuzi were grouped into three clusters according to their chemical fingerprints. UPLC-PDA was used to generate metabolomic fingerprints. and quantitative thermo-kinetic parameters obtained from the thermogenic curves of mitochondria metabolic activity were analyzed using PCA. The metabolomics fingerprint-efficacy relationship for fuzi was established using canonical correlation analysis. The study showed that the origin of fuzi samples had an impact on the metabolomic fingerprints and biological activity of this CHM, and it suggests a possible quality control for fuzi that could be extended to other CHM.

Similarly, chemometrics have been applied to study decoction times and their impact on metabolomic profiles and toxicity in *Radix Aconiti Lateralis Preparata* samples. In this instance, UPLC-qTOF-MS was used to generate the metabolomic fingerprint. The data was processed by partial, least square-discriminant analysis (PLS-DA) to compare the difference among these samples. This methodology determined that the metabolites in decoctions varied, depending on the duration of decoction with aconitine, mesaconitine and hypaconitine present in higher concentrations between 2 and 10 min, while the contents of the monoester-diterpenoid alkaloids, such as benzoylaconine, benzoylmesaconine, and benzoylhypaconine, increased during the first 60 min and then stabilized. This study confirmed that understanding the impact of processing conditions for *Radix Aconiti Lateralis Preparata,* and, by extension, other CHMs, is essential for attenuating toxicity and increasing efficiency. It also confirms that a chemometric approach is appropriate for investigating the effects of such parameters.

The correlation of metabolomic fingerprint and biological activity of finished herbal injections has also been analyzed by Zhang et al. (2014b). In this study, a biological fingerprint for the quality control of herbal injections of Shuang-Huang-Lian lyophilized powder (SHL) was investigated. SHL is an herbal injection that has been widely reported to cause adverse reactions. In the study, cellular responses were continuously monitored by time-dependent cell response profiles (TCRP) of RBL-2H3 cells. Seven regular SHL samples were compared with 21 artificial abnormal samples that had been chemically altered by subjection to environmental factors such as light, temperature, and oxidation. In this study, chemometric analysis clearly discriminated different groups of abnormal samples; this quality control approach can be applied to other herbal injections. There are over 130 types of herbal injections in clinical use in China for some 400,000,000 patients each year, with total sales of more than US $4 billion per year.

Chemometrics can also be applied to identify active constituents. In a recent study on the antitrypanosomal activities of highly variable extracts from *Juglans* spp., hydrojuglone glucoside was predicted and confirmed as the active constituent by correlating LC-MS fingerprints with their biological activity using PLS (Ellendorff et al. 2015). This approach appears especially promising for the research on preparations used in TCMs, where detailed traditional rules for their production imply that more or less active preparations can be produced, depending on how rigidly these rules are followed (Sheridan et al. 2012).

A variety of reviews have looked at various applications and approaches to the application of chemometrics, and confirm the importance of this approach to the ongoing quality control of existing and new herbal formulas. Bansal et al. (2014) focus on important analytical techniques, chemometric tools, and interpretation of results in quality and efficacy evaluation of HMP. Wolfender et al. (2015) have reviewed the application of chromatographic techniques, and Gad (2013) the role of chemometrics in the authentication of HMs. Wang and Yu (2015) have specifically covered the use of non-destructive near-IR for species authentication and discrimination of the geographical origins of HMPs, while Rohman et al. (2014) have evaluated the application of vibrational spectroscopy in combination with chemometrics in the authentication of CHM.

Future Challenges

The popularity of HMs has increased worldwide, and as a result there are many challenges facing their global production and use. If we consider the production chain of HMP, the first real challenge lies in the continued and sustainable supply of quality herbal material. As we have seen, herbal material is substantially sourced through wild harvesting, with only a limited volume of material and a small number of species being produced by cultivation. Without the rigorous implementation of GACP and GMP, there is a risk of overharvesting threatened species, and inferior, contaminated, or adulterated products being introduced to the market. Directly resulting from this are potentially serious adverse health consequences for the consumer. A key challenge for regulation in this area is the harmonization of international guidelines, categorization, standards, and requirements. The adoption of the WHO guidelines, and the use and elaboration of additional international monographs addressing the quality and safety of HMPs, will address this shortfall (Zhang et al. 2012; Knöss and Chinou 2015).

The advances in modern analytical methodology and pharmaceutical techniques have contributed significantly to determining the quality of HMP and also to teasing out the complex interactions and synergisms that occur in HMs. However, studies are limited in terms of the number of species and in global distribution. The continuation of metabolomic and chemometric investigations into the relationships between composition and biological activity, and the important role of concomitant compounds will facilitate the optimization of existing HMPs and also empower the design of new formulations and ensure modernized products that retain the biological attributes of the parent HMPs.

There is a need for continued clinical research to support the safety and efficacy of the quality of HMPs that will then be available to support registration of HMPs in various regions. To date, there are limited robust clinical data relating to common single HMs, and more so for multicomponent medicines. However, there is extensive clinical evidence for some of the most popular HMs; for example, there have been over 400 clinical trials carried out on *Ginkgo biloba*, looking at its use in a variety of clinical settings. In 2007 it was recognized as the most popular natural product in the U.S. (Barnes et al. 2004, 2008; updated 2015). St. John's wort is also a much-investigated HM that has been subject to clinical trials. However, a meta-analysis of clinical trial results in 2005 called for more trials to compare extracts, placebos, and conventional antidepressants in test and control patient populations with and without depression (Linde et al. 2005). For other HMs, robust clinical data related to quality herbal material will contribute to accessing marketing authorization and the introduction of HMs to the European region that currently do not comply with the well-established use or traditional use time requirements (European Commission 2004). Guidelines for HM have been drawn up by the EMEA (2015), and Flower et al. (2011) have made recommendations concerning randomized controlled trials investigating CHMs (2012).

Ultimately, the greatest challenge to HMs is that of taking its place in a global integrated medical system, as the underlying requirements for doing this are rigorous proof of quality, safety, and efficacy (WHO 2005). The international regulatory environment and many involved national governments have embraced these fundamental requirements and the move toward harmonization of standards will address some of the challenges to this growing area. The scientific community continues to address the shortfalls in understanding the complex relationships in HMs and HMPs.

Conclusions

Without a doubt, herbal medicines and their various markets represent a rapidly growing global industry. The need for internationally established and harmonious standards for the quality of starting materials and finished products is paramount to ensuring the quality and safety of these complex products. Due to the numerous factors that can impact the chemical profile of HMP, quality needs to be built in at the very beginning of the production chain with standards established and governed by GACP and GMP. There is an ongoing and pressing need to understand, in detail, the complex relationships between changes in chemical profiles and biological activity and the importance of synergisms in the final bioactivity of an HM. Rapid developments in analytical methodology and hyphenated chromatographic and spectroscopic techniques, coupled with chemometric analysis and high throughput methods for measuring bioactivity will be very important for developing a greater understanding of these relationships. Such understanding will contribute to optimized herbal products and modernized formulations that will still retain the favorable properties of the original HM.

References

Aarland RC, Rivera-Cabrera F, Pérez-Flores LJ, de Léon Sánchez FD, Mendoza-Espinoza JA (2014) Relevance of chemical standardization and innocuousness in the process of development of herbal medicines: a review. Asian J Plant Sci 13:1–7. doi:10.3923/ajps.2014.1.7

Abascal K, Ganora L, Yarnell E (2005) The effect of freeze-drying and its implications for botanical medicine: a review. Phytother Res 19:655–660. doi:10.1002/ptr.1651

Ajazuddin SS, Alexander A, Qureshi A, Kumari L, Vaishnav P et al (2014) Role of herbal bioactives as a potential bioavailability enhancer for active pharmaceutical ingredients. Fitoterapia 97:1–14. doi:10.1016/j.fitote.2014.05.005

Ang CYW, Hu L, Heinze TM, Cui Y, Freeman JP, Kozak K, Luo W (2004) Instability of St. John's wort (*Hypericum perforatum* L.) and degradation of hyperforin in aqueous solutions and functional beverage. Agric Food Chem 52:6156–6164. doi:10.1021/jf0490596

Atzorn R, Weiler EW, Zenk MH (1981) Formation and distribution of sennosides in *Cassia angustifolia*, as determined by a sensitive and specific radioimmunoassay. Planta Med 41:1–14

Awang DVC (1997) Quality control and good manufacturing practices: safety and efficacy of commercial herbals. Food Drug Law Inst 52:341–344

Bailey DG (2010) Fruit juice inhibition of uptake transport: a new type of food-drug interaction. Br J Clin Pharmacol 70(5):645–655

Bailey DG, Dresser G, Malcolm J, Arnold O (2012) Grapefruit-medication interactions: forbidden fruit or avoidable consequences? CMAJ. doi:10.1503/cmaj.120951

Bansal A, Chhabra V, Rawal RK, Sharman S (2014) Chemometrics: a new scenario in herbal drug standardization. J Pharm Anal 4(4):223–233

Barnes PM, Powell-Griner E, McFann K, Nahin RL (2004) Complementary and alternative medicine use among adults: United States, 2002. Advance data from vital and health statistics; no 343. National Center for Health Statistics, Hyattsville

Barnes PM, Bloom B, Nahin R (2008) Complementary and alternative medicine use among adults and children. United States. 2007. http://www.cdc.gov/nchs/data/nhsr/nhsr012.pdf

Bauer R, Franz G (2010) Modern European monographs for quality control of Chinese herbs. Planta Med 76:2004–2011, 10.15806

Bennett BC, Balick MJ (2014) Does the name really matter? The importance of botanical nomenclature and plant taxonomy in biomedical research. J Ethnopharmacol 152:387–392. doi:10.1016/j.jep.2013.11.042

Berenbaum MC (1989) What is synergy? Pharmacol Rev 41:93–141

Bergeron C, Gafner S, Batcha LL, Angerhofer CK (2002) Stabilization of caffeic acid derivatives in *Echinacea purpurea* L. Glycerin extract. J Agric Food Chem 50:3967–3970. doi:10.1021/jf011582m

Biopharma (2013) http://www.biopharmalink.com/companies/phyto.htm

Bisson J, McAlpine JB, Friesen JB, Chen S-N, Graham J, Pauli GF (2015). Can invalid bioactives undermine natural product-based drug discovery? J Med Chem 59:1671–1690. doi:10.1021/acs.jmedchem.5b01009

Bouin N, Wierer M (2014) Quality standards of the European Pharmacopoeia. J Ethnopharmacol 158(Part B):454–45. doi:10.1016/j.jep.2014.07.020

Bruni R, Sacchetti G (2009) Factors affecting polyphenol biosynthesis in wild and field grown St. John's wort (*Hypericum perforatum* L. Hypericaceae/Guttiferae). Molecules 14:682–725. doi:10.3390/molecules14020682

Bucar F, Wube A, Schmid M (2013) Natural product isolation – how to get from biological material to pure compounds. Nat Prod Rep 30:525–545. doi:10.1039/c3np20106f

Busse W (2000) The significance of quality for efficacy and safety of herbal medicinal products. Drug Inf J 34:15–23. doi:10.1177/009286150003400102

Butterweck V, Petereit F, Winterhoff H, Nahrstedt A (1998) Solubilized hypericin and pseudohypericin from *Hypericum perforatum* exert antidepressant activity in the forced swimming test. Planta Med 64:291–294. doi:10.1055/s-2006-957437

Calixto JB (2000) Efficacy, safety, quality control, marketing and regulatory guidelines for herbal medicines (phytotherapeutic agents). Braz J Med Biol Res 33:179–189

Chan K, Leung KSC-Y, Zhao SS (2009) Harmonization of monographic standards is needed to ensure the quality of Chinese medicinal materials. Chin Med 4:18. doi:10.1186/1749-8546-4-18, http://www.cmjournal.org/content/4/1/18

Chan K, Shaw D, Simmonds MSJ, Leon CJ, Xu Q et al (2012) Good practice in reviewing and publishing studies on herbal medicine and Chinese materia medica. J Ethnopharmacol 140:469–475. doi:10.1016/j.jep.2012.01.038

Chand GS, Shenoy S, Mita K (2013) A review on role of *Samskara* (special processing) in food and pharmaceutical preparations. Int Ayurvedic Med J 1(5):1–10. http://www.iamj.in/images/upload/01.05.10_IAMJ.pdf

Chau FT, Chan HY, Cheung CY, Xu CJ, Liang Y, Kvalheim OM (2009) Recipe for uncovering the bioactive components in herbal medicine. Anal Chem 81:7217–7225

Chau F, Fung K, Koon C et al (2011) Bioactive components in herbal medicine experimental approaches. In: Benzie IFF, Wachtel-Galor S (eds) Herbal medicine: biomolecular and clinical aspects, 2nd edn. CRC Press, Boca Raton, Chapter 20. Available from: http://www.ncbi.nlm.nih.gov/books/NBK92759/

Chen C, Jie Yuan J, Li X-J, Shen Z-B, Yu D-H, Zhu J-F, Zeng F-L (2013) Chemometrics-based approach to feature selection of chromatographic profiles and its application to search active fraction of herbal medicine. Chem Biol Drug Des 81:688–694. doi:10.1111/cbdd.12114

Chenery PN (2009) Plants, colloids and tinctures – nature's pharmaceutics, 2nd edn. Herbal Research Notes, Town Park Farm

Choi RCY, Gao QT, Cheung AWH, Zhu JTT, Lau FTC et al (2011a) A Chinese herbal decoction, *Danggui Buxue Tang*, stimulates proliferation, differentiation and gene expression of cultured osteosarcoma cells: genomic approach to reveal specific gene activation. Evid Based Complement Alternat Med 2011:1–13. doi:10.1093/ecam/nen085

Choi YH, van Spronsen J, Dai Y, Verberne M, Hollmann F (2011b) Are natural deep eutective solvents the missing link in understanding cellular metabolism and physiology? Plant Physiol 156:1701–1705. doi:10.1104/99.111.178426

Cordell GA (2014) Phytochemistry and traditional medicine – the revolution continues. Phytochem Lett 10:28–40, 10.1016

Council of Europe 9 (2015) European pharmacopoeia, 8th edn. Strasbourg: Council of Europe.

Dai Y, Verpoorte R, Choi YH (2014) Natural deep eutectic solvents providing enhanced stability of natural colorants from safflower (*Carthamus tinctorius*). Food Chem 159:116–121

Dong TTX, Zhao KJ, Gao QT, Ji ZN, Zhu TT et al (2006) Chemical and biological assessment of a Chinese herbal decoction containing *Radix Astragali* and *Radix Angelicae Sinesis*: determination of drug ration in having optimized properties. J Agric Food Chem 54:2767–2774

Dudhatra GB, Mody SK, Awale MM, Patel HB, Modi CM et al (2012) A comprehensive review on pharmacotherapeutics of herbal bioenhancers. Sci World J 2012:637953. doi:10.1100/2012/637953

Dziri S, Casabianca H, Hanchi B, Hosni K (2014) Composition of garlic oil (*Allium sativum* L.) as influenced by drying method. J Essent Oil Res 26:91–96. doi:10.1080/10412905.2013.868329

Eder M, Mehnert W (2000) Pflanzliche Begleitstoffe – wertvolle Hilfsstoffe oder überflüssiger Ballast? Pharm Unserer Zeit 29:377–384, 10.1002/1615-1003 (200012)29:6<377::AID-PAUZ377>3.0.CO;2-#

Ellendorff T (2014) Entdeckung bioaktiver Naturstoffe unter Verwendung der multivariaten Datenanalyze. Ph.D. thesis, University of Münster, Germany

Ellendorff T, Brun R, Kaizer M, Sendker J, Schmidt TJ (2015) PLS-prediction and confirmation of hydrojuglone glucoside as the antitrypanosomal constituent of *Juglans Spp*. Molecules 20:10082–10094. doi:10.3390/molecules200610082

EMEA (2006) Guideline on Good Agricultural and Collection Practice (GACP) for starting materials of herbal origin. http://www.ema.europa.eu/docs/en_GB/document_library/Scientific_guideline/2009/09/WC500003362.pdf

EMEA (2011) Guideline on quality of herbal medicinal products/traditional herbal medicinal products. Committee on Herbal Medicinal Products (HMPC). EMA/HMPC/201116/2005 Rev. 2

EMEA (2015) European Union herbal monographs http://www.ema.europa.eu/ema/index.jsp?curl=pages/regulation/document_listing/document_listing_000212.jsp&mid=WC0b01ac058003380a

EU (2008) EudraLex. The rules governing medicinal products in the European Union. Volume 4 EU guidelines to good manufacturing practice medicinal products for human and veterinary use. Annex 7 manufacture of herbal medicinal products. http://ec.europa.eu/health/files/eudralex/vol-4/vol4_an7_2008_09_en.pdf

EUROPAM (2006) Guidelines for Good Agricultural and Wild Collection Practice (GACP) of medicinal and aromatic plants. http://www.europam.net/documents/gacp/EUROPAM_GACP_7.3.pdf

European Commission (2004) Directive 2004/24/EC of the European parliament and of the council of 31 March 2004 amending, as regards traditional herbal medicinal products. Directive 2001/83EC of the Community Code Relating to Medicinal Products for Human use. http://eur-lex.europa.eu/LexUrizerv/LexUrizerv.do?uri=OJ:L:2004:136:0085:0090:en:PDF

Fan XH, Cheng YY, Ye ZL, Lin RC, Qian ZZ (2006) Multiple chromatographic fingerprinting and its application to the quality control of herbal medicines. Anal Chim Acta 555:217–224. doi:10.1016/j.aca.2005.09.037

Flower A, Witt C, Liu JP, Ulrich-Merzenich G, Yu H, Lewith G (2011) Guidelines for randomized controlled trials investigating Chinese herbal medicine. J Ethnopharmacol 140(3):550–554. doi:10.1016/j.jep.2011.12.017

Flück H (1968) Haltbarkeit und haltbarmachung der arzneimittel IV. Arzneidrogen. Pharm Acta Helv 43:1–15

Gad HA (2013) Application of chemometrics in authentication of herbal medicines: a review: chemometry and herbal medicines. Phytochem Anal 24(1):1–24

Gaedcke F (2003) Herstell- und Qualitätsaspekte pflanzlicher Extrakte. Pharm Unserer Zeit 32:192–201. doi:10.1002/pauz.200390064

Gaedcke F, Steinhoff B (2000) Phytopharmaka – Wissenschaftliche und rechtliche Grundlagen für die Entwicklung, Standardisierung und Zulassung in Deutschland und Europa. Wissenschaftliche Verlagsgesellschaft, Stuttgart

Gafner S, Bergeron C (2005) The challenges of chemical stability testing of herbal extracts in finished products using state-of-the-art analytical methodologies. Curr Pharm Anal 1:203–215

Gao QT, Cheung JKH, Li J, Chu GKY, Duan R et al (2006) A Chinese herbal decoction, *Danggui Buxue Tang*, prepared from *Radix Astragali* and *Radix Anglicae Sinensis* stimulates the immune response. Planta Med 72:1227–1231. doi:10.1055/s-2006-947186

Gertsch J (2011) Botanical drugs, synergy, and network pharmacology: forth and back to intelligent mixtures. Planta Med 77:1086–1098. doi:10.1055/s-0030-1270904

Goodarzi M, Russell PJ, Vander Heyden Y (2013) Similarity analyses of chromatographic herbal fingerprints: a review. Anal Chim Acta 804:16–28. doi:10.1016/j.aca.2013.09.017

Goppel M, Franz G (2004) Stability control of senna leaves and senna extracts. Planta Med 70:432–436. doi:10.1055/s-2004-818971

Govindaraghavan S (2008) Quality assurance of herbal raw materials in supply chain: challenges and opportunities. J Diet Suppl 5:176–212. doi:10.1080/193390210802332810
Güçlü-Üstündağ Ö, Mazza G (2007) Saponins: properties, applications and processing. Crit Rev Food Sci Nutr 47:231–258. doi:10.1080/10408390600698197
Guo D, Wu WY, Ye M, Liu X, Cordell GA (2014) A holistic approach to the quality control of traditional Chinese medicines. Science 347(6219 Suppl):S29–S31
Gurley BJ (2012) Pharmacokinetic herb-drug-interactions (part 1): origins, mechanisms, and the impact of botanical dietary supplements. Planta Med 78:1478–1489. doi:10.1055/s-0031-1298273
He TT, Ung COL, Hu H, Wang Y-T (2015) Good manufacturing practice (GMP) regulation of herbal medicine in comparative research: China GMP, cGMP, WHO-GMP, PIC/S and EU-GMP. Eur J Integr Med 7(1):55–66
Hu C, Xu G (2014) Metabolomics and traditional Chinese medicine. Trends Anal Chem 61:207–214
Hu J, Wu Z, Yan J, Pang W, Liang D, Xu X (2009) A promising approach for understanding the mechanism of Traditional Chinese Medicine by the aggregation morphology. J Ethnopharmacol 123:267–274. doi:10.1016/j.jep.2009.03.007
WHO (2008) The International Pharmacopoeia. World Health Organization, Geneva, Dept. of Essential Medicines and Pharmaceutical Policies.
Jaiswal Y, Liang Z, Yong P, Chen H, Zhao Z (2013) A comparative study on the traditional Indian Shodhana and Chinese processing methods for aconite roots by characterization and determination of the major components. Chem Cent J 7:169
Janeš D (2010) Assessment report on *Echinacea purpurea* (L.) Moench radix. Committee on Herbal Medicinal Products (HMPC). EMA/HMPC/577786/2008
Jiang X, Oh IH, Lee SG, Choi HK (2013) The application of metabolomics to processed traditional Chinese medicine. J Korean Soc Appl Biol Chem 56:475–481. doi:10.1007/s13765-013-3075-z
Junior JOCS, Costa RMR, Teixara FM, Barbosa WLR (2014) Processing and quality control of herbal drugs and their derivatives. In: Shoyaman Y (ed) Quality control of herbal medicines and related areas. InTech Rijeka, Crotia. ISBN 978-953-307-682-9. doi:10.5772/1141
Kalwij JM (2012) Review of 'The Plant List', a working list of all plant species. J Veg Sci 23:998–1002. doi:10.1111/j.1654-1103.2012.01407.x
Knöss W (2007) Assessment report for Rhubarb (*Rhei radix*). Committee on Herbal Medicinal Products (HMPC). EMEA/HMPC/189626/2007
Knöss W, Chinou I (2015) Traditional herbal medicines in the European Union: implementing standardization and regulation. Science 350(6262 Suppl):S59–S60. doi:10.1126/science.350.6262.871-c
Kreis W (2007) Enzyme bei der gewinnung von drogen und der herstellung von phytopharmaka. In: Hänsel R, Sticher (eds) Pharmakognosie Phytopharmazie, 8th edn. Springer Medizin Verlag, Heidelberg, pp 286–291
Länger R (2014) Draft assessment report on *Capsicum annuum* L. var. minimum (Miller) Heizer and small fruited varieties of *Capsicum frutescens* L., fructus. Committee on Herbal Medicinal Products (HMPC). EMA/HMPC/674138/2013
Li S, Han Q, Qioa C, Song J, Cheng CL, Xu H (2008) Chemical markers for the quality control of herbal medicines: an overview. Chin Med 3:7. doi:10.1186/1749-8546-3-7
Li X, Chen Y, Lai Y, Yang Q, Hu H, Wang Y (2014) Sustainable utilization of Traditional Chinese Medicine resources: systematic evaluation on different production modes. Evid Based Complement Alternat Med. Article 218901. http://dx.doi.org/10.1155/2015/218901
Linde K, Berner M, Egger M, Murlow C (2005) St John's wort for depression meta-analysis of randomized controlled trials – meta-analysis of randomized controlled trials. Br J Psychiatry 18(6):99–107. doi:10.1192/bjp.186.2.99
Liu Z (2009) Diterpene glycosides as neutral solubilizers. U.S. patent US 20110033525 A1
Liu Z (2010) Terpene glycosides and their combinations as solubilizing agents. U.S. patent U.S. 8551507 B2

Ma XH, Zheng CJ, Han LY, Xie B, Jia J et al (2009) Synergistic therapeutic actions of herbal ingredients and their mechanisms from molecular interaction and network perspectives. Drug Discov Today 14:579–588. doi:10.1016/j.drudis.2009.03.012

Mak DHF, Chiu PY, Dong TTX, Tsim KWK, Ko KM (2006) *Dang-Gui Buxue Tang* produces a more potent cardioprotective effect than its component herb extracts and enhances glutathione status in rat heart mitochondria and erythrocytes. Phytother Res 20:561–567. doi:10.1002/ptr.1904

Maltese F, van der Kooy F, Verpoorte R (2009) Solvent derived artifacts in natural product chemistry. Nat Prod Commun 4:447–454

McNeill J, Barrie FR, Bruck WR, Demoulin V, Greuter W et al. (2012) International code of nomenclature for algae, fungi, and plants (Melbourne code). Regnum Vegetabile 154. www.iapt-taxon.org/nomen/main.php?page=title. Accessed on 10 Apr 2015

Moore BD, Andrew RL, Külheim C, Foley WJ (2014) Explaining intraspecific diversity in plant secondary metabolites in an ecological context. New Phytol 201:733–750. doi:10.1111/nph.12526

Müller J, Heindl A (2006) Drying of medicinal plants. In: Bogers RJ, Craker LE, Lange D (eds) *Medicinal and Aromatic Plants: Agricultural, Commercial, Ecological, Legal, Pharmacological and Social Aspects*. Springer Netherlands, pp 237–252. ISBN: 978-1-4020-544-8.

Murch SJ, Saxena PK (2006) St. John's wort (*Hypericum perforatum* L.): challenges and strategies for production of chemically-consistent plants. Can J Plant Sci 86:765–771. doi:10.4141/P05-179

Murch SJ, Haq K, Rupasinghe V, Saxena PK (2003) Nickel contamination affects growth and secondary metabolite composition of St. John's wort (*Hypericum perforatum* L.). Environ Exp Bot 49:251–257, doi: unknown

Nahrstedt A (1989) Phytopharmaka: zubereitungsformen und inhaltsstoffe. Z Phytotherapie 10:83–86

Nahrstedt A, Butterweck V (2010) Lessons learned from herbal medicinal products: the example of St. John's wort. J Nat Prod 73:1015–1021. doi:10.1021/np1000329

Nöldner M, Schötz K (2002) Rutin is essential for the antidepressant activity of *Hypericum perforatum* extracts in the forced swimming test. Planta Med 68(7):577–580. doi:10.1055/s-2002-32908

Nortier JL, Martinez MC, Schmeizer HH, Arlt VM, Bieler CA, Petein M et al (2000) Urothelial carcinoma associated with the use of a Chinese herb (*Aristolochia fangchi*). N Engl J Med 342(23):1686–1692

Nualkaew S, Gritsanapan W, Petereit F, Nahrstedt A (2004) New fatty acid esters originate during storage by the interaction of components in prasaplai, a Thai traditional medicine. Planta Med 70:1243–1246. doi:10.1055/s-2004-835861

Nüsslein B, Kurzmann M, Bauer R, Kreis W (2000) Enzymatic degradation of cichoric acid in *Echinacea purpurea* preparations. J Nat Prod 63:1615–1618. doi:10.2021/np0002839

Okem A, Southway C, Stirk WA, Street RA, Finnie JF, Van Staden J (2015) Effect of cadmium and aluminum on growth, metabolite content and biological activity in *Drimia elata* (Jacq.) hyacinthaceae. S Afr J Bot 98:142–147

Patel AR, Seijen-ten-Hoorn J, Heussen PCM, Drost E, Hazekamp J, Velikov KP (2012) Straightforward preparation of organic colloidal particles by harnessing spontaneous non-covalent interactions of active molecules from natural origin. J Colloid Interface Sci 374:150–156. doi:10.1016/j.jcis.2012.01.056

Paton A (2009) Biodiversity informatics and the plant conservation baseline. Trends Plant Sci 14:629–637. doi:10.1016/j.tplants.2009.08.007

Pavol M, Repčak M, Ciccarelli D, Garbari F (2001) *Hypericum perforatum* L. – chemotype without rutin from Italy. Biochem Syst Ecol 29:659–661. doi:10.1016/S0305-1978(00)00094-6

Prothon F, Ahrné L, Sjöholm I (2003) Mechanisms and prevention of plant tissue collapse during dehydration: a critical review. Crit Rev Food Sci Nutr 43:447–479

Qi Z, Kelley E (2014) The WHO traditional medicine strategy 2014–2023: a perspective. Science 346(6216 Suppl):S5–S6

Rascio N, Rocca NL (2005) Resurrection plants: the puzzle of surviving extreme vegetative desiccation. Crit Rev Plant Sci 24:209–225. doi:10.1080/07352680591008583

Rivera D, Allkin R, Obón C, Alcaraz F, Verpoorte R, Heinrich M (2014) What is in a name? The need for accurate scientific nomenclature for plants. J Ethnopharmacol 152:393–402. doi:10.1016/j.jep.2013.12.022

Rocha RP, Melo EC, Radünz LL (2011) Influence of drying process on the quality of medicinal plants: a review. J Med Plants Res 5:7076–7084. doi:10.5897/JMPRx11.001

Rohman A, Nugroho A, Lukitaningsih E, Sudjadi (2014) Application of vibrational spectroscopy in combination with chemometrics techniques for authentication of herbal medicine. Appl Spectrosc Rev 49(8):603–613. doi:10.1080/05704928.2014.882347

Seidler-Łożykowska K (2003) Secondary metabolites content of *Hypericum* sp. in different stages and plant parts. In: Ernst E (ed) *Hypericum*: the genus hypericum. Taylor and Francis, London, pp 100–105. ISBN 0-415-36954-1

Selmar D, Kleinwächter M (2013) Influencing the product quality by deliberately applying drought stress during the cultivation of medicinal plants. Ind Crop Prod 42:558–566. doi:10.1016/j.indcrop.2012.06.020

Sendker J (2006) Zur entstehung primärer amidglykoside aus cyanogenen glykosiden. Ph.D. thesis, University of Münster, Germany

Sendker J, Nahrstedt A (2009) Generation of primary amides from cyanogenic glucosides. Phytochemistry 70:388–393. doi:10.1016/j.phytochem.2008.11.008

Sher H, Ali H, Rehman S (2012) Identification and conservation of important plant areas (Ipas) for the distribution of medicinal, aromatic and economic plants in the Hindukush-Himalaya mountain range. Pak J Bot 44:187–194

Sheridan H, Krenn L, Jiang R, Sutherland I, Ignatova S, Marmann A et al (2012) The potential of metabolic fingerprinting as a tool for the modernization of TCM preparations. J Ethnopharmacol 140:482–491. doi:10.1016/j.jep.2012.01.050

Sheridan H, Kopp B, Krenn L, Guo D, Sendker J (2015) Traditional Chinese herbal medicinal preparation: invoking the butterfly effect. Science 350(6262 Suppl):S64–S66. doi:10.1126/science.350.6262.871-c

Shoba G, Joy D, Joseph T, Majeed M, Rajendran R, Srinivas PSSR (1998) Influence of piperine on the pharmacokinetics of curcumin in animals and human volunteers. Planta Med 64:353–356. doi:10.1055/s-2006-957450

Singhuber J, Zhu M, Prinz S, Kopp B (2009) *Aconitum* in traditional Chinese medicine – a valuable drug or an unpredictable risk? J Ethnopharmacol 126:18–30. doi:10.1016/j.jep.2009.07.031

Siow YL, Gong Y, Au-Yeung KKW, Woo CWH, Choy PC (2005) Emerging issues in traditional Chinese medicine. Can J Physiol Pharmacol 83:321–334. doi:10.1139/y05-029

State Pharmacopoeia Committee (2010) Pharmacopoeia of the People's Republic of China. China Medical Science and Technology Press, Beijing

Sticher O (2007) Isoprenoide als inhaltsstoffe. In: Hänsel R, Sticher (eds) Pharmakognosie Phytopharmazie, 8th edn. Springer Medizin Verlag, Heidelberg, pp 838–846

Sun Y, Feng F, Yu X (2012) Pharmacokinetics of geniposide in Zhi-Zi-Hou-Pu decoction and in different combinations of its constituent herbs. Phytother Res 26(1):67–72. doi:10.1002/ptr.3516

Tanaka T, Zhang H, Jiang ZH, Koundo I (1997) Relationship between hydrophobicity and structure of hydrolysable tannins, and association of tannins with crude drug constituents in aqueous solution. Chem Pharm Bull 45:1891–1897. doi:10.1248/cpb.45.1891

Tanko H, Carrier DJ, Duan L, Clausen E (2005) Pre- and post-harvest processing of medicinal plants. Plant Genet Resour 3:304–313, 10.1079/PGR200569

The Plant List (2013) Version 1.1. Published on the Internet; http://www.theplantlist.org/. Accessed on 10.04.2015

Tistaert C, Thierry L, Szandragh A, Dejaegher B, Fan G, Frédérich M, Hyeden YV (2011) Quality control of Citri reticulatae pericarpium: exploratory analysis and discrimination. Anal Chim Acta 705(2011):111–122

Tschiggerl C, Bucar F (2011) Influence of saponin plants on the volatile fraction of thyme in herbal teas. Fitoterapia 82:903–910. doi:10.1016/j.fitote.2011.05.006

Umek A, Kreft S, Kartnig T, Heydel B (1999) Quantitative phytochemical analyses of six *hypericum* species growing in Slovenia. Planta Med 65:388–390, doi: unknown

Uptodate (2015) http://www.uptodate.com/contents/clinical-use-of-ginkgo-biloba?source=search_result&search=ginko+biloba&selectedTitle=1~41

Verpoorte R, Choi YH, Mustafa NR, Kim HK (2008) Metabolomics: back to basics. Phytochem Rev 7:525–537

Wagner H, Ulrich-Merzenich G (2009) Synergy research: approaching a new generation of phytopharmaceuticals. Phytomedicine 16(2–3):97–110. doi:10.1016/j.phymed.2008.12.018

Walthelm U, Dittrich K, Gelbrich G, Schöpke T (2001) Effects of saponins on the water solubility of different model compounds. Planta Med 67:49–54. doi:10.1055/s-2001-10876

Wang P, Yu Z (2015) Review paper: species authentication and geographical origin discrimination of herbal medicines by near infrared spectroscopy: a review. J Pharm Anal. doi:10.1016/j.jpha.2015.04.001

Wang Y, Wang X, Cheng Y (2006) A computational approach to botanical drug design by modelling quantitative composition-activity relationship. Chem Biol Drug Des 68(3):166–172

Watanabe K, Fujino H, Morita T, Kasai R, Tanaka O (1988) Solubilization of saponins of *bupleuri radix* with ginseng saponins: cooperative effect of dammarane saponins. Planta Med 54:405–409. doi:10.1055/s-2006-962482

Werner C, Merz B (2007) Assessment report on *Rhamnus frangula* L., cortex. Committee on Herbal Medicinal Products (HMPC). EMEA/HMPC/76306/2006

WHO (2000) General Guidelines for Methodologies on Research and Evaluation of Traditional Medicine, WHO/EDM/TRM/2000.1. World Health Organization, Geneva, http://apps.who.int/iris/bitstream/10665/66783/1/WHO_EDM_TRM_2000.1.pdf

WHO (2001) Index of pharmacopoeias. http://apps.who.int/medicinedocs/en/d/Jwhozip19e/

WHO (2003a) WHO Guidelines on Good Agricultural and Collection Practices (GACP) for Medicinal Plants. World Health Organization, Geneva, http://apps.who.int/iris/bitstream/10665/42783/1/9241546271.pdf

WHO (2003b) Guide to Good Storage Practices for Pharmaceuticals, WHO Technical Report Series, No. 908. Annex 9. World Health Organization, Geneva, http://apps.who.int/medicinedocs/en/d/Js5517e/

WHO (2005) National Policy on Traditional Medicine and Regulation of Herbal Medicines. Report of a WHO global survey. http://whqlibdoc.who.int/publications/2005/9241593237.pdf

WHO (2007) WHO Guidelines on Good Manufacturing Practice (GMP) for Herbal Medicines. http://apps.who.int/medicinedocs/documents/s14215e/s14215e.pdf

WHO (2013) Traditional Medicine Strategy: 2014–2023. http://apps.who.int/iris/bitstream/10665/92455/1/9789241506090_eng.pdf

WHO (2015) http://apps.who.int/medicinedocs/en/d/Js2200e/

Wichtl M (1970) Wirkstoffveränderung bei der drogenbereitung (trocknung der arzneipflanzen). Die Pharm 25:692–697

Wikene KO, Bruzell E, Tønnesen HH (2015) Improved antibacterial phototoxicity of a neutral prophyrin in natural deep eutectic solvents. J Photochem Photobiol B Biol. doi:10.1016/j.jphotobiol.2015.04.022, In Press

Williamson EM (2001) Synergy and other interactions in phytomedicines. Phytomedicine 8:401–409, doi: unknown

Wolfender JL, Marti G, Thomas A, Bertrand (2015) Current approaches and challenges for the metabolite profiling of complex natural extracts. J Chromatogr A 1382:136–164. doi:10.1016/j.chroma.2014.10.091

Xiao PG (1986) Medicinal plants: the Chinese approach. World Health Forum 6:84–85, http://whqlibdoc.who.int/whf/1986/vol7-no1/WHF_1986_7%281%29_p84-85.pdf

Yakovishin LA, Grishkovets VI, Belash DY, Yarovoy IR (2011) Molecular complexation of ivy and licorice saponins with some drugs of aromatic nature (trans: Russian). Scientific notes of Taurida National University Series Biology, Chemistry 24:4–10. doi: unknown

Yakovishin LA, Lekar AV, Vetrova EV, Borizenko NI, Borizenko SN, Grishkovets VI (2012) Molecular complexes of the triterpene glycosides with L-tyrosine and their biological activity. Biopolymers Cell 28:62–67. doi:10.7124/bc.0002E, Translated from Russian

Yang Y, Zhang Z, Li S, Ye X, Li X, He K (2014) Synergy of herb extracts: pharmacokinetics and pharmacodynamics basis. Fitoterapia 92:133–147. doi:10.1016/j.fitote.2013.10.010

Yeong-Deug Y, Chang IM (2004) An overview of traditional Chinese herbal formulae and a proposal of a new code system for expressing the formula titles. Evid Based Complement Alternat Med 1:125–132. doi:10.1093/ecam/neh019

Yi LZ, Liang YZ, Wu H, Yuan DL (2009) The analysis of *Radix Angelicae Sinensis* (Danggui). J Chromatogr A 1216:1991–2001. doi:10.4103/0973-1296.111255

Yuan R, Lin Y (2000) Traditional Chinese medicine: an approach to scientific proof and clinical validation. Pharmacol Ther 86:191–198. doi:10.1038/sj.bjp.0706945

Zhang B, Peng Y, Zhang Z, Liu H, Qi Y, Liu S, Xiao P (2010) GAP production of TCM herbs in China. Planta Med 76:1948–1955. doi:10.1055/s-0030-1250527

Zhang J, Wider B, Shang H, Li X, Ernst E (2012) Quality of herbal medicines: challenges and solutions. Complement Ther Med 20:100–106. doi:10.1016/j.ctim.2011.09.004

Zhang WL, Chen JP, Lam KYC, Zhan JYX, Yao P et al (2014a) Hydrolysis of glycosidic flavonoids during the preparation of *Danggui Buxue Tang*: an outcome of moderate boiling of Chinese herbal mixture. Evid Based Complement Alternat Med. doi:10.1155/2014/608721, 11 pages

Zhang L, Ma L, Feng W, Zhang C, Sheng F, Zhang Y et al (2014b) Quality fluctuation detection of an herbal injection based on biological fingerprint combined with chemical fingerprint. Anal Bioanal Chem 406(20):5009–5018. doi:10.1007/s00216-014-7918-1

Zhao Z, Liang Z, Chan K, Lu G, Lee E, Chen H, Li L (2010) A unique issue in the standardization of Chinese materia medica: processing. Planta Med 76:1975–1986. doi:10.1055/s-0030-1250522

Zhao Z, Guo P, Brand E (2012) The formation of *daodi* quality materials. J Ethnopharmacol 140:476–481. doi:10.1016/j.jep.2012.01.048

Zhou XH, Kasi R, Yoshikawa M, Kitagawa I, Tanaka O (1991) Solubilization of saponins of *bupleuri radix* with ginseng saponins: effect of malonyl-ginsenosides on water solubility of saikosaponin-b_1. Chem Pharm Bull 39:1250–1252

Zhou J, Gao G, Chu Q, Wang H, Rao P, Ke L (2014) Chromatographic isolation of nanoparticles from Ma-Xing-Shi-Tang decoction and their characterization. J Ethnopharmacol 151:1116–1123. doi:10.1016/j.jep.2013.12.029

Zhu Y, Zheng Q, Sun Z, Chen Z, Zhao Y et al (2014) Fingerprint–efficacy study of *Radix Aconiti Lateralis Preparata* (Fuzi) in quality control of Chinese herbal medicine. J Therm Anal Calorim 118(3):1763–1772. doi:10.1371/journal.pone.0086539

Zhuang Y, Yan J, Zhu W, Chen L, Liang D, Xu X (2008) Can the aggregation be a new approach for understanding the mechanism of traditional Chinese medicine? J Ethnopharmacol 117:378–384. doi:10.1016/j.jep.2008.02.017

Chapter 4
Toxicokinetics of Herbal Products

Olavi Pelkonen and Jorma T. Ahokas

Abstract Pharmacokinetics (PK) and toxicokinetics (TK) mean essentially the same thing, only the final effect of the studied substance differs. In this chapter, the abbreviation "TK" is usually used to acknowledge the title of the book (toxicology), but also the abbreviation "PK" is used, depending on the context. TK is an essential part of the characterization of a conventional pharmaceutical and, besides its "intrinsic scientific" value, TK constitutes a backdrop for understanding and delineating a substance's in vivo potency, potential toxicities, and particular clinical conditions. The same TK principles should apply to herbal medicinal products; however, these products are complex chemical mixtures, with tens or hundreds of major and minor components belonging to a variety of chemical groups and classes, making it rather difficult to study their TK, both in theory and in practice. The TK of an herbal product should address both the time course of its active constituents, and the impact of the various components on the TK processing (metabolism and transport) of its own constituents and simultaneously administered pharmaceuticals.

This chapter describes some of the major areas that should be addressed when investigating the PK/TK of herbal medicinal products. Appropriate analytical methods exist to address major TK issues, despite the complex composition of herbal products. However, the success of these studies depends on pharmacodynamic and mechanistic studies to decide which of the many components should be targeted for the ADME (absorption, distribution, metabolism, and excretion) characterization. The prevailing tenet in the area of herbal products is that the "whole product" is responsible for the therapeutic action. However, such a statement is scientifically inadequate and therefore not really helpful. The dissection of contributing components and their interactions with respect to both therapeutic effects and potential toxicities requires the application of advanced analytical and high-content

O. Pelkonen (✉)
Department of Pharmacology and Toxicology, Institute of Biomedicine,
University of Oulu, Aapistie 5 B, FIN-90014 Oulu, Finland
e-mail: olavi.pelkonen@oulu.fi

J.T. Ahokas
School of Medical Sciences, RMIT-University,
Bundoora Campus, PO Box 71, Bundoora, VIC 3083, Australia
e-mail: Jorma.Ahokas@rmit.edu.au

O. Pelkonen et al. (eds.), *Toxicology of Herbal Products*,
DOI 10.1007/978-3-319-43806-1_4

technologies, including "omics" methods, computational modelling and simulation approaches, and, most of all, systems biological thinking.

Keywords Toxicokinetics • ADME • Absorption distribution • Metabolism • Excretion

Basic Concepts and Processes of Toxicokinetics of Single Substances

Toxicokinetic Processes

The primary reason for pharmacokinetic studies of pharmaceutical agents is to enable their rational use. The clinical use of drugs depends on knowing the bioavailability of the drug and its absorption, distribution, and elimination characteristics. The elimination characteristics are only fully understood with adequate knowledge of the metabolism of the active constituents. Metabolic characterization also includes the identification of active and/or possibly toxic metabolic products or intermediates.

The very cornerstone of pharmacology and toxicology is that there is a direct relationship between the concentration of an active constituent of a therapeutic product at the site of action and the extent of the effect. However, after a given dose, the concentration at the site of action is extensively modulated by several pharmacokinetic factors (Fig. 4.1). With therapeutically used drugs, the pharmacokinetic principles are well developed and the predictions of concentration at the site of action are good. Similarly, the factors contributing to the inter-individual variability are well established. The most significant sources of variability stem from formulation, body build, and metabolism and/or excretion (Fig. 4.1). These factors affect the ultimate concentration at the active site resulting from a given administered dose.

With a therapeutically used drug, the problem setting is relatively clear-cut, in that one is dealing with a very well characterized active constituent and its metabolites. The information regarding every therapeutically used compound is obtained by a systematic study of absorption, distribution, and elimination kinetics of the active constituent(s). As well as the kinetic parameters, the metabolism of the compound is determined. This characterization includes determining the metabolite pattern and assignment of the enzymes (CYP and conjugating enzymes) responsible for the metabolism. Further studies involve determination of active transport of the compound and its metabolites in and out of various organs.

Once a detailed pharmacokinetic database has been established for a pharmaceutical agent, it is possible to extend the studies to predict most drug-drug and drug-food interactions, or interactions with any other agent. The most prominent cause of significant interactions involves CYP-based interactions. Nevertheless, PK may be non-linear (e.g., saturable enzymatic or transport systems) and so may not reliably indicate the TK at supra-pharmacological dosages. Also, some genetic outliers may

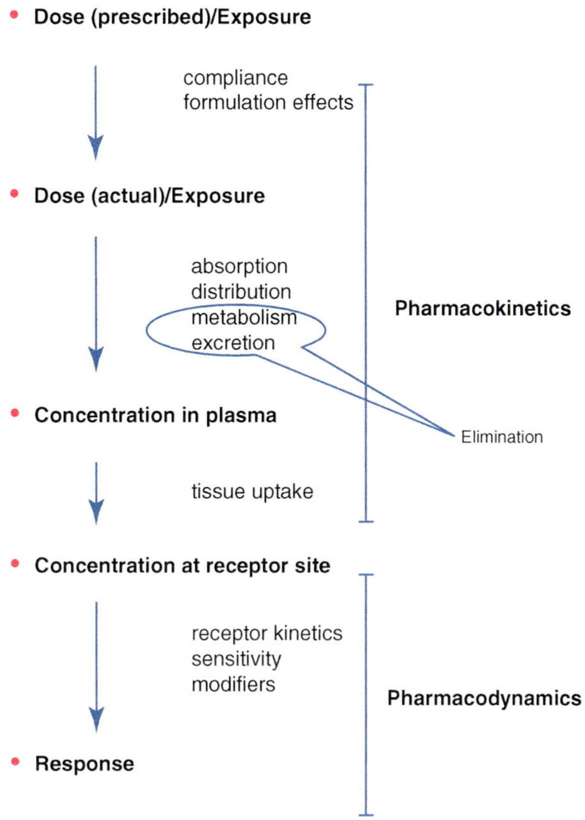

Fig. 4.1 A schematic view to delineate processes between dose/exposure and the ultimate response (therapeutic or toxicological)

be missed by preclinical and clinical studies, resulting in unexpected (idiosyncratic) toxicities or interactions.

Inter-individual variability of pharmacokinetics is an important cause of concern in therapeutic failure or adverse effects caused by excessive concentration. Such variability can be caused by genetic differences or various organ-specific diseases.

What Is Pharmacokinetics/Toxicokinetics of Herbal Medicinal Preparations?

While pharmacokinetic studies of therapeutically used drugs can be conducted with relative ease, the difficulties with complementary preparations are significant. The difficulties do not stem from lack of available methodology, but from the nature of the herbal products and the basic philosophy of using herbal products in health care. If the whole product in its entirety is considered to be the active "medicine," then what is the meaning of pharmacokinetics? However, current systems pharmacology thinking assumes that there are several active components in the whole preparation, targeting various processes in a therapeutically meaningful manner. As the active constituents are often not characterized or even known, how could pharmacokinetics then be studied? Furthermore, the composition of crude herbal products is highly variable and their standardization cannot be easily effected, resulting in variable doses and ratios of (possibly) active components. One is faced with so many unanswered and unanswerable questions that problems seem almost unsurmountable.

As described above, the study of pharmacokinetics/toxicokinetics of a single identifiable substance is very clear, whereas with crude or partially processed plant products, the concept becomes complex and in some aspects meaningless. Especially with ill-defined products and products with questionable efficacy, it becomes a question of kinetics of "what," and consequently kinetic studies are without a clear purpose. In the next section, an attempt is made to categorize various possibilities to define pharmaco/toxicokinetics of an herbal medicinal substance.

Which Components of an Herbal Medicinal Substance Should Be Analyzed Regarding TK?

General Considerations

Commercial stakeholders of herbal medicinal products usually state that the whole herbal substance is the active principle of herbal medicinal product (Lu et al. 2015). This statement is often used to downgrade the significance of toxicities of isolated ingredients or as a hypothetical claim that the whole is more than the sum of its parts (or in the case of toxicity, less than the sum of the parts). In the area of TK, this statement has been used as an excuse NOT to perform kinetic studies of isolated ingredients, because the results of such studies cannot be extrapolated to the situation in which the whole herbal medicinal product is being used.

It is fair to say that recent developments in "omics" approaches and network or systems pharmacology and toxicology have given at least some credibility to those earlier concepts that the whole is more than the sum of its parts. In particular, studies of traditional Chinese medicines (TCM) have started to produce, even if it is still very tentative, credible data to justify systems views on complex medicinal preparations (see, for example, reviews of Wang et al. 2009; Xu et al. 2012). These systems views are based on detailed studies of comprehensive analytical techniques, medium-throughput bioassays, target interactions, and integrated computational tools – i.e., actual experimental and computational approaches (see this chapter and Chap. 5).

Pharmacodynamic and Toxicodynamic Considerations

In order to overcome the fundamental questions about which components to measure, pharmacodynamic studies and efficacy trials of complementary medicines should precede extensive kinetic studies. These studies should be performed with adequate knowledge of the composition of the product and the preclinical screen of pharmacodynamics in order to provide a preliminary understanding of which components may or may not be linked with the therapeutic responses or adverse outcomes observed during the clinical studies. Products with proven efficacy and/or observed adverse effects should be automatically considered for pharmacokinetic studies. With respect to these products, it is within our scientific capability to identify active constituents that can be characterized with respect to their pharmacokinetic behavior. The modern metabolomic approach may offer an alternative or additional way of tackling the behavior and pharmacokinetic impact of complex mixtures.

Thus, it seems reasonable to state that clinically or experimentally observed beneficial or adverse effects and their tentative linkages with identified components of the mixture should be a prerequisite and starting point for kinetic studies. Without the pharmaco(toxico)dynamic background knowledge, it is useless to embark on kinetic studies. However, since many herbal medicinal products are on the market and are being used without proof of efficacy, it may still be necessary to conduct certain TK studies to address selected safety concerns with these products. Such concerns include interference with pharmacokinetics of concurrently administered drugs.

Analytical and Biomarker Considerations

Quality control is naturally a *sine qua non* consideration for herbal medicinal products, because the composition and pharmaceutical properties of a product have to be within certain pre-determined tolerances. For quality control purposes, usually one or several major components are selected and their concentrations in the finished product have to be determined on a routine basis (see Chap. 3). If a component is a toxic one, the concentration should be under a certain limit value, as is the case with aconite alkaloids.

A totally different question is whether there is a need to measure a reference constituent (a marker substance used for quality control) in the body just for bioavailability or therapeutic monitoring reasons? Obviously this matter needs further elaboration.

Toxicological Considerations

Often safety concerns arise from a known presence of a toxic substance in the product. Table 4.1 lists some examples of plant-derived substances known to elicit various toxicities. An obvious problem is that these substances are present in greatly varied amounts, together with greatly varied compositions of other constituents in plant-derived products. It is clear that if there is a strong suspicion of toxicological potential of a component, its risks should be adequately evaluated. However, risk assessment of a single component may be a tricky task because of potential interactions with other components, either in the product or in the body.

Table 4.1 Examples of substances as constituents in plant-derived products that have been linked with adverse effects

Substance (typical plant)	Adverse effect	Relevant toxicokinetic characteristics	References
Aristolochic acid (*Aristolochia* genus)	Nephrotoxic, carcinogenic	metabolic activation to DNA-reactive metabolites	See Chaps. 9 and 13
Pulegone/menthofuran (*Mentha pulegium* L.)	Hepatotoxic, carcinogenic (animals)	Glutathione conjugation; metabolic activation in the liver	Nelson (1995)
Estragole (*Foeniculum vulgare* Mill.)	Hepatotoxic, carcinogenic (animals)	Metabolic activation by oxidation and sulphate conjugation	Rietjens et al (2005)
Thujone (*Artemisia absinthium* L.)	Neurotoxic	Metabolic detoxification	Pelkonen et al (2013)
Rhein (*Polygonum multiflorum* Thunb.[a])	Cytotoxic, mutagenic	Metabolic fate in the colon and the body	Lin et al (2015)
Aconitine (*Aconitum sp*)	High acute toxicity	Metabolic detoxification	Singhuber et al (2009)

[a]Is in fact a synonym of *Reynoutria multiflora* (Thunb.) Moldenke

Matrix-Derived Effect

Within a complex herbal substance or preparation, there is a large number of potential components that may affect the toxicity of other components. For example, the induction or inhibition of metabolic or transport processes by a "perpetrator" component may drastically change the concentration of a target (also referred to as "victim") component. Plant products also often contain various antioxidant or pro-oxidant compounds that may affect the in vivo effects of other components in the same product, if they reach sufficient concentrations in the body compartments. However, it is clear that toxicological characteristics of single substances cannot be neglected even if they are in a complex mixture. There are many past examples of toxicological effects of single components administered as constituents of an herbal product. Usually the actual harmful effects of most toxic compounds have been detected in humans as sporadic or epidemic case reports (Table 4.1). It can be easily agreed that there may be various interactions between components of complex mixtures; these interactions can be inhibitory, additive, or synergistic in nature and should be taken into consideration if such effects have been observed or demonstrated. It is also clear that the toxicity study of a well-characterized mixture itself provides the most reliable results when it comes to most adverse outcomes. However, genotoxicity or carcinogenicity, for example, are two outcomes that are difficult to detect in conventional toxicity tests (or by pharmacovigilance), and they can have severe adverse outcomes that must be addressed even if the evidence is based on a single component study.

Toxicology of complex mixtures has been a topic of recent reviews (Meek et al. 2011). Furthermore, these reviews specifically addressed the matrix-derived combination effects, although a fair number of examples have come from food and environmental research (Rietjens et al. 2015; van der Berg et al. 2013).

Mechanisms Behind Matrix-Derived Effects

Matrix-derived effects may be detected, impacting at all levels of ADME processes, with some examples listed in Table 4.2. It seems obvious that a comprehensive evaluation of matrix effects of herbal products will become an important topic in future studies of herbal medicinal products (see below for methods).

Considering gastrointestinal absorption, the rate and the extent of release of a substance from the herbal matrix may be highly variable as adsorbents, tensioactive

Table 4.2 Examples of potential sites and mechanisms of matrix-derived effects on ADME characteristics and systemic exposure of components of herbal medicinal products. For background reviews, see Wagner and Ulrich-Merzenich (2009) and Rietjens et al (2015)

Site/mechanism	Consequence of interaction	Example	Reference
Release from matrix and solubilization in water	Potentially improved bioavailability	Hypericin by procyanidins	Nahrstedt and Butterweck (2010)
			See Chap. 3 for other examples
Inhibition of transporters (P-glycoprotein) in the gut wall	Enhanced basolateral transport (enhanced bioavailability)	Hypericin by flavonoid glycosides and proanthocyanidins	Nahrstedt and Butterweck (2010), Yang et al (2014), and Sevior (2012)
Metabolism in the gut wall	Increased/decreased bioavailability	Potentially e.g., mutual interactions of CYP3A4 substrates and inhibitors	Ajazuddin et al (2014) and Yang et al (2014)
			See Chap. 3 for other examples
Metabolism in the liver	Inhibition of metabolic activation	Estragole and methyleugenol by nevadensin	van den Berg et al (2013)
			Alhusainy et al (2014)
Metabolism in the liver	Induction/inhibition of metabolic activation or detoxication	Potentially e.g., mutual interactions of CYP3A4 substrates and inhibitors	Sevior (2012) and Rietjens et al (2015)
Excretion into the bile	Inhibition of biliary transporters	Isorhamnetin (Gingko component) on MATE1	Kawasaki et al (2014)
Binding to plasma proteins	Displacement from binding sites	Potentially, e.g., albumin and alpha-1-acid glycoprotein	Sevior (2012)
Excretion in the kidney	Inhibition of tubular transporters	Isorhamnetin (Gingko component) on MATE1	Kawasaki et al (2014)

or dispersing agents may accompany the studied substance. Transporters in the gut wall may be inhibited, which may lead to increased bioavailability when efflux transporters such as P-glycoprotein at the apical membrane are inhibited. The consequences of transporter modifications depend on both their localization in the cell, i.e., whether they are apical or baso-lateral, and their function, i.e., whether the herbal constituents are ligands or modulators of influx or efflux transporters. The same type of interaction can be seen for biliary transport affecting excretion and entero-hepatic cycling. Xenobiotic-metabolizing enzymes in the gut wall may also affect the bioavailability of various components of herbal medicines.

Very little is known about the distribution of constituents of herbal medicinal products. It is hypothetically possible that multiple components bind to the same plasma proteins or tissue proteins and thus affect each other's free (i.e., diffusible) concentrations. There are also some examples of displacing concurrently administered drugs by herbal constituents, thus causing an increase in the concentration of the free fraction of a drug (Sevior 2012). In general, such interactions are not clinically relevant as the increase in free fraction is usually compensated for – except in some cases of hepatic or renal failure. Because some organ barriers in the body, such as placenta or blood-brain barriers, contain both influx and efflux transporters, it is possible that the competition of herbal components at those transporters results in changes of barrier penetration, but again these possibilities are mostly hypothetical.

In terms of biotransformation, various hepatic drug metabolizing enzymes provide the principal routes for detoxification and metabolic activation of components of herbal medicines as well as determining the elimination of metabolizable components. Theoretically, matrix effects based on hepatic metabolism are expected to be possibly the most frequent mechanism due to a large number of potentially metabolizable components in any herbal medicinal product. Although examples of matrix effects based on metabolism are increasingly published (Table 4.2), most of them are observed in animal studies and "real-life" examples in human use of herbal medicines are scarce.

Interactions Between Components

Pharmacokinetic interactions (as discussed in Chap. 5) can be studied with herbal products without knowing the chemical composition of the product. Crude extracts/products can be subjected to studies where the induction and inhibition of pharmacokinetic processes are assessed; this involves the study of the effect of herbal products on drug metabolizing enzymes and transporters. Also, displacement of drugs from protein binding sites can be investigated, and, of course, variability in the herbal products is an ongoing problem. Differential variation of constituents that may affect any one of the pharmacokinetic processes is problematic, as it cannot be compensated for, even by monitoring a single or a few selected marker compounds in the herbal product. However, some benefit can be gained by screening crude herbal products for their impact on pharmacokinetics. It is possible to add some value to this approach if the goal of the studies is to deal with drugs that have a narrow therapeutic index. Matching the therapeutic indications of drugs with the therapeutic indications of herbal products is of substantial value in focusing these studies (this is discussed in detail in Sevior 2012).

Possibilities for Solving the TK of Herbal Medicinal Products

A frequently expressed claim is that the efficacy and safety of herbal medicines is thought to be due to a number of active components exerting their effects on multiple targets, possibly also acting synergistically or antagonistically. Consequently, it is important to assess how the kinetic behavior is dictated in complex mixtures. Similarly, we need to assess whether there are so-called "matrix effects" and interactions between components affecting the fate and effects of the total mixture. At least at present, it is difficult or practically impossible to elucidate such complex possibilities in clinical studies; consequently, it is advisable to start with in vitro studies, which are routinely used in the pre-clinical ADME studies of conventional pharmaceuticals.

Analytical Methods

As herbal medicinal products are complex plant-derived mixtures, the primary goal – and also the prerequisite for further studies – is to employ analytical methods that can identify the major and minor components and, when necessary, quantitate the components of concern. Various chromatographic separation methods, linked with mass spectrometry, are the methods of choice, because at least a tentative composition and relative quantification of the major components can be achieved relatively easily within a reasonable time frame. Furthermore, the information on components is beneficial for various non-clinical and clinical pharmacodynamics studies. However, these studies have to be conducted according to basic requirements; otherwise, they yield impressive but totally useless and misleading data (Sansone et al. 2007; see also Pelkonen et al. 2012). Indeed the unequivocal identification of a component, as well as absolute quantitation, need additional analytical methods, such as NMR and synthesis of reference compounds. The need for such methods and studies, however, has to be justified by other requirements, such as the elucidation of pharmacodynamic or toxicological characteristics of active components, or the use of a component as an analytical marker or as a biomonitoring marker.

In Vitro and In Vivo Methods for Studying TK

In Table 4.3, some possibilities for studying TK and ADME processes are presented. Almost all these methods were originally developed and intended for pre-clinical studies during drug development of conventional medicines. However, these methods should be used also to study herbal products and their components, even if modifications are probably required just for the sake of the complexity of herbal products. Firstly, the metabolism and metabolic interactions of complex mixtures and their main components are studied in in vitro systems, preferably human-derived preparations such as subcellular organelles (microsomes) and hepatocytes. In the cellular systems, it is also possible, at least in principle, to study associations between the metabolic and kinetic behavior of an herbal preparation and/or its components and cellular effects and toxic outcomes and markers. These effects and markers include selected in vitro toxicity outcomes (e.g.,, cytotoxicity in hepatocytes) or specific markers of mechanisms of action of toxic compounds (e.g., mitochondrial toxicity, reactive intermediates). Subsequently, the findings and predictions most relevant to clinical investigations in human volunteers or patients and, in selected cases, in vivo experimental animals are carried out. The goal is to develop a viable scheme for elucidating major ADME properties of complex herbal mixtures in in vitro systems so that the results can be extended to in vivo for comparison (see the section on poly-pharmacokinetics). *In vitro* testing systems are already being used to study metabolic interactions between herbal medicines and conventional medicines (see Sevior 2012; Pozadski et al. 2013), but other ADME-associated characteristics have not been extensively studied. It is not currently known to what extent the in vitro approach designed for single drugs can be applied for the studies of herbal products; therefore, the above-mentioned scheme is mostly conjectural, but it is believed to be a feasible starting point for investigations directed towards building an evidence-based scientific and regulatory TK dossier for herbal medicinal products.

Poly-pharmacokinetics

As described earlier in this chapter, determining pharmacokinetics of a complex herbal medicinal product, especially in clinical settings in humans, is a daunting task. Recently, Chinese scientists suggested that the advent of comprehensive profiling technologies offer new opportunities for understanding multicomponent pharmacokinetics (Lan and Jia 2010; Lan et al. 2013; Jia et al. 2015). Firstly, metabolomics, coupled with multivariate statistical tools to study the metabolism of xenobiotics, delineate the complicated variations in multiple metabolites of exogenous origin within the landscape of herbal medicine-dietary exposure-gut microbial-host metabolic

Table 4.3 *In vitro* and in vivo approaches to studying ADME (absorption, distribution, metabolism, excretion) processes of herbal medicinal products

Study specifics	Purpose	Methods	Outcome
Herbal substance/ preparation	Identification and quantification of major and minor components	Metabolomics	"Xenometabolome" of the preparation (composition of the preparation; principal 'active' components)
Subcellular organelles (microsomes), primary and permanent cells in culture	Metabolism, identification and quantitation of metabolites	Conventional analytics, metabolomics	*In vitro* "xenometabolome" (after metabolism)
		Metabolism of individual principal components	Rate of metabolism (clearance), metabolic profile of individual components
Major and minor constituents of the herbal substance	Identification of metabolizing enzymes	Recombinant enzymes, enzyme-selective inhibitors, antibodies	Enzymes catalyzing different pathways (especially of "active" components)
Intestinal cells (Caco-2, etc.), hepatocytes in culture	Permeation, transporters, metabolism	LC-MS, inhibitors of transporters	Rate of permeation, contribution of transporters, bioavailability of individual components
Cryopreserved primary hepatocytes, HepaRG cells	Metabolism	Metabolomics	Prediction of clearance and metabolic routes of "active" components
Experimental animals	"Poly-pharmacokinetics"	Metabolomics	In vivo "xenometabolome" in experimental animals
Administration of preparation or components	In vivo kinetics of herbal product and/or components	Conventional analytics	Bioavailability
Volunteers, patients	"Poly-pharmacokinetics"	Metabolomics	In vivo "xenometabolome" in humans
Administration of preparation or components	In vivo kinetics of herbal product and/or components	Conventional analytics	Bioavailability

Modified from Pelkonen et al. (2012). Copyright 2012 with permission from Elsevier
LC-MS liquid chromatography-mass spectrometry

machinery interactions, and unravel the underlying mechanisms in terms of systems biology. Secondly, metabolomics of body fluids offer simultaneously a view on endogenous metabolites altered in response to the treatment. Thirdly, integration of the bioavailable ("systemic") PK profile of the herbal product with an endogenous metabolomic profile may indicate herbal-target-effect relationships of clinical and treatment outcomes. In the words of the authors (Lan et al. 2013): "Acquisition of a complete and dynamic panel of pharmacokinetic parameters for multicomponent dosage regimens to achieve desired therapeutic efficacies is essential to minimize toxicity, reduce overdosing and drug complications, keep healthcare costs at a minimum, and, ultimately, increase patient compliance and quality of life."

Intuitively, the poly-PK integrated with pharmacodynamics seems to be an appropriate approach to study the PK/TK of complex herbal products. It is, however, obvious that the presented scheme is just the first attempt to formulate a feasible research agenda to elucidate the PK/TK of complex herbal products. It also requires up-to-date, i.e., expensive and labor-intensive analytical and computational tools for metabolomic elucidation of the resulting complex biological fluids that encompass (1) the absorbed components of the herbal product; (2) their metabolites; and (3) endogenous metabolites (e.g., glucose, cytokines, prostaglandins, etc.), which may be affected by the effects of herbal products. Analogous to conventional medicines, these studies should be carried out at different herb dosages and administration regimens, in groups of sufficient numbers of patients to be representative of the possibly encountered variabilities due to for example, genetics (CYP profiles), age, hepatic and renal function status, clinical condition, etc. Still, at least theoretically, there is a risk of missing important information from possibly low-level but very active or toxic components of the herbal mixture.

Conclusions and Perspectives

It should be obvious that the elucidation of the TK of a single component of an herbal medicinal product can follow the established paths when the compound is studied as a single isolated chemical. The elucidation of its fate as a component of an herbal medicine needs a consideration of various matrix-related effects, e.g., potential changes in bioavailability, metabolism, etc., which should be studied when a suspicion arises that other components would affect its TK significantly.

On the other hand, a comprehensive evaluation of the TK of the whole herbal medicine via the above-mentioned traditional single-substance pathway seems not to be feasible or even worthy of efforts for all components. Instead, the significance of TK elucidation of any single component should be reflected on the basis of its activity in the herbal medicinal product, whether in the area of efficacy or of safety. It is possible to screen the product itself and its constituents at group or single substance levels by various in vitro and ex vivo techniques to give a tentative view of potentially active components. However, the next consideration is whether the active concentration is actually reached in in vivo conditions. This question can be solved only by in vivo (human) studies in which the actual systemic pharmacokinetics of the active components is elucidated. This in vivo step requires a comprehensive approach, e.g., poly-pharmacokinetic integration with pharmacodynamics. The above considerations are formally outlined in Fig. 4.2.

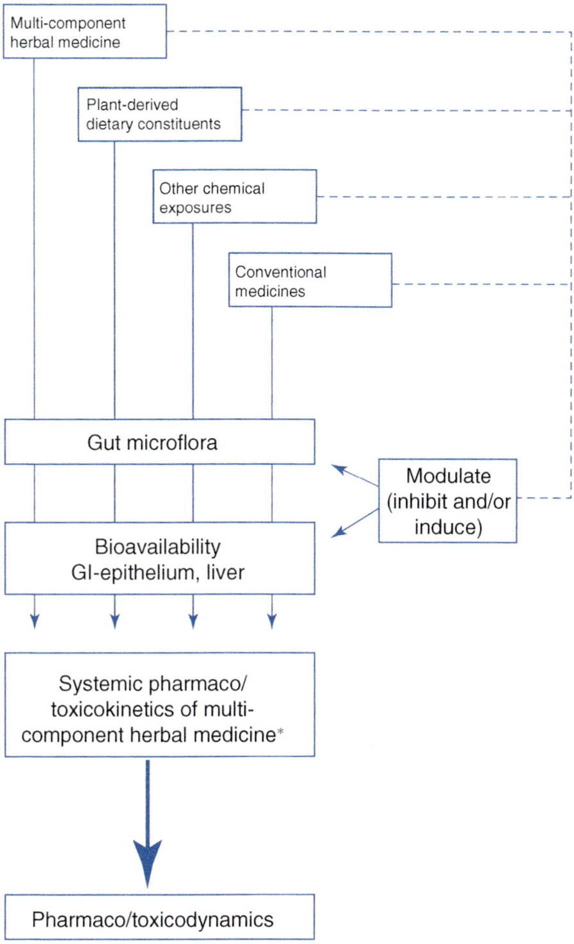

Fig. 4.2 A simplified scheme of the factors to be taken into consideration when elucidating the pharmaco-toxicokinetics of a multi-component herbal medicine. This represents a complex example of *polypharmacokinetics that can only be resolved in a satisfactory manner when the active and/or toxic components have been identified

References

Ajazuddin, Alexander A, Qureshi A, Kumari L, Vaishnav P, Sharma M, Saraf S, Saraf S (2014) Role of herbal bioactives as a potential bioavailability enhancer for active pharmaceutical ingredients. Fitoterapia 97:1–14. doi:10.1016/j.fitote.2014.05.005

Alhusainy W, Williams GM, Jeffrey AM, Iatropoulos MJ, Taylor S, Adams TB, Rietjens IM (2014) The natural basil flavonoid nevadensin protects against a methyleugenol-induced marker of hepatocarcinogenicity in male F344 rat. Food Chem Toxicol 74:28–34. doi:10.1016/j.fct.2014.08.016

Jia W, Fan TP, Wang X, Xie G (2015) The polypharmacokinetics of herbal medicines. Science 350(6262 Suppl):S76–S78

Kawasaki T, Ito H, Omote H (2014) Components of foods inhibit a drug exporter, human multidrug and toxin extrusion transporter 1. Biol Pharm Bull 37(2):292–297

Lan K, Jia W (2010) An integrated metabolomics and pharmacokinetics strategy for multi-component drugs evaluation. Curr Drug Metab 11(1):105–114

Lan K, Xie G, Jia W (2013) Towards polypharmacokinetics: pharmacokinetics of multicomponent drugs and herbal medicines using a metabolomics approach. Evid Based Complement Alternat Med 2013, 819147. doi:10.1155/2013/819147

Lin L, Ni B, Lin H, Zhang M, Li X, Yin X, Qu C, Ni J (2015) Traditional usages, botany, phytochemistry, pharmacology and toxicology of Polygonum multiflorum Thunb.: a review. J Ethnopharmacol 159:158–183. doi:10.1016/j.jep.2014.11.009

Lu LL, Liu XH, Leung ELH, Wang Y, Shi J, Hu M, Liu L, Liu ZQ (2015) The bioavailability barrier and personalized traditional Chinese medicine. Science 350(6262 Suppl):S79–S81

Meek ME, Boobis AR, Crofton KM, Heinemeyer G, Raaij MV, Vickers C (2011). Risk assessment of combined exposure to multiple chemicals: a WHO/IPCS framework. Regul Toxicol Pharmacol. [Epub ahead of print] PMID:21466831

Nahrstedt A, Butterweck V (2010) Lessons learned from herbal medicinal products: the example of St. John's Wort. J Nat Prod 73(5):1015–1021. doi:10.1021/np1000329

Nelson SD (1995) Mechanisms of the formation and disposition of reactive metabolites that can cause acute liver injury. Drug Metab Rev 27(1–2):147–177

Pelkonen O, Pasanen M, Lindon JC, Chan K, Zhao L, Deal G, Xu Q, Fan TP (2012) Omics and its potential impact on R&D and regulation of complex herbal products. J Ethnopharmacol 140:587–593

Pelkonen O, Abass KM, Wiesner J (2013) Thujone and thujone-containing herbal medicinal and botanical products: toxicological assessment. Regul Toxicol Pharmacol 65:100–107

Posadzki P, Watson L, Ernst E (2013) Herb-drug interactions: an overview of systematic reviews. Br J Clin Pharmacol 75:603–618

Rietjens IM, Boersma MG, van der Woude H, Jeurissen SM, Schutte ME, Alink GM (2005) Flavonoids and alkenylbenzenes: mechanisms of mutagenic action and carcinogenic risk. Mutat Res 574(1–2):124–138

Rietjens IMSM, Tyrakowska B, van den Berg SJPL, Soffers AEMF, Punt A (2015) Matrix-derived combination effects influencing absorption, distribution, metabolism and excretion (ADME) of food-borne toxic compounds: implications for risk assessment. Toxicol Res 4:23–35

Sansone SA, Fan T, Goodacre R, Griffin JL, Hardy NW, Kaddurah-Daouk R et al (2007) The metabolomics standards initiative. Nat Biotechnol 25:846–848

Sevior D (2012) Complementary products and drug interactions screening for the potential to cause pharmacokinetic interactions. PhD Thesis, RMIT Toxicology, School of Medical Sciences, RMIT University, Melbourne, Australia

Singhuber J, Zhu M, Prinz S, Kopp B (2009) Aconitum in traditional Chinese medicine: a valuable drug or an unpredictable risk? J Ethnopharmacol 126(1):18–30. doi:10.1016/j.jep.2009.07.031

van den Berg SJ, Klaus V, Alhusainy W, Rietjens IM (2013) Matrix-derived combination effect and risk assessment for estragole from basil-containing plant food supplements (PFS). Food Chem Toxicol 62:32–40

Wagner H, Ulrich-Merzenich G (2009) Synergy research: approaching a new generation of phytopharmaceuticals. Phytomedicine 16(2–3):97–110. doi:10.1016/j.phymed.2008.12.018

Wang J, van der Heijden R, Spruit S, Hankermeier T, Chan K, van der Greef J, Xu G, Wang M (2009) Quality and safety of Chinese herbal medicines guided by a systems biology perspective. J Ethnopharmacol 126(1):31–41

Xu Q, Qu F, Pelkonen O (2012) Network pharmacology and traditional Chinese medicine. In: Sakagami H (ed) Alternative medicine. Intech, Rijeka, pp 277–297

Yang Y, Zhang Z, Li S, Ye X, Li X, He K (2014) Synergy effects of herb extracts: pharmacokinetics and pharmacodynamic basis. Fitoterapia 92:133–147. doi:10.1016/j.fitote.2013.10.010

Chapter 5
Interactions Between Conventional and Herbal Medicinal Products

Danielle Sevior and Jorma Ahokas

Abstract Drug interactions are a commonplace occurrence, and with the majority of pharmaceuticals, such interactions are well characterized. These interactions may be beneficial, with augmentation of the effects of one of the agents used. However, interactions with adverse outcomes are also common. Harmful interactions can result in therapeutic failure of a drug or in a toxic outcome. Just as two drugs can interact, so can complementary products and drugs; such interactions are referred to as "herb-drug interactions" and, like all drug interactions, can be potentially harmful. Herb-drug interactions present their unique additional complications due to the complex and variable nature of the products and their somewhat random use.

Investigations into herb-drug interactions are challenging due to the very nature of the product; many components that may be active or inactive when isolated may behave very differently when in a mixture. Additionally, there is great product variability and currently, regulations requiring standardization or demonstration of product safety, are carried out mainly in developed parts of the world. Also, the methodology used to study herb-drug interactions can have a significant impact on the results obtained, and an understanding of these methods and the extrapolation of the data to human impact is required.

In this chapter we discuss herb-drug interactions of potential or reported clinical significance.

Keywords Herb-drug interactions • Investigation methods • Pharmacokinetics • Adverse event reporting

---

D. Sevior
Department of Pharmacology and Toxicology, University of Oulu, Oulu, Finland

School of Medical Sciences, RMIT-University, Bundoora, VIC, Australia

J. Ahokas (✉)
School of Medical Sciences, RMIT-University, Bundoora Campus,
PO Box 71, Bundoora, VIC 3083, Australia
e-mail: jorma.ahokas@icloud.com

O. Pelkonen et al. (eds.), *Toxicology of Herbal Products*,
DOI 10.1007/978-3-319-43806-1_5

Introduction

Drug interactions can significantly alter the effect of a drug. This may manifest as increased or decreased effectiveness of the drug, or an atypical effect. Drug-drug and food-drug interactions are often well documented, and many resources, including Web-based tools, exist to help practitioners and consumers to be aware of, and to avoid, common interactions. For the well-documented and extensively studied interactions, the mechanisms are often identified or suspected, and it is possible to predict the probable outcomes with reasonable certainty.

This is not necessarily the situation with herb-drug interactions, which are becoming an increasingly important area of interest, as their market share and use continue to grow. In the USA, their reported use has been estimated to be 18 % of the adult population (Barnes et al. 2008). Similar figures have been reported in Australia, with 25 % reportedly using complementary products. In the UK, the use of herbal products has been reported to be as high as 22 % (Thomas et al. 2001).

The World Health Organization (WHO) reports that the use of traditional medicine accounts for 80–95 % of primary health care in Africa and Asia, although this figure includes other forms of traditional medicine, such as acupuncture and homeopathy (World Health Organization 2008). The global market for these products was estimated at US $83 billion annually in 2008 (Robinson and Zhang 2011). Demand for herbal products worldwide has increased at an annual rate of 8 % between 1994 and 2001, and according to WHO forecasts, the global herbal market will be worth $5 trillion by the year 2050.

Irrespective of the outcome of the interaction (not all interactions are harmful), the unpredictable and often unknown nature of these interactions is cause for concern. Ideally these products should undergo the standard safety testing that is required during pre-clinical investigations into new therapeutic agents and continues throughout the market life of a product. Currently, pre-clinical testing and regulatory investigations that are carried out on pharmaceutical agents are not required (although in some countries testing is required if a therapeutic claim is made), and post-market surveillance is limited and inconsistent. Complicating the issue further, each country and the various regulators have their own requirements.

Nevertheless, in many parts of the world, herbal products continue to be generally unregulated and untested for quality, efficacy, and more importantly, safety. In that way, most of the products on the global market are untested and used by the general public without medical supervision. Post-marketing surveillance and adverse drug-interaction reporting are hampered by this global market that involves diverse and conflicting proprietary trade names, Internet sales, and unlicensed practitioners.

Under European medicines legislation (Directive 2004/24/EC), medicinal products containing herbal substances/preparations must fall within one of the following three categories in order to reach the market:

- A product can be classified under traditional medicinal use provisions ("traditional use"), accepted on the basis of sufficient safety data and plausible efficacy: the product is granted a traditional use registration.

- A product can be classified under well-established medicinal use provisions ("well-established use"). This is demonstrated with sufficient safety and efficacy data. As a result, the product is granted a marketing authorization.
- A product can be authorized after the evaluation of a marketing authorization application consisting of only product-specific safety and efficacy data ("full dossier"). As a result, the product is granted marketing authorization.

Since September 2015, the European Medical Agency (EMA) has been responsible for monitoring a number of substances and selected medical literature to identify suspected adverse reactions with medicines authorized in the European Union, and for entering the relevant information into the EudraVigilance database. The list of substances being monitored includes many herbal products, including valerian, gingko and angelica.

In the United States, the National Center for Complementary and Alternative Medicine (NCCAM), a center within the National Institutes of Health (NIH), advises the U.S. Food and Drug Administration (FDA) on issues relating to complementary products. The regulation of these products is under the control of the FDA, and many of them fall under the regulation of the Dietary Supplement Health and Education Act (DSHEA) of 1994. This act regulates products that are intended to supplement the diet and includes vitamins, minerals, herbs, or other botanicals, amino acids, and substances such as enzymes, organ tissues, glandular, and metabolites. Under this act, the manufacturer is responsible for ensuring that the ingredient is safe before it is marketed, although products do not need to be registered with the FDA or gain approval prior to being marketed. Manufacturers must make sure that product label information is truthful and not misleading, and that products are manufactured under Good Manufacturing Practices (GMP). The FDA is responsible for taking action against any unsafe dietary supplement product after it reaches the market.

The frequency of side effects due to herbal products is not known, as the current systems in place are inadequate; the U.S. Department of Health and Human Services has estimated that only 1 % of events are detected (General 2001). It is proposed that consumers are less likely to consider an adverse event linked to a herbal product that is generally perceived as "natural and safe" (Eisenberg et al. 1998).

Investigations on Herb-Drug Interactions

To carry out investigations into complementary products involves additional challenges to what is already a complex area of study. Like all drug interactions, herb-drug interactions may result from a pharmacokinetic or a pharmacodynamic interaction. The complexity is exacerbated when you consider the variable nature of these herbal and complementary products. The qualitative and quantitative composition of important components can vary greatly. This variability was recognized well over a century ago with the cardiac glycosides containing foxglove, *Digitalis purpurea* L. (Breckenridge 2006). The growth conditions, seasonal and geographic variability, the processing, and the storage of the product can alter the composition and concentration of the constituents (Braun and Cohen 2010). The eventual extraction and final compounding is also variable as the products are used in creams, teas, tablets, and capsules, all of which may lead to the patient's exposure to various compounds and varying concentrations from the same starting material. An additional difficulty with complementary products is that multiple constituents often exhibit biological activity.

The manufacturers' claim about the main ingredient may also be misleading, as it may not be the constituent responsible for an interaction. Furthermore, the claimed concentrations are often inaccurate. In a review of 25 ginseng products marketed in the USA, all were found to be correctly labeled with regards to the genus of the plant used (*Panax* or *Eleutherococcus*), but with regards to the claimed concentration of the marker compounds, a 15- to 36-fold variability of ginsenosides and 43- to 200-fold variability of eleutherosides was reported (Harkey et al. 2001).

In a study of 880 individual preparations marketed in the USA, only 43 % were consistent with labeling: another 20 % were consistent with respect to ingredients, but not dosage. A total of 37 % were either not consistent with respect to ingredients, or the label information was so vague that determination of ingredients was impossible (Garrard et al. 2003). This highlights the misleading nature of the product labeling provided by the manufacturers, which results in confusion among customers. While the products may have multiple active components, an additional complication in safety evaluation or testing is that a pharmacologically inactive component may be responsible for any herb-drug interaction.

A simple search for the term "herb-drug interaction" in the NCBI database yields over 1,700 hits (database accessed January 2016). Many of these investigations focus on isolated components tested using in vitro systems, and while these studies are crucial for understanding the underlying mechanism, they may not always translate into clinically significant interactions and are generally not supported by evidence-based clinical trials. Many of the claimed herb-drug interactions reported are based on theoretical suspicion or anecdotal case reports. On the other hand, in some cases, despite clear clinical evidence of a herb-drug interaction, the underlying mechanism of action is poorly understood.

It is important to remember that, as with drug-drug interactions, the risk of an interaction occurring and being clinically relevant must always be considered. This is particularly so when the interaction involves a drug with a narrow therapeutic window, such as warfarin, or a drug for which therapeutic failure would be life-threatening. In these situations, the potential for an interaction must be weighed against any potential benefit of using a herbal product.

Methods for Investigating Herb-Drug Interactions

A rational approach to interaction studies can be based on understanding the mechanisms of interaction or a relative risk assessment. The mechanistic studies can be broadly categorized into pharmacodynamic and pharmacokinetic interactions.

The risk assessment approach, on the other hand, relates to the likelihood of two compounds being consumed concurrently, and clinical significance of relevant interaction.

The concurrent use of complementary products with prescription medications is an issue for prescribing practitioners as consumers conceal the use of complementary products both unintentionally and intentionally. The unintentional failure to reveal their use is a consequence of not considering these products relevant. They may intentionally conceal their use for fear of being judged by medical practitioners. Medical professionals must assure their patients that they are asking them specifically about their use of herbal products, and encourage them to discuss these products as part of their treatment.

Criteria for Prioritization (Ranking Based on Risk)

With the large number of products available to the general public and the variability in their composition and formulation, undertaking interaction studies in each circumstance may not be practical. Therefore, a system of prioritizing products for study should be considered. In order to best identify the significant interactions, several factors can be considered. In each of the following circumstances, an interaction is likely to produce a significant risk to the patient.

- Likelihood of concurrent use with therapeutic agents based on therapeutic claims:
  - Many complementary products make broad therapeutic claims or claims of certain health benefits. If these claims indicate that a product is likely to be concurrently used with a therapeutic agent – for example a product to treat nausea or boost the immune system – something that patients receiving chemotherapy may likely want, then there is a high risk of concurrent use and the potential for any interaction should be investigated.
- Likelihood of a complementary product being used concurrently with drugs with a narrow therapeutic index:
  - In cases where a therapeutic agent known to have a narrow therapeutic index, such as warfarin, is administered, then products likely to be concurrently used should be tested for interaction potential.
- Circumstances for which therapeutic failure would be significant or life-threatening:
  - In cases where the patient requires therapeutic treatment for a life- threatening disease, such as cancer or HIV, then products that are likely to be taken concurrently should be investigated for their interaction potential.

A product that definitely meets these criteria is St. John's wort (*Hypericum perforatum* L.), which is one of the most-studied herbal products and is responsible for

Table 5.1 Potential herb-drug interactions with St. John's wort (*Hypericum perforatum* L.)

Drug class	Example (drug/s)	Interaction noted	Proposed mechanism of action
Anticancer agents	Imatinib	Decreased blood concentration and AUC for therapeutic agent	Induction of CYP3A4
Anti-HIV agents	Indinavir, saquinavir, and ritonavir	Reduced blood concentration and decreased effectiveness of HIV suppression	Induction of CYP3A4 and P-glycoprotein
Antimicrobial agents	Voriconazole, Erythromycin	Increased clearance of the therapeutic agent	Induction of CYP3A4
Cardiovascular agents	Warfarin and digoxin	Reduced anticoagulant effect of warfarin and blood concentration of digoxin with possible resulting heart failure	Induction of CYP2C9 and P-glycoprotein
SSRIs	Fluoxetine and fluvoxamine	Increase in serotonergic effects	Pharmacodynamic
Hypoglycemic agents	Gliclazide	Increased clearance of therapeutic agent	Independent of CYP2C9 phenotype
Immuno-modulators	Cyclosporin and acrolimus	Reduced concentration of contraceptive and risk of transplant rejection	Induction of CYP3A4 and P-glycoprotein
Oral contraceptives	Ethinyl estradiol	Reduced concentration of contraceptive and risk of pregnancy	Induction of CYP1A2 and CYP3A4
Anti-convulsants	Phenobarbitone and phenytoin	Reduced blood concentration and increased risk of seizures	Induction of CYP3A4

both pharmacokinetic and pharmacodynamic interactions (Henderson et al. 2002; Zhou et al. 2003, 2004a, b; Gurley et al. 2005a). Many of the interactions are clinically significant, causing therapeutic failure with life-threatening outcomes; several key interactions are summarized in Table 5.1. The therapeutic claims of St. John's wort – including treatment for depression and anxiety – indicates that it could be concurrently used with antidepressant medications.

Methodology for Assessment

Considerations When Studying Complex Mixtures

Many supporters of these products suggest that they cannot be studied with the same protocols as therapeutic agents, as this does not allow for the complexity of the product to be considered. It is claimed that data obtained by isolating and standardizing the components are not reflective of the true holistic and complex dosages required for herbal products. The claim is that if these products are studied as isolated components, then it is not reflective of what the consumer is exposed to (Benzie and Wachtel-Galor 2011). Thus, one of the biggest challenges in studying complementary products for their interaction potential is in the preparation of the extract(s) to be tested.

Alcohol- and water-based extractions will isolate different components that are present in the product, which may produce very different results when testing for activity and interactions. Also, low-polarity compounds would be absent from such extracts, whereas they could be solubilized in preparations for traditional use and so be bioavailable to the consumers.

In vitro studies into Dong Quai (*Angelica sinensis* (Oliv.) Diels) with both aqueous and methanolic extracts, showed the variability in results that are seen with various extraction methods. Using these extracts to study the inhibition of cytochrome P450 enzymes, aqueous extracts resulted in an IC_{50} <100 μg/ml for the nine CYP enzymes tested for inhibition (Sevior et al. 2010; Sevior 2012). The methanolic extracts caused significant inhibition of three isoenzymes: CYP2A6, CYP2B6 and CYP2C19 with IC_{50}'s of 94.7, 11.4 and 14.0 μg/ml, respectively. Inhibition by Dong Quai may be significant, since well-known CYP2C19 inhibitors, including isoniazid (K_i=25.4 μM), are likely to cause adverse clinical interactions when administered concurrently with CYP2C19 substrates such as diazepam or phenytoin (Kay et al. 1985; Ochs et al. 1981).

Advanced separation and isolation using techniques, such as supercritical fluid extraction, microwave and ultrasonication-assisted extraction, are now used and have been shown to influence the composition of the extracts (Zhang et al. 2015; Anubala et al. 2014). The reality is that these techniques are also used in the preparation of products offered to the consumers, thus changing their composition.

The inactivation of active components when in mixture, compared to isolated compound studies, can be suspected whenever the in vitro studies are not supported by clinical findings. *In vitro* investigation of valerian (*Valeriana officinalis* L.) reported that (1) commercial extracts caused inhibition of CYP3A4 (Lefebvre and Foster 2004; Hellum and Nilsen 2008), and (2) methanolic extracts inhibited additional Cytochrome P450 enzymes, including CYP1A2 and CYP2C19 (Sevior et al. 2010); but in vivo investigations did not correlate with these findings (Gurley et al. 2005b).

Reporting of Adverse Drug Reactions

Case reports are one way to detect potential interactions, but post-marketing surveillance and patient reporting is not required under current regulations for products marketed as nutrients or food supplements. A comprehensive database to collate suspected herb-drug interactions may provide researchers with guidance helping to focus studies, but until both practitioners and consumers understand the risk of these interactions, case studies will continue to be limited in their use. They nevertheless constitute important warning signals to detect harmful combinations (Shaw et al. 2012; Zhang et al. 2012).

Table 5.2 Web-based resources for herb-drug interactions

Website	Availability	Information referenced	Notes
http://who.int/en/	Free	Limited	Limited information on interactions and adverse events
http://www.fda.gov	Free	Case studies and literature referenced	Safety reporting portal for consumers and practitioners to report suspected adverse events
http://www.ema.europa.eu/ema/	Free	Case studies and literature referenced	Allows reports for suspected adverse interactions and publishes a downloadable bulletin
http://www.ema.europa.eu/ema/index.jsp?curl=pages/regulation/general/general_content_000633.jsp	Free	All known product information	Excel spread sheet that contains lists of all products being monitored for suspected adverse reactions
https://www.tga.gov.au	Free	Case studies and literature referenced	Allows reports for suspected adverse interactions and publishes a downloadable bulletin
http://www.mims.com.au	Subscription required	Case studies and literature referenced	Information regularly updated. Search feature to identify potential interactions
http://reference.medscape.com	Free	No references listed	Ability to check for interactions between two agents

At present, suspected interactions can be found using various online tools (Table 5.2). Although these resources are limited and provide varying levels in the quality of the information, they may act as a guide to practitioners and researchers.

In Vitro

In vitro studies are often relatively inexpensive and rapid and can be upscaled as a way of screening many compounds in a relatively short time. Such studies are often used during the pre-clinical phase of drug development to assess the potential for a compound to bind to therapeutic targets and pharmacokinetic profiles. Pharmacokinetic profiling of test compounds includes elucidation of metabolites and interference by other drugs. This same information can be obtained for herbal products by in vitro studies; similarly these studies can identify potential herb-drug interactions and assist in the identification of the component(s) responsible.

In vitro studies may be used as the basis for extrapolation to in vivo studies, but this can be complex and relies on assumptions regarding the behavior of the compound in vivo, and is more difficult when working with complex herbal products. Without the information regarding the metabolic clearance, metabolite formation, and stability, in vitro-*in vivo* extrapolation is practically impossible (Wilk-Zasadna et al. 2015).

In Silico

With the development of *in silico* technology, it may eventually be possible to identify the components present in complementary products that are likely to be involved in interactions. If the composition of a product is known, the compounds can be subjected to quantitative structure-activity relationship (QSAR) studies in order to predict behavior with therapeutic targets and metabolizing enzymes, such as cytochrome P450.

The 'omics-based technologies, which include NMR and LC-MS-based metabolomics, may provide significant advances in the studies of herbal products. The promise of these technologies not only involves the identification of herb-drug interactions with the identification of biomarkers, but also the identification and tracking of the active constituents and establishing chemical variability (Yang et al. 2012; Sheridan et al. 2012). These advanced analytical and highly sensitive techniques can track biomarkers that could reveal both adverse effects and interactions in vivo.

Evidence for using gene based 'omics technologies with interesting outcomes has involved products such as kava kava (*Piper methysticum*) (Guo et al. 2009, 2010b). In these studies, kava kava was administered to rats; subsequent genome-wide gene expression revealed differences in the expression of drug metabolizing enzymes in the exposed group, and alterations in the mitochondrial function and oxidative stress response pathways.

Microarrays have also been used to identify gene expression changes. *Ginkgo biloba* L. has been shown to alter a number of genes in a dose-dependent manner (Guo et al. 2010a). Of the 31,802 genes investigated, 2,011 were altered. This included an increase in the expression of alcohol dehydrogenase and cytochrome P450 family 1 and a decrease in the expression of aldehyde dehydrogenase and cytochrome P450 family 2.

These "omics" and computational technologies are rapidly expanding and may act as a complement to the standard in vitro and in vivo tests. These studies do, however, require strict quality control and reliable and repeatable methods to analyze the large amounts of data that are generated.

Pharmacokinetic Studies

Isolating individual components of herbal preparations is one approach that can be used to assist in the determination of the pharmacokinetic profiles of these products, although this is time-consuming, and in the end may not provide the answers required as these products are not taken as isolated components, but as complex mixtures.

Practitioners of herbal medicine do not consider the multitude of components of complementary products in isolation. Instead, the multiple components present are often considered integral to the action, with "a concerted pharmacological intervention of multiple compounds interacting with multiple targets and possessing mutually interdependent activities that are required for an optimal effect" (Chan 1995). For this reason, proponents of herbal products often discredit studies involving isolated components; however, these studies still provide valuable information and insights into the specific components possibly responsible for an interaction.

Absorption/Distribution

Interactions of herbal products with the absorption and distribution of clinical drugs have the potential to alter the concentration of the drug in the systemic circulation.

Interaction with Transporters

The extensive family of transporters is responsible for the flux of a wide range of compounds in and out of cells. Consequently, they are important in the absorption, redistribution, and elimination of drugs. Any change in the relative abundance or activity of transporters can have consequences. These transporters can be induced or inhibited by a wide variety of compounds. The pharmacokinetic behavior of drugs that are substrates for transporters can be influenced by concurrently administered compounds including conventional medicines, foods, and complementary products that function as inhibitors or inducers of transporter function.

To date, several complementary and alternative products have been identified as substrates, inhibitors, and/or inducers of P-glycoprotein (P-gp), which is a major efflux transporter in many organs, such as the intestine, liver, kidney, and brain; P-gp notably opposes the intestinal absorption of xenobiotics. One of the most clinically significant examples is St. John's wort (*Hypericum perforatum* L.), which is commonly used to treat mild depression (Nathan 1999).

In vitro and in vivo studies have indicated that St. John's wort induces intestinal P-gp (Hennessy et al. 2002; Perloff et al. 2001). These findings have been consistent in healthy volunteers who consumed the product for 14 days, with a subsequent 1.4--to 1.5-fold increase in their expression of P-gp (Dürr et al. 2000). This induction of P-gp can be expected to enhance the clearance of therapeutic agents from the circulation of patients. In the case of patients undergoing treatment with the chemotherapeutic drugs etoposide and doxorubicin, there is a risk of a decrease in the therapeutic effect and therapeutic failure (Hennessy and Spiers 2007).

Milk thistle is suspected to inhibit P-gp, and studies have shown that silymarin, a mixture of flavolignanes, major components of milk thistle, inhibits the transport of P-gp substrates such as calcein into the cell in in vitro experiments, although in vivo studies with healthy human volunteers determined no statistically significant effect on digoxin pharmacokinetics with milk thistle (900 mg a day) (Gurley et al. 2006). Studies such as this with milk thistle highlight the problems that can occur when isolated components are investigated for drug-interactions and the need for additional clinical data.

Interaction with Protein Binding

Binding to plasma proteins can influence the distribution, metabolism, and excretion of many endogenous and exogenous compounds. *In vivo*, drug molecules are either bound to plasma and tissue proteins and plasma lipids, or are free. It is generally accepted that the free molecules represent the fraction of a drug that is able to interact with the therapeutic target to produce an effect.

The fraction of a drug or ligand that is bound to protein can change significantly due to co-administered drugs/ligands. Thus the ability to bind to protein and to displace a previously bound compound from protein is an important consideration in drug development and in the prediction of drug interactions. Several human plasma proteins, such as albumin and alpha-1-acid glycoprotein, play a key role in binding drugs.

One significant example of herbal constituents binding to a plasma protein and causing a potential herb-drug interaction is seen with albumin and Danshen (a traditional Chinese medicine prepared from the root of *Salvia miltiorrhiza* Bunge). Studies claim that Danshen components "are 50-70 % bound to albumin, and can displace salicylate from its binding site" (Gupta et al. 2002); this study appears to be alarming due to the potential for concurrent use of Danshen (recommended for the treatment of "stagnation of blood flow") with a drug such as warfarin, which has a narrow therapeutic index. While the methods used in this study are valid, the interpretation by these authors is highly misleading and highlights the need for caution when assessing the literature for potential interactions; the authors should specify that "at the Danshen level used, 50-70 % of salicylate binding sites are occupied by as yet unidentified components of Danshen." Limitations of this study include using non-physiological concentrations of albumin and very high concentrations of Danshen, measuring the binding and subsequent displacement of salicylate instead of digoxin, and assuming the Danshen would displace the digoxin from the albumin, rather than a direct measurement of binding.

Drug interactions due to protein-binding and displacement highlight the importance of considering the in vitro and in vivo results in the clinical setting. Let us assume that two compounds (A and B) are administered concurrently (or when an herb and a drug are co-administered). If compound A is displaced by compound B, an increase in the free concentration of compound A results, but the increase in free compound A may not correlate with an increase in compound A at the receptor site because A will also be available for redistribution to the rest of the body. Any increase in free A following redistribution will also be available for elimination. Even for low clearance drugs, where intrinsic clearance of free compounds is the only determinant of mean steady-state free drug concentration, free concentration for A will return to the pre-B level. Thus any increase in the pharmacological effect of A will generally be transient and cannot be sustained; this means that the interaction may become significant if the transient increase in the free compound is for a drug with a very narrow therapeutic index.

Predicting clinically significant interactions from in vitro data is further complicated by the possibility for drug metabolites to bind to plasma proteins and potentially displace previously bound compounds. As the metabolic stability of a molecule and the rate of the biotransformation can affect its toxic potential, disposition in the body and eventual excretion (Wilk-Zasadna et al. 2015; Coecke et al. 2013), in vitro studies should consider the possible metabolites that may also be biologically active in vivo.

Metabolism

Drug interactions involving metabolism are very important as they are often the cause of adverse outcomes.

The cytochrome P450 (CYP) system is comprised of a family of heme-containing (hemoprotein) enzymes with closely related isoforms classified as "mixed function oxidases." This enzymatic system is found in almost all organisms, including mammals, bacteria, and plants,being crucial for the oxidative, per-oxidative, and reductive metabolism of exogenous and endogenous compounds. More than >11,000 CYPs have been identified, and new CYPS are continuously being added (Nelson 2009).

To date, there have been 107 genes identified as encoding for CYPs in humans, with 18 families and 45 subfamilies. Of the genes identified, 57 have been isolated, identified, and classified, and most appear to be expressed primarily in the endoplasmic reticulum. Of the CYPs with known catalytic activities, 14 are involved in steroidogenesis, 4 in the metabolism of vitamins, 5 in eicosanoid metabolism, 4 have fatty acids as their substrates, and 15 catalyze transformation of xenobiotic chemicals (Guengerich 2005).

The investigation of interactions involving the drug-metabolizing enzymes, and in particular the CYP group, is a key area of study as it has been shown that ≈75 % of all drugs can be metabolized by three CYPs (CYP3A4, CYP2D6, and CYP2C9) (Rendic 2002), and a set of six to seven CYPs (CYP1A1, CYP1A2, CYP1B1, CYP2A6, CYP2B6, CYP2C8, CYP2C9, CYP2C19, CYP2D6 CYP2E1 and CYP3A4) account for 90–95 % of all drug metabolism (Evans and Relling 1999). CYPs are involved in the metabolism of 78 % of the 200 top-selling prescription medications in the U.S. (Zanger et al. 2008).

Complementary products are usually complex mixtures containing many potentially pharmacologically active components, notably essential oils, tannins, flavonoids, anthraquinones, alkaloids or polyphenols (Liu et al. 2006; Zhou et al.. 2004a, b, 2005; Izzo and Ernst 2009). Several complementary products, such as St. John's wort (*Hypericum perforatum* L.), garlic (*Allium sativum* L.), liquorice (*Glycyrrhiza glabra* L.), and ginseng (*Panax ginseng* C.A.Mey.) have been proven to interact with CYPs. They act either as inhibitors or inducers of more or less specific CYPs, with St. John's wort known to induce CYP3A4 (Henderson et al. 2002).

Furanocoumarins found in several herbal medicines, including Chinese angelica (*Angelica sinensis*) and also in grapefruit (*Citrus* × *paradisi* Macfad.), are known to be potent inhibitors of CYP1A2 and CYP3A4, and therefore the cause of drug interactions (Tassaneeyakul et al. 2000). Products suspected of containing furanocoumarins – and therefore likely to cause a drug interaction – can be screened using standard HPLC analysis, allowing the rapid identification of potential drug interactions without the need for more complex experiments.

Furanocoumarins isolated from grapefruit juice are mechanism-based inhibitors of CYP3A4 (Tassaneeyakul et al. 2000), and resveratrol from *Vitis vinifera* L., have been shown to be mechanism-based inhibitors of CYP1A2 and CYP3A4 (Chan and Delucchi 2000). The analysis of 30 herbal plants from Indonesia showed their ability to inhibit CYP2D6 and CYP3A4 (Subehan et al. 2006). With an increase in the incubation time, five plants showed more than a 30 % increase in mechanism-based inhibition of CYP3A4, and three showed a 30 % increase in mechanism-based inhibition of CYP2D6.

CYP2E1 inhibitors from natural sources include compounds originating from garlic (diallyl sulfide, diallyl sulfoxide, and diallyl sulfone) that were shown to competitively inhibit CYP2E1 (Gurley et al. 2005a; Brady et al. 1991). This was confirmed by two clinical trials, as revealed by the decreased 6-hydroxychlorzoxazone/chlorzoxazone serum ratios (Gurley et al. 2005a).

In vivo studies in humans initially found that black cohosh (*Actaea racemosa* L.) may weakly inhibit CYP2D6 (Gurley et al. 2005b), but this observation was negated by later studies conducted by the same group (Gurley et al. 2008). In both these studies, the method for determining inhibition is the same but the product investigated was different. The 2005 study used a black cohosh that was standardized to 0.2 % triterpene glycosides per tablet, while the 2008 study used a product that was standardized to 2.5 % triterpene glycosides per tablet. This finding suggests that while the triterpene glycosides are often reported as the principal component of these preparations, they may not be the components that cause CYP inhibition; as such other components are not yet identified and may be present or absent; depending on the extraction conditions, this complicates safety assessments.

The induction of CYPs can have a major impact on drug metabolism, pharmacokinetic behavior, drug-drug interactions, on the toxicity of foreign chemicals, and on the activity and disposition of endogenous hormones (Conney 1982). It is now known that a wide range of chemicals are capable of causing induction; these include therapeutic agents, pesticides, food additives, industrial chemicals, dietary constituents, natural products, and environmental pollutants (Pelkonen et al. 2008).

Pharmacodynamic Interactions

Pharmacodynamic interactions with herbal products are not as well studied or identified. Early reports suggested that garlic and ginkgo interacted with anticoagulants and caused increased bleeding in consumers. For gingko, adverse events were particularly severe for aspirin (spontaneous hyphema), warfarin (intracerebral hemorrhage) and ibuprofen (comatose state with an intracerebral mass bleeding from which the patient died). However, recent trials did not confirm these effects (Izzo and Ernst 2009).

Ginger (*Zingiber officinale*) was also suspected to interact with warfarin but, as with ginkgo and garlic, in vitro studies indicated that it could inhibit platelet aggregation, and in vivo studies with healthy human volunteers showed that co-administration of warfarin with ginseng did not alter the pharmacokinetics or the pharmacodynamics of warfarin (Jiang et al. 2005).

Warfarin is a complex target of interacting substances. Apart from pharmacokinetic interactions, warfarin therapy is subject to any dietary product containing vitamin K; these include many vegetables and green tea.

Clinically significant interactions have shown that when green tea is consumed along with warfarin, the antagonism may result in dramatic reductions of a patient's INR value (Taylor and Wilt 1999).

Clinical Considerations

Predicting adverse herb-drug interactions in humans is complicated by the influence of genes, disease state, and age on the expression of many proteins and enzymes that are involved in interactions. As the liver contains the greatest concentration of CYPs, diseases affecting the liver, including cirrhosis, alcoholic liver disease, hepatitis, and hepatocarcinoma, all alter the level and activities of the CYPs and therefore the drug metabolism. CYP activity may also be decreased due to altered hepatic blood flow and hypoalbuminemia.

While the changes in the drug metabolizing enzymes that occur with age and in disease states are well documented, there are also changes in the serum protein levels. Alpha-1-acid glycoprotein (AGP) is a protein that is known to bind mainly to basic drugs in the serum. A critically important characteristic of AGP is the change in plasma levels that occurs with conditions such as inflammation (e.g., arthritis) and chronic disease (e.g,, cancer). In these patients, expression is increased two- to fivefold (Fournier et al. 2000). An increase in AGP also occurs following myocardial infarction (Johansson et al. 1972) and surgery (Voulgari et al. 1982), while lower levels are seen during pregnancy (Perucca and Crema 1982), in thyroid disease (Feely et al. 1981), and in patients with liver cirrhosis (Serbouce-Hougel et al. 1981). Such wild fluctuations must be considered when the potential for drug interactions are investigated with regards to AGP.

Conclusion

The inescapable reality is that most patients resorting to the use of herbal products are likely to be treated with prescription or over-the-counter drugs (MacLennan et al. 2002; Eisenberg et al. 1998). Additionally, many patients see multiple therapists along with medical practitioners.

Safety studies and interaction studies of herbal products are lacking and are urgently required to ensure consumer safety. Several steps should be mandatory regarding the standardization and quality of these products.

- Worldwide databases need to be established with full product monographs to ensure correct product identification;
- the composition of products should be determined and compounds known to cause drug interactions flagged.

To monitor the safety of these products, the cooperation between traditional practitioners, pharmacists, medical professionals, and consumers needs to improve to ensure that accurate reports of adverse events are identified. Due to the complex nature of these products and the specific challenges associated with them, pharmacovigilance needs to be expanded to incorporate all medications and herbal products, whatever their marketing status, and consumers need to be educated to consider complementary products in this light. Mandatory labelling requirements can be set so that consumers are warned of potential interactions.

Reported suspected interactions with complementary products should include additional information, such as the origin of the product, the presence of any contaminant and adulterants, and the name attributed to the product (traditional, botanical, common). If this information were available to regulators worldwide, then consistent risk assessments could be carried out and the risk to patients would be decreased.

Studying herbal products as pure compounds is relatively uncomplicated; however, complementary products present a number of difficult problems that must be solved in order to conduct any meaningful studies with them. These problems include:

- Product selection
- Extraction method
- Metabolic profiling and constituent concentration

Finally, the study of drug interactions is complicated by the variation in response by individual subjects; causes of this variability include ethnicity, age, genetic factors, disease state (in particular, kidney and liver failure), and pregnancy.

References

Anubala S, Sekar R, Nagaiah K (2014) Development and validation of an analytical method for the separation and determination of major bioactive curcuminoids in curcuma longa rhizomes and herbal products using non-aqueous capillary electrophoresis. Talanta 123:10–17

Barnes P, Bloom B, Nahin R (2008) Complementary and alternative medicine use among adults and children: United States, 2007. National Health Statistics Report. U.S. Department of Health and Human Services

Benzie IFF, Wachtel-Galor S (2011) Herbal medicine: bimolecular and clinical aspects. CRC Press/Taylor & Francis, Florida

Brady JF, Xiao F, Wang MH, Li Y, Ning SM, Gapac JM, Yang CS (1991) Effects of disulfiram on hepatic P450IIE1, other microsomal enzymes, and hepatotoxicity in rats. Toxicol Appl Pharmacol 108:366–373

Braun L, Cohen M (2010) Herbs & natural supplements – an evidence-based guide. Elsevier Australia (Sydney)

Breckenridge A (2006) William Withering's legacy – for the good of the patient. Clin Med 6:393–397

Chan K (1995) Progress in traditional Chinese medicine. Trends Pharmacol 16:182–187

Chan WK, Delucchi AB (2000) Resveratrol, a red wine constituent, is a mechanism-based inactivator of cytochrome P450 3A4. Life Sci 67:3103–3112

Coecke S, Pelkonen O, Leite S, Bernauer U, Bessems JGM, Bois FY, Gundert-Remy U, Loizou G, Testai E, Zaldivar JM (2013) Toxicokientics as a key to the integrated toxicity risk assessment based primarily on non-animal approaches. Toxicol In Vitro 27:1570–1577

Conney AH (1982) Induction of microsomal enzymes by foreign chemicals and carcinogenesis by polycyclic aromatic hydrocarbons: G.H.A. Clowes memorial lecture. Cancer Res 42:4875–4917

Dürr D, Stieger B, Kullak-Ublick GA, Rentsch KM, Steinert HC, Meier PJ, Fattinger K (2000) St John's wort induces intestinal P-glycoprtoein/MDR1 and intestinal and hepatic CYP3A4. Clin Pharmacol Ther 68:598–604

Eisenberg DM, Davis RB, Ettner SL, Appel S, Wilkey S, Van Rompay MI, Kessler RC (1998) Trends in alternative medicine use in the United States, 1990–1997. J Am Med Assoc 280:1569–1575

Evans WE, Relling MV (1999) Pharmacogenomics: translating functional genomics into rational therapeutics. Science 286:487–491

Feely J, Stevenson I, Crooks J (1981) Alteres plasma protein binding of drugs in thyroid disease. Clin Pharmacokinet 6:298–305

Fournier T, Medjoubi-N N, Porquet D (2000) Alpha-1- acid glycoprotein. Biochim Biophys Acta 1482:157–171

Garrard J, Harms S, Eberly LE, Matiak A (2003) Variations in product choices of frequently purchased herbs. Arch Intern Med 163:2290–2295

Office of Inspector General (2001) Adverse event reporting for dietary supplements. An inadequate safety valve. Department of Health and Human Services - USA. Boston Regional Office, OEI-01-00-00180. Available at: http://www.dhhs.gov/oig/oei

Guengerich FP (2005) Human cytochrome P450 enzymes. In: Ortiz De Montellano PR (ed) Cytochrome 450. Structure, mechanism and biochemistry, 3rd edn. Kluwer Academic/Plenum Publishers, New York

Guo L, Li QZ, Xia Q, Dial S, Chan PC, Fu P (2009) Analysis of gene expression changes of drug metabolizing enzymes in the livers of F344 rats following oral treatment with kava extract. Food Chem Toxicol 47:433–442

Guo L, Mei N, Lioa W, Chan PC, Fu PP (2010a) Ginkgo biloba extract induces gene expression changes in xenobiotics metabolism and the Myc-centered network. OMICS 14:75–90

Guo L, Shi Q, Dial S, Xia Q, Mei N, Li QZ, Chan PC, Fu P (2010b) Gene expression profiling in male B6C3F1 mouse livers exposed to kava identifies changes in drug metabolizing genes and potential mechanisms linked to kava toxicity. Food Chem Toxicol 48:686–696

Gupta D, Jalali M, Wells A, Dasgupta A (2002) Drug-herb interactions: unexpected suppression of free danshen concentrations by salicylate. J Clin Lab Anal 16:293–294

Gurley BJ, Gardner SF, Hubbard MA, Williams DK, Gentry WB, Cui Y, Ang CY (2005a) Clinical assessment of effects of botanical supplementation on cytochrome P450 phenotypes in the elderly: St. John's wort, garlic oil, Panax ginseng and Ginkgo biloba. Drugs Aging 22:525–539

Gurley BJ, Gardner SF, Hubbard MA, Williams DK, Gentry WB, Khan IA, Shah A (2005b) In vivo effects of goldenseal, kava kava, black cohosh and valerian on human cytochrome P450 1A2, 2D6, 2E1 and 3A4 phenotypes. Clin Pharmacol Ther 77:415–426

Gurley BJ, Barone GW, Williams DK, Carrier J, Breen P, Yates CR, Song PF, Hubbard MA, Tong Y, Cheboyina S (2006) Effect of milk thistle (Silybum marianum) and black cohosh (Cimicifugaq racemosa) supplementation on digoxin pharmacokinetics in humans. Drug Metab Dispos 34:69–74

Gurley BJ, Swain A, Hubbard MA, Williams DK, Barone GW, Hartsfield F, Tong Y, Carrier DJ, Cheboyina S, Battu SK (2008) Clinical assessment of CYP2D6-mediated herb-drug interactions in humans: effects of milk thistle, black cohosh, goldenseal, kava kava, St. John's wort and echinacea. Mol Nutr Food Res 52:1–9

Harkey MR, Henderson GL, Gershwin ME, Stern JS, Hackman RM (2001) Variability in commercial ginseng products: an analysis of 25. Clin Nutr 73:1101–1106

Hellum BH, Nilsen OG (2008) In vitro inhibition of CYP3A4 metabolism and P-glycoprotein mediated transport by trade herbal products. Basic Clin Pharmacol Toxicol 102:466–475

Henderson L, Yue QY, Bergquist C, Gerden B, Arlett P (2002) St. John's wort (hypericum perforatum): drug interactions and clinical outcomes. Br J Clin Pharmacol 54:349–356

Hennessy M, Spiers JP (2007) A primer on the mechanics of P-glycportein the multidrug transporter. Pharmacol Res 55:1–15

Hennessy M, Kelleher D, Spiers JP, Barry M, Kavanagh P, Back D, Mulcahy F, Feely J (2002) St. Johns wort increases expression of P-glycoprotein: implications for drug interactions. Br J Pharmacol 53:75–82

Izzo AA, Ernst E (2009) Interactions between herbal medicines and prescribed drugs. An updated systematic review. Drugs 69:1777–1798

Jiang X, Williams KM, Liauw WS, Ammit AJ, Roufogalis BO, Duke CC, Day RO, Mclachlan AJ (2005) Effect of ginkgo and ginger on the pharmacokientics and pharmacodynamics of warfarin in healthy subjects. Br J Clin Pharmacol 59:425–432

Johansson B, Kindmark C, Trell E, Wollheim F (1972) Sequential changes of plasma proteins after myocardial infarction. Scand J Clin Lab Investig 124:117–126

Kay L, Kampmann JP, Svendsen TL, Vergman B, Hansen JEM, Skovsted L, Kristensen M (1985) Influence of rifampicin and isoniazid on the kinetics of phenytoin. Br J Clin Pharmacol 20:323–326

Lefebvre T, Foster BC (2004) In vitro activity of commercial valerian root extracts against human cytochrome P450 3A4. J Pharmacol Pharm Sci 7:265–273

Liu KH, Kim MJ, Jeon BH, Shon JH, Cha IJ, Cho KH, Lee SS, Shin JG (2006) Inhibition of human cytochrome P450 isoforms and NADPH-CYP reductase in vitro by 15 herbal medicines, including *Epimedii herba*. J Clin Pharm Ther 31:83–91

Maclennan AH, Wilson DH, Taylor AW (2002) The escalating cost and prevalence of alternative medicine. Preventative Med 35:166–173

Nathan P (1999) The experimental and clinical pharmacology of St. John's Wort (hypericum perforatum L.). Mol Psychiatry 4:333–338

Nelson DR (2009) The cytochrome P450 homepage. Hum Genomics 4(1):59–65

Ochs HR, Greenblatt DJ, Roberts GM, Dengler HJ (1981) Diazepam interaction with antituberculosis drugs. Clin Pharmacol Ther 29:671–678

Pelkonen OP, Turpeinen M, Hakkola J, Honkakoski P, Hukkanen J, Raunio H (2008) Inhibition and induction of human cytochrome P450 enzymes: current status. Arch Toxicol 82:667–715

Perloff MD, Von Moltke LL, Störmer E, Shader RI, Greenblatt DJ (2001) St. John's wort: an in vitro analysis of P-glycoprotein induction due to extended exposure. Br J Pharmacol 134:1601–1608

Perucca E, Crema A (1982) Plasma protein binding of drugs during pregnancy. Clin Pharmacokinet 7:336–352

Rendic S (2002) Summary of information on human CYP enzymes: human P450 metabolism data. Drug Metab Rev 34:83–448

Robinson MM, Zhang X (2011) Traditional medicines: global situation, issues and challenges. In: WHO/EMP/MIE (ed) The world medicines situation 2011. World Health Organization, Geneva

Serbouce-Hougel S, Durand G, Corbii M, Agneray J, Feger J (1981) Alterations in relative proportions of microheterogenous forms of human alpha 1-acid glycoprotein in liver disease. J Hepatol 2:245–252

Sevior DK (2012) Complementary products and drug interactions – screening for the potential to cause pharmacokinetic interactions. Doctor of Philosophy, RMIT University

Sevior D, Hokkanen J, Tolonen A, Abass K, Tursas L, Pelkonen O, Ahokas JT (2010) Rapid screening of commercially available herbal products for the inhibition of major human hepatic cytochrome P450 enzymes using the N-in-One cocktail. Xenobiotica 40:245–254

Shaw D, Graeme L, Pierre D, Elizabeth W, Kelvin C (2012) Pharmacovigilance of herbal medicine. J Ethnopharmacol 140:513–518

Sheridan H, Krenn L, Jiang R, Sutherland I, Ignatova S, Marmann A, Liang X, Sendker J (2012) The potential of metabolic fingerprinting as a tool for the modernization of TCM preparations. J Ethnopharmacol 140:482–491

Subehan A, Usia T, Iwata H, Kadota S, Tezuka Y (2006) Mechanism-based inhibition of CYP3A4 and CYP2D6 by Indonesian medicinal plants. J Ethnopharmacol 105:449–455

Tassaneeyakul W, Guo LQ, Fukuda K, Ohta T, Yamazoe Y (2000) Inhibition selectivity of grapefruit juice components on human cytochromes P450. Arch Biochem Biophys 378:356–363

Taylor JR, Wilt VM (1999) Probable antagonism of warfarin by green tea. Ann Pharmacother 33:426–428

Thomas KJ, Nicholl JP, Coleman P (2001) Use and expenditure on complementary medicine in England: a population based survey. Complement Ther Med 9:2–11

Voulgari F, Cummins P, Gardecki T, Beeching N, Stone P, Stuart J (1982) Serum levels of acute phase and cardiac proteins after myocardial infarction, surgery, and infection. Br Heart J 1982:4

Wilk-Zasadna I, Bernasconi C, Pelkonen O, Coecke S (2015) Biotransformation in vitro: an essential consideration in the quantitative in vitro-to-iv-vivo extrapolation (QIVIVE) of toxicity data. Toxicology 332:8–19

World Health Organization (2008) Traditional medicine. WHO. http://www.who.int/mediacentre/factsheets/2003/fs134/en/

Yang SO, Shin YS, Hyun SH, Cho S, Bang KH, Lee D, Choi SP, Choi HK (2012) NMR-based metabolic profiling and differentiation of ginseng roots according to cultivation ages. J Pharm Biomed Anal 58:19–26

Zanger UM, Turpeinen M, Klein K, Schwab M (2008) Functional pharmacogenetics/genomics of human cytochromes P450 involved in drug biotransformation. Anal Bioanal Chem 392:1093–1108

Zhang L, Yan J, Liu X, Ye Z, Yang X, Meyboom R, Chan K, Shaw D, Duez P (2012) Pharmacovigilance practice and risk control of traditional Chinese medicine drugs in China: current status and future perspective. J Ethnopharmacol 140:519–525

Zhang Y, Liu C, Qi Y, Li Y, Wang L, Ren J, Tang Y (2015) Efficient combination of circulating ultrasound-assisted extraction partition chromatography for extraction and on-line separation of chemical constituents from Stellera chamaejasme L. Phytochem Anal 26:301–309

Zhou S, Gao Y, Jiang W, Huang M, Xu A, Paxton JW (2003) Interactions of herbs with cytochrome P450. Drug Metab Rev 35:35–98

Zhou S, Lim L, Chowbray B (2004a) Herbal modulation of P-glycoprotein. Drug Metab Rev 36:57–104

Zhou SF, Koh HL, Gao Y, Gong ZY, Lee EJD (2004b) Herbal bioactivation: the good, the bad and the ugly. Life Sci 74:935–968

Zhou S, Huang M, Xu A, Yang H, Duan W, Paxton JW (2005) Prediction of herb-drug metabolic interactions: a simulation study. Phytother Res 19:464–471

Chapter 6
Safety Pharmacology of Herbal Products

Gert Laekeman

Abstract This chapter aims at a comprehensive review on safety pharmacology. Foxglove and the discovery of acetylsalicylic acid are given as historical examples emerging from benefit-risk balance. Before going to practical examples of *in vitro* research, a theoretical approach is given to competitive and non-competitive antagonism. Results of experimental work are presented in order to show how early toxic activity can be detected and can even be related to mechanisms of action (e.g., the indole alkaloids of *Pterotaberna inconspicua* and the sesquiterpene lactone vernolepin from *Vernonia amygdalina*). If substances can be detected as responsible agents for a biological activity of complex mixtures, the safety approach can be simplified (e.g., eugenol in powder of *Myristica fragrans*). Pharmacokinetic considerations are important when taking into consideration safety and therapeutic issues of red yeast rice. A pharmacokinetic schedule is given and illustrated with herbal examples. Finally, a framework of questions is proposed to cover quality as well as pre-clinical and clinical safety.

Keywords Pharmacology • Safety • History • Checklist • Pharmacokinetics • Pharmacodynamics

Safety Pharmacology: An Introduction

Although there seems to be no internationally accepted definition of safety pharmacology, Pugsley et al. (2008) define the concept as follows: "... Safety Pharmacology is the discipline that seeks to predict whether a drug (in the widest sense of the word), if administered to human (or animal) populations, is likely to be found unsafe, and its professional mandate is to prevent such an occurrence . . ."

In its *Guidance for Industry*, the Food and Drug Administration defines the objectives of safety pharmacology studies: (1) to identify undesirable pharmacodynamics properties of a substance that may have relevance to its human

G. Laekeman
Clinical Pharmacology and Pharmacotherapy, Faculty of Pharmaceutical Sciences, KU Leuven, Leuven, Belgium
e-mail: gert.laekeman@kuleuven.be

O. Pelkonen et al. (eds.), *Toxicology of Herbal Products*,
DOI 10.1007/978-3-319-43806-1_6

safety; (2) to evaluate adverse pharmacodynamic and/or pathophysiological effects of a substance observed in toxicology and/or clinical studies; and (3) to investigate the mechanism of the adverse pharmacodynamic effects observed and/or suspected (ICH 2001).

The experimental approach of safety pharmacology can be designed according to the hypothesized properties of the substance or preparation. The purpose is not to address an exhaustive list, but a few examples can be illustrative.

A therapeutic class to which the principle to be tested belongs, permits foreseeing possible toxic effects, e.g., an antispasmodic activity on smooth muscle cells can result in irreversible paralysis. However, adverse effects are sometimes surprising: for example, substances with an antipsychotic action can cause QT-prolongation. The last example points to the importance of taking vital organs and systems to the testing. Pugsley et al. (2008) present an overview of a multidisciplinary integration required to evaluate the safety profile of new chemical entities in safety pharmacology. According to their view, the central nervous activity, as well as respiratory and cardiovascular functions should be covered.

Ligand binding of molecules to receptors studies give factual evidence for specificity and affinity, but do not permit making predictions on the intrinsic activity that is related to the cascade of events triggered by the binding. The same can be said about enzyme-inhibiting assays. Moreover, experimental models should also take into account that substances can be metabolized, and that the safety of metabolites should be tested as well.

As a result of the above-mentioned considerations, pharmacological testing of safety remains a challenge. Reliable results can only be obtained by triangulation, i.e., combining various techniques and models, which often means combining in vitro and in vivo testing. For example, the European Medicines Agency (EMA) has established a Joint Expert Group for Reduction, Replacement and Refinement (JEG 3Rs) in order to reduce the number of animals in experimental pharmacological procedures (EMA 2012). The JEG3R examines possibilities and limitations of the 3R principle. In the near future, guiding documents will be available on the EMA website.

This chapter aims at a comprehensive view of historical and actual examples of how to get an estimate of the benefit-risk balance. It also presents a primary framework of binary questions that should be feasible for safety when considering herbal practice.

Historical Considerations

Throughout history, the beneficial or toxic effects of many plant materials must have been recognized by trial and error. Attempts were made to develop methods in order to turn pharmacological activity into therapeutic practice. Experimentation and observation suffered from the difficulty of characterizing real actions of natural substances and mixtures and making the difference between a real effect and a coincidence.

Around the end of the eighteenth century, developments in chemistry helped master the way of making preparations in a reproducible way. The time for standardization was still to come, but the notion of dose-response was something within reach.

Digitalis Purpurea as an Example

William Withering (1741–1799) was one of the first practitioners carrying out "phase one" experiments *avant la lettre* by administering increasing dosages of leaves of *Digitalis purpurea* to patients and studying their cardiac function. Firstly, he collected and prepared the leaves of the purple foxglove (*Digitalis purpurea*) to obtain a product of reasonable consistency. Secondly, he identified most of the adverse effects of digitalis and how toxicity could be minimized by dose reduction. Thirdly, he investigated the dose-response characteristics of digitalis, with respect to both slowing the heart rate and inducing diuresis. He also showed that some individuals were more sensitive and responsive to digitalis than others (Breckenridge 2006) (Fig. 6.1).

The lessons that Withering learned from his studies of digitalis are still relevant today. He dealt – probably for the first time in history – with the basic properties of medicines: quality, safety, and effectiveness.

Reasonable Consistency: The Quality Issue

A quote can be taken from William Withering's *Account of the Foxglove and Some of Its Medical Uses* (London: CGJ and Robinson 1785): cited by Breckenridge (2006)

> I was well aware of the uncertainty which must attend on the exhibition of the root of a biennial plant and therefore continued to use the leaves. These I found to vary much at different seasons of the year, but I found that if gathered at one time of year, namely when it

Fig. 6.1 *Digitalis pupurea* L. or foxglove, the plant from which William Withering evaluated therapeutic effects and safety, using the leaves in clinical experiments (Image taken from http://www.kuleuven-kulak.be/bioweb/?page=guide&lang=nl. Accessed on April 30, 2015)

> was in its flowering state and carefully dried, the dose could be determined as exactly as any other medicine. The more I saw of the great powers of this plant, the more it seemed necessary to bring the doses to the greatest degree of accuracy.

A clearer historical reference to the nature and quality of herbal raw materials can probably not be made. Until now, the composition of various preparations has been discussed. The definition of the preparations is an important issue in the making of herbal monographs within the Herbal Medicinal Product Committee of the European Medicines Agency (EMA). Very recently, the possible consequences of differences in the composition of *Serenoa repens* (saw palmetto) extracts (hexane, ethanol, and critical CO_2) were discussed. Advanced chemical analytical techniques go hand in hand with the evaluation of pharmacological and safety evaluation (Habib and Wyllie 2004; Scaglione et al. 2008, Scaglione et al. 2012; Booker et al. 2014; De Monte et al. 2014; De Combarieu et al. 2015.

Adverse Events: The Safety Issue

All cardioactive steroids share the property of being potent and highly specific inhibitors of the intrinsic Na1K1-ATPase membrane protein. This enzyme comprises the cellular "sodium pump" in which membrane ion translocation is coupled with the hydrolysis of a high-energy ATP phosphate. Despite their more than 200 years in use, the safety of cardiac glycosides still remains a matter of debate (Hauptman and Kelly 1999). Withering gained more experience with the use of digitalis, which helped to lower adverse effects. It is not clear whether or not this was related to various dosage regimens or to the use of various preparations of digitalis that probably had varying degrees of bioavailability (Breckenridge 2006).

Until now there has been an ongoing debate on the safety of digoxin. Results of a meta-analysis suggest that digoxin use is associated with a greater risk of mortality in patients with atrial fibrillation, regardless of concomitant heart failure. Indeed, shortening of the refractory period may enhance the risk of cardiac arrhythmia. Moreover, signs for under-treatment are similar to those of overtreatment (Ouyang et al. 2015).

Dose-Response Characteristics: The Efficacy Issue

Withering considered foxglove to be a useful therapeutic tool in case of edema and weakness of the heart. Clinical research shows that digoxin reduces hospitalization for worsening heart failure (HF) when patients are admitted with sinus heart rhythm and ejection fraction lower than 45 %. International guidelines recommend that digoxin be considered an add-on therapy in patients with symptomatic heart failure when insufficient therapeutic results are obtained with optimal doses of angiotensin-converting enzyme inhibitors or angiotensin receptor blockers, β-blockers, and mineralocorticoid receptor antagonists. However, there is a lack of definitive evidence from randomized studies for the efficacy of digoxin as add-on therapy in this context. Questions also remain regarding the optimal dosage of digoxin. Contemporary data for digoxin in HF are derived from observational studies, and the findings are conflicting. Despite two centuries' experience with cardiac glycosides to treat HF, fundamental questions remain unanswered (Chaggar et al. 2015).

Observing and Understanding: The Key to Progress

From the current perspective, it seems self-evident that isolation and identification of active principles is the key to understanding the efficacy and safety of herbal medicines. In this respect, it is astonishing to know that at the beginning of the nineteenth century, scientists could isolate and identify quite complex structures, such as morphine from *Papaver somniferum* Friedrich Sertürner in 1804 (Schmidt 1985) and quinine from *Cinchona officinalis* Pierre Joseph Pelletier and Joseph Bienaimé Caventou in 1820 (Kyle & Shampe 1974). These isolations began from factual evidence based on observed efficient and safe use of the herbal preparations.

The definition of the symptoms of inflammation by *rubor*, *calor*, *dolor*, and *tumor* by Aurus Cornelius Celsus (ca 25 B.C. to ca 50 A.D.) helped make a therapeutic approach to willow leaves as a useful tool already in Roman times. It was the Rev. Edward Stone (1702–1768) who successfully tested willow as a possible alternative for the expensive *Cinchona* in case of fever. German scientists succeeded in isolating salicylic acid from meadowsweet (*Filipendula ulmaria*) Löwig and Weidman in 1839; unfortunately, the substance was revealed to have strong stomach-churning acidity. One hundred years after Edward Stone, Charles Gerhardt (1816–1856), a French professor of chemistry, synthesized acetylsalicylic acid for the first time, in an attempt to find a solution to the gastric irritation caused by salicylic acid. However, he found the synthesis procedure too complicated and left the findings in the drawer. The finding of a more suitable way to synthesize acetylsalicylic acid was claimed by Felix Hoffman (1868–1946) and Arthur Eichengrün (1867–1949), and the drug turned out to be the first commercial success of a herbal-derived chemical substance with feasible therapeutic properties and – for that time – an acceptable safety profile (Jeffreys 2005).

For more than 2000 years, salicylic acid-containing plants have been used in traditional herbal medicinal practice. The acetylsalicylic acid that emerged from this use has been commercialized for more than 70 years, before the mechanism of action was discovered by Sir John Vane (1927–2004) (Vane 1971). For this discovery, Vane was awarded a Nobel Prize in Physiology or Medicine in 1982.

Pharmacodynamics and Safety Issues

Pharmacology is the scientific discipline that studies the interaction of a substance with the body, whereas pharmacokinetics deals with the action of the body on a substance. In a clinical environment, there is a tendency to replace pharmacology by pharmacodynamics. Pharmacokinetics and pharmacodynamics go in pair to the so-called PK/PD approach, in which an attempt is made to translate the insight in the way the body deals with a medicine into a therapeutic perspective.

Having an insight into this process means knowing the pathway that the substances are following. The more specific and the more selective these substances are, the better they can be traced, and the better their safety can be controlled. An approach is made to competitive and non-competitive antagonism and to selectivity and specificity as important issues in the evaluation of safety and pharmacological perspectives.

Competitive and Non-competitive Antagonism: Theoretical Framework

A clear dose – or concentration relationship – of a natural substance or a plant extract is proof for pharmacological validation. Experimental data obtained with only one concentration or one dose should not be considered as proof of a pharmacological action. The description below is based on the models developed by Arunlakshana and Schild (1959) and Van Rossum (1963).

Agonists

A substance or an extract can behave as an agonist in an experimental model. In order to study real agonistic activity, the effects on an isolated organ are suitable for investigation. *In vitro* ligand binding to isolated receptors is a useful approach to the affinity of a substance to a receptor, but not to the intrinsic activity emerging from the binding of the substance with the receptor. An approach to affinity, as well as to intrinsic activity, can be made by considering the agonist "A" and the receptor "R" as two reacting principles. Once the agonist-receptor complex has been formed, an effect "E" can be the result.

$$A + R \rightleftharpoons AR \longrightarrow E$$

A is the agonist, whereas R is the receptor. The effect E depends upon the affinity of the agonist A for the receptor and the intrinsic activity, i.e., the intensity with which the effect is generated. It can be written as the equation below. In this equation, α stands for the intrinsic activity.

$$AR = \alpha \times E$$

The model fits for a one-by-one relationship, hypothesizing the interaction of one molecule with one receptor. The stronger the intrinsic activity, the higher the effect. Affinity and intrinsic activity are specific for an agonist-receptor interaction.

Agonists can be compared by calculating the pD_2 defined as the negative logarithm of the molar concentration leading to 50 % of the maximal effect. The pD_2 can only be calculated when pure substances are tested. Plant extracts must be considered to be complex mixtures. Instead of molar concentrations of a pure compound, only EC_{50} values can be calculated, and these can be used to compare mutual activity. The EC_{50} value gives the concentration for which 50 % of the maximal effect is obtained.

Competitive Antagonist

Suppose an antagonist B is interfering with the activity of an agonist A. Again, the interaction of agonist, antagonist, and receptor R can be theorized by using a chemical approach. In the case of a competitive antagonist, this antagonist B competes for the same receptor as the agonist A; the affinity of both B and A for the receptor will have important consequences for the resulting effect E. The simplest approach is the one for which the antagonist has zero intrinsic activity – which is usually the case.

When a competitive antagonist binds reversibly to a receptor, increasing concentrations of agonist will chase it from the receptor. However, if the binding of B with the receptor is too strong and increasing concentrations of agonist cannot reverse the situation, this might be a sign of irreversibility and, as a consequence, of toxicity. In the case of competitive antagonism, concentration-effect curves will shift to the right and the same maximal effect will be obtained. Obtaining the maximal effect is a guarantee of non-toxicity. When the antagonist is washed out, the concentration-effect curves must be comparable to those obtained before the antagonist was brought in contact with the receptor.

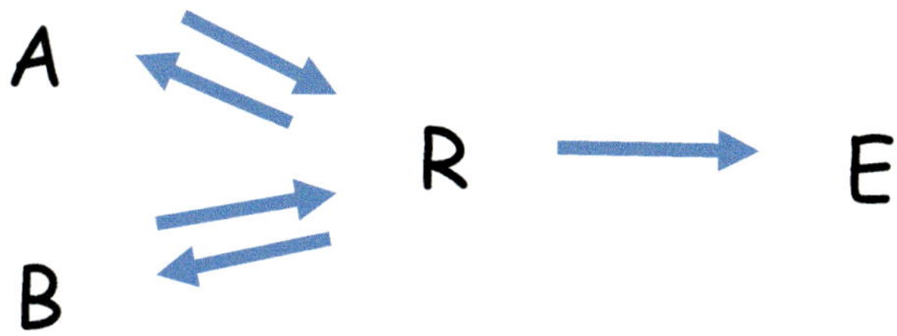

In order to compare competitive antagonists, the pA_2 has been introduced. It is the negative logarithm of the molar concentration of an antagonist for which the concentration of the agonist must be doubled in order to obtain the same effect as without an antagonist.

Non-competitive Antagonist

When the antagonist Z acts in a non-competitive way, it is hypothesized that it does not bind on the same place on the receptor as the agonist. The interaction between agonist and receptor takes place without hindrance, i.e., the affinity is not influenced.

Z will bind to another binding site, R, and the complex formed will negatively influence the effect E by interfering with the intrinsic activity. As a result, the maximum effect of the concentration-response curve will be lowered. The lowering is concentration-dependent and must be reversible once the antagonist has been washed out.

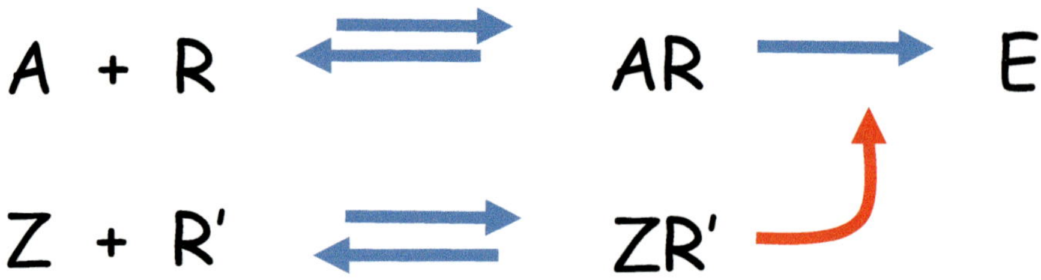

In order to compare non-competitive antagonists, the pD'_2 has been introduced. It is the negative logarithm of the molar concentration of the antagonist for which the maximal effect of the agonist is reduced to 50 %, as compared to the maximal effect without antagonist.

Natural Substances as Examples of Competitive and Non-competitive Antagonism

The theoretical framework of competitive and non-competitive antagonism is illustrated by the activity of indole alkaloids isolated from the leaves of *Pterotaberna inconspicua* Stapf. (*Apocynaceae*). This plant species grows as a shrub in Central Africa; its leaves are used in Congolese traditional medicines to treat hypertension, gastro-intestinal upsets, and several kinds of aches. The alkaloid content of the leaves was explored, having led to identification of 2-acylindole alkaloids of which the antihistaminic properties were studied on the isolated guinea-pig (Bakana et al. 1985).

The isolated ileum of the guinea pig offers opportunities to use many types of agonists that lead to contractions of the organ. One animal offers enough material to test properties of substances for at least one day. Several agonists can be used, giving typical patterns of contractions. Figure 6.2 gives examples of agonists that play a role in the human body; these can be administered to an ileum mounted in an isolated organ bath with *in vitro* conditions optimized by using a physiological mixture of electrolytes and glucose, at a temperature of 37 °C and oxygenated with a mixture of 95 % O_2 + 5 % CO_2. There is a nice concentration-response relationship that can be visualized by administering cumulative concentrations until a maximal response has been reached.

As can be seen in Fig. 6.2, contraction patterns differ from one agonist to another. Acetylcholine, bradykinine, and histamine give a sustained contraction, whereas contractions by the other agonists are rather short-lasting. Tachyphylaxis is observed for serotonin: even after washing, contractions with the same concentration fade.

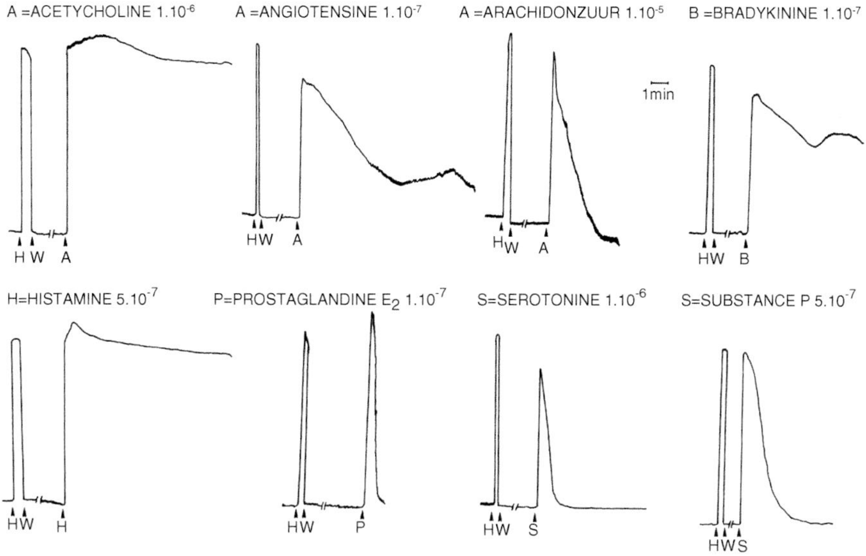

Fig. 6.2 Examples of contractions of different agonists on the isolated ileum of the guinea pig. All concentrations are expressed as g/ml (final concentrations in the isolated organ bath). Arachidonzuur = arachidonic acid (Laekeman 1980). *H* histamine 5.10^{-7} g/ml as initial positive control for contractility. *W* washing out histamine. Agonists were given within 10 min of the initial contraction by histamine

Epimethuenine and metheunine were the 2-acylindole alkaloids isolated from the leaves of *P. inconspicua*. Both substances are indole alkaloids with a different steric position of the substituents on C16 (Fig. 6.3).

From the cumulative dose response curves, a pD_2 of the agonist histamine of 6.13 +/− 0.05 was calculated (mean ± SEM; n = 24). This value corresponds with the values found in the literature (Brownlee and Johnson et al. 1963; Van Rossum 1963), whereas methuenine showed a non-competitive type of antagonism (depression of the maximal response). This led to calculation of $pD'_2 = 5.13 \pm 0.14$ (+/- SEM) for methuenine. The type of antagonism found for epimethuenine was a competitive one (shifting the response by the agonist to the right). For epimethuenine, the pA2 equalled 6.55 ± 0.08 (± SEM) (Fig. 6.4).

Methuenine also depressed the maximum response by acetylcholine as an agonist, although the effect was inferior to that seen on histamine contractions, hypothesizing a certain selectivity towards muscarine and histamine receptors. Epimethuenine did not influence the concentration-response curve of acetylcholine (Fig. 6.5).

All inhibitions seen were reversible: maximal contractions by the agonists could be reached again, after washing the alkaloids from the isolated organ, by renewing the physiological fluid in the isolated organ bath. These findings clearly show the existence of specificity for natural products. Reversibility of the antagonism is a guarantee for the absence of toxicity on the organ used.

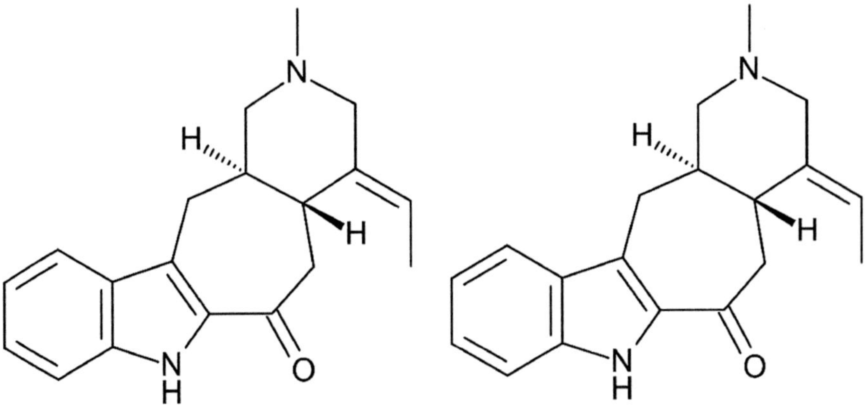

Fig. 6.3 2-acylindole alkaloids isolated from the leaves of *P. inconspicua*. Left: epimethuenine; right: methuenine

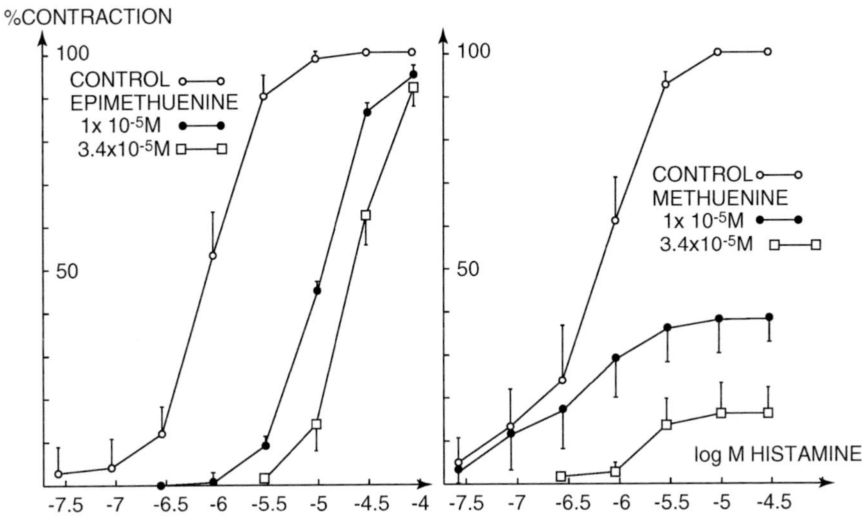

Fig. 6.4 Cumulative concentration-response curves with histamine on the isolated ileum of the guinea pig. Competitive antagonism obtained with epimethuenine and non-competitive antagonism by methuenine, 2-acylindole alkaloids isolated from *P. inconspicua*. Concentrations of alkaloids are given in Molar. Each point represents the mean ± SEM of at least five experiments, except for 3.4×10^{-5} M 16-epimethuenine where three experiments were carried out (Bakana et al. 1985)

Early Detection of Toxicity by an Experimental Pharmacological Approach

The isolated ileum of the guinea pig is a useful model for making predictions about the toxicity of natural compounds. In the case of methuenine and epimethuenine, the inhibiting activity was reversible, which is a guarantee of non-toxicity of the concentrations used.

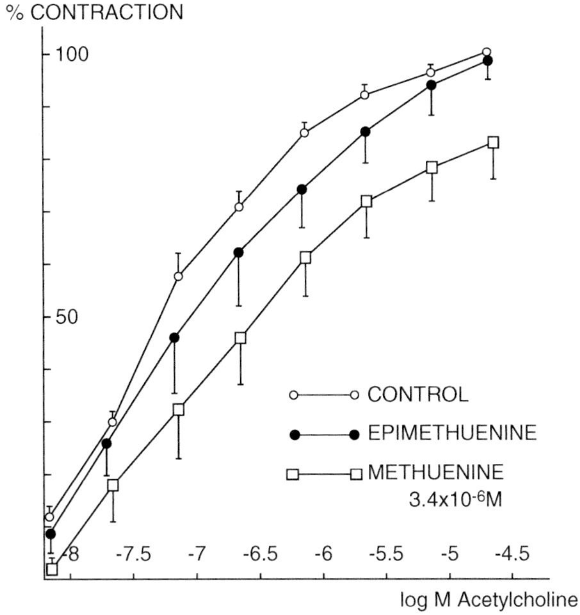

Fig. 6.5 Cumulative concentration-response curves with acetylcholine on the isolated ileum of the guinea pig, influenced by methuenine and 16-epimethuenine. Each point represents the mean ± SEM of six experiments (Bakana et al. 1985)

Another effect is seen with vernolepin, a sesquiterpene lactone isolated from *Vernonia amygdalina* Del. *(Asteraceae)* (Fig. 6.6). The plant is well known in African countries such as Cameroon, Guinea, Ghana, and Sao-Touré. It is also known in Angola, Congo, Ethiopia, Kenya, Rwanda, Tanzania, Uganda, Zambia, and Zimbabwe. The fruits of the plant are eaten; the leaves are used for various conditions, from infectious diseases to stimulation of circulation.

When vernolepin was tested on acetylcholine-induced contractions of the isolated guinea pig ileum, no inhibition was seen after an incubation time of less than 15 min. Enhancing the incubation time, however, resulted in an inhibition of the maximal response. After 60 min., contractions completely stopped. The total inhibition was irreversible, i.e., repeated washing was not able to restore contractions with any agonist, and the organ seemed completely paralyzed. This is proof of the fact that contact time can also be an important factor to take into account (Fig. 6.7) (Laekeman et al. 1983).

Interference with Neurotransmission

The coaxial stimulation of the guinea pig ileum gives insight into intestinal cholinergic neurotransmission. By the presence of neuronal plexi and their common embryonal origin, the isolated ileum of the guinea pig and the brain can be considered to be related. Dr. Paul Janssen, the founder of Janssen Pharmaceutica, once called the isolated guinea pig ileum a "mini central nervous system." Indeed, most of the neurotransmitters that are expressed in the central nervous system are

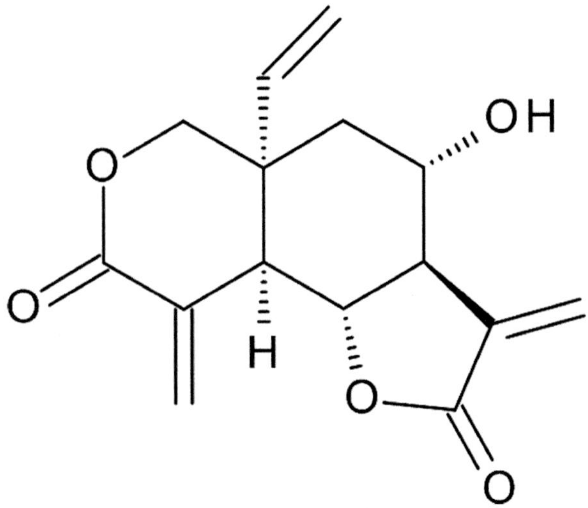

Fig. 6.6 Vernolepin isolated from *V. amygdalina*

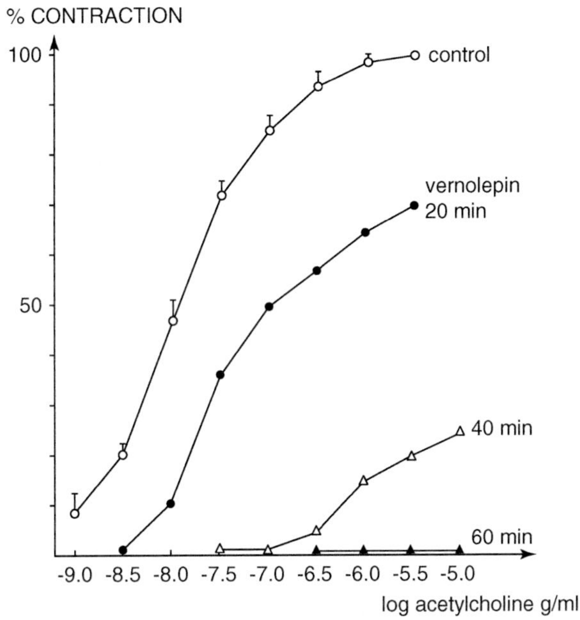

Fig. 6.7 Cumulative concentration-response curves with acetylcholine on the guinea pig ileum, influenced by vernolepin. A fixed concentration of 1.8×10^{-5}M was applied with the incubation time as a variable (Laekeman 1988)

represented within subclasses of the phenotypically diverse neurons of the enteric nervous system. The function of the intestine is influenced by nervous impulses, but the plexi of Auerbach and Meissner in the isolated ileum of the guinea pig can function independently (Nijenhuis et al. 2011). To trigger intestinal activity, acetylcholine is an important intestinal neurotransmitter, acting on local muscarine-2 receptors (Tayebati et al. 1999).

In order to study the influence of substances and complex mixtures, pieces of ileum are mounted over an electrode in an isolated organ bath. Every 10 s an electric pulse is delivered through the organ, leading to the release of small amounts of acetylcholine. This release causes a twitch contraction that lasts for less than one second, because the acetylcholine that is released is immediately hydrolyzed by cholinesterase. The quantity of acetylcholine per twitch is unmeasurably low. The coaxially stimulated ileum can be used to test antimuscarinic properties, but can be studied at the same possible interference with neuronal release of acetylcholine (Fig. 6.8) (Day and Vane 1963).

When testing the effect of methuenine and epimethuenine, it can be expected that the former will more likely inhibit contractions caused by acetylcholine released from the neuronal plexi in the ileum. But when twitch contractions are less prominent under the influence of natural substances, it must be determined whether the inhibition is caused by an inhibition of the release of acetylcholine, by a muscarinic antagonism, or by a non-specific activity.

Methuenine inhibited the coaxial stimulation more, and also inhibited contractions by exogenous acetylcholine. As it was shown that epimethuenine had a clear antihistaminic property, it inhibited twitch contractions only slightly. Prostaglandin E2 was able to partially restore the twitch contractions suppressed by methuenine. The action of methuenine could not be reversed by naloxone, which is additional proof that methuenine does not interfere with the neurotransmission (Van Nueten et al. 1976).

When applied to the coaxially stimulated, isolated ileum, vernolepin led to an initial stimulation of the twitch contractions and a time-dependent inhibition. It was proven that exogenous acetylcholine was also inhibited. As for the cumulative dose-response curves, the inhibition was irreversible, i.e., contractions could not be restored after washing (Fig. 6.9).

Morphine specifically inhibits cholinergic neurotransmission. Its action targets only the electrically evoked twitch contractions, whereas contraction by exogenous acetylcholine is not influenced. Moreover, the inhibiting activity of morphine is reversed by naloxone. Interpretation of the results obtained with coaxial stimulation can teach us more about specificity and reversibility of actions by natural substances (Roquebert et al. 1971).

Platelet Aggregation: Mechanisms *In Vitro* Predicting Safety In Vivo

Human or animal (e.g., rabbit) blood platelets are a useful experimental tool for studying the interaction of natural substances with an intracellular system. Platelet-rich plasma is obtained by centrifuging citrated blood samples at a lower speed. The aggregation of platelets can be triggered by several agents, a.o. arachidonic acid (AA), adenosine diphosphate (ADP), or collagen. Aggregation takes place in a small tube, filled with stirred platelet-rich plasma. The tube is placed between a light source and a photometer; as long as the platelet rich plasma is homogeneous,

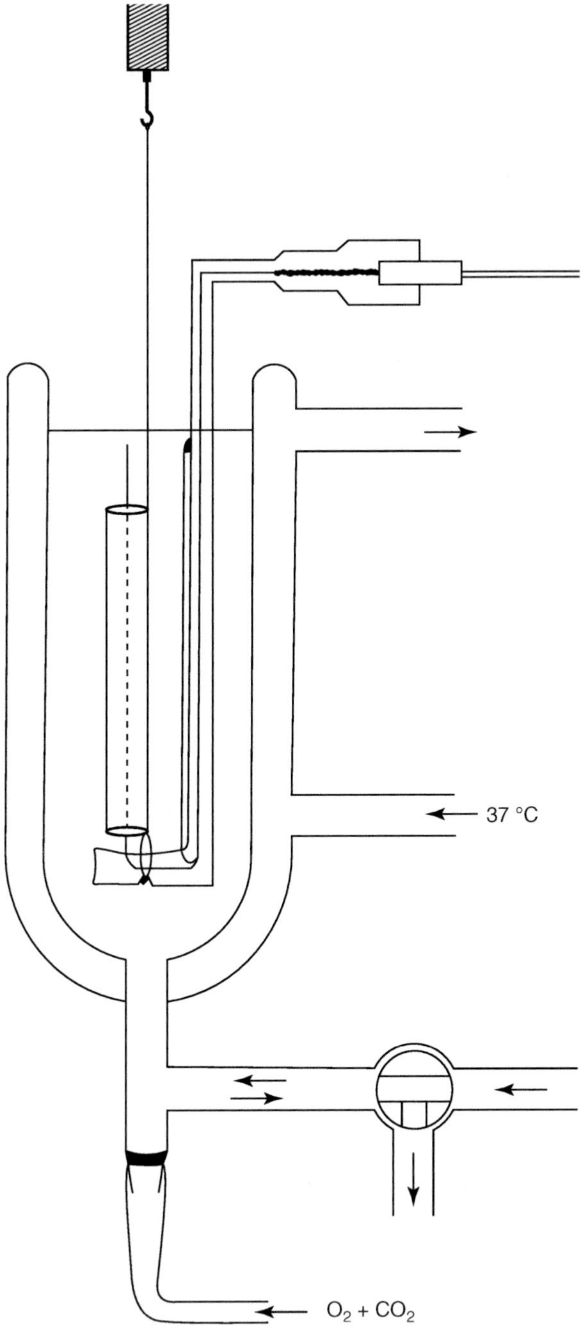

Fig. 6.8 Setup of the coaxial stimulation of the guinea pig ileum in an isolated organ bath. Twitch contractions are detected using an isometric transducer and signals are transmitted for recording

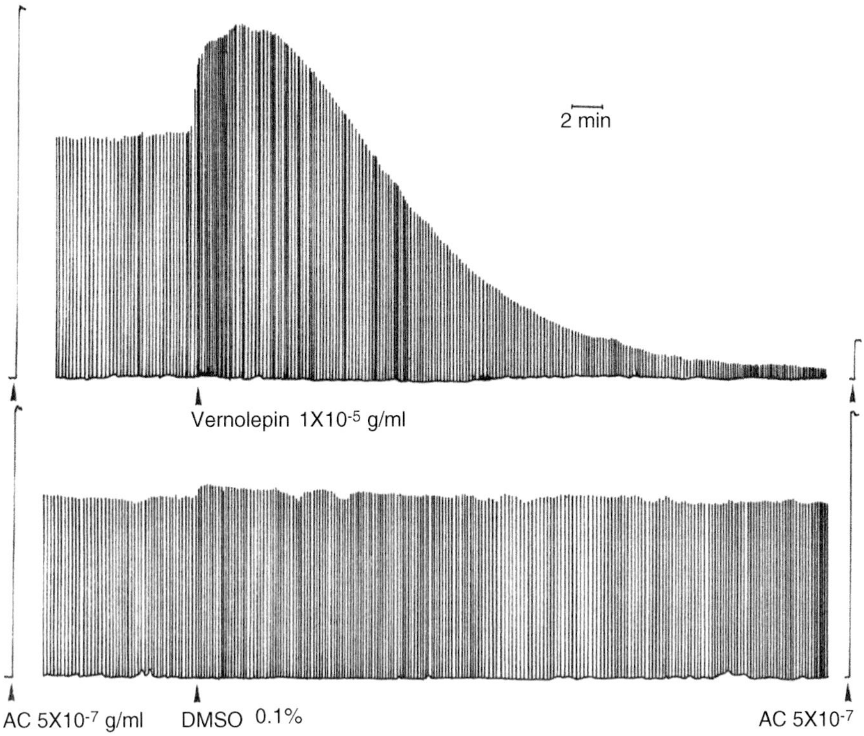

Fig. 6.9 Original tracing of a coaxially stimulated guinea pig ileum. Substances added to the isolated organ bath: *AC* acetylcholine, *DMSO* dimethylsulfoxide as solvent control, vernolepin (Laekeman 1988)

light transmission is hampered. When pro-aggregating agents are added, lumps of platelets are formed and sink to the bottom of the tube, as the stirring is not sufficient to keep the lumps afloat. Light transmission is enhanced when more aggregating lumps are formed, until all platelets have been coagulated (Born 1962).

Platelet aggregation triggered by AA, ADP and collagen was inhibited by vernolepin, and this inhibition was time-dependent (Fig. 6.10). The aggregation seemed to be reversible, at least after short incubation times, with the pattern passing by a maximum, close to the maximal aggregation, and then returning to baseline. When re-challenged, the platelets no longer responded to the aggregating agents, which seemed like they were protected against clotting.

In order to see what happened on a subcellular level, electron microscopy was performed at distinct phases before and during the aggregation process. Aliquots of the platelet suspension were stained with glutaraldehyde, the solution was decanted, and platelets were mixed with an ice-cold cacodylate-sucrose buffer, centrifuged and dehydrated with ethanol. Platelets were subsequently prepared for electron microscopy by embedding them in Epon-Epon in order to prepare ultrathin sections stained with uranyl acetate and lead citrate (Laekeman et al. 1985).

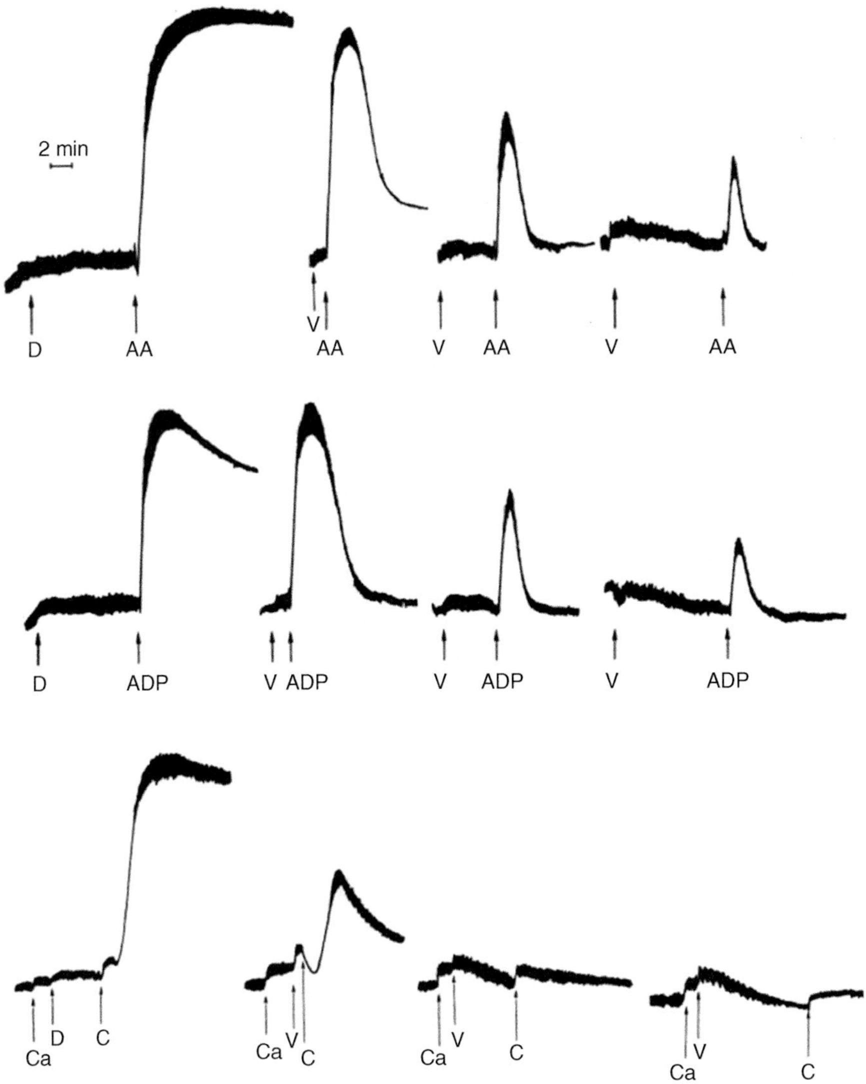

Fig. 6.10 The effect of vernolepin on rabbit platelets aggregated with various aggregating agents. Recording based on light transmission (Laekeman 1988). *AA* arachidonic acid (1×10^{-4} g/ml). *ADP* adenosinediphosphate (4×10^{-6} g/ml). *C* collagen (1×10^{-5} g/ml). *Ca* $CaCl_2$ (2 mM). *D* dimethylsulfoxide (0.1 %). *V* vernolepin (1×10^{-5} g/ml). All concentrations are final concentrations in the cuvette

Subcellular structures were seen by enlarging them more than 10,000 times. It was seen that although platelets tended to stick together by forming typical pseudopods, the aggregation did not take place. Platelets remained in suspension individually, but they did not return to their flattened discoid form, and some pseudopods

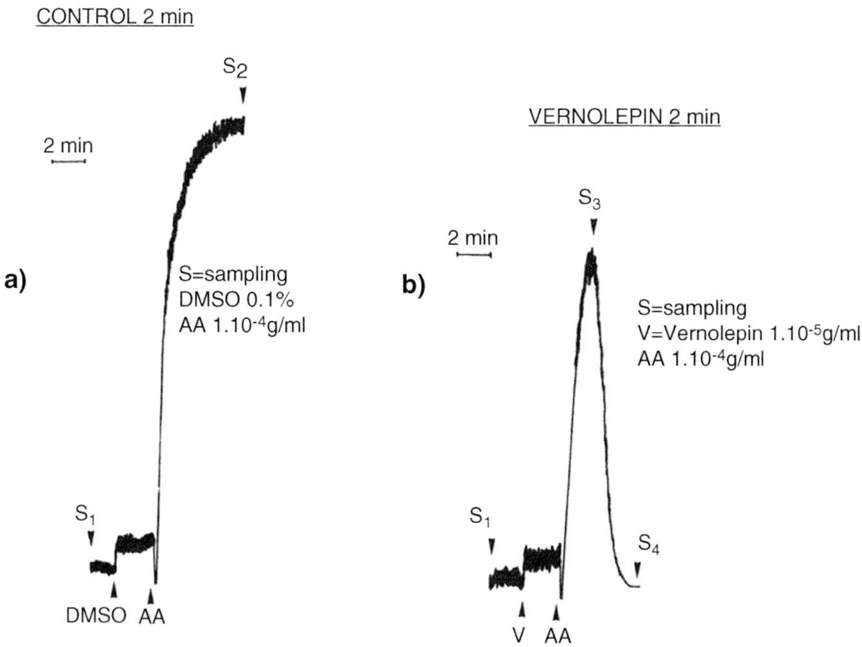

Fig. 6.11 Sampling of platelets for electron microscopy (Laekeman et al. 1985) published with permission of Springer. *AA* arachidonic acid (1×10^{-4} g/ml)

were kept. The intracellular granules were still visible, but not redistributed over the cellular surface (Figs. 6.11 and 6.12).

When platelets were incubated with vernolepin alone, a certain shape change occurred. The majority of platelets were swollen, and some of them had pseudopods. The main activity of vernolepin seemed to be concentrated on the platelet membrane. The microtubular system plays an important role in the aggregation process (Crawford 1976). An interaction of vernolepin at this level could be hypothesized. From subsequent *in vivo* experiments on rats, it could be seen that vernolepin had too small a therapeutic index to create perspectives. Intraperitoneal doses, from 10 mg/kg on up, resulted in lethal outcomes (Fig. 6.13).

The Relationship Between Secondary Metabolites and Pharmacological Activity: Simplification for Safety

The ripe seeds of *Myristica fragrans* Houtt. (family *Myristicaceae*) or nutmeg are used for digestive purposes. In China, the powdered nut is used as a warming and astringent remedy against dysentery, especially in children and the elderly. In Indochina, the powdered seeds are mixed with boiled rice against dysentery, anorexia, and colic (Perry 1980).

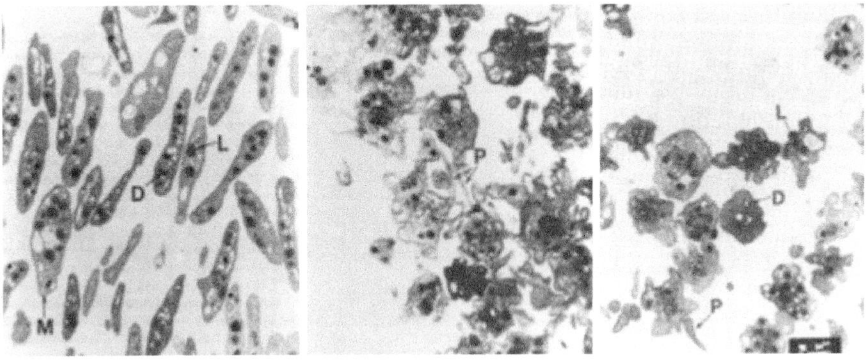

Fig. 6.12 Electron microscopic pictures of rabbit platelets activated by arachidonic acid. *Left* (=S_1 on Fig. 6.11): normal platelets in resting situation (*D* dense granule or body, *L* large granule, *M* microtubular system; enlargement 5,050 × 2.2). *Middle* (=S_3 on Fig. 6.11): platelets activated with arachidonic acid with formation of pseudopods (=P; enlargement 6,900 × 1.6). *Right* (=S_4 on Fig. 6.11): platelets back to baseline after nearly aggregation (*D* dense granule or body; *L* large granule, *P* pseudopods enlargement 6,900 × 1.6) (Laekeman et al. 1985). Published with permission of Springer

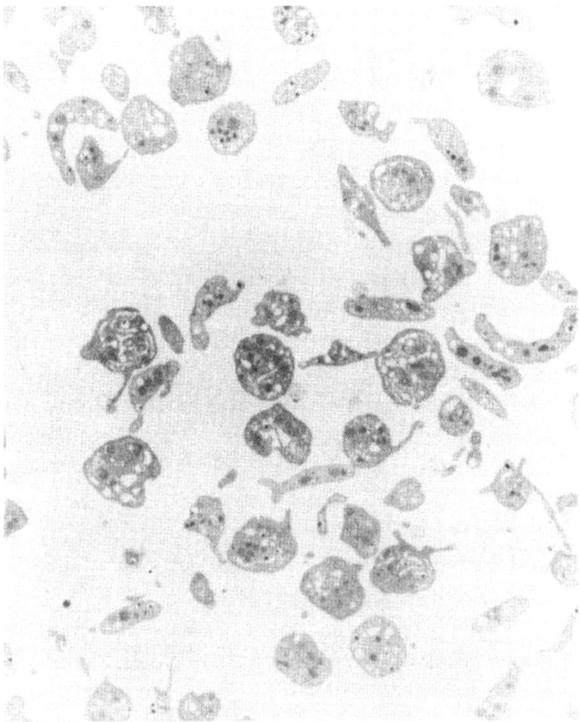

Fig. 6.13 Shape change of non-activated rabbit blood platelets after exposure to vernolepin 1×10^{-5} g/ml (enlargement 2,290 × 2.2) (Laekeman et al. 1985). Published with permission of Springer

Crude extracts of nutmeg inhibit the formation of prostaglandin-like compounds in vitro. Rasheed et al. (1984) reported strong evidence that the essential oil is the most important part of nutmeg with regard to pharmacological activity. They found eugenol and isoeugenol to be the most important compounds with regard to pharmacological activity in vitro. Bennett et al. (1988) have widened knowledge on the biological activity of eugenol.

Janssens et al. (1990) investigated the pharmacological activity of aromatic compounds of commercially available nutmeg oil. The composition of the oil was characterized by gas chromatography, with an OV-17 glass capillary column and equipped with a flame ionization detector. The temperature of the column gradually rose from 100 to 200 °C. 3-t-butyl-4-hydroxanisole was added as an internal standard. An identical response was assumed for eugenol and isoeugenol; terpinen-4-ol and terpineol; safrole, myristicin, and elemicin. Peak areas were integrated. Fractions to be tested were prepared, using silicagel column chromatography with benzene, chloroform, and ethyl acetate as eluents. Fractions were pharmacologically tested.

In vitro rabbit blood platelets' aggregation was used to quantify the pharmacological activity of nutmeg oil samples (Born 1962). Aggregation curves were recorded as described earlier in this chapter. Arachidonic acid was used as a pro-aggregating agent, and polysorbate 80 (tween 80) was used as an emulsifying agent for the essential oil samples and standard compounds.

The qualitative composition of essential oil samples was roughly checked via thin layer chromatography. The following components could be compared with standard spots: safrole, myristicin, eugenol and/or isoeugenol, terpinen-4-ol, elemicin, and alpha-terpineol. A more detailed quantification of the oils was carried out by GC/MS.

Only five batches were found to be active: (1) fraction A contained large amounts of myristicin and a little safrole; (2) fraction B contained equal amounts of isoeugenol and myristicin; (3) fraction C contained methyleugenol and a trace of eugenol and isoeugenol; (4) fraction D contained myristicin, a small amount of eugenol but no isoeugenol; and (5) fraction E contained elemicin. Other volatile monoterpenes, such as alpha- and betapinene, were also found, but these showed no pharmacological activity. The activity of methyleugenol was largely inferior to that of eugenol and isoeugenol.

Eugenol and isoeugenol are the most potent constituents of nutmeg oil for the inhibition of platelet aggregation induced by arachidonic acid; all other constituents were 100-1,000 times less potent. No IC_{50} could be determined for linalool, alpha- and betapinene, alpha terpineol, terpinene4-ol and camphene, due to lack of activity. There was no potentiating action of one compound on another, when equal amounts of eugenol, safrole, myristicin and a-terpineol were mixed. In Table 6.1 the IC_{50} values are given in molar concentrations in order to compare the activity of the different compounds on a molecular basis.

Table 6.1 Platelet aggregation IC_{50} values of nutmeg oil constituents and indomethacin (positive control) (Janssens et al. 1990)

Product	IC_{50} (M)
Eugenol	3.0×10^{-7}
Isoeugenol	7.2×10^{-7}
Elemicin	3.6×10^{-4}
Myristicin	2.5×10^{-4}
Safrole	1.1×10^{-4}
Indomethacin	2.2×10^{-7}

Table 6.2 Comparison between experimentally obtained IC_{50} values (g/ml) of nutmeg oil and calculated activity based on eugenol and isoeugenol contents in the oils (Janssens et al. 1990)

Oil batch	Experimental $IC_{50} \pm$ SEM ($\times 10^{-5}$ g/ml)	Calculated IC_{50} (95 % CI) ($\times 10^{-5}$ g/ml
Donck	1.17 ± 0.63	0.86 (0.67–1.17)
Liebig	1.35 ± 0.89	1.21 (1.17–1.39)
Universal flavor	1.59 ± 1.16	0.90 (0.79–1.05)

When rough calculations are made as to the pharmacological activity of different oils using the concentrations of eugenol and isoeugenol in the oils as a starting point, comparison can be made between the activity of the oil and its most active compounds. Theoretical IC_{50} values were calculated, as if the oils were only composed of eugenol and isoeugenol diluted with pharmacologically inert material. These calculations resulted in values that could be compared to the actual IC_{50} values obtained for the real oils using *in vitro* blood platelet aggregation; the comparisons are shown in Table 6.2.

The results indicate the usefulness of the identification of highly active substances and of comparing their share in pharmacological activity of the complex mixture. It can be concluded that the matrix effect is not valid for nutmeg essential oil, when investigating possible anti-inflammatory action, eugenol and iso-eugenol as the most prominent active substances. At the same time, it showed that substances for which there is a safety concern, like safrole, can be eliminated without losing pharmacological activity (EMA 2014).

Pharmacokinetics and Safety Issues

The Case of Red Yeast Rice Extracts

The importance of pharmacokinetics can be illustrated by the actual development of red yeast rice preparations. Extracts of red yeast rice are used to lower cholesterol in mild hypercholesterolemia. Preparations of red yeast rice have been used for more than 2000 years in China; long-term cardiovascular beneficial effects have been claimed, even in patients at risk (Lu et al. 2008). Monacoline K, which is the most important active substance in red yeast rice extracts, has been identified as the inspiring natural substance that led to the discovery of simvastatin (Fig. 6.14).

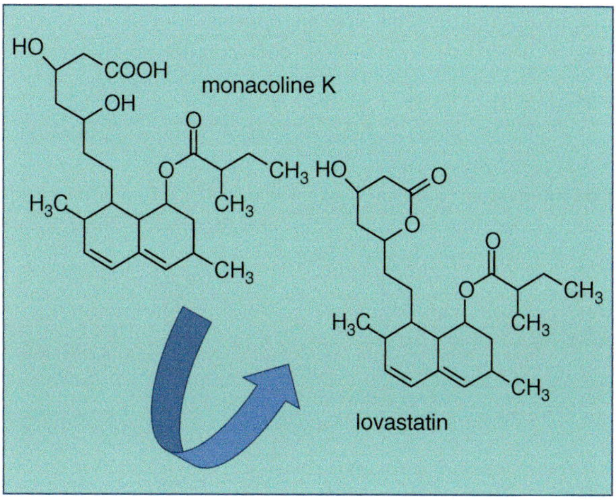

Fig. 6.14 Structures of monacoline K and lovastatin present in red yeast rice

There has been a debate with regard to the relationship between the composition of extracts and the benefit-risk ratio, with extrapolation to all statins. Indeed, monacoline K can be in its lactone form (lovastatin), or its open ring structure (with OH- and COOH-groups free = active configuration). The latter has a more hydrophilic character than the former. There is much (commercial) speculation about the prevailing configuration in extracts available on the market. Most of the preparations are food supplements, and there are no valid comparative data on possible undesirable effects of one extract versus another; consequently, it is difficult to address a risk-benefit balance of various extracts. The fact that mono-preparations as well as combinations are on the market hampers straightforward evaluation, so there is little underpinning evidence for making predictions. Due to the lack of data on monacoline K, research on commercialized statins may be inspiring. It seems that desolvation and resolvation of the statins before and after binding to the OH-Methyl-Glutaryl Coenzyme A Reductase (HMGCR) are the major determinants of the energetics of the binding process. An analysis of the amphiphilic nature of lovastatin anion, acid, and lactone, and their abilities to cross the blood brain barrier, has indicated that this process may be dominated by desolvation and resolvation effects as well. Crossing the blood brain barrier may influence central nervous system cholesterol physiology with unknown consequences. (Fong 2014).

Applying these hypotheses to a pharmacokinetic model can inspire further investigation of benefit-risk aspects related to the prevalence of one form or another in commercialized extracts. The findings should be linked to clinical observations. Figure 6.15 can help identify important issues. Herbal examples of absorption-distribution-metabolism-excretion processes can be seen.

Absorption: The Case of Anthraquinones-containing Plants

Anthraquinones can be present as C-O or C-C glycosides and have to be considered as prodrugs. A bacterial conversion is necessary to produce the bioactive forms. Anthracenic glycosides are not absorbed in the small intestine; they can resist

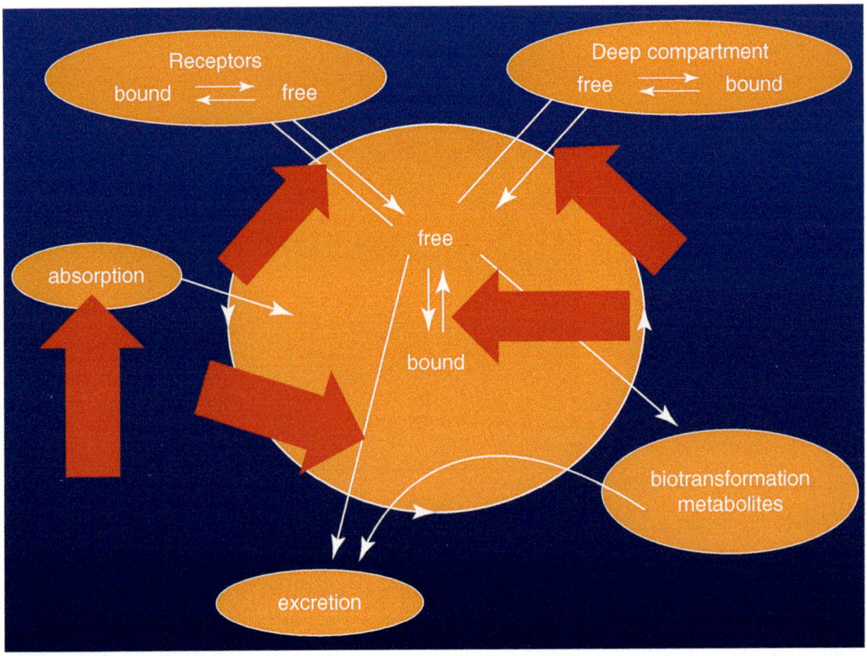

Fig. 6.15 Red arrows indicate possible critical points in the transfer of hydrophilic and lipophilic forms of statin-like compounds in red yeast rice extracts. Several processes ask for a transfer from hydrophilic to lipophilic configuration and vice versa: (1) absorption through the enterocytes; (2) binding to the plasma proteins; (3) migration to the deep compartment (more particularly, skeletal muscle cells); (4) binding to the HMGCR; (5) biotransformation and excretion

stomach acid and intestinal glycosidases and arrive at the colon intact, where they are converted to aglycones that can irritate the mucosal layers and invert the absorption of water and electrolytes. It is argued that plant species mainly containing anthrone C-glycosides (present for example in *Rhamnus purshianus* D.C.) or dianthrone O-glycosides (for example in *Cassia senna* L. and *Cassia angustifolia* Vahl) are preferable over species that contain more aglycones, which can be already absorbed in the small intestine (for example preparations from *Aloe* species that can contain *Aloe* emodins originating from the pericycle cells and adjacent leaf parenchyma) (De Witte et al. 2008; De Witte and Lemli 1990).

Biotransformation: Hepatotoxicity of Concentrated Camellia Extracts

Preparations of *Camellia sinensis* (L.) Kuntze are considered to be safe. Their traditional use has been approved by the EMA, and a European Union monograph was made after the assessment by the Herbal Medicinal Product Committee (Grigoras and Purdel 2014). However, the safety of orally taken preparations depends on the concentration of polyphenols circulating through the liver. A case of acute hepatitis in a 46-year-old woman was related to the use of a highly concentrated *Camellia* extract made with 80 % ethanol (Vial et al. 2003). Mazzanti et al. (2015) report on 19 cases of hepatotoxicity due to green tea preparations. The high content of epigallocathechines may be the most probable cause, although factors related to the patient are also thought to be involved.

Binding to Proteins: Digoxin Binding Ab

Digoxin isolated from *Digitalis purpurea* L. leaves is a molecule with a particular pharmacokinetic profile. The distribution volume is estimated to be 510 L. This is rather high and means that there is an important intracellular penetration with concentrations in the heart muscle more than 30 times the plasma concentrations. The circulating digoxin is only for 30 % bound to plasma proteins. In the case of intoxications or overdose, hemodialysis makes no sense (Lanoxin® SPC 2015). Digoxin-specific antibody fragments (digoxin-Fab) are widely regarded as a safe and effective treatment for the management of digoxin poisoning. Although there are no randomized clinical trials, reports on more than 2,000 cases of acute or chronic poisoning show a therapeutic response from 50 % to 80–90 %. The time for reversal of digoxin toxicity varies between 30 and 45 min. Studies with pharmacokinetic data show that free digoxin concentration fell to almost zero within a few minutes following the administration of digoxin-Fab, which is a good example of applied safety pharmacokinetics (Chan and Buckley 2014).

Receptor Binding: Berberin As an Emerging Phytochemical Substance

PCSK9, or Protoprotein Convertase Subtilisine/Kexine type 9, downregulates the LDL-receptor. The protein can be seen as a negative factor hampering the elimination of LDL-cholesterol. Berberin, an isoquinoline alkaloid isolated from *Berberis* and *Hydrastis* species, lowered circulating PCSK9 concentrations and hepatic PCSK9 mRNA levels. The mechanism seems to be related to overexpressing ubiquitin, which in turn inhibits activation of the Hepatocyte Nuclear Factor 1-alpha (HNF1-alpha), a factor essential to PCSK9 expression. Berberin is a good example of a substance that binds to receptors at a subcellular level, with a mechanism of action closely related to recent developments in lipid lowering. These discoveries are contributing to clinical studies with plant species containing berberin, but a thorough risk-benefit analysis must go hand in hand with therapeutic development (Dong et al. 2015).

Excretion: The Example of Arbutin

The leaves of *Arctostaphylos uva ursi* (L.) Spreng or bearberry are granted a traditional use status by EMA, and an assessment of this plant has been carried out (Heroutová 2012). The toxicity of hydroquinone is an apparent risk related to its use. However, the amount of hydroquinone corresponding to the recommended dose of bearberry leaf extract is considered to be safe for short-term use. In order to remain in a safe area, it is important to obtain a maximal yield within the dosage range applied. In a pilot study, coated tablets with bearberry extract were administered together with 10 g of sodium hydrogen carbonate. The pH of the urine changed from 6.5 to 7.4, and in one case to pH 8 for 1 h. Free hydroquinone was found in a therapeutic concentration in urine only if the pH was alkaline (pH 8) (Frohne 1970).

Safety Frame

Until now, no algorithm has been available for safety evaluation in general, and safety evaluation on pharmacological grounds in particular. The question of whether or not an herbal medicinal product is safe may be answered by using a multifocal approach.

As already shown by William Withering, the triad of "quality-safety-efficacy" plays a crucial role (Breckenridge 2006). Aspects related to quality and efficacy are dealt with in other chapters. Safety can be subdivided into several topics. The research question to be answered when considering the framework below can be formulated as "To what extent can a framework be considered to evaluate the safety of herbal products in a reproducible and comparative way and contribute to risk-benefit analysis?"

Ideally, the quality issue can be covered by Pharmacopoeia monographs. There is close co-operation between the Herbal Medicinal Product Committee within the EMA and Group 13B and 13C within the European Pharmacopoeia. The quality of the herbal material should be taken into consideration before beginning pharmacological investigations (Table 6.3).

As proven by a few examples, an experimental pharmacological approach focused on safety should take into account a number of the issues listed in Table 6.4, which deals with content and methodology. When extrapolating pharmacological,

Table 6.3 A series of questions related to the quality of herbal materials. The materials should yield as many green fields as possible

Question	Answer	
	NO	YES
1. Is the herbal medicinal product subject of a European Pharmacopoeia monograph?		
2. Is the herbal medicinal product subject of a monograph in an accessible Pharmacopoeia other than European Pharmacopoeia?		
3. Is an unambiguous macroscopic identification of the herbal medicinal product possible?		
4. Is an unambiguous chemical characterization of the herbal medicinal product possible?		
5. Is adulteration/contamination of the herbal medicinal product possible/probable?		

pharmacodynamic, and pharmacokinetic findings to clinical practice, a series of questions could be examined as presented in Table 6.5.

Conclusions

Safety pharmacology needs a practical approach, beginning from fundamental pharmacodynamics and pharmacokinetic properties of substances alone, or as a component in a complex matrix. An *in vitro* approach seems to be the most suitable way to begin. Irreversibility of the observed effects must be interpreted as an early warning for toxicity. *In vitro* techniques allow the study of types of antagonisms and mechanisms of action. Safety is linked to well-documented pharmacodynamics; in complex mixtures, it is useful to try to attribute pharmacological effects to individual substances. Pharmacokinetics of complex mixtures remains tricky, especially when families of related substances are present.

The proposed safety framework may help evaluate a safety profile of a substance or a complex mixture.

Table 6.4 A series of questions related to a safety pharmacology approach. The materials should yield as many green fields as possible

Question	Answer	
	NO	YES
1. Is the preparation tested related to the traditionally used preparation according to an ethnopharmacological approach?		
2. Does a concentration-response relationship exist in an *in vitro* experimental pharmacological approach?		
3. Does a dose-response relationship exist in an *in vivo* experimental pharmacological approach?		
4. Does an effect seen in an *in vitro* pharmacological approach correspond to an identifiable mechanism of action?		
5. Does an effect seen in an *in vivo* pharmacological approach correspond to an identifiable mechanism of action?		
6. Is there a comparison between the concentrations needed for a pharmacological effect *in vitro* and the doses needed for a pharmacological effect *in vivo*?		
7. Does an effect seen in an *in vitro* experimental model correspond to a possibly toxic action?[1]		
8. Does a relationship exist between an *in vitro* or *in vivo* pharmacological effect and identified natural substances?		
9. Does an affinity exist for certain types of CYP-isoenzymes or PgP-proteins?		
10. Can undesirable effects be hypothesized, starting from experimental pharmacological findings?		
11. Does the experimental pharmacological approach contribute to the underpinning of existing therapeutic indications?		
12. Does the experimental pharmacological approach open new perspectives for new therapeutic indications?		

Table 6.5 A series of questions related to clinical safety aspects of herbal medicines. The materials should yield as many green fields as possible

Question	Answer	
	NO	YES
1. Are non-serious undesirable effects[a] with a therapeutic posology of the herbal medicinal product reported in literature or reference sources?		
2. Were serious undersirable effects[b] with a therapeutic posology of the herbal medicinal product reported in literature or reference sources with a well-documented history[c]?		
3. Were serious undesirable effects with or without specifications of the therapeutic posology of the herbal medicinal product reported in literature or reference sources in general?		
4. Were one or more intoxications due to (in)voluntary overdose with the herbal medicinal product reported in literature or reference sources as documented case studies?		
5. Were one or more intoxications due to (in)voluntary overdose with the herbal medicinal product reported in literature or reference sources in general?		
6. Is (are) there (a) well defined therapeutic indication(s) for the herbal medicinal product according to international standards[d]?		
7. Can the symptoms of the conditions covered by the claimed indications be related to serious health deficiencies creating the need for medical supervision?		
8. Are there constitutional groups[e] at risk by using the herbal substance?		
9. Are drug-drug interactions of the herbal medicinal product with other medicines reported in literature or reference sources as case studies?		
10. Does there exist a plausibility for drug-drug interactions of the herbal medicinal product?		
11. Are there patient groups[f] at risk when considering the use of the herbal medicinal product?		
12. Can the herbal medicinal product be replaced by other medicines with stronger therapeutic activity and less undesirable effects?		
13. Can the herbal medicinal product be replaced by other medicines with stronger therapeutic activity but more undesirable effects?		
14. Is the use of the herbal medicinal product known in western traditional medicine?		
15. Is the herbal medicinal product studied in one or more placebo controlled clinical trials?		
16. Is the herbal medicinal product subject of one or more meta-analysis or reviews published in international scientific journals?		
17. Is the herbal medicinal product on the market in EU countries since 30 years or more?		

[a] Non serious undesirable effects = not leading to hospitalisation and reversible.

[b] Serious side effects = leading to hospitalization or dead

[c] Reference sources are accessible written sources, pharmacovigilance reports included.
In clinical studies there exists a need for immediate reporting serious side effects in the pharmacovigilance cascade reported to the ethical committee when submitting the protocol of the study. Four conditions have simultaneously to be present in order to report: there must be *drug* involved (= one or more well defined medicinal products, among them the herbal medicinal product dealt with), there must be an *event* (= well documented phenomenon happening to the patient), there must be a *source* (= a well-defined source: first order = patient her/himself or people immediately involved in the event) and there must be a *patient* (= a well-defined person taking among others the herbal medicinal product)

[d] These therapeutic indication(s) are referring to international standards, e.g. ICD-10 criteria (http://www.who.int/classifications/apps/icd/icd10online/) and/or internationally recognised diagnostic references (e.g. DSM-IV criteria for mood disorders; NYHA for cardiac insufficiency; Vahlensieck for BPH).

[e] Constitutional groups = children, pregnant and lactating women and elderly

[f] Patient groups are considered as being at risk by the pathology they are suffering from

References

Arunlakshana O, Schild M (1959) Some quantitative uses of drugs antagonists. Br J Pharmacol 14:48–58

Bakana P, Laekeman GM, Totte J, Herman AG, Vlietinck AJ (1985) Stereochemical considerations in relation to the pharmacological activity of Pterotaberna alkaloids. J Nat Prod 48:766–771

Bennett A, Stamford IF, Tavares IA et al (1988) The biological activity of eugenol, a major constituent of nutmeg (*Myristica fragrans*): studies on prostaglandins, the intestines and other tissues. Phytother Res 2:124–130

Booker A, Suter A, Krnjic A et al (2014) A phytochemical comparison of saw palmetto products using gas chromatography and 1H nuclear magnetic resonance spectroscopy metabolomic profiling. J Pharm Pharmacogn 66:811–822

Born GVR (1962) Aggregation of blood platelets by adenosine diphosphate and its reversal. Nature 194:927–929

Breckenridge A (2006) William Withering's legacy – for the good of the patient. Clin Med 6: 393–397

Brownlee G, Johnson ES (1965) The release of acetylcholine from the isolated ileum of the guinea-pig induced by 5-hydroxytryptamine and dimethylphenylpiperazine. Br J Pharmacol 24: 689–700.

Chaggar PS, Shaw SM, Williams SG (2015) Is foxglove effective in heart failure? Cardiovasc Ther. doi:10.1111/1755-5922 [Epub ahead of print]

Chan BS, Buckley NA (2014) Digoxin-specific antibody fragments in the treatment of digoxin toxicity. Clin Toxicol 52:824–836

Crawford N (1976) Platelete microfilaments and microtubules. In: Gordon JL (Ed) Platelets in biology and pathology. North-Holland Publishing Company, Amsterdam pp. 121–157.

Day M, Vane JR (1963) An analysis of direct and indirect actions of drugs on the isolated guinea-pig ileum. Br J Pharmacol Chemother 20:150–170

de Combarieu E, Martinelli EM, Pace R, Sardone N (2015) Metabolomics study of Saw palmetto extracts based on ^{1}H NMR spectroscopy. Fitoterapia 102:56–60

De Monte C, Carradori S, Granese A, Di Pierro GB, Leonardo C, De Nunzio C (2014) Modern extraction techniques and their impact on the pharmacological profile of *Serenoa repens* extracts for the treatment of lower urinary tract symptoms. BMC Urol 14:63–74

De Witte P, Lemli L (1990) The metabolism of anthranoid laxatives. Hepatogastroenterology 37:601–605

De Witte P, Demeyer K, Delmulle L et al (2008) Pharmacokinetics. In: Delmulle L, Demeyer K (eds) Anthraquinones containing plants reconsidered. Standaard Uitgeverij, Antwerpen, pp 99–107

Dong B, Li H, Singh AB, Cao A, Liu J (2015) Inhibition of PCSK9 transcription by berberine involves down regulation of hepatic HNF1-alpha protein expression through the ubiquitin-proteasome pathway. J Biol Chem 290:4047–4058

EMA (European Medicines Agency) (2012) Recommendation to marketing authorisation holders, highlighting the need to ensure compliance with 3Rs methods described in the European Pharmacopoeia. Committee for Medicinal Products for Human Use (CHMP) and Committee for Medicinal Products for Veterinary Use (CVMP), EMA

EMA (European Medicines Agency) (2014) Public statement on the use of herbal medicinal products containing estragole. Committee on Herbal Medicinal Products, EMA

Fong CW (2014) Statins in therapy: understanding their hydrophilicity, lipophilicity, binding to 3-hydroxy-3-methylglutaryl-CoA reductase, ability to cross the blood brain barrier and metabolic stability based on electrostatic molecular orbital studies. Eur J Med Chem 85:661–674

Frohne D (1970) Untersuchungen zur Frage der harndesinfizierenden Wirkungen von Bärentraubenblatt-Extrakten. Planta Med 18:1–25

Grigoras N, Purdel C (2014) Assessment report on *Camellia sinensis* (L.) Kuntze, non fermentatum folium. European Medicines Agency. http://www.ema.europa.eu/ema/

Habib FK, Wyllie MG (2004) Not all brands are created equal: a comparison of selected components of different brands of *Serenoa repens.*. Prostate Cancer Prostatic Dis 7:195–200

Hauptmann PJ, Kelly RA (1999) Digitalis. Circulation 99: 1265–1270.

Heroutová M (2012) Assessment report on *Arctostaphylos uva-ursi* (L.) Spreng., folium. EMA. http://www.ema.europa.eu/ema/

ICH (2001) Guidance for Industry: S7A safety pharmacology studies for human pharmaceuticals. U.S. Department of Health and Human Services, Food and Drug Administration, Center for Drug Evaluation and Research (CDER), Center for Biologics Evaluation and Research (CBER)

Janssens J, Laekeman GM, Pieters LAC, Totte J, Herman AG, Vlietinck AJ (1990) Nutmeg oil: identification and quantification of its most active constituents as inhibitors of platelet aggregation. J Ethnopharmacol 29:179–188

Jeffreys D (2005) The remarkable story of a wonder drug Aspirin. Bloomsbury, London, pp 5–74

Kyle R, Shampe M (1974). Discoverers of quinine. JAMA. 229 (4): 462.

Laekeman G (1980) Pharmacological study of the Schulz-Dale reaction. University of Antwerp, PhD thesis

Laekeman G (1988) Phytopharmacological study of vernolepin and eugenol. University of Antwerp, PhD aggregation thesis

Laekeman GM, Mertens J, Totte J, Bult H, Vlietinck AJ, Herman AG (1983) Isolation and pharmacological characterization of vernolepin. J Nat Prod 46:161–169

Laekeman GM, De Clerck F, Vlietinck AJ, Herman AG (1985) Vernolepin: an antiplatelet compound of natural origin. Naunyn Schmiedeberg's Arch Pharmacol 331:108–113

Lanoxin (2015) Summary of Product Characteristics last update June 2015

Löwig C, Weidmann S (n.d.) Beiträge zur organischen Chemie (Contributions to organic chemistry); "III. Untersuchungen mit dem destillierten Wasser der Blüthen von *Spiraea Ulmaria*" (III. Investigations of the water distilled from the blossoms of *Spiraea ulmaria*). Annalen der Physik und Chemie 46:45–91; 57–83

Lu Z, Kou W, Du B et al (2008) Effect of Xuezhikang, an extract from red yeast Chinese rice, on coronary events in a Chinese population with previous myocardial infarction. Am J Cardiol 101:1689–1693

Mazzanti G, Di Sotto A, Vitalone A (2015) Hepatotoxicity of green tea: an update. Arch Toxicol 89:1175–1191

Nijenhuis CM, ter Horst PGJ, de Jong-van den Berg LTW, Wilffert B (2011) Disturbed development of the enteric nervous system after *in utero* exposure of selective serotonin re-uptake inhibitors and tricyclic antidepressants. Part 1: literature review. Br J Clin Pharmacol 73:16–26

Ouyang AJ, Lv YN, Zhong HL, Wen JH, Wei XH, Peng HW, Zhou J, Liu LL (2015) Meta-analysis of digoxin use and risk of mortality in patients with atrial fibrillation. Am J Cardiol 115(7):901–916

Perry LM (1980) Medicinal plants of East and Southeast Asia: attributed properties and uses. MIT Press, Cambridge, MA, pp 141–142, 153 and 227–228

Pugsley MK, Authier S, Curtis MJ (2008) Review: frontiers in pharmacology. Principles of safety pharmacology. Br J Pharmacol 154:1382–1399

Rasheed A, Laekeman G, Totte J, Vlietinck AJ, Herman AG (1984) Eugenol and prostaglandin biosynthesis. NEJM 310:50–51

Roquebert J, Dufour P, Canellas J (1971) Action of some morphine analgesics on the effects of coaxial electric stimulation of the isolated guinea-pig ileum. C R Seances Soc Biol Fil 165:577–580

Scaglione F, Lucini V, Pannacci M, Caronno A, Leone C (2008) Comparison of the potency of different brands of *Serenoa Repens* extract on 5alpha-reductase Types I and II in prostatic co-cultured epithelial and fibroblast cells. Pharmacology 82:270–275

Scaglione F, Lucini V, Pannacci M, Dugnani S, Leone C (2012) Comparison of the potency of 10 different brands of *Serenoa repens.*. Eur Rev Med Pharmacol Sci 16:569–574

Schmitz, R (1985). Friedrich Wilhelm Sertürner and the discovery of morphine". Pharmacy in history. 27 (2): 61–74.

Tayebati SK, Piergentili A, Natale D, Amenta F (1999) Evaluation of an agonist index: affinity ratio for compounds active on muscarinic cholinergic M2 receptors. J Auton Pharmacol 19:77–84

Van Nueten JM, Janssens AJ, Fontaine J (1976) Unexpected reversal effects of naloxone on the guinea pig ileum. Life Sci 18:803–809

Van Rossum JM (1963) Cumulative dose-response curves II. Technique for the making of dose-response curves in isolated organs and the evaluation of drug parameters. Arch Int Pharmacodyn 143:299–330

Vane J (1971) Inhibition of prostaglandin synthesis as a mechanism of action for aspirin-like drugs. Nat New Biol 231:232–235

Vial T, Bernard G, Lewden B, Dumortier J, Descotes J (2003) Hépatite aiguë imputable à l'Exolise® (*Camellia sinensis*). Gastroenterol Clin Biol 27(12):1166

Chapter 7
Systems Network Pharmaco-Toxicology in the Study of Herbal Medicines

Alessandro Buriani, Stefano Fortinguerra, Maria Carrara, and Olavi Pelkonen

Abstract Analytic "'omic" techniques and systems biology driven bioinformatics have increasingly been a game changer in the study of herbal drugs, thanks to the simultaneous detection of entire molecular families in a given biological system, and the ability to collect, classify, network, and visualize a large number of analytical data through bioinformatics. The genomics area has been at the vanguard of this evolution. Other "'omic" techniques, such as proteomics and metabonomics, are providing a fast-growing body of data both on biological targets and on phytocomplexes and their interactions. This has favored a more global view of biological processes, describing how perturbations can influence the steady state of a large number of the components of the system and their relations, changing the system as a whole. It is thus apparent that biological responses induced by phytocomplexes represent the net output of changes in the properties of a very large number of molecules, all acting in an interdependent fashion to form a highly connected network.

"'Omic" techniques and systems biology are applied in herbal medicine at various levels, and provide novel strategies that can be exploited both for herbal drug research and medical use, in applications ranging from drug quality control to patient stratification. Network pharmaco-toxicology represents one of the most important applications of this new approach. Building up networks of molecular interactions between phytocomplex components and pharmaco-toxicological processes can provide a powerful predictive tool in herbal medicine. There is an increasing number of Web-based systems biology platforms, continuously fed with "'omics"

A. Buriani (✉) • S. Fortinguerra
Center Maria Paola Belloni Regazzo for Pharmacogenomics and Personalized Medicine, Gruppo Data Medica, Padova, Italy

Department of Pharmacological and Pharmaceutical Sciences, University of Padova, Padova, Italy
e-mail: alessandro.buriani@gmail.com

M. Carrara
Department of Pharmacological and Pharmaceutical Sciences, University of Padova, Padova, Italy

O. Pelkonen
Department of Pharmacology and Toxicology, University of Oulu, Oulu, Finland

O. Pelkonen et al. (eds.), *Toxicology of Herbal Products*,
DOI 10.1007/978-3-319-43806-1_7

data, providing a view of the complete biological system modulated by a given drug that can be used for predictive pharmacology and toxicology. Systems toxicology promises to be the best context for providing a mechanistic understanding of toxicological effects, thus allowing the prediction of responses to phytochemicals.

Keywords 'omics • Pharmacogenomics • Genomics • Whole genome sequencing • Natural drugs • Proteomics • Metabonomics • Metabolomics • Microbiome • Systems biology • Network pharmacology • Network toxicology • Herbal medicines • Phytocomplex • Holistic • Traditional Chinese medicine • Ayurveda • Jamu • Kampo • Traditional Iranian medicine

Abbreviations

ADME	Adsorption, distribution, metabolism, excretion
ADMET	Adsorption, distribution, metabolism, excretion and toxicology
AOP	Adverse outcome pathway
AST	Addition and subtraction theory
CMAP	Connectivity map
DCHD	Da Chaihu decoction
DILI	Drug-induced liver injury
FDA	Food and Drug Administration
HMDB	Human metabolome database
HMSP	Herbal medicine systems pharmacology
HS	UK National Health System
KEGG	Kyoto Encyclopedia of Genes and Genomes
LBVS	Ligand-based virtual screening
LINCS	Library of integrated network-based cellular signatures
MHD	Ma-huang decoction
MS	Mass spectrometry
NGS	Next generation sequencing
NIH	(United States) National Institutes of Health
NMR	Nuclear magnetic resonance
PEA	Probability ensemble approach
PGP	Personal genome project
PoT	Pathways of toxicity
PRM	Parallel reaction monitoring
QSARs	Quantitative structure-activity relationships
SBVS	Structure-based virtual screening
SNPs	Single nucleotide polymorphisms
SRM	Selected reaction monitoring
TCM	Traditional Chinese medicine

TCMSP	Traditional Chinese medicine systems pharmacology database
VS	Virtual screening
WES	Whole exome sequencing
XCHD	Xiao Chaihu decoction
WGS	Whole genome sequencing

Introduction

It is probably safe to say that the introduction in the laboratory of the "'omic" techniques has provided a new, unexpected strategy to overcome several major problems that have historically hampered the pharmaco-toxicological study of phytocomplexes, thus limiting their introduction in Western mainstream medical practice. The new technical ability of simultaneously identifying global multi-molecular patterns in biological systems has provided a more suitable tool for studying herbal drugs, which are inherently multicomponent and act in a multi-targeted fashion. At the same time, the development of appropriate bioinformatics tools (databases, software) to handle the new complexity of biological data flowing in from "'omic" techniques, increasingly allows the building of knowledge-based networks in order to better understand the biological phenomena in their wholeness, picturing simultaneously each single dynamic interaction between biomolecules, and between biomolecules and drugs, thus allowing a whole vision of the components of biological effects and their crosstalk.

Genomic, proteomic, metabonomic, and the other "'omic" techniques are revealing the molecular multitude within biological circuits, while systems biology bioinformatics can identify the large number of direct and indirect relations and interactions, simultaneously taking into account many different aspects, in a dynamic multilayer, three-dimensional molecular network of molecules, functions, and conditions, building up interactive associations between molecules and causal relationships between molecules and biological events, and between these and pathophysiological conditions. The study of the medical use of phytocomplexes thus seems to have found in network pharmaco-toxicology the perfect discipline for its development, a fact confirmed by the unprecedented and increasingly large numbers of scientific results on the molecular networks implicated in the pharmacology of herbal drugs.

This new approach, both molecular and holistic at the same time, can overcome the limits of a reductionist approach to studying complex systems, such as the phytocomplex and its multi-target effects, and it is providing a fast-growing body of systems biology-oriented knowledge that is also paving the way for a new molecular approach to traditional medicines like Traditional Chinese Medicine (TCM), Ayurveda, and Traditional Iranian Medicine (TIM). Systems biology and "'omics" are thus providing an unprecedented opportunity to demystify these traditional medicines, thus making their rational introduction into the mainstream medical practice a realistic goal.

The Systems Biology World of "'Omics" Techniques

The Dawn of the Postgenomic Era, from Genes to Proteins and Metabolites; The High Throughput 'Omic Techniques

At least two main scientific evolutions in the last 20 years have made a major contribution towards unveiling and understanding the complexity of biological events: the simultaneous detection of entire molecular families in a given biological system, and the bioinformatics' capability of collecting, classifying, networking and visualizing large sets of data. This has totally changed our approach to any biological field – especially pharmacology and toxicology – with their background firmly based on a consolidated reductionist "one molecule, one target ,one effect" approach. The new paradigm requires a more global, holistic view, where the focus backtracks from the drug-target molecular interaction to comprise the entire biological system affected, seen as a set of intermingled networks, allowing a more comprehensive view of what's behind the effect in its dynamic, plural, and three-dimensional context (molecules-targets, targets-pathways, pathways-effect), and its capability to affect health and disease.

Genomics and Transcriptomics

Genomics, the so-called "mother of all 'omics," is still today the leading family of techniques, especially in its clinical application. Over the past 15 years, genomics has achieved landmark results, from the completion of the Human Genome Project, to the HapMap Project, which produced a genome-wide database of human genetic variation for use in genetic association studies of common diseases. Genome-wide association studies have been developed to identify multifactorial disease-predisposing variants and large-scale whole-genome and whole-exome sequencing studies have been introduced (Zheng et al. 2015). Numerous genetic susceptibility loci have been identified to be associated with many complex diseases via genome-wide association studies, including a variety of cancers (Haiman et al. 2011; Jia et al. 2013; Lan et al. 2012; Shi et al. 2011; Thomas et al. 2008), bipolar disorder, coronary artery disease, Crohn's disease, hypertension, rheumatoid arthritis, type I and type II diabetes (The Wellcome Trust Case Control Consortium 2007; Franke et al. 2010; Voight et al. 2010; Deloukas et al. 2013), inflammatory bowel disease (Duerr et al. 2006), obesity (Wen et al. 2012), and others. Next-generation sequencing (NGS) has greatly improved the techniques, bringing them to a higher level, and can even be used to identify disease-associated low-frequency alleles in the case of complex diseases (Palotie et al. 2013; Manolio and Collins 2009).

In the clinic, both predictive genomics and pharmaco-toxicogenomics are mostly based on the analysis of specific single nucleotide polymorphisms (SNPs), some of the most frequently occurring genetic variations in the human genome, that can be associated with pathologically significant characteristics, such as disease susceptibility, differential sensitivity to drugs, and altered enzymatic activities (Guo et al. 2014). Today the ultimate frontier in genomics is probably the sequence of the whole personal DNA, which has become more and more affordable for the public thanks to the rapid evolution of analytical techniques and associated bioinformatics tools (Heather and Chain 2015). Whole genome sequencing (WGS) is increasingly applied in the clinic, and despite its current limitations, can allow the identification of clinically actionable genetic information, allowing early medical intervention (Dewey et al. 2014; Glusman et al. 2015). A major collaborative project aimed at sequencing the genome of a pool of 100,000 healthy individuals is on, and will be used to set up and optimize a clinical protocol. Genomics England, a company wholly owned and funded by the U.K. Department of Health, was set up to deliver this project that by 2017 will have sequenced 100,000 whole genomes from patients of the U.K. National Health System (NHS) (http://www.genomicsengland.co.uk), aiming at, among others, setting up a genomic medicine service for the NHS (Siva 2015). Other national and international multicenter collaborations are being organized with the aim of sharing databases continuously fed with genetic information from the clinic, such as the multicenter Human Variome Project, whose aim is to collect, share, and interpret human genetic variation data from routine clinical practice and research (Smith et al. 2015). In a smaller scale example, the personalized genomic disease risk of a group of 81 volunteers was evaluated by whole exome sequencing (WES) using next generation sequencing which, by integrating the patient's genomic data with medical records and pedigree data, allowed a clear link between personal disease histories and causative disease genes (Gonzalez-Garay et al. 2013). Another important example is the non-anonymous, U.S.- based Personal Genome Project (PGP) which, thanks to its public access and participatory research model, besides its contribution to understanding the relationships between genomes and phenotypes, promotes participant education and compliance as well as method sharing and other community-driven benefits (Ball et al. 2012, 2014; Chen et al. 2014).

Downstream from the genotype, the analysis of transcripts (transcriptomics) has been developed, focusing more on what is actually transcribed following specific stimuli. Unlike DNA analysis, transcriptomics allows a closer look at epigenetic phenomena and takes into account events that can affect DNA expression (Crosetto et al. 2015). Transcriptomic techniques have been applied in the clinical context, and thanks to next-generation sequencing, transcriptome-wide structure determinations can be performed in association with RNA structure probing to elucidate RNA structure, which is a pivotal aspect of all its functions (Kwok et al. 2015). Among epigenetic mechanisms of gene-expression control, regulation by non-coding RNAs, such as microRNAs, certainly plays a central role. Given their importance in fine-tuning gene expression, microRNAs can be considered part of the epigenome, and have been associated with disease susceptibilities, opening new prospects for the development of new diagnostic and therapeutic opportunities (Jirtle and Skinner 2007).

Finally, with the increasing knowledge that we are not alone in our own bodies, but possess unique sets of microbiota that actively and decisively participate in our daily exchange with the external environment, genomics has again provided the technical means for classifying and analyzing the variegated world of saprophytes living with us, especially in the intestines. The term "metagenomics" indicates the analysis of the DNA of entire populations of microorganisms. Thanks to metagenomics, a technique that made possible the Human Microbiome Project (Pennisi 2007; Turnbaugh et al. 2007), it is possible to identify our gut microbiome from feces samples and follow its evolutions with respect not just to intestinal health, but to systemic conditions and reactions to food and drugs as well, given the central role of the microbiome in transforming and metabolizing dietary ingredients and even drugs (Ji and Nielsen 2015; D'argenio and Salvatore 2015).

Proteomics

Proteomic techniques have undergone a formidable evolution thanks to the technical and computational advances of mass spectrometry-based proteomics, which has simplified the identification and quantification of proteins in a given biological sample, giving way to next-generation proteomics, thus leading to the establishment of fast-growing proteomics resources and repositories (Altelaar et al. 2013; Chen et al. 2015b). Besides the general mechanistic advancement in describing the association of protein phenotypes with human health and diseases, proteomics represents a significant source of protein biomarker candidates that can be used in the clinic (Drabovicha et al. 2015).

The proteome represents the full complement of the proteins in a cell, organ, or organism, and proteomics is a systematic approach to characterizing all, or an enriched subset of proteins therein. Measuring changes in levels and modifications of proteins is applied to diagnostics, drug discovery, and investigating toxic events. Resolving the molecular details of proteome variation in the various tissues and organs of the human body will greatly increase our knowledge of human biology and disease (Uhlén et al. 2015). Proteomic data are of particular value in pharmacotoxicology because the proteome is an important mediator of altered biological responses as a consequence of exposure to active substances such as phytocomplexes (Sturla et al. 2014; Pelkonen et al. 2012). Increases or decreases in protein levels may be a direct consequence of corresponding mRNA-expression changes, but increases or decreases in protein function may also be influenced by post-translational modifications (Sturla et al. 2014; Morris et al. 2014; Uhlén et al. 2015). For example, protein phosphorylation, which can be further addressed by high-throughput phosphoproteomics, enables the characterization of molecular events proximal to disease-related signaling mechanisms and is particularly interesting for its potential pharmaco-toxicological applications (Morris et al. 2014; Kinoshita et al. 2015).

Mass spectrometry (MS) is widely considered to be the central technology for modern proteomics, mainly because of its unsurpassed sensitivity and throughput. Thanks to the accuracy of MS, peptides in the sub-femtomolar range can be detected in biological samples with a mass accuracy of less than 10 ppm. This level of accuracy is necessary to compare proteins between samples derived from altered and control systems. Isotope tagging for relative and absolute quantification (iTRAQ) is used in comparative proteomics for pharmaco-toxicological purposes because it enables the relative quantification of protein species between samples in a non-targeted fashion (Sturla et al. 2014; Guo et al. 2015b). This method can be further complemented with a targeted method of even greater accuracy, selected reaction monitoring (SRM), especially when an accurate quantification of a specified set of peptides/proteins across multiple samples is required. Technically, SRM enables the precise quantification of predefined proteins by measuring peptides produced by the controlled enzymatic digestion of the proteome as surrogates to their corresponding proteins in triple quadrupole MS (Titz et al. 2014). Because it is a targeted approach, any proteomics analysis by SRM requires the a priori selection of the proteins to quantify. The list of selected proteins is then processed with bioinformatic tools to identify at least two proteolytic peptides that optimally represent the protein and distinguish it from all others. This step is followed by several optimization and validation steps to ensure unique identification and accurate quantification; the method enables a multiplexed approach by which tens or hundreds of proteins can be quantified in a single MS run with absolute molecular specificity (Picotti and Aebersold 2012). A new MS-based targeted approach, called "parallel reaction monitoring" (PRM), that is centered on the use of next-generation, quadrupole-equipped high-resolution and accurate mass instruments, has been developed. This approach is closely related to SRM, but allows the automated measurement of all fragmentation products of given peptides and produces high-quality data that can be easily interpreted and provide mechanistic details of pharmacology and toxicology pathways (Peterson et al. 2012).

Metabonomics

Metabonomics, or metabolomics, measures the metabolic response of biological systems to internal or external stimuli, aiming at understanding the systemic change in complex multicellular systems (Nicholson and Lindon 2008). Metabonomics involves a comprehensive and quantitative analysis of all metabolites or low molecular weight organic or inorganic chemicals that are products or substrates of enzyme-mediated processes, in a given biological sample (Li et al. 2011). Metabonomics is unique because it can be used to define amounts of internalized xenobiotic chemicals and their biotransformation products, with such an analytical power that today it is possible to characterize even the most complex mixtures such as the food metabolome (Scalbert et al. 2014). The analytical method is also used to investigate the changes of the perturbed endogenous metabolome, which represents the ultimate change in the levels of chemical species resulting from molecular perturbations at the genomic and proteomic levels (Robertson 2005; Bouhifd et al. 2013; Ramirez et al. 2013). In the first case, an understanding of the kinetic behavior of xenobiotics and their metabolites is necessary to identify candidate clinical biomarkers. The second application involves both identification of endogenous metabolites and quantification of changes in their abundance. From a technical perspective, metabonomics most commonly involves nuclear magnetic resonance (NMR) spectroscopy and/or MS techniques, in both untargeted and targeted analytical strategies. Profiling of a metabonome may entail global detection and relative quantification of a large number of metabolites without a priori knowledge (Robertson 2005; Ideker et al. 2001).One of the aspects that makes the metabonome very attractive for the development of new pharmaco-toxicological and diagnostic tools, especially when determined in excretions like urine, is its being downstream from the genome, the proteome, the exposome (the cumulative measure of environmental influences and associated biological responses throughout its lifespan (Miller and Jones 2014), and all the other "'omes," thus representing the endpoint of all the biological networks perturbations, and it is the end result of whichever biological effects, negative and positive feedbacks, redundancies, buffers and synergies have occurred, each of them dependent from specific genotypic and phenotypic individual characteristics, and from the endogenous or exogenous elements challenging their homeostasis (Li et al. 2011). Therefore, the metabonome can be considered the closest thing to a final biological signature that can be traced in its dynamics from the early events to the last stages, providing a potentially formidable tool – not just for early detection of disease states, but also for its complete monitoring (Zhang et al. 2015a), including treatment pharmacokinetics, pharmacodynamics, and toxicological aspects. Despite the numerous important results coming from all medical research fields, where it is being increasingly exploited, the use of metabonomics as a tool in the diagnostic routine is still limited, but it is easily predicted that it will rapidly become popular in the clinic in the next few years, especially considering the special efforts and resources that have been funnelled towards this accomplishment (Zanetti et al. 2014). (A more detailed view on metabolomics and its applications in the area of herbal medicines is presented in Chap. 8 of this book.)

A Systems Biology Approach for Exploiting 'Omics Techniques

For almost 20 years, systems biology has adopted a system-level understanding and a holistic vision of all biologic phenomena, based on the analysis of molecular networks in their dynamic interactions, and of the cross-talks among these highly interconnected pathways, thereby distinguishing itself from the reductionism-oriented scientific and medical mainstream (Li et al. 2015a; Ideker et al. 2001; Mitra et al. 2013; Hood et al. 2004; Csermely et al. 2013). The advent of 'omics techniques has given a novel impulse to systems biology, allowing the simultaneous detection of entire classes of interrelated molecules (Nicholson and Wilson 2003). Together with the advances in bioinformatics that have provided more user-friendly tools, from 'omics-fed databases to multivariate analytical tools, and to association and visualization software, the systems approach has been rapidly spreading in all biomedical fields (Ghosh et al. 2011; Joyce and Pallson 2006). Today it is safe to say that it is not possible to exploit 'omic data without a systems biology approach and, conversely, it is not feasible to use a systems biology approach without using 'omic techniques; the two have inextricably merged and their advancement is mutually dependent (Toga et al. 2015; Liu 2009; Ahn et al. 2006).

The shift of focus from single molecules to entire molecular classes and their networks has contributed to a more global view of biological processes and has shown how perturbations in biological systems are never restricted to a single molecule, or even a small group of molecules. Instead, any given perturbation influences the steady state of a large number of the components of the system and their relations, changing the system as a whole. Biomolecules form tightly integrated networks and biological responses derive from the behavior of such networks (Ma'ayan 2011). It is thus apparent that biological responses induced by phytocomplexes represent the net output of changes in the properties of a very large number of molecules, all acting in an interdependent fashion to form a highly connected network (Ma'ayan 2011; Barabàsi and Oltvai 2004).

The new perspective, unlike the traditional reductionist approach, allows a view of biological systems as a whole and has probably been one of the most exciting cultural developments deriving from the mutual empowerment of 'omics and systems biology (Ma'ayan 2011; Kitano 2002; Barabàsi and Oltvai 2004). Understanding the topological and functional organization of the various molecules in the system at temporally distinct periods and the dynamics thereof, is central to obtaining a deeper insight into biological phenomena. Large-scale data sets generated with 'omics, comprising gene expression profiles, proteome profiles, metabonome profiles, microbiota profiles, lipidome profiles and interactomes, are being produced as a result of the newfound interest in the systems approach (Ideker et al. 2001; Mitra et al. 2013).

Biological phenomena are thus recognized as representing the integrated outcome of a complex interplay between the individual constituents of the system. The analyses must be global, integrative and dynamic, ultimately allowing the exploration of new dimensions of biologically meaningful data from patients and providing relevant information on health and disease for the individual. Step-by-step systems biology will allow a continuously improving analysis of large networks that describe the properties of entire genomes, of the proteome and the corresponding interactome, the comprehensive mapping and functional integration of metabolic pathways and the combination of all of these systems at different scales of biological organization, thus moving traditional biology and medicine towards information science (Barabàsi and Oltvai 2004; Ideker et al. 2001).

Ideally, by providing an understanding of disease at the molecular level, systems medicine will eventually be able to predict when an organ will become diseased or when a perturbation in a biological network could progress to disease (Barabàsi et al. 2011). Today this systems approach has indeed contributed to a rapid evolution in each biomedical field that would have been unthinkable just few years ago. While next-generation scientists are being appropriately trained, the present generation is being introduced to the new approach; after initial anxiety about challenging the very basic reductionist dogma, they are now starting to ride the new wave in most biomedical fields (Wang et al. 2010; Shen et al. 2015; Posma et al. 2014; Zou et al. 2015).

Chemogenomics, Bridging the Chemical and the Pharmaco-Toxicological Space

The management of libraries of biological information coming from 'omics has found in bioinformatics its natural companion, providing appropriate databases for data storage, software for consultation, and platforms for *in silico* screening of compounds and the identification of their molecular targets (Bredel and Jacoby 2004; Brown et al. 2014). At the same time, advances in medicinal chemistry and the rapid growth of the available crystallographic information on proteins has advanced *in silico* research methodologies, both for molecular modeling and chemoinformatics, improving the prediction of pharmaco-toxicological effects based on the structural information of target-ligand complexes (Medina-Franco et al. 2014; Barlow et al. 2012). Today structural information is increasingly available and several online resources can provide structural characterization of the available binding pockets, their interactions with ligands, and their effects (Ilatovskiy et al. 2013; Wang and Xie 2014; Brown et al. 2014). *In silico* methodologies based on pockets/ligands interactions can be subdivided into compound docking, performed for a single compound and a single binding pocket; compound screening, where different compounds have to be docked and scored in a single binding pocket; compound profiling, where a single compound is docked and scored in the binding pockets of various proteins of pharmaco-toxicological interest; and compound binding affinity prediction, where compounds are docked and scored in one binding pocket represented by more than one structure (Ilatovskiy et al. 2013). Following the identification of the molecular target, and thus the foreseeable pharmaco-toxicological action(s), key absorption, distribution, metabolism, excretion, and toxicology (ADMET) properties need to be determined. *In silico* model systems can be used to suggest metabolite patterns from a candidate drug by virtual means, and the methods to predict metabolite formation can be divided into two main classes: comprehensive (global), and specific (local).

As for ligand-target modeling, *in silico* ADMET profiling has seen a rapid advance, and physiologically based pharmacokinetic models are becoming a key tool in pharmaco-toxicological prediction studies (Pelkonen et al. 2009). These *in silico* methods include both comprehensive and specific approaches. Comprehensive methods, also called expert systems, mimic human reasoning and formalize existing knowledge on ADMET pathways. Specific methods generally apply to specific metabolic enzymes and include ligand-based and target-based techniques. The former are based on the known biological activities of the ligand and can derive ligand-target(s) interaction models. The target-based models can evaluate the molecular interactions between the three-dimensional structure of the ligand and the target, thus predicting an affinity-based activity. Today, ligand and target-based methods are often combined (Pelkonen et al. 2011).

In silico models for pharmaco-toxicological research have thus provided excellent platforms for the exploitation of data coming from 'omic techniques, which can then be fed to such *in silico* experimental platforms, continuously enriching them with biological information on the direct effects of one molecule on a given substrate, as well as on downstream and sidestream effects, their metabolic fate, and their toxic potential, making them potent tools for pharmaco-toxicological prediction. The use of informatic platforms exploiting knowledge from medicinal chemistry and 'omics data for *in silico* screening and prediction, gave rise to the term "chemogenomics," an *in silico* pharmaco-toxicological approach taking into account the structural molecular aspects, as well as the associated knowledge from genomics, but also from proteomics and other 'omics. Today, methods for the identification of potential drugs by computational methods represents a crucial step in early-stage drug discovery, such as virtual screening (VS), a type of compound database searching approach used to find novel compounds with a certain biological activity, similar to existing ligands, or for unexplored putative drug targets.

As mentioned above, VS chemogenomic experiments can be divided into two main approaches: ligand-based (LBVS) and structure-based (SBVS), with some also indicated as "target-based." In the first case, structure-activity data from a set of known biological effects are used to identify compounds for experimental evaluation. The protein structure of interest is available in chemical libraries with biological information and the compound is explored by docking it into the active site of the target using dedicated software. Examples of LBVS are similarity and substructure searching and quantitative structure-activity relationships (QSAR). SBVS, on the other hand, utilizes the amino-acid sequence (sequence-based comparisons), or the three-dimensional (3-D) structure (structure-based comparisons) of the biological target to dock the candidate molecules and rank them based on their predicted binding affinity or complementarity to the binding site. In this paradigm, receptors are no longer viewed as single entities but grouped into sets of related proteins or receptor families and as such can be explored in a systematic manner (Lavecchia and Di Giovanni 2013; Klabunde 2007; Rognan 2007; Nantasenamat and Prachayasittikul 2015).

Chemogenomics can thus provide a profound insight into relevant biological targets or pathways, and can be seen as an emerging interdisciplinary field aiming at identifying all possible ligands for all possible targets. Besides computational approaches such as docking, similarity searching, and pharmacophore modeling that can identify new ligands for known targets (VS), putative targets for known ligands can be identified (target fishing). This methodology can lead to the generation of the global pharmaco-toxicological profile of a compound and actually lies at the basis of the polypharmacological profile of bioactive compounds (Medina-Franco et al. 2013). The prediction of full bioprofiles can uncover unknown pharmacological activities or indicate potential off-targets of a compound that may cause unwanted side effects. Thanks to the versatility and the wide knowledge-based chemogenomics approach, it is possible to evaluate compounds on entire families of related proteins or on a full metabolic pathway. This is particularly important, considering that many, if not all drugs bind to several targets (polypharmacology), and becomes fundamental for the correct evaluation of mixtures of compounds – such as in the case of the phytocomplexes in herbal drugs, where a multi-target mediated effect is always present. Computational chemogenomic approaches can thus enable a systematic proteome-wide assessment of protein-ligand interactions and the correlations of molecular interactions with biological effects and clinical outcomes, thus supporting and facilitating a shift in the conventional one-drug, one-target drug discovery process to the new paradigm of polypharmacology (Lavecchia and Di Giovanni 2013; Medina-Franco et al. 2013; Xu et al. 2012; Brown et al. 2014), a paradigm that is changing the approach to herbal drugs.

Thanks to the increased ability to investigate multi-target biological contexts, pharmacologically active phytocomplexes administered with herbal drugs, whose activity is the synthesis of the effects of mixtures of bioactive compounds, can be considered to be a pharmacological whole, and their study not limited just to the simple identification and isolation of druggable active compounds. Thanks to chemogenomics, the multifaceted nature of phytocomplex drugs can be fully appreciated, exploited, and controlled. Application of chemogenomics to toxicology is particularly valuable for its ability to predict the potential toxicity of chemicals in an adequately reliable manner before humans (or any other living organisms) are exposed to them, thus making such *in silico* tools an extraordinary resource for the prediction of main and side effects of single drugs as well as mixtures (Pelkonen 2010).

There is a rapidly growing number of compound databases annotated with bioactivity that represent an important source for mining complex ligand-target relationships as well as toxicological and kinetic aspects. Examples of major databases available to the public are ChEMBL, PubChem, GLIDA, PDSP, BindingDB, IUPHARdb and DrugBank, while examples of public platforms for data integration and mining are the Open Pharmacological Concept Triple Store (Open PHACTS) project, SILIRID (Simple Ligand-Receptor Interaction Descriptor), the PharmaTrek Web explorer, "Structure-Activity Relationship (SAR) Matrix" (SARM) methodology, the ChemMapper and the PreDPI-Ki server. These can systematically explore the integrated pharmaco-toxicological knowledge, allowing the evaluation and prediction of ligand-target interactions in a qualitative and/or quantitative way (Barlow et al. 2012; Carrascosa et al. 2012; Medina-Franco et al. 2013; Paricharak et al. 2015; Gong et al. 2013; Cao et al. 2013; Chupakhin et al. 2014; Gupta-Ostermann and Bajorath 2014). Most of these platforms use Bayesian target prediction algorithms for qualitative polypharmacological profiling, thus giving a prediction of the interaction between a compound and a panel of targets. Structure-activity relationship techniques can then be used to provide quantitative bioactivity predictions. The two approaches can also be merged in integrated *in silico* platforms for target prediction and proteochemometric modelling (PCM) for the prediction of compound polypharmacology, potency, and affinity, and also for anticipating adverse drug effects (Paricharak et al. 2015). Given the large number of chemogenomics databases and platforms, there is a growing need to integrate data from the many different sources. Semantic Web technology allowing interoperability needs to be developed, an upgrade facilitated by the recent transition from desktop-based to cloud-based applications with open standards and open source software. Examples of interoperative platforms via cloud computing services are being developed by the Blue Obelisk and the Open Pharmacological Concepts Triple Store consortiums (Nantasenamat and Prachayasittikul 2015).

As an example of the use of these platforms in polypharmacology, the Web-based server PreDPI-Ki was designed to identify drug-target associations by developing a chemogenomics approach using integrated molecular features and Kis. This approach provides a predictive model that can differentiate drug-target interactions with strong binding affinity from those with weak binding affinity (10 μM Ki value is used as the critical threshold value to discriminate). Once the significant drug(s)-target(s) interactions have been determined, the software graphically visualizes the interactions with edges between circles (drugs) and triangles (targets). Given that a drug can associate to more than one target, and different targets can associate with different drugs, closely related members of the target gene family will show significant drug promiscuity and give rise to complex clinical pharmacology. The final result is a network reconstruction of the associations between drugs and targets that can comprise secondary and side effects and can be applied to the study of phyto-complexes (Cao et al. 2013). This network approach, thoroughly illustrated in the next paragraph, is often used, and ultimately aims at building the biological network of proteins and their regulators in a large-scale structure-activity relationships graphic mode, making it simpler, more accessible, and more intuitive to utilize such *in silico* analytical tools, especially when dealing with herbal drugs, where the complexity of multiple targets is further complicated by the presence of multiple compounds.

Back to the Future: New Systems Pharmaco-Toxicology Meets Old Herbal Medicines

'Omics, Systems Biology and Their Applications to Pharmaco-Toxicology

Pharmacology and toxicology are fields where 'omic techniques have been successfully used from the very beginning, and in each 'omic discipline, both fields have consolidated their own interest areas. Pharmacogenomics and toxicogenomics exploit the whole range of genomic approaches, from SNPs to WGS, to DNA signatures, and to mRNA arrays. Specific genetic variants can be predictive of pharmacological efficacy or toxic side effects, while certain changes in expression of genes for several receptors and enzyme families can be predictive of toxicity, and induction of expression of drug metabolizing enzymes and transporters can give indications about ADME characteristics. Gene expression profiling has been used to assess drug safety and to predict toxicity, leading to the production of dedicated toxicogenomics arrays (Lord et al. 2006; Taboureau et al. 2012), which can reveal molecular signatures of drug-induced toxicity when analyzed in the context of the appropriate database (Zhang et al. 2014). Toxicoproteomics, using mostly protein chips or advanced MS technology, aims at identifying and characterizing the pathological responses to specific toxicants at the protein level and, like toxicogenomics, can be used to identify toxicity biomarkers (Gao et al. 2009). Toxicological applications have been successfully used for the identification and characterization of networks of toxicologically relevant proteins and peptides, as well as their post-translational modifications, also thanks to available network analysis software that allows for import, annotation, visualization, and basic analysis of molecular interactions and networks (Titz et al. 2014). In systems biology, a network is formed by nodes, their connections (edges) and high density collections of nodes (modules) organized according to a hierarchy in the properties of nodes, in that some of them (hubs) are more central, expressing higher numbers of connections with respect to the rest. Bridging nodes connect two other nodes or modules in the network. It is thus clear how with increasing complexity, bioinformatics tools need to be developed with matching increased logical capability. Systems biology aims at integration of biological complexity at all levels of biological organizations, whether cell, organ, organism, or a human being. Building more and more complex networks will ideally eventually lead to the *in silico* replica of a whole human body, an objective that has been hypothesized to reach its 90% by 2038 (Xu et al. 2012).

Large amounts of experimental data need to be collected in order to appropriately feed up systems biology databases. Among "'omics" applied to pharmaco-toxicology, one of the most promising is probably pharmaco-toxicometabonomics, considering the relative simplicity of the assays, the availability of fluid samples like urines, and the possibility of obtaining comprehensive information on biological phenomena downstream from any internal or external perturbation. Pharmaco-metabonomics allows the identification and analysis of contributions to the overall drug response from the environment, diet or the microbiome, something that pharmacogenomics cannot take into account, thus promising better chances for personalized treatment and early toxicological prediction (Clayton et al. 2006). Metabolic profiling can offer rapid, non-invasive pharmaco-toxicological information both for identifying individual best treatment conditions and for monitoring therapeutic regimens. Pharmacometabonomics has had a wide range of applications, from CYP-mediated drug metabolism to the identification of specific drug-responsive phenotypes, while toxicometabonomics has been used in the whole range of toxicological testing, from in vitro to drug-mediated organ toxicity (James 2013; Wang et al. 2010). In systems pharmaco-toxicology, metabonomics may be accompanied by concomitant transcriptomic and proteomic measurements to provide the full context of the exposure to exogenous molecules. Integrated analysis of these diverse data types is important for enabling a full understanding of the mechanistic events driving metabonomic changes (Van den Hof et al. 2015). The latest in the family is pharmacomicrobiomics, which aims at analyzing drug-microbiome interactions and the impact of human microbiome variations on pharmacodynamics and pharmacokinetics (Elrakaiby et al. 2014).

Given its superiority in investigating complex biological networks, systems biology can be considered the elective approach to analyze the entire biological context of drug effects both in the pharmacological and toxicological fields. For a long time, a focus on specific molecular targets of single drugs has driven the core of pharmaco-toxicological research, highlighting pharmacological events in basic molecular experimental models whose biological relevance would not sustain the increasing complexity of higher systems at the cellular or organism levels. In the path to drug characterization and biomarkers identification, most molecular interactions of pharmacological interest are not necessarily meaningful when considered within the entirety of the living organism. Thousands of molecules can inhibit an enzyme in vitro or bind effectively to a membrane receptor, but fail to reproduce the effect when challenged in cellular or in vivo systems, and fail to translate into clinical applications. Systems biology, being a network science that by definition does not isolate molecular events, has a better chance of achieving results that are reproducible in the clinical validation path, and it has rapidly gained ground among pharmacologists and toxicologists (Butcher et al. 2004; Ravindranath et al. 2015; Bai et al. 2014). Today, systems pharmaco-toxicology is taking up the major challenge of clinical toxicology and pharmacology to map out, understand – and model in quantifiable terms – the topological and dynamic properties of the various networks that control the behavior of the pharmaco-toxicologically relevant biological systems. The realization that biological responses stem from aggregate properties of underlying molecular networks, signifies a major drive towards a fast development in the application of complex network theory, and facilitates advances towards uncovering the organizing principles that govern the formation and evolution of complex biological networks by looking at the global relationships and patterns of interactions among the myriad molecular constituents (Barabàsi and Oltvai 2004).

There is an increasing number of Web-based systems biology platforms, continuously fed with 'omics data, that can be used for predictive pharmacology and toxicology. They provide a view of the complete biological system that can be modulated by a given compound, while pathway profiling toolkits that allow for predicting optimal treatments for patients are increasingly being proposed for clinical application (Shen et al. 2015; Martínez et al. 2014; Ekins et al. 2005; Hamon et al. 2014; Gika et al. 2014). The use of comprehensive databases can even allow the identification of unknown potential links between chemicals and human diseases, or toxicological effects, by feeding the chemical-protein and protein-disease associations networks with experimental data (Kongsbak et al. 2014; Rouquié et al. 2015). Recently a major collective project was organized to define and characterize what the main pathways involved in toxicological effects are, in order to produce dedicated databases and software that can be used as an internationally shared methodology for identifying toxicants and their effects on significant biological pathways. This strategy moves away from traditional animal testing, which provides only limited mechanistic information, and towards a pathway-oriented approach. The outcome of the project will be the development of "pathways of toxicity" (PoT), defined as "a molecular definition of cellular processes shown to mediate adverse outcomes of toxicants," leading to the completion of the "human toxome." This formidable key tool for toxicant identification and prediction of toxicity, based on systems biology knowledge of how genes, proteins, and metabolites interact in molecular networks, is developing at a rapid pace, and it is conceivable that in the near future the interrogation of PoTs will became a leading effective strategy in systems toxicology, both for the clinic and in drug design (Kleensang et al. 2014; Bouhifd et al. 2014).

Another proposed strategy in the context of systems toxicology aims at building dynamic adverse outcome pathway (AOP) models starting from biological networks (Ideker et al. 2001). An AOP is a model that describes a process as key events (nodes) and the causal relationships between them (edges). This is achieved in three broad steps: The first is the identification of toxicant-related causal biological networks and their links with the organ-level responses. In the second step, dynamic models linking the exposure with the toxic responses can be developed. The third step enables the simulation of population-level effects of an exposure. The AOP model can identify biomarkers of exposure, effect, or susceptibility, which can then be used for risk prediction and to monitor exposure to toxicants (Ideker et al. 2001). Briefly, once a critical amount of qualitative and quantitative information regarding molecular responses is collected with transcriptomic, proteomic, and metabolomic techniques, data are managed, integrated, and processed through dedicated computational workflows that allow biological pathway analysis, aiming to develop predictive *in silico* models that can be used in risk assessment and prediction. Publicly accessible databases of biological networks and pathways exist, such as the Kyoto Encyclopedia of Genes and Genomes (KEGG) (Manyam et al. 2015), but they often consist of rather static representations of biological processes, while they cannot be used for representing the adverse event mechanisms following a specific exposure.

The transition from static to computable biological network models is thus a necessary step for analyzing efficiently toxicological effects, a step that has to proceed to a virtual organism model. An example is the DILIsym modelling software, a mathematical multi-scale representation of drug-induced liver injury (DILI), to carry out an in vitro to in vivo extrapolation, allowing the exploration of the inter-individual variability in response to potential hepatotoxic substances (Howell et al. 2012). In summary, systems toxicology promises to be the best context for providing a mechanistic understanding of toxicological effects, thus allowing the prediction of responses to substances, including phytochemicals (Knudsen et al. 2015). (See Fig. 7.1 for a synthetic representation of systems toxicology.)

Fig. 7.1 A synthetic representation of systems toxicology, which is aimed at decoding the toxicological blueprint of active substances that interact with living systems. It resides at the intersection of systems biology with toxicology and chemistry, and integrates classic toxicology approaches with network models and quantitative measurements of molecular and functional changes occurring across multiple levels of biological organization. The multidisciplinary systems toxicology approach combines principles of chemistry, computer science, engineering, mathematics, and physics with high-content experimental data obtained at the molecular, cellular, organ, organism, and population levels to characterize and evaluate interactions between potential hazards and the components of a biological system. It is aimed at developing a detailed mechanistic as well as quantitative and dynamic understanding of toxicological processes, permitting prediction and accurate simulation of complex (emergent) adverse outcomes. Thereby, the approach provides a basis for translation between model systems (*in vivo* and *in vitro*) and study systems (e.g., human, ecosystem). systems toxicology, therefore, has an ultimate potential for extrapolating from early and highly sensitive quantifiable molecular and cellular events to medium- and long-term outcomes at the organism level, and its application could be part of a new paradigm for risk assessment (Artwork by Samantha J. Elmhurst (www.livingart.org.uk)). Figure is reproduced with the kind permission of Sturla et al. (2014))

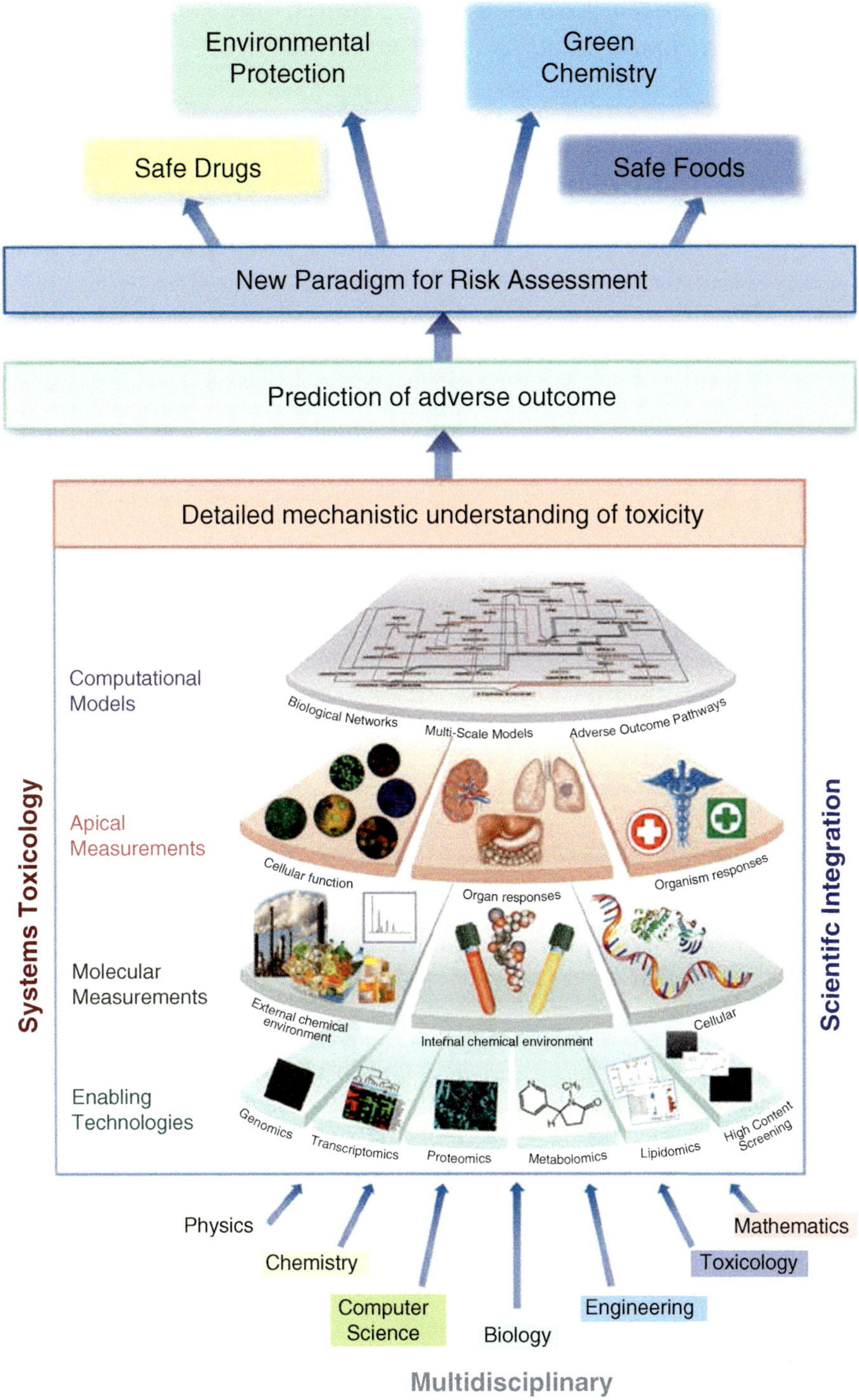
Environmental Protection
Green Chemistry
Safe Drugs
Safe Foods
New Paradigm for Risk Assessment
Prediction of adverse outcome
Detailed mechanistic understanding of toxicity
Computational Models
Biological Networks
Multi-Scale Models
Adverse Outcome Pathways
Apical Measurements
Cellular function
Organ responses
Organism responses
Molecular Measurements
External chemical environment
Internal chemical environment
Cellular
Enabling Technologies
Genomics
Transcriptomics
Proteomics
Metabolomics
Lipidomics
High Content Screening
Systems Toxicology
Scientifc Integration
Physics
Chemistry
Computer Science
Biology
Engineering
Toxicology
Mathematics
Multidisciplinary

Network Pharmaco-Toxicology and "'Omics," the Perfect Match for the Phytocomplex and Its Multi-target Activity

The pharmaco-toxicological interest in phytocomplexes is on the rise, not just as alternative or non-conventional therapies, but also as a source of novel therapeutic pharmaceuticals. A study analyzing new medicines approved by the U.S. Food and Drug Administration (FDA) over the last 30 years revealed that 34 % of medicines based on small molecules were natural products or direct derivatives of natural products. Moreover, approximately 15 % of the drug interventions in the United States ClinicalTrials.gov database are plant-related (Harvey et al. 2015). Identification of the natural sources of herbal medicines, and the variability of their components, are some of the intrinsic problems of phytotherapy, and they require standardization. High-throughput, information-rich 'omic assays can be used to fingerprint herbs and botanical extracts, improving the reproducibility and standardization of biological effects. DNA analysis is being used to categorize medicinal herbs by sequencing, assembling, and annotating their genomes, and by analyzing their genes' functions. Standardized DNA bar-coding identification systems are available and are revolutionizing the practice of herbal identification, utilizing the concept of "one sequence, one species." Herbal genetic information is thus being accumulated, and several herb-related databases have been developed, including genomic and transcriptomic information (http://herbalgenomics.org – http://medicinalplantgenomics.msu.edu) (Chen et al. 2015a). Genotyping single plants, however, is not sufficient for the standardization of phytocomplex preparations, given that many herbal medicine traditions are based on the use of herbal mixtures, such as Traditional Chinese Medicine (TCM) formulas, where admixtures of multiple plants, with components that work synergistically to achieve the therapeutic effects, are used. Establishing a chemical and biological quality standard for such complex TCM preparations might require a comprehensive analytical approach, integrating chemical, metabolic, and biological methods, where DNA barcoding is integrated with metabolomic analysis of the extract allowing the chemical profiling and quantification of all bioactive constituents (Harvey et al. 2015; Guo et al. 2015a). Besides quality standardization, the phytocomplex profile can be applied, before its clinical use, to check the known associations of its components with biological molecular targets of pharmaco-toxicological interest, using the appropriate databases (see the paragraph below on this subject, as well as section "'Omics, Systems Biology and Their Applications to Pharmaco-Toxicology").

The use of 'omic techniques is particularly appropriate for analyzing the biological effects of herbal drugs, which elicit multiple simultaneous perturbations on primary and secondary molecular targets. 'Omic techniques allow examining simultaneous molecular effects and, with the help of bio-informatics, it is possible to look at such effects with a global view on the biological system affected. Analytical models can decode large quantity of raw data, allowing correlations of the multiple components of phytocomplexes with their biological targets and effects. This approach is even more relevant when using multi-herbal mixtures, as in TCM in China, Traditional Iranian Medicine in Iran, Kampo Medicine in Japan, Ayurveda, Siddha and Unani in India, or Jamu in Indonesia, where plants are used as blended herbal medicines and formulas are used that comprise mixtures of mixtures, with each herbal component supposed to exert its specific role, either as an effector, an enhancer, or a mitigator. As an example, in TCM Jun-Chen-Zuo-Shi and Qiqing (seven ways of pairing compatible herbs) are the basic theories guiding the combination of different herbal medicines in Fufang (TCM formulas), based on the properties and constituents of each herb. The Jun (emperor) is the principal phytocomplex targeting the major symptom of the disease. The Chen (minister) herbs synergize with Jun to strengthen its therapeutic effects, the Zuo (assistant) medicinal reduces or eliminates possible adverse or toxic effects of the Jun and/or Chen components, while the Shi (courier) herbs facilitate the adsorption, distribution, and delivery to the target of the active components (Zhao et al. 2015b). It is thus clear that the number of molecular components of pharmaco-toxicological interest can be very high in a single preparation, and identifying them and their multitudes of molecular direct or indirect targets with consistency was virtually unimaginable until the advent of 'omic techniques (Ding et al. 2015).

The complexity of the herbal molecular composition and of the biological perturbations to take into consideration are particularly complex in Ayurveda, where the number of herbal components can be much higher than in TCM in a single formulation, and where even nutritional components are often an integral part of the therapeutic prescription. In Ayurveda, individual Prakriti, a person's constitution type, is defined; this corresponds to a profile conceptually close to the genetic variability affecting susceptibility to disease and response to xenobiotics. This affects the way an individual will react to drugs or food and deeply affects the medical practice in Ayurveda, which evolved around the concept of maintaining a balance of three physiological entities, called Tri-doshas, namely, Vata, Pitta, and Kapha. Ayurveda merges foods (Pathya or Ahara) and drugs (Ausadha) into the concept of therapeutics, to maintain harmonization of the Doshas or physiological factors (Fauzi et al. 2013). Nutrition thus has a central role in Ayurvedic therapy in that food needs to be carefully chosen so it can interact with the Ayurvedic phytocomplex or with its targets, creating a context where pharmaco-toxicogenomics needs to merge with nutrigenomics (Banerjeea et al. 2015). Chemogenomics approaches to rationalizing the mode-of-action of TCM and Ayurvedic medicines may explain side effects, and it provides a resource for new molecular entities with possibly higher efficacy in the clinic than those identified by single-target biochemical assays (Fauzi et al. 2013).

The application of 'omics techniques thus seems inherently appropriate and even necessary for the assessment of efficacy and safety of herbal medicines, using a holistic strategy to evaluate the whole medicine in the context of multiple biological targets, rather than separating target specific responses to each specific component of the mixture (Pelkonen et al. 2012; Efferth and Koch 2011; Ouedraogo et al. 2012). 'Omic techniques, allowing a simultaneous observation of classes of molecules in a given system, are thus the main analytical drivers of a systems approach to herbal medicines. Accordingly, information-rich assays can be applied to the fine description of a complex formulation, a multi-target biological context as well as to the observation of their interactions and the subsequent perturbation of the biological equilibrium (Buriani et al. 2012). Treatments with phytocomplexes can be laid out, characterized and monitored with genomics, proteomics and metabonomics. While all the 'omic techniques are firmly pushing their way in pharmacognostical research and application, metabonomics seems to be gaining ground with respect to the others, as mentioned previously. This is probably due to the information-rich outcomes and to the simplicity of the assay, which allows direct and detailed analysis of large numbers of biological samples which, like urine, can be frequently and easily obtained. Metabonomics measures the metabolic profiling of the systems and provides the most holistic picture on the effects of drugs, but since the metabonomics, genomics, and proteomics analyze different molecules during drug treatment, they can complement each other and can be used as an integrated pharmaco-toxicological tool (Jiang et al. 2015). Combined with bioinformatics and statistical analysis, this can provide information on the altered molecular mechanisms, downstream effects, and targeted pathways.

Taken together, systems biology and 'omic techniques provide a unique opportunity to understand the complex biological perturbations induced by phytocomplexes, since several active ingredients in one prescription are aimed at numerous targets and work together to provide therapeutic benefits (Buriani et al. 2012; Lao et al. 2014). This new vision goes beyond single molecule pharmacology and target specificity, embraces the entire equilibrium of a biological system undergoing simultaneous perturbations, and can be considered elective to investigate the effects of multi-chemical mixtures on a plurality of biological targets, affording new opportunities to study biological effects holistically, rather than by following the classical reductionistic approach. The multitude of experimental data obtained need to be handled both in terms of multi-chemical identification and pathophysiological correlations.

Specific software and databases have been developed in order to exploit 'omics data and allow correlation of the multiple components of a given phytocomplex with its biological effects. This systems biology approach integrates powerful information-rich technologies, and computational tools and knowledge bases, making it possible to establish links between molecular patterns, biological functions, and a wide range of human diseases and pharmacological interventions. In particular, *in silico* tools are used to assist in molecular identification, and databases such as the METLIN and MassBank for metabolomics, are available to assist in this task. Other databases are used for molecular network building, pathway identification and associations, allowing data mining, integrated correlations, and modeling of biochemical pathways, thus providing an integrated platform for translational medicine (Barabàsi and Oltvai 2004). Among the many databases available, Recon2 represents the state-of-the-art human metabolic network reconstruction, with networks consisting of some 1789 enzyme-encoding genes, 7440 reactions, and 2626 unique metabolites distributed over eight cellular compartments (Kell and Goodacre 2014). The Human Metabolome Database (HMDB) is another frequently used medical bioinformatics and chemo-informatics database, with thousands of metabolite entries and tools for viewing, sorting metabolites and pathways, and customized, clickable metabolic maps, as well as information on disease (Kouskoumvekaki and Panagiotou 2011). Some databases are dedicated to specific fields like T3TB, considered the Toxic Exposome Database, providing descriptions, mechanisms of action, and information on toxins and toxin-targets (Wishart et al. 2015).

Many different databases and software dedicated to the pharmaco-toxicology of herbal drugs are also available (Barlow et al. 2012), and TCM drug formulation is probably one of the specific areas receiving the most attention. A special effort is being made for the construction of a TCM drugs-targets-diseases network aiming at the elucidation of the scientific base of TCM. Linking the multiple components that play principal, complementary and assistant therapeutic roles in TCM formulas to principal, complementary, and assistant targets in a disease network, heavily relies on 'omic platforms as well as computational tools. Among the many TCM databases, the "Connectivity Map" (CMAP), a database widely used for the identification of functional connections between drugs, genes and diseases, has been successfully used to examine the biological effects of Chinese formulas (Wen et al. 2011). Among others, the Chem-TCM Database, the World Traditional and Natural Medicine Patent Database, the Herbal Ingredient Target Database, and the TCM Information Database Alternative Medicine have been recently developed (Xu et al. 2012). Furthermore, the traditional Chinese medicine systems pharmacology database and analysis platform (TCMSP), freely available at http://sm.nwsuaf.edu.cn/lsp/tcmsp.php, was built based on the framework of systems pharmacology for herbal medicines. It consists of all the 499 Chinese herbs registered in the Chinese pharmacopoeia, with 29,384 ingredients, 3,311 targets, and 837 associated diseases. ADME-related properties can be evaluated. TCMSP also provides drug targets and diseases related to each active compound, showing the compound-target and target-disease networks, currently with more than 84,260 compound-target pairs and 2387 target-disease pairs. Last but not least, the TCMSP website is more than a data repository; it contains tools for visualization and analysis of TCM (Ru et al. 2014).

Another popular TCM information database is TCMGeneDIT, providing data collected from public databases and giving association information related to TCM and associated genes, while TCM database@Taiwan and TCMOnline are among the largest TCM herb databases (Gu and Chen 2014). There are other software platforms for network and pathway analysis (e.g., STRING, KEGG) to visualize the functional contexts of TCM, and recently, in an important development, the U.S. National Institutes of Health (NIH) launched a program called library of integrated network-based cellular signatures (LINCS) (Lao et al. 2014). Outside TCM there are other examples of databases and dedicated software that have been developed, such as the KNApSAcK Family database systems for metabolomics and related areas, which have been applied to Indonesian blended herbal medicines (Jamu) (Afendi et al. 2013). In summary, the large amount of experimental data generated by high-throughput techniques is available through various public repositories. Understanding transcriptional regulation, molecular interaction networks, metabolic pathways, and associations with disease conditions is rapidly expanding, thanks to the innovative systems biology approach, information-rich omics technology and dedicated databases, and this rapidly evolving context can provide a molecular characterization of the effects of phytocomplexes, thus supporting the use of herbal medicines in the clinical context.

Making Sense of the Phytocomplex Within Biological Networks, Systems, and Beyond

The efficiency and effectiveness shown by the bioinformatic applications of network pharmaco-toxicology has placed it at the forefront of current advances in systems biology. By integrating molecular knowledge and systems approaches as well as computational and experimental methods, network pharmacology and its holistic approach has revealed its potential to act as the next generation mode of drug research, with obvious advantages in all pharmaco-toxicological applications (Xu et al. 2012).

Characterized by holistic theory and rich experience in multicomponent therapeutics, TCM herbal formulas offer clear examples of a systems control of complex diseases, and indeed the introduction of network pharmaco-toxicology in TCM is proving its power in the identification of bioactive ingredients and endogenous/exogenous biomarkers, providing an increasing amount of data on mechanisms of action and exploring scientific evidence of numerous herbs and herbal formulas. In TCM network pharmacology, complex TCM drugs themselves can be envisioned as a molecular network interacting with "network targets," helping the body to regain its lost balance. Importantly, as mentioned in the preceding paragraph, some components of TCM drugs are not aimed at "network targets", but at other drug components, or their secondary targets, so as to alleviate side effects, improve activity of the principal drug component, improve absorption, and/or facilitate delivery of the principal drug to the targeted disease areas (Xu et al. 2012; Li and Zhang 2013; Gu and Chen 2014; Li et al. 2015d). Systems approaches have emerged that are specialized in investigating how herbs interact with the human body from a molecular level to the organism level – for instance, the integrated herbal medicine systems pharmacology (HmSP) platform. This network pharmacology methodology allows the systematization of current and traditional knowledge of herbal medicines, and has been applied to dissect basic TCM theories (e.g., yin-yang, qi-blood, herbal synergy) and to the development of new drugs (Ru et al. 2014; Li et al. 2015c). While leading the way in reinterpreting TCM, network pharmacology shows how this same paradigm should be applied to pharmaco-toxicology as a whole and that all drug activities should be reinterpreted in a more holistic fashion. Many diseases, especially chronic ones, are initiated and perpetrated via dysregulation of multiple pathways, even when the primary trigger is the mutation in a central gene associated with an endogenous or exogenous insult. The application of network analysis on human diseases has made it increasingly clear that chronic diseases are associated with changes in the expression of a large number of genes, proteins, and metabolites, involving a large number of modules or functional units, while at the same time showing considerable overlaps of important genes and network modules (Barabàsi and Oltvai 2004). The implications of this complexity are that the idea of single-targeted drugs may be completely inadequate to remedy a complex dysregulation, and the efficiency of any drug could be highly dependent on importance (centrality) of the target ("node" or "edge") in the disease network.

The One Drug, One Target Paradigm Versus Polypharmacology

Even if the current paradigm has been "one target (or disease/symptom) – one drug," pharmacologists and toxicologists have always known that practically all drugs have multiple effects – some desirable, and others apparently indifferent or harmful. The "one target, one drug" paradigm created the illusion of a "magic bullet" that was eagerly adopted, although some scientists pointed out that even such "magic bullets" have pharmacokinetics-associated problems (Xu et al. 2012). Now it is becoming increasingly apparent that biological systems are complex, redundant, homeostatic and resilient to perturbations, and most diseases exhibit much wider perturbations and variations than once thought. A new discipline, loosely termed "polypharmacology," has been gaining ground both conceptually and experimentally. Current drug-protein interactions and chemogenomic studies have shown that many drugs interact with two or more targets with reasonably close affinities. Databases of FDA-approved drugs and their targets (and effects) have been used to create networks of multiple drug-target interactions, obtaining a polypharmacological network that is mapped within the biological network to reveal multiple actions of drugs on multiple targets and multiple diseases (Xu et al. 2012). Many of the *in silico* findings have later been actually validated experimentally, contributing to the notion that network approaches can be effectively used to identify and characterize on-target and off-target toxicity of pharmaceuticals, including phytocomplexes (Li et al. 2015c; Li and Zhang 2013; Gu and Chen 2014). Using this polypharmacological approach, systems pharmacology frameworks on computational models have been developed to predict drug combinations, such as the probability ensemble approach (PEA), to analyze both the efficacy and adverse effects of drug combinations. PEA can also quantitatively evaluate the efficacy and toxicity and detect the therapeutic indications for drug combinations (Huang et al. 2014).

Systems biology is thus quickly spreading within the biomedical community, but it is probably running even faster in the TCM environment where, thanks to its holistic vision, it is considered to be the most appropriate approach to explain – with scientifically sound evidence – the millennia old Chinese medical tradition. In a relatively short time, systems pharmaco-toxicology has gained a prominent position at the forefront of TCM research (Li et al. 2015b, d; Ma et al. 2015; Wang et al. 2015; Zhao et al. 2015a; Zheng et al. 2015).

An Example: Ma Huang

Taking as an example a TCM formula, Ma Huang decoction (also known as Ephedra Decoction, MHD), a systems pharmacology method integrating pharmacokinetic analysis, drug targeting, and drug-target-disease network, was developed by Yao et al. (2013). MHD is an ancient TCM formula described in the Treatise on Cold Pathogenic Diseases during the Chinese Eastern Han Dynasty. It is composed of four herbal drugs: Herba Ephedrae (Ma-huang), Ramulus Cinnamomi (Gui-zhi), Semen Armeniacae Amarum (Xing-ren), and Radix Glycyrrhizae (Gan-cao). The aim of the study was to validate the supposed herbal rules that had guided the composition of the TCM formula. Six main steps were taken; the first involved the literature and online databases' search for the identification of the chemicals present in all four herbs, in order to build a dedicated molecular dataset. Differences and similarities between herbs were then investigated, based on their molecular composition. Drug-likeness was evaluated, oral bioavailability screening was carried out, and a thorough evaluation of predicted interactions with ADME key proteins was completed. The identification of bioavailable compound-target interactions was then carried out with an in-house comprehensive *in silico* model, integrating a large scale of chemical, genomic, and pharmacological data. In the last step, Drug-Target-Disease Network construction and analysis was finally performed. The resulting network confirmed the cooperation of the four herbs towards the pharmacological effect. In particular, the various supposed roles of each herb in the prescription formula were highlighted, with Herba Ephedrae clearly emerging as the principal effector, and the others cooperating, either by synergy or by enhancing the bioavailability of the active components (Yao et al. 2013) (see Fig. 7.2).

Complex Herbal Medicines: Building a Systems Pharmacology/Toxicology Approach

In another study, the addition and subtraction theory (AST) of TCM was addressed with a systems biology approach (Li et al. 2014). According to AST, which plays a core role in individualized TCM medicine, adding or removing one or more herbal medicines, or changing the dosage from an original "foundational formula" can be used to modify a formula into an improved one, according to the patient's specific needs. Two classical TCM prescriptions were used; the first was a Xiao Chaihu decoction (XCHD), composed of Radix bupleuri (Chaihu), Radix scutellariae (Huangqin), Rhizoma pinelliae (Banxia), Rhizoma zingiberis recens (Shengjiang), Fructus jujubae (Dazao), licorice (Gancao) and Panax ginseng (Renshen). The second prescription was Da Chaihu decoction (DCHD), which is derived from XCHD, but with the elimination of P. ginseng and licorice, and the addition of Fructus aurantii immaturus (Zhishi) and Paeonia lactiflora (Shaoyao). The five common herbs of the two prescriptions are the "foundational formula," while the other two herbs are "additive herbs." Following the initial identification of all the ingredients

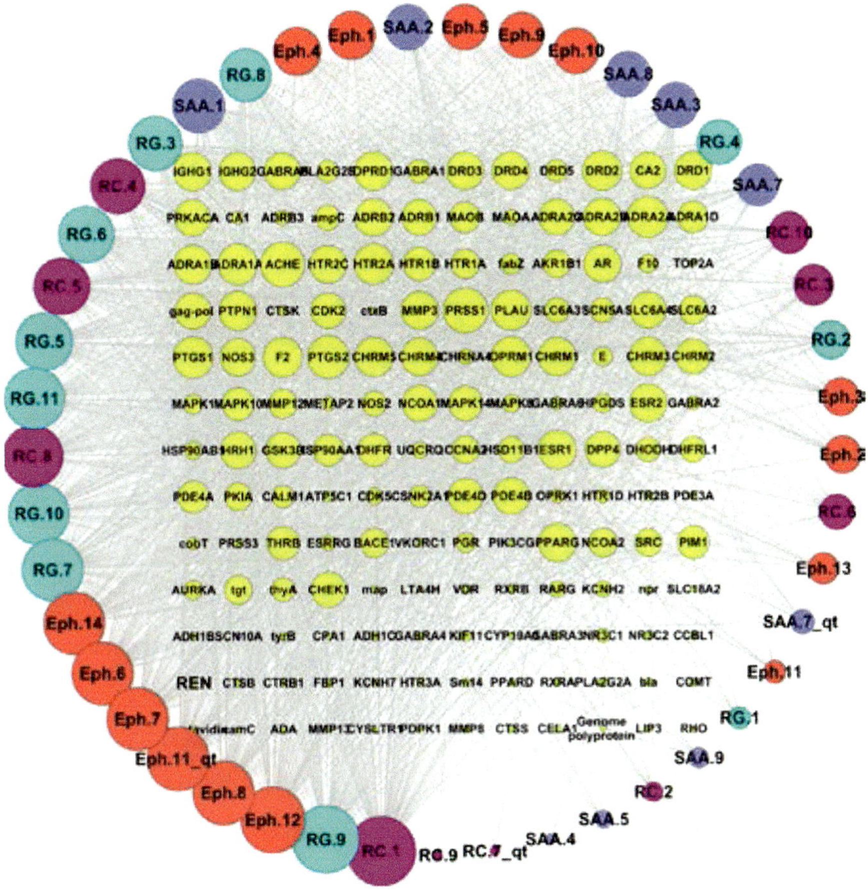

Fig. 7.2 Example of drug-target and target-disease networks for a TCM formula. A systems pharmacology method integrating pharmacokinetic analysis, drug targeting, and drug-target-disease network is developed to dissect the rule embedded in the representative TCM herbal formula Ma-huang decoction, made up of four herbs (*Eph* Herba Ephedrae, *RC* Ramulus Cinnamomi, *SAA* Semen Armeniacae Amarum, *RG* Radix Glycyrrhizae). Above graph of Drug–Target Network. Eph (*red*), RC (magenta), SAA (*purple*), RG (*cyan*) and targets (*yellow*). Candidate herbal components are connected to their related targets, and the network analysis method is used for analysing properties of the network. The cD–T Network consists of 92 nodes and 1,049 edges, with 36 candidate herbal components (drugs) and 56 potential targets. The area of the node is proportional to the degree of interaction. The *yellow circles* are the common molecular targets of all four herbs, illustrating the synergistic effect in the mixture. Below: graph of Target-Disease Network. The T-D Network linking potential molecular targets and diseases is constructed for exploring the protein interactions and the therapeutic targets for diseases. It has been proven that different diseases might share common symptoms, and thus potentially be cured by the same formula; in other words, one formula might be used to treat multiple diseases. Target proteins (*circle*, *jasper*) are connected to various kinds of diseases (*square*, *purple*), 35 of which have at least one link to other diseases (Figures are reproduced with the kind permission of Yao et al. (2013). Please refer to original article for a more detailed explanation)

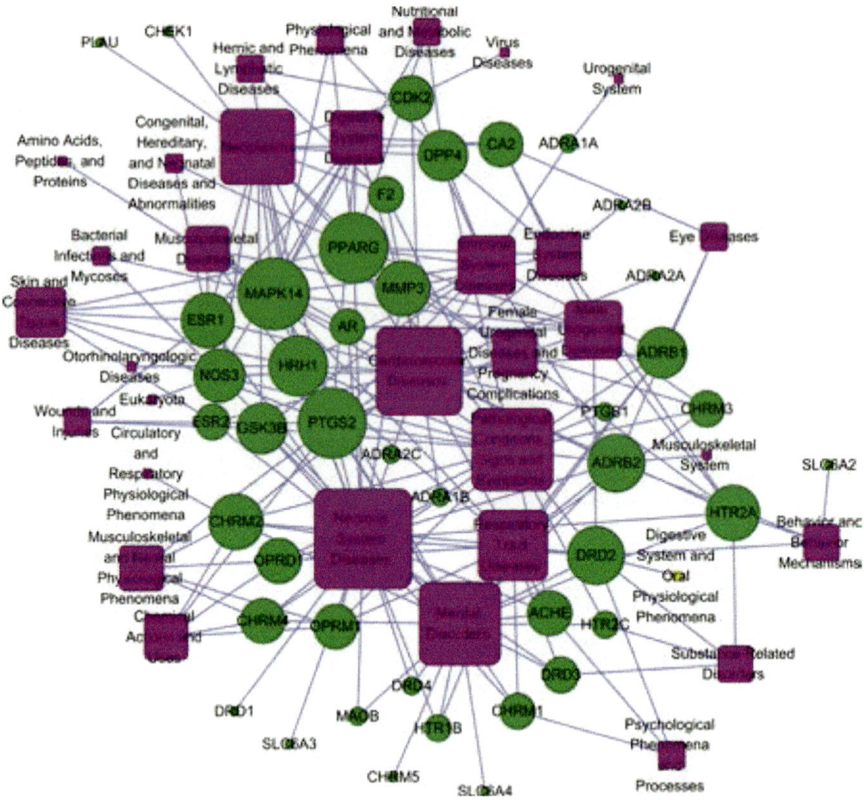

Fig. 7.2 (continued)

from the herbs, a database was built. Oral bioavailability and drug half-life were then evaluated and, finally, the potential targets were predicted. Pharmacological data were integrated into drug-target and target-disease networks to outline the mechanisms involved in the herb functions and to verify the addition and subtraction theory. Using the network approach, key molecular targets and interactions involved in XCHD and DCHD were indeed identified. In particular, the results showed that the "fundamental formula" was responsible for the major therapeutic effects, whereas the "additive herbs" were shown to synergistically enhance the treatment outcomes by targeting the same or complementary proteins (Li et al. 2014) (see Fig. 7.3).

In other examples TCM specific holistic pathophysiologic concepts as well as their treatment, were addressed with the same systems approach, by building specific drug-target-disease networks. Herbal formulas described in TCM to be used to "eliminate blood stasis" were shown to possess blood vessels' dilating activity, thus improving the microcirculation, while "qi-enhancing" herbs were shown to enhance energy metabolism (Zhou and Wang 2013). Single TCM herbs have also been characterized using systems pharmacology, providing rather complete networks for each indication of a given herbal drug, networks that de facto have become an integral part of a drug's portfolio. As an example the anti-inflammatory Folium Eriobotryae, the dried leaves of Eriobotrya japonica (Thunb.) Lindl., were characterized using drug-target-disease pathways focusing on inflammation. Briefly, 11 compounds of this herb were identified with favorable pharmacokinetic properties and were predicted to have anti-inflammatory activity. Their targets were then iden-

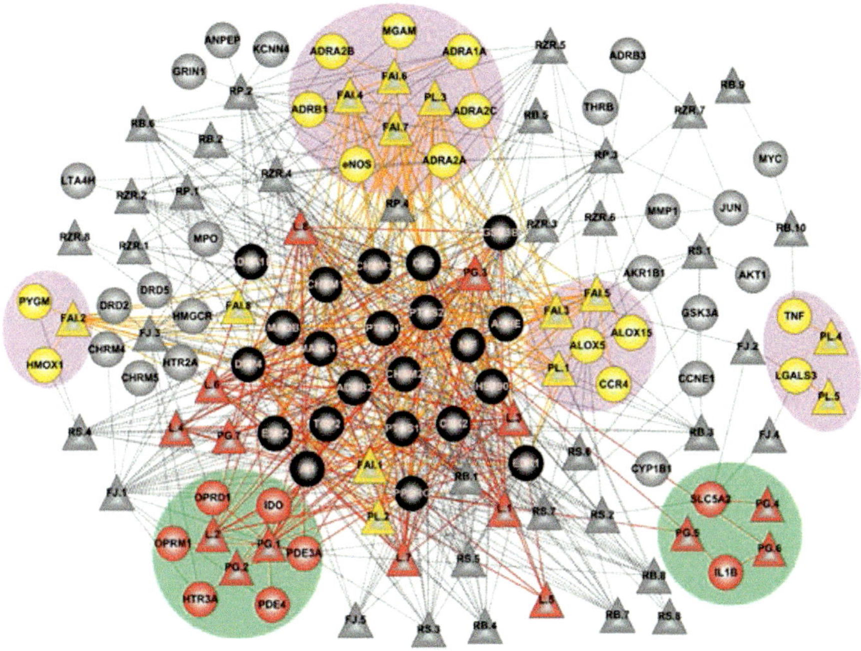

Fig. 7.3 Example of application of systems biology approach to the TCM formulas XCHD and DCHD. See text for explanation. (*RB* Radix bupleuri, *RS* Radix scutellariae, *RP* Rhizoma pinelliae, *RZR* Rhizoma zingiberis recens, *FJ* Fructus jujubae, *L* Licorice, *PG* Panax ginseng, *FAI* Fructus aurantii immaturus, *PL* Paeonia lactiflora). Above: Drug–target network. Drug–target interactions are depicted as connecting lines between drugs (*compounds*, *triangles*) and targets (*circles*). The black nodes (*circles*) represent targets shared by all herbs. Drugs belonging to the L and PG are indicated in *red* (highlighted in *green* background), and their corresponding targets and the linked lines are also indicated in the same color. Drugs belonging to PL and FAI are indicated in *yellow* (highlighted in *purple* background), and their linked targets and the corresponding lines are also in *yellow*. Drugs in the fundamental herbs (see text) and corresponding targets are colored in gray, and linked by *gray dashed lines*. Below Target-disease network. Targets (*circles*) and diseases (*squares*). Where *red* and *yellow circles* (highlighted in *green* and *purple* background) are proteins strengthened by XCHD additive herbs and DCHD additive herbs, respectively. And *gray circles* are proteins targeted by the fundamental herbs. The *red squares* indicate the diseases targeted by XCHD additive herbs and/or fundamental herbs. The *squares* in *yellow squares* indicate the diseases targeted by DCHD additive herbs and/or fundamental herbs. The shared diseases display a split color code (Figures are reproduced with copyright permission from Li et al. (2014). The figure is given just as an example of systems biology application to TCM, for detailed explanation of specific acronisms and codes of single compounds and targets, please refer to the original article)

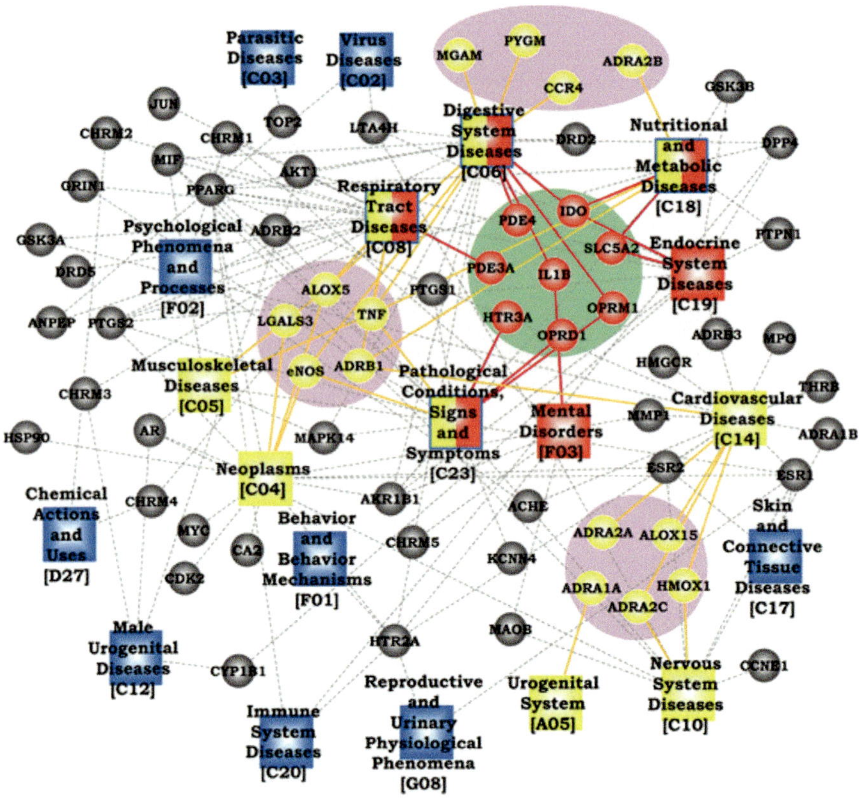

Fig. 7.3 (continued)

tified in 43 inflammation-associated proteins – some of them particularly important, like COX2, ALOX5, PPARG, TNF and RELA, implicating the MAPK signalling pathway, the rheumatoid arthritis pathway, and NF-kB signaling pathway, all meaningful pathways for an anti-inflammatory activity (Zhang et al. 2015b).

Indian Ayurveda provides further examples of how traditionally used phytocomplexes can be characterized by using a network pharmaco-toxicology approach. In Ayurveda, it is actually believed that the synergistic effect of combined extracts of many plants is more beneficial than the extract of a single plant, so combined extracts of plants are normally used. An example of an interesting network approach application was recently provided (Fayaz et al. 2014), focusing on anti-diabetic phytochemicals traditionally used in Ayurveda. Initially, a rigorous step-by-step network analysis was performed to establish: the antidiabetic plants-active compounds network, the active compounds-protein targets network, the protein targets-pathways network, and the pathways-disease network. Then a series of observations were made and a model was extrapolated with a series of predictive general criteria, to estimate the effectiveness of antidiabetic combinations of chemicals and herbs that could be used as a strategy for preparing new effective anti-diabetic formulations (Fayaz et al. 2014). Steps similar to those described in the examples above could be used to standardize a methodology for setting up therapeutic protocols for using phytocomplexes in the clinic, in order to facilitate their therapeutic use in the routine.

Conclusions

The above analysis shows clearly that "'omic" techniques and systems biology are increasingly applied in herbal medicines at various levels and promise to transform the scientific framework to be exploited in research and clinical use. The use of genomic technique applications in classification and quality control of herbal medicines has already advanced to a considerable extent, and metabolomic promises to become a spearheading approach in the methodological repertoire of herbal medicinal investigations.

Network pharmaco-toxicology represents one of the most important applications of the new approaches described in this chapter. Building up networks of molecular interactions between phytocomplex components and pharmaco-toxicological events can provide a powerful predictive tool in herbal medicine. There is an increasing number of Web-based systems biology platforms, continuously fed with "'omics" data, providing a view of the complete biological system modulated by a given medicine. Systems toxicology may become the best conceptual framework to provide mechanistic insights into herbal-associated adverse effects, thus allowing the prediction of responses to phytochemicals.

In summary. the large amount of experimental data generated by high-throughput techniques is available through various public repositories. Information about transcriptional regulation, molecular interaction networks, metabolic pathways, and associations with disease conditions is rapidly expanding, thus providing the right platform for the molecular characterization of the effects of phytocomplexes. Systems biology is spreading fast in the biomedical community, but probably even faster in the holistic medicine community where it is felt to be the most appropriate approach to explain millennia-old medical traditions with scientifically sound evidence. Positive results are coming in with increasing frequency, especially in TCM research, and in a relatively short time, systems pharmaco-toxicology has gained a prominent position at the forefront of herbal research.

Acknowledgements We would like to thank Roberta Sato at the Library of the Department of Pharmaceutical and Pharmacological Sciences of the University of Padova for her technical assistance with database search and bibliography, and Mariagnese Barbera for text revision.

Conflict of Interest Alessandro Buriani and Stefano Fortinguerra are in charge of the Personalized Medicine service of the Gruppo Data Medica Padova.

References

Afendi FM, Ono N, Nakamura Y et al (2013) Data mining methods for omics and knowledge of crude medicinal plants toward big data biology. Comput Struct Biotechnol J 4:e201301010. doi:10.5936/csbj.201301010

Ahn AC, Tewari M, Poon CS et al (2006) The clinical applications of a systems approach. PLoS Med 3:956–960

Altelaar AF, Munoz J, Heck AJ (2013) Next-generation proteomics: towards an integrative view of proteome dynamics. Nat Rev Genet 14:35–48. doi:10.1038/nrg3356

Bai JP, Fontana RJ, Price ND et al (2014) Systems pharmacology modeling: an approach to improving drug safety. Biopharm Drug Dispos 35:1–14. doi:10.1002/bdd.1871

Ball MP, Thakuria JV, Zaranek AW et al (2012) A public resource facilitating clinical use of genomes. Proc Natl Acad Sci USA 109:11920–11927. doi:10.1073/pnas.1201904109

Ball MP, Bobe JR, Chou MF et al (2014) Harvard personal genome project: lessons from participatory public research. Genome Med 6:10. doi:10.1186/gm527

Banerjeea S, Debnathb P, Kumar Debnathc P (2015) Ayurnutrigenomics: Ayurveda-inspired personalized nutrition from inception to evidence. J Tradit Complement Med. doi:10.1016/j.jtcme.2014.12.009

Barabàsi AL, Oltvai ZN (2004) Network biology: understanding the cell's functional organization. Nat Rev Genet 5:101–113. doi:10.1038/nrg1272

Barabàsi AL, Gulbahce N, Loscalzo J (2011) Network medicine: a network-based approach to human disease. Nat Rev Genet 12:56–68. doi:10.1038/nrg2918

Barlow DJ, Buriani A, Ehrman T et al (2012) In-silico studies in Chinese herbal medicines' research: evaluation of in-silico methodologies and phytochemical data sources, and a review of research to date. J Ethnopharmacol 140:526–534. doi:10.1016/j.jep.2012.01.041

Bouhifd M, Hartung T, Hogberg HT et al (2013) Toxicometabolomics. J Appl Toxicol 33:1365–1383. doi:10.1002/jat.2874

Bouhifd M, Hogberg HT, Kleensang A et al (2014) Mapping the human toxome by systems toxicology. Basic Clin Pharmacol Toxicol 115:24–31. doi:10.1111/bcpt.12198

Bredel M, Jacoby E (2004) Chemogenomics: an emerging strategy for rapid target and drug discovery. Nat Rev Genet 5:262–275. doi:10.1038/nrg1317

Brown JB, Okuno Y, Marcou G et al (2014) Computational chemogenomics: is it more than inductive transfer? J Comput Aided Mol Des 28:597–618. doi:10.1007/s10822-014-9743-1

Buriani A, Garcia-Bermejo ML, Bosisio E et al (2012) Omic techniques in systems biology approaches to traditional Chinese medicine research: present and future. J Ethnopharmacol 140:535–544. doi:10.1016/j.jep.2012.01.055

Butcher EC, Berg EL, Kunkel EJ (2004) Systems biology in drug discovery. Nat Biotechnol 22:1253–1259

Cao DS, Liang YZ, Deng Z et al (2013) Genome-scale screening of drug-target associations relevant to Ki using a chemogenomics approach. PLoS One 8:e57680. doi:10.1371/journal.pone.0057680

Carrascosa MC, Massaguer OL, Mestres J (2012) PharmaTrek: a semantic Web explorer for open innovation in multitarget drug discovery. Mol Inform 31:537–541. doi:10.1002/minf.201200070

Chen YC, Douville C, Wang C et al (2014) A probabilistic model to predict clinical phenotypic traits from genome sequencing. PLoS Comput Biol 10:e1003825. doi:10.1371/journal.pcbi.1003825

Chen S, Song J, Sun C et al (2015a) Herbal genomics: examining the biology of traditional medicines. Science 347:S27–S29

Chen T, Zhao J, Ma J et al (2015b) Web resources for mass spectrometry-based proteomics. Genomics Proteomics Bioinforma 13:36–39. doi:10.1016/j.gpb.2015.01.004

Chupakhin V, Marcou G, Gaspar H et al (2014) Simple ligand-receptor interaction descriptor (SILIRID) for alignment-free binding site comparison. Comput Struct Biotechnol J 10:33–37. doi:10.1016/j.csbj.2014.05.004

Clayton TA, Lindon JC, Cloarec O et al (2006) Pharmaco-metabonomic phenotyping and personalized drug treatment. Nature 440:1073–1077
Crosetto N, Bienko M, van Oudenaarden A (2015) Spatially resolved transcriptomics and beyond. Nat Rev Genet 16:57–66. doi:10.1038/nrg3832
Csermely P, Korcsmáros T, Kiss HJ (2013) Structure and dynamics of molecular networks: a novel paradigm of drug discovery a comprehensive review. Pharmacol Ther 138:333–408. doi:10.1016/j.pharmthera.2013.01.016
D'Argenio V, Salvatore F (2015) The role of the gut microbiome in the healthy adult status. Clin Chim Acta. doi:10.1016/j.cca.2015.01.003
Deloukas P, Kanoni S, Willenborg C et al (2013) Large-scale association analysis identifies new risk loci for coronary artery disease. Nat Genet 45:25–33. doi:10.1038/ng.2480
Dewey FE, Grove ME, Pan C et al (2014) Clinical interpretation and implications of whole-genome sequencing. JAMA 311:1035–1045. doi:10.1001/jama.2014.1717
Ding F, Zhang Q, Ung COL et al (2015) An analysis of chemical ingredients network of Chinese herbal formulae for the treatment of coronary heart disease. PLoS One 10:e0116441. doi:10.1371/journal.pone.0116441
Drabovicha AP, Martínez-Morilloa E, Diamandisa EP (2015) Toward an integrated pipeline for protein biomarker development. Biochim Biophys Acta 1854:677–686. doi:10.1016/j.bbapap.2014.09.006
Duerr RH, Taylor KD, Brant SR et al (2006) Genome-wide association study identifies IL23R as an inflammatory bowel disease gene. Science 314:1461–1463. doi:10.1126/science.1135245
Efferth T, Koch E (2011) Complex interactions between phytochemicals. The multi-target therapeutic concept of phytotherapy. Curr Drug Targets 12:122–123. doi:10.2174/138945011793591626
Ekins S, Nikolsky Y, Nikolskaya T (2005) Techniques: application of systems biology to absorption, distribution, metabolism, excretion, and toxicity. Trends Pharmacol Sci 26:202–209
Elrakaiby M, Dutilh BE, Rizkallah MR et al (2014) Pharmacomicrobiomics: the impact of human microbiome variations on systems pharmacology and personalized therapeutics. OMICS 18:402–414. doi:10.1089/omi.2014.0018
Fauzi FM, Koutsoukas A, Lowe R et al (2013) Chemogenomics approach to razionalizing the mode-of-action of traditional Chinese and Ayurvedic medicines. J Chem Inf Model 53:661–673. doi:10.1021/ci3005513
Fayaz SM, Suvanish Kumar VS, Rajanikant KG (2014) Finding needles in a haystack: application of network analysis and target enrichment studies for the identification of potential anti-diabetic phytochemicals. PLoS One 9:e112911. doi:10.1371/journal.pone.0112911
Franke A, McGoven DPB, Barrett JC et al (2010) Genome-wide meta-analysis increases to 71 the number of confirmed Crohn's disease susceptibility loci. Nat Genet 42:1118–1125. doi:10.1038/ng.717
Gao Y, Holland RD, Yu LR (2009) Quantitative proteomics for drug toxicity. Brief Func Genom Proteomic 8:158–166. doi:10.1093/bfgp/elp006
Ghosh S, Matsuoka Y, Asai Y et al (2011) Software for systems biology: from tools to integrated platforms. Nat Rev Genet 12:821–832. doi:10.1038/nrg3096
Gika HG, Wilson ID, Theodoridis GA (2014) LC-MS-based holistic metabolic profiling. Problems, limitations, advantages, and future perspectives. J Chromatogr B Anal Technol Biomed Life Sci 966:1–6. doi:10.1016/j.jchromb.2014.01.054
Glusman G, Severson A, Dhankani V et al (2015) Identification of copy number variants in whole-genome data using reference coverage profiles. Front Genet 6:45. doi:10.3389/fgene.2015.00045
Gong J, Cai C, Liu X et al (2013) ChemMapper: a versatile web server for exploring pharmacology and chemical structure association based on molecular 3D similarity method. Bioinformatics 29:1827–1829. doi:10.1093/bioinformatics/btt270
Gonzalez-Garay ML, McGuire AL, Pereira S et al (2013) Personalized genomic disease risk of volunteers. Proc Natl Acad Sci USA 110:16957–16962. doi:10.1073/pnas.1315934110
Gu P, Chen H (2014) Modern bioinformatics meets traditional Chinese medicine. Brief Bioinform 15:984–1003. doi:10.1093/bib/bbt063

Guo L, Du Y, Chang S et al (2014) rSNPBase: a database for curated regulatory SNPs. Nucleic Acids Res 42(Database issue):D1033–D1039. doi:10.1093/nar/gkt1167

Guo DA, Wu WY, Ye M et al (2015a) A holistic approach to the quality control of traditional Chinese medicines. Science 347:S29–S31

Guo Z, Zhang Y, Zou L et al (2015b) A proteomic analysis of individual and gender variations in normal human urine and cerebrospinal fluid using iTRAQ quantification. PLoS One 10:e0133270. doi:10.1371/journal.pone.0133270

Gupta-Ostermann D, Bajorath J (2014) The 'SAR Matrix' method and its extensions for applications in medicinal chemistry and chemogenomics. Version 2 F1000Res 3:113. doi:10.12688/f1000research.4185.2

Haiman CA, Chen GK, Vachon CM et al (2011) A common variant at the TERT-CLPTM1L locus is associated with estrogen receptor-negative breast cancer. Nat Genet 43:1210–1214. doi:10.1038/ng.985

Hamon J, Jennings P, Bois FY (2014) Systems biology modeling of omics data: effect of cyclosporine a on the Nrf2 pathway in human renal cells. BMC Syst Biol 8:76. doi:10.1186/1752-0509-8-76

Harvey AL, Edrada-Ebel R, Quinn RJ (2015) The re-emergence of natural products for drug discovery in the genomics era. Nat Rev Drug Discov 14:111–129. doi:10.1038/nrd4510

Heather JM, Chain B (2015) The sequence of sequencers: the history of sequencing DNA. Genomics 10:1–8. doi:10.1016/j.ygeno.2015.11.003

Hood L, Heath JR, Phelps ME et al (2004) Systems biology and new technologies enable predictive and preventative medicine. Science 306:640–643

Howell BA, Yang Y, Kumar R et al (2012) In vitro to in vivo extrapolation and species response comparisons for drug-induced liver injury (DILI) using DILIsym: a mechanistic, mathematical model of DILI. J Pharmacokinet Pharmacodyn 39:527–541. doi:10.1007/s10928-012-9266-0

Huang C, Zheng C, Li Y et al (2014) Systems pharmacology in drug discovery and therapeutic insight for herbal medicines. Brief Bioinform 15:710–733. doi:10.1093/bib/bbt035

Ideker T, Galitsky T, Hood L (2001) A new approach to decoding life: systems biology. Annu Rev Genomics Hum Genet 2:343–372

Ilatovskiy AV, Abagyan R, Kufareva I (2013) Quantum mechanics approaches to drug research in the era of structural chemogenomics. Int J Quantum Chem 113:1669–1675. doi:10.1002/qua.24400

James LP (2013) Metabolomics: integration of a new "Omics" with clinical pharmacology. Clin Pharmacol Ther 94:547–551. doi:10.1038/clpt.2013.166

Ji B, Nielsen J (2015) New insight into the gut microbiome through metagenomics. Adv Genomics Genet 2015–5:77–91. doi:10.2147/AGG.S57215

Jia WH, Zhang B, Matsuo K et al (2013) Genome-wide association analysis in East Asians identify new susceptibility loci for colorectal cancer. Nat Genet 45:191–196. doi:10.1038/ng.2505

Jiang X, Lv B, Li P et al (2015) Bioactivity-integrated UPLC/Q-TOF-MS of Danhong injection to identify NF-kB inhibitors and anti-inflammatory targets based on endothelial cell culture and network pharmacology. J Ethnopharmacol 174:270–276. doi:10.1016/j.jep.2015.08.026

Jirtle RL, Skinner MK (2007) Environmental epigenomics and disease susceptibility. Nat Rev Genet 8:253–262

Joyce AR, Pallson BØ (2006) The model organism as a system: integrating "omics" data sets. Nat Rev Mol Cell Biol 7:198–210

Kell DB, Goodacre R (2014) Metabolomics and systems pharmacology: why and how to model the human metabolic network for drug discovery. Drug Discov Today 19:171–182. doi:10.1016/j.drudis.2013.07.014

Kinoshita E, Kinoshita-Kikuta E, Koike T (2015) Advances in Phos-tag-based methodologies for separation and detection of the phosphoproteome. Biochim Biophys Acta 1854:601–608. doi:10.1016/j.bbapap.2014.10.004

Kitano H (2002) Computational systems biology. Nature 420:206–210. doi:10.1038/nature1254

Klabunde T (2007) Chemogenomic approaches to drug discovery: similar receptors bind similar ligands. Br J Pharmacol 152:5–7. doi:10.1038/sj.bjp.0707308

Kleensang A, Maertens A, Rosenberg M et al (2014) t4 workshop report: pathways of toxicity. ALTEX 31:53–61. doi:10.14573/altex.1309261
Knudsen TB, Keller DA, Sander M et al (2015) FutureTox II: in vitro data and in silico models for predictive toxicology. Toxicol Sci 143:256–267. doi:10.1093/toxsci/kfu234
Kongsbak K, Hadrup N, Audouze K et al (2014) Applicability of computational systems biology in toxicology. Basic Clin Pharmacol Toxicol 115:45–49. doi:10.1111/bcpt.12216
Kouskoumvekaki I, Panagiotou G (2011) Navigating the human metabolome for biomarker identification and design of pharmaceutical molecules. J Biomed Biotechnol. doi:10.1155/2011/525497
Kwok CK, Tang Y, Assmann SM et al (2015) The RNA structurome: transcriptome-wide structure probing with next-generation sequencing. Trends Biochem Sci 40:221–232. doi:10.1016/j.tibs.2015.02.005
Lan Q, Hsiung CA, Matsuo K et al (2012) Genome-wide association analysis identifies new lung cancer susceptibility loci in never-smoking women in Asia. Nat Genet 44:1330–1335. doi:10.1038/ng.2456
Lao Y, Wang X, Xu N et al (2014) Application of proteomics to determine the mechanism of action of traditional Chinese medicine. remedies. J Ethnopharmacol 155:1–8. doi:10.1016/j.jep.2014.05.022
Lavecchia A, Di Giovanni C (2013) Virtual screening strategies in drug discovery: a critical review. Curr Med Chem 20:2839–2860. doi:10.2174/09298673113209990001
Li S, Zhang B (2013) Traditional Chinese medicine network pharmacology: theory, methodology and application. Chin J Nat Med 11:110–120. doi:10.1016/S1875-5364(13)60037-0
Li F, Lu J, Ma X (2011) Profiling the reactive metabolites of xenobiotics using metabolomic technologies. Chem Res Toxicol 24:744–751. doi:10.1021/tx200033v
Li B, Tao W, Zheng C et al (2014) Systems pharmacology-based approach for dissecting the addition and subtraction theory of traditional Chinese medicine: an example using Xiao-Chaihu-Decoction and Da-Chaihu-Decoction. Comput Biol Med 53:19–29. doi:10.1016/j.compbiomed.2014.05.007
Li B, Zhang Y, Yu Y et al (2015a) Quantitative assessment of gene expression network module-validation methods. Sci Rep 5:15258. doi:10.1038/srep15258
Li J, Zhao P, Li Y et al (2015b) Systems pharmacology-based dissection of mechanisms of Chinese medicinal formula Bufei Yishen as an effective treatment for chronic obstructive pulmonary disease. Sci Rep 5:15290. doi:10.1038/srep15290
Li P, Huang C, Fu Y et al (2015c) Large-scale exploration and analysis of drug combinations. Bioinformatics. doi:10.1093/bioinformatics/btv080
Li Y, Zhang J, Zhang L et al (2015d) Systems pharmacology to decipher the combinational anti-migraine effects of Tianshu formula. J Ethnopharmacol 174:45–56. doi:10.1016/j.jep.2015.07.043
Liu ET (2009) Integrative biology – a strategy for systems biomedicine. Nat Rev Genet 10:64–68
Lord PG, Nie A, McMillian M (2006) Application of genomics in preclinical drug safety evaluation. Basic Clin Pharmacol Toxicol 98:537–546
Ma YM, Zhang XZ, Su ZZ et al (2015) Insight into the molecular mechanism of a herbal injection by integrating network pharmacology and in vitro. J Ethnopharmacol 173:91–99. doi:10.1016/j.jep.2015.07.016
Ma'ayan A (2011) Introduction to network analysis in systems biology. Sci Signal 4:tr5. doi:10.1126/scisignal.2001965
Manolio TA, Collins FS (2009) The HapMap and genome-wide association studies in diagnosis and therapy. Annu Rev Med 60:443–456. doi:10.1146/annurev.med.60.061907.093117
Manyam G, Birerdinc A, Baranova A (2015) KPP: KEGG pathway painter. BMC Syst Biol 9:S3. doi:10.1186/1752-0509-9-S2-S3
Martínez V, Navarro C, Cano C et al (2014) DrugNet: network-based drug-disease prioritization by integrating heterogeneous data. Artif Intell Med 63:41–49. doi:10.1016/j.artmed.2014.11.003
Medina-Franco JL, Giulianotti MA, Welmaker GS et al (2013) Shifting from the single- to the multitarget paradigm in drug discovery. Drug Discov Today 18:495–501. doi:10.1016/j.drudis.2013.01.008

Medina-Franco JL, Méndez-Lucio O, Martinez-Mayorga K (2014) The interplay between molecular modeling and chemoinformatics to characterize protein-ligand and protein-protein interactions landscapes for drug discovery. Adv Protein Chem Struct Biol 96:1–37. doi:10.1016/bs.apcsb.2014.06.001

Miller GW, Jones DP (2014) The nature of nurture: refining the definition of the exposome. Toxicol Sci 137:1–2. doi:10.1093/toxsci/kft251

Mitra K, Carvunis AR, Ramesh SK et al (2013) Integrative approaches for finding modular structure in biological networks. Nat Rev Genet 14:719–732. doi:10.1038/nrg3552

Morris MK, Chi A, Melas IN et al (2014) Phosphoproteomics in drug discovery. Drug Discov Today 19:425–432. doi:10.1016/j.drudis.2013.10.010

Nantasenamat C, Prachayasittikul V (2015) Maximizing computational tools for successful drug discovery. Expert Opin Drug Discov 10:321–329. doi:10.1517/17460441.2015.1016497

Nicholson JK, Lindon JC (2008) Systems biology: metabonomics. Nature 455:1054–1056. doi:10.1038/4551054a

Nicholson JK, Wilson ID (2003) Understanding "global" systems biology: metabonomics and the continuum of metabolism. Nat Rev Drug Discov 2:668–676

Ouedraogo M1, Baudoux T, Stévigny C et al (2012) Review of current and "omics" methods for assessing the toxicity (genotoxicity, teratogenicity and nephrotoxicity) of herbal medicines and mushrooms. J Ethnopharmacol 140:492–512. doi:10.1016/j.jep.2012.01.059

Palotie A, Widén E, Ripatti S (2013) From genetic discovery to future personalized health research. New Biotechnol 30:291–295

Paricharak S, Cortés-Ciriano I, IJzerman AP et al (2015) Proteochemometric modelling coupled to in silico target prediction: an integrated approach for the simultaneous prediction of polypharmacology and binding affinity/potency of small molecules. J Cheminform 15:7–15. doi:10.1186/s13321-015-0063-9

Pelkonen O (2010) Predictive toxicity: grand challenges. Front Pharmacol 1:3. doi:10.3389/fphar.2010.00003

Pelkonen O, Tolonen A, Korjamo T et al (2009) From known knowns to known unknowns: predicting in vivo drug metabolites. Bioanalysis 1:393–414. doi:10.4155/bio.09.32

Pelkonen O, Turpeinen M, Raunio H (2011) In vivo-in vitro-in silico pharmacokinetic modelling in drug development: current status and future directions. Clin Pharmacokinet 50:483–491. doi:10.2165/11592400-000000000-00000

Pelkonen O, Pasanen M, Lindon JC et al (2012) Omics and its potential impact on R&D and regulation of complex herbal products. J Ethnopharmacol 140:587–593. doi:10.1016/j.jep.2012.01.035

Pennisi E (2007) Metagenomics massive microbial sequence project proposed. Science 315:1781

Peterson AC, Russell JD, Bailey DJ et al (2012) Parallell reaction monitoring for high resolution and high mass accuracy quantitative, targeted proteomics. Moll Cell Proteomics 11:1475–1488. doi:10.1074/mcp.O112.020131

Picotti P, Aebersold R (2012) Selected reaction monitoring–based proteomics: workflows, potential, pitfalls and future directions. Nat Methods 9:555–566. doi:10.1038/nmeth.2015

Posma JM, Robinette SL, Holmes E et al (2014) Metabonetworks, an interactive Matlab-based toolbox for creating, customizing and exploring subnetworks from KEGG. Bioinformatics 30:893–895. doi:10.1093/bioinformatics/btt612

Ramirez T, Daneshian M, Kamp H et al (2013) Metabolomics in toxicology and preclinical research. ALTEX 30:209–225

Ravindranath AC, Perualila-Tan N, Kasim A et al (2015) Connecting gene expression data from connectivity map and in silico target predictions for small molecule mechanism-of-action analysis. Mol Biosyst 11:86–96. doi:10.1039/c4mb00328d

Robertson DG (2005) Metabonomics in toxicology: a review. Toxicol Sci 85:809–822. doi:10.1093/toxsci/kfi102

Rognan D (2007) Chemogenomic approaches to rational drug design. Br J Pharmacol 152:38–52. doi:10.1038/sj.bjp.0707307

Rouquié D, Heneweer M, Botham J et al (2015) Contribution of new technologies to characterization and prediction of adverse effects. Crit Rev Toxicol 45:172–183. doi:10.3109/10408444.2014.986054
Ru J, Li P, Wang J et al (2014) TCMSP: a database of systems pharmacology for drug discovery from herbal medicines. J Cheminform 6:13. doi:10.1186/1758-2946-6-13
Scalbert A, Brennan L, Manach C et al (2014) The food metabolome: a window over dietary exposure. Am J Clin Nutr 99:1286–1308. doi:10.3945/ajcn.113.076133
Shen Y, Rahman M, Piccolo SR et al (2015) ASSIGN: context-specific genomic profiling of multiple heterogeneous biological pathways. Bioinformatics 31:1745–1753. doi:10.1093/bioinformatics/btv031
Shi Y, Hu Z, Wu C et al (2011) A genome-wide association study identifies new susceptibility loci for non-cardia gastric cancer at 3q13.31 and 5p13.1. Nat Genet 43:1215–1218. doi:10.1038/ng.978
Siva N (2015) UK gears up to decode 100,000 genomes from NHS patients. Lancet 385:103–104. doi:10.1016/S0140-6736(14)62453-3
Smith TD, Vihinen M, on Behalf of the Human Variome Project (2015) Standard development at the Human Variome Project. Database. doi:10.1093/database/bav024, pii: bav024
Sturla SJ, Boobis AR, FitzGerald RE et al (2014) Systems toxicology: from basic research to risk assessment. Chem Res Toxicol 27:314–329. doi:10.1021/tx400410s
Taboureau O, Hersey A, Audouze K et al (2012) Toxicogenomics investigation under the eTOX project. J Pharmacogenomics Pharmacoproteomics S7:001. doi:10.4172/2153-0645.S7-001
The Wellcome Trust Case Control Consortium (2007) Genome-wide association study of 14,000 cases of seven common diseases and 3,000 shared controls. Nature 447:661–678. doi:10.1038/nature05911
Thomas G, Jacobs KB, Yeager M et al (2008) Multiple loci identified in a genome-wide association study of prostate cancer. Nat Genet 40:310–315. doi:10.1038/ng.91
Titz B, Elamin A, Martin F et al (2014) Proteomics for systems toxicology. Comput Struct Biotechnol J 11:73–90. doi:10.1016/j.csbj.2014.08.004
Toga AW, Foster I, Kesselman C et al (2015) Big biomedical data as the key resource for discovery science. J Am Med Inform Assoc 22:1126–1131. doi:10.1093/jamia/ocv077
Turnbaugh PJ, Ley RE, Hamady M et al (2007) The human microbiome project. Nature 449:804–810
Uhlén M, Fagerberg L, Hallström BM et al (2015) Tissue-based map of the human proteome. Science 347:6220. doi:10.1126/science.1260419
Van den Hof WF, Ruiz-Aracama A, Van Summeren A et al (2015) Integrating multiple 'omics to unravel mechanisms of Cyclosporin A induced hepatotoxicity in vitro. Toxicol In Vitro 29:489–501. doi:10.1016/j.tiv.2014.12.016
Voight BF, Scott LJ, Steinthorsdottir V et al (2010) Twelve type 2 diabetes susceptibility loci identified through large-scale association analysis. Nat Genet 42:579–589. doi:10.1038/ng.609
Wang L, Xie XQ (2014) Computational target fishing: what should chemogenomics researchers expect for the future of in silico drug design and discovery? Future Med Chem 6:247–249. doi:10.4155/fmc.14.5
Wang K, Lee I, Carlson G et al (2010) Systems biology and the discovery of diagnostic biomarkers. Dis Markers 28:199–207. doi:10.3233/DMA-2010-0697
Wang S, Tong Y, Ng TB et al (2015) Network pharmacological identification of active compounds and potential actions of Erxian decoction in alleviating menopause-related symptoms. Chin Med 10:19. doi:10.1186/s13020-015-0051-z
Wen Z, Wang Z, Wang S et al (2011) Discovery of molecular mechanisms of traditional Chinese medicinal formula Si-Wu-Tang using gene expression microarray and connectivity map. PLoS One 6:e18278. doi:10.1371/journal.pone.0018278
Wen W, Cho YS, Zheng W et al (2012) Meta-analysis identifies common variants associated with body mass index in East Asians. Nat Genet 44:307–311. doi:10.1038/ng.1087
Wishart D, Arndt D, Pon A et al (2015) T3DB: the toxic exposome database. Nucleic Acids Res 43:D928–D934. doi:10.1093/nar/gku1004

Xu Q, Qu F, Pelkonen O (2012) Network pharmacology and traditional Chinese medicine In: Sakagami H (ed) Alternative medicine, Intech, (ISBN:978-953-51-0903-7), pp 277–297. doi.org/10.5772/53868

Yao Y, Zhang X, Wang Z et al (2013) Deciphering the combination principles of traditional Chinese medicine from a systems pharmacology perspective based on Ma-huang decoction. J Ethnopharmacol 150:619–638. doi:10.1016/j.jep.2013.09.018

Zanetti KA, Mette E, Maruvada P et al (2014) The future of metabolomic profiling in population-based research: opportunities and challenges. J Anal Bioanal Tech 5:203. doi:10.4172/2155-9872

Zeng P, Zhao Y, Qian C et al (2015) Statistical analysis for genome-wide association study. J Biomed Res 29:285–297. doi:10.7555/JBR.29.20140007

Zhang JD, Berntenis N, Roth A et al (2014) Data mining reveals a network of early-response genes as a consensus signature of drug-induced in vitro and in vivo toxicity. Pharmacogenomics J 14:208–216. doi:10.1038/tpj.2013.39

Zhang A, Sun H, Yan G et al (2015a) Metabolomics for biomarker discovery: moving to the clinic. Biomed Res Int 2015:354671. doi:10.1155/2015/354671

Zhang J, Li Y, Chen SS et al (2015b) Systems pharmacology dissection of the anti-inflammatory mechanism for the medicinal herb Folium eriobotryae. Int J Mol Sci 16:2913–2941. doi:10.3390/ijms16022913

Zhao F, Li G, Yang Y et al (2015a) A network pharmacology approach to determine active ingredients and rationality of herb combinations of Modified-Simiaowan for treatment of gout. J Ethnopharmacol 168:1–16. doi:10.1016/j.jep.2015.03.035

Zhao X, Zheng X, Fan TP et al (2015b) A novel drug discovery strategy inspired by traditional medicine philosophies. Science 347:S38–S40

Zheng CS, Fu CL, Pan CB et al (2015) Deciphering the underlying mechanism of Diesun Miaofang in traumatic injury from systems pharmacology perspective. Mol Med Rep 12:1769–1776. doi:10.3892/mmr.2015.3638

Zhou W, Wang Y (2013) A network-based analysis of the types of coronary artery disease from traditional Chinese medicine perspective: potential for therapeutics and drug discovery. J Ethnopharmacol 151:66–77. doi:10.1016/j.jep.2013.11.007

Zou D, Ma L, Yu J et al (2015) Biological databases for human research. Genomics Proteomics Bioinforma 13:55–63. doi:10.1016/j.gpb.2015.01.006

Chapter 8
Metabolomics Applications in Herbal Medicine

Kati Hanhineva and Markku Pasanen

Abstract Metabolomics analytics focuses on the concomitant measuring of a vast array of small molecule metabolites in a biological sample. These approaches are heavily technique-driven, and have recently been increasingly utilized within the context of herbal medicine. Metabolomics analytics has been used to identify composition or origin of an herbal supplement, or to investigate clinical safety and efficacy responses in animals and humans following treatment with herbal products. Two main approaches – mass spectrometry preceded by either liquid or gas chromatographic separation, and nuclear magnetic resonance – are the most widely used techniques, and the development of instrumentation will be the driving force in terms of analytical accuracy when developing "end-points" or "biomarkers" for the clinical use of herbals or any other therapy.

Keywords Clinical trial • Drug interaction • Mass spectrometry • Metabolism • Nuclear magnetic resonance • Phytocomplex • Toxicity

Abbreviations

ADME	Absorption, distribution, metabolism, excretion
GC/TOF MS	Gas chromatography-time of flight mass spectrometry
HPLC	High performance liquid chromatography
MS	Mass spectrometry
NMR	Nuclear magnetic resonance
PLS-DA	Partial least squares discriminant analysis

K. Hanhineva (✉)
Institute of Public Health and Clinical Nutrition, University of Eastern Finland, P.O. Box 1627, 70211 Kuopio, Finland
e-mail: kati.hanhineva@uef.fi

M. Pasanen
School of Pharmacy, University of Eastern Finland, P.O. Box 1627, 70211 Kuopio, Finland

O. Pelkonen et al. (eds.), *Toxicology of Herbal Products*,
DOI 10.1007/978-3-319-43806-1_8

UPLC/Q-TOF MS	Ultrahigh-performance liquid chromatography-quadrupole time-of-flight mass spectrometry
TCM	Traditional Chinese medicines

Introduction

Metabolomics (metabonomics/metabolic profiling) is – together with transcriptomics and proteomics – an integral component of the systems biology concept (Nicholson et al. 2004; Baker 2011; Blow 2008; Patti et al. 2012). Metabolomics analyses focus on investigating the composition of low molecular weight biomolecules and correlating their patterns and concentrations with phenotypic observations, typically after various perturbations depending on the study focus (e.g., diet, medical/supplement treatment, stress, physical exercise, state of growth, or diseases). A metabolomics study setup involves the extraction of metabolites, analysis in a suitable platform (typically MS or NMR), data acquisition, collection, preprocessing, and statistical evaluation (Fig. 8.1). The metabolite alterations caused by the perturbations are revealed with statistical and chemometric evaluations, including unexpected changes, which are neglected in traditional hypothesis-driven targeted analyses. Such wide-scale approaches have been enabled by the improvements in analytical technologies within the past decade that provide for an extremely accurate, sensitive, and high-throughput approach to explore the metabolite content in virtually any biological material. This capacity can be utilized to measure thousands of analytes concomitantly from minimal sample material in non-targeted metabolite profiling setups. For these reasons, the non-targeted metabolite examination is particularly useful in addressing research questions involving myriad chemical species, such as in the case of herbal medicine.

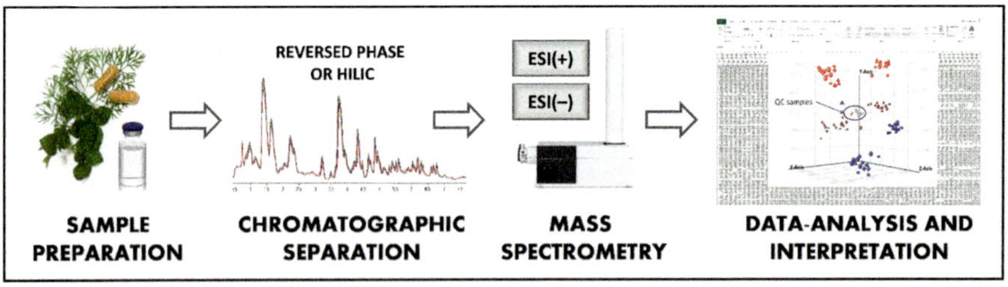

Fig. 8.1 Outline of the LC-MS based metabolomics analysis. The sample preparation typically involves straightforward one-step solvent extraction, followed by injection to the chromatographic separation and mass spectrometric analysis. The data are collected to large data matrices typically consisting of several thousand chemical entities, followed by pre-processing, statistical evaluation, metabolite identification, and biological interpretation utilizing various vendor-specific and open-source software

Herbal products or phytocomplexes used in herbal medicine typically contain complex mixtures of phytochemicals that can be characterized with large-scale metabolomics approaches more efficiently than by conventional targeted single-analyte methods. In particular, the non-targeted metabolite profiling approaches enable the examination of thousands of metabolic features on a single analytical experiment, and therefore offer a powerful tool for both the characterization of the phytochemical content in planta or in the used product, as well as in human borne samples when assessing the metabolic impact of the herbal products (Yuliana et al. 2013; AiHua et al. 2010; Zhang et al. 2012; Liu and Wang 2014). It offers a very wide window to observe the metabolite levels, which is inevitable for assessing the synergistic effects of the phytochemical mixtures on various cellular pathways. Metabolomics adopts a "top-down" strategy to reflect the function of organisms from terminal symptoms of metabolic network and understand metabolic changes of a complete system caused by various treatments in a holistic context.

Owing to its ability to systematically monitor vast metabolite content concomitantly, and thus provide a dynamic picture of the phenotype, one of the key areas where metabolomics is predicted to enable major achievements is drug discovery from traditional Chinese medicine. Traditional Chinese medicines (TCM) are a rich source of potential compounds for drug development, and the hypothesis-free metabolite profiling approach allows the analysis of a priori unknown chemicals, unlike targeted chemical analyses. Likewise, metabolomics allows for discovering biomarkers and perturbed pathways that can clarify the pharmacological mechanisms of traditional Chinese medicines, therefore offering the opportunity to scientifically express the meaning of evidence-based Chinese medicine (Zhang et al. 2012; Quan et al. 2014; Cao et al. 2015). Likewise, the possibility to utilize metabolite profiling approaches in the modernization of TCM preparations is foreseen, as the preparation of the traditional herbal products is complicated in practice. The optimization of modern extraction procedures may be done after more detailed knowledge about the traditional extracts' chemical profiles and their impact on biological activity is achieved with the aid of metabolite profiling, in order to obtain modernized extracts that contain the whole range of compounds relevant for the efficacy of the traditional application (Sheridan et al. 2012). Also, the possibility of using metabolomics analytics in the ADME evaluation of herbal products is foreseen, due to the ability of metabolomics approaches to capture a vast number of compounds and their potential products of metabolism in the evaluation of the pharmacokinetics and pharmacological mechanisms. The chemical diversity of compounds present in herbal products and the complex interactions they may have within cellular metabolism are not captured within traditional single-analyte biochemical measurements, and thus the capacity of metabolomics techniques is foreseen to aid in the analysis of mechanistic links between herbal-derived phytochemicals and cellular metabolic processes (Lan and Xie 2013; Xin et al. 2011; Wang and Chen 2013).

The two prevailing technologies utilized in metabolomics evaluation in general, as in herbal medicine, are mass spectrometry (MS) (Allwood and Goodacre 2010), and nuclear magnetic resonance (NMR) (Ludwig and Viant 2010; Schripsema

2010). While NMR has the advantage of providing actual quantities of analytes in focus and elucidating structures of unknown compounds, the major benefit from MS approaches is its superior sensitivity and therefore the ability to measure compounds in minimal, picomolar concentrations. Owing to this capacity to focus on low-level compounds, MS is typically more widely utilized in metabolomics experiments when addressing the phytochemical composition of plant-borne material, such as herbal supplements (Allwood and Goodacre 2010; Yang et al. 2012).

Metabolomics Approaches in the Characterization of the Phytochemical Content of Herbal Products

Until now, the area of herbal medicine where metabolomics approaches have been utilized most intensively by far is the characterization of the composition of bioactive phytochemicals in herbal plants and products made from them (Table 8.1). In particular, metabolite profiling has been proven useful in the quality assurance of herbal products. The authentication of herbal raw materials and products made from them is necessary in order to ensure the safety and efficacy of the products. The non-targeted metabolite profiling with large-scale data collection, followed by chemo-metric evaluation, has been proven to be a powerful tool for classifying even very closely related species that can otherwise be difficult to distinguish, based on morphological characteristics only. The metabolomics approach has been used as a classifier for seven *Lonicera* species flower buds, and the chemo-metric model created based on the metabolomics data showed good prediction performance (Gao et al. 2012). Likewise, the metabolite profile of curcuma extracts determined using gas chromatography-time of flight mass spectrometry (GC/TOF MS) and ultrahigh-performance liquid chromatography-quadrupole time-of-flight mass spectrometry (UPLC/Q-TOF MS) enabled characterizing differences between *Curcuma aromatica* and *C. longa*

Table 8.1 Metabolomics applications in the analysis of herbal plants/products

Substance	Study focus	Method	Reference
Licorice species Echinacea species *Echinacea pallida* (Nutt.)	Analysis of phytochemical composition	MS and NMR MS MS	Farag et al. (2012) Hou et al. (2010) Pellati et al. (2012)
Lonicera Asian and American ginseng *Panax ginseng* and *Panax quinquefolius* Korean, Chinese and American ginseng	Quality assurance of herbal products	MS NMR MS MS	Gao et al. (2012) Zhao et al. (2015) Park et al. (2014) Yang et al. (2013)
Curcuma *Schisandra chinensis* Goji berries *Angelica gigas*	Identification of geographical origin of herbs	MS MS MS NMR and MS	Lee et al. (2014) Zhang et al. (2014) Bondia-Pons et al. (2014) Kim et al. (2011)

MS mass spectrometry, *NMR* nuclear magnetic resonance

grown in South Korea. The identified metabolites included several curcuminoids and terpenoids, and the metabolite profile allowed discriminating curcuma samples according to species or geographical origin (Lee et al. 2014). Also, the metabolite patterns of goji berries from various geographic origins were analyzed by LC-qTOF-MS metabolite profiling (Bondia-Pons et al. 2014). The different geographic origins clearly differentiate the replicative samples in principal component analysis, as shown in Fig. 8.2.

Ginseng is one of the most widely studied medicinal herbs, and publications describing its metabolomics applications are rapidly appearing. An NMR metabolomics approach was undertaken on 31 batches of ginseng from Chinese stores and assessed with a multi-step principal component analysis. Distinctive differences between Asian ginseng and American ginseng were found mainly in the levels of sucrose, glucose, arginine, choline, and 2-oxoglutarate and malate. Additionally, main differences between wild and cultivated ginseng were identified as ginsenosides (Zhao et al. 2015a). Also, the discrimination between *Panax ginseng* and *P. quinquefolius* has been proven feasible with the UPLC-QTOF-MS-based metabolic profiling method. This approach is very important since these two species have similar chemical and physical properties and the characteristics of their appearance are very similar, but the therapeutic effects are different (Park et al. 2014). Similarily, in another study, *Panax ginseng* Meyer (Korean origin and Chinese origin of Korean ginseng) and *P. quinquefolius* (American ginseng) were analyzed to investigate patterns in major metabolites using HPLC-based metabolic profiling. Partial least squares discriminant analysis (PLS-DA) showed a clear separation between *Panax*

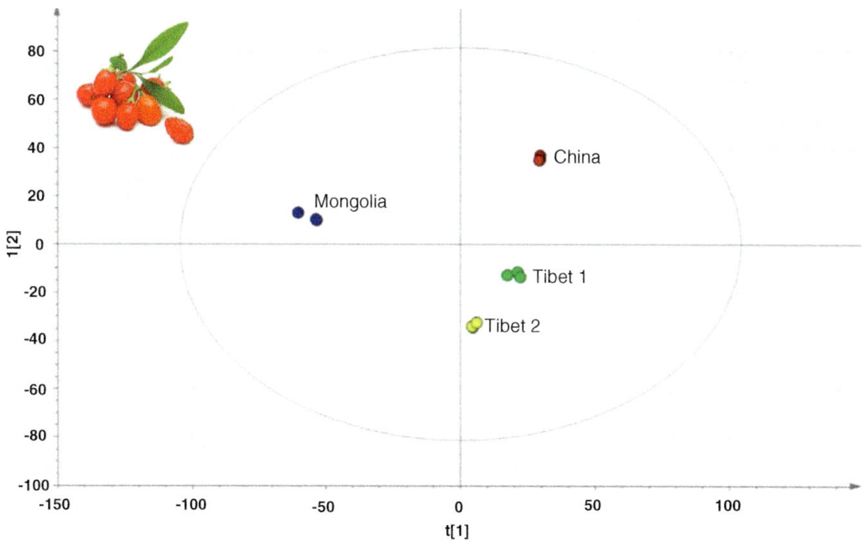

Fig. 8.2 Principal component analysis score plot derived from LC-qTOF-MS data using negative electrospray ionization of 12 Goji berry samples. Each circle represents a sample: Mongolian (*blue*), Chinese (*red*), Tibetan 1 (*green*), Tibetan 2 (*yellow*) (Reprinted from *Food Research International*, Volume 63: 132–138 Copyright 2014, with permission from Elsevier)

species and/or origins from different countries in the PLS-DA score plots, with various ginsenosides as the main differential compounds in the analysis (Yang et al. 2013). All the studies concluded that metabolite profiling can be used to undertake quality control of *Panax* products.

The geographical origin of Wu Wei Zi (*Schisandra chinensis*), an important herbal medicine mainly distributed in northeast China, was evaluated, based on a comprehensive metabolite profiling approach using GC-TOF-MS, ultra-performance LC (UPLC) quadrupole TOF (QTOF) MS. The different phytochemical composition of Wu Wei Zi from different areas including Heilongjiang, Liaoning, Jilin, and Shanxi of China was resolved with the robust and reliable method (Zhang et al. 2014). Furthermore, a combination of nuclear magnetic resonance (NMR) spectroscopy and ultraperformance liquid chromatography-mass spectrometry (UPLC-MS) followed by multivariate data analyses including principal component analysis and orthogonal partial least squares-discriminant analysis was used to characterize *Angelica gigas* obtained from other geographical regions in Korea (Kim et al. 2011).

In addition to the assessment of geographical origins or other quality control approaches, metabolite profiling techniques are widely used for the general, explorative characterization of the phytochemical composition of various herbal products. For example, various licorice species were characterized by large-scale metabolic profiling techniques to gain a broader insight into *Glycyrrhiza* species chemical composition, including *Glycyrrhiza glabra*, *G. uralensis*, *G. inflata* and *G. echinata*, which contained a plethora of phytochemicals, including terpenoids, saponins, flavonoids, polyamines, and polysaccharides (Farag et al. 2012).

Among the most widely used herbal supplements worldwide are various *Echinacea* preparations; however, the composition of different *Echinacea* plant species in the commercial *Echinacea* products is typically not well defined. A comparative metabolomics study by HPLC-MS was used to show that the three most used medicinal *Echinacea* species – *Echinacea purpurea*, *E. pallida*, and *E. angustifolia* – can be classified by the metabolite content, with alkamides and phenolic compounds as its main constituents (Hou et al. 2010). Another study focused on the detailed phytochemical characterization of *E. pallida* (Nutt.) root extracts and dietary supplements with a combination of HPLC with diode array and electrospray ionization-mass spectrometry (ESI-MS) detection (with ion trap and triple quadrupole mass analyzers). The quantitative analysis showed great variability in the number of the many bioactive compounds, including echinacoside and polyacetylenes and polyenes (tetradec-(8Z)-ene-11,13-diyn-2-one, pentedeca-(8Z,11Z)-dien-2-one and pentadec-(8Z)-en-2-one) (Pellati et al. 2012).

Metabolomics Investigations of In Vitro Studies in Herbal Medicine

Several studies, with numerous herbal medicine preparations or purified extracts of bioactive compounds from herbal products, have been published with various in vitro approaches using different animal or human-derived testing platforms,

Table 8.2 Metabolomics investigations of in vitro studies in herbal medicine

Substance	Study focus	Method	Reference
Echinacea	Role of alkylamides in the inhibition of CYP3A4	NMR and MS	Modarai et al. (2010)
Polygonum capitatum	Anti-tumor activity of ellagitannin fraction	MS	Ma et al. (2014)
Pyrrolizidine alkaloids	Metabolism of pyrrolizidine alkaloids	MS	Fashe et al. (2014), (2015a)
			Fashe et al. (2015a)

MS mass spectrometry, *NMR* nuclear magnetic resonance

including exposure of cell cultures (Table 8.2). Nowadays, increasingly, these studies use metabolite profiling techniques to gain a wider view on the metabolism of the herbal-derived chemicals, generation of metabolites, and cellular metabolism perturbations caused by the herb being studied.

Herbals have raised concern about drug-drug interactions in polypharmacy. One classic example is hyperforin fraction, which is isolated from St. John's Wort, which has proved to be a potent inhibitor of the human CYP3A4 enzyme and responsible for clinically significant interactions (Hohmann et al. 2015; Rahimi and Abdollahi 2012). Based on it, several approaches to identifying metabolites and subsequent interaction studies have been taken. For example, the metabolomic profiling of liquid *Echinacea* containing medicinal products has identified two alkylamides (dodeca-2 E,4 E,8 Z,10 E/Z-tetraenoic acid, and a new compound (putative molecular formula $C_{18}H_{36}NO(+)$), which are responsible for inhibiting CYP3A4 (Modarai 2010); however, its clinical significance is still unresolved. In another study using rat and human subcellular organelles and rat primary hepatocytes, altogether eight metabolites of ellagitannin polyphenols from *Polygonum capitatum* were successfully identified in the microsomes and isolated for further evaluations of pharmacologically active compounds (Ma et al. 2014).

Hepatotoxic and genotoxic pyrrolizidine alkaloids are often components and contaminants in herbal medicinal products and food or food supplements (EFSA CONTAM Panel 2011; EMA/HMPC/893108/2011 Committee on Herbal Medicinal Products 2014). Mimicking metabolomics platforms, a thorough metabolite profile evaluation in human subcellular organelles and LC/MS analysis introduced a new glutathione reactive metabolite, ((3H-pyrrolizin-7-yl) methanol) that can be formed from all pyrrolizidine alkaloids (Fashe et al. 2014, 2015a). Additionally, metabolite profiles between pyrrolizidine alkaloids sensitive and resistant species were different; more GSH-reactive metabolites were formed in sensitive species than resistant ones (Fashe et al. 2015b). On the other hand, although pyrrolizidine alkaloids are extensively metabolized and bioactivated by human CYP3A4 (Fashe et al. 2015a), based on the low quantity of parent compounds in any herbal or other source of exposure, they will not be responsible for any clinically relevant drug interactions, but toxic concerns still remain. Therefore, detailed in vitro approaches to identify generated metabolites and cellular responses using different test platforms of human origin will be of great value when establishing safety margins for herbal medicines.

Metabolomics Investigations of Herbal Medicine Studies Using Animal Experiments

The analysis of the biological/metabolic impact of herbal treatment often involves the use of animal experiments. Such analyses have superiority when compared to human trials in that they offer the possibility to examine the metabolic events perturbed with the herbal treatment also in various organs, rather than solely biofluid samples as is the case in human trials. In these experiments, the non-targeted metabolite profiling approaches are useful in providing detailed information about the herbal-derived compounds that potentially harbor the various organs, as well as monitoring the effect on the endogenous metabolite levels, which may aid in understanding the actual effect of the herbal treatment on a wider scale than possible when focusing on plasma (Table 8.3).

Metabonomic profiling in an experimental setting has been carried out, for example, in the examination of the chemopreventive effect of American ginseng (*Panax quinquefolius* L.). The serum metabolic alterations after treatment with ginseng suggested that the chemopreventive effects are exerted by anti-inflammatory and antioxidant mechanisms, as attenuation of impaired amino acids, carbohydrates, and lipid metabolism was observed in the GC-TOF-MS metabolomics analysis (Xie et al. 2015). An LC-qTOF-MS approach has been undertaken on a mouse model of systemic lupus erythematosus that was treated with jieduquyuziyin. The

Table 8.3 Metabolomics applications in the animal studies of herbal medicine

Substance	Study focus	Method	Reference
American ginseng	Chemoprevention in mouse	MS	Xie et al. (2015)
Jieduquyuziyin prescription	Effect on systemic lupus erythematosus in mouse	MS	Ding et al. (2014)
Chotosan	Effect on type 2 diabetes-induced dementia in mouse	MS	Niu et al. (2015)
Fu Fang Jin Jing	Effect on hypoxia and anxiety in mouse	NMR	Liu et al. (2013)
Da-Cheng-Qi	Effect on acute pancreatitis in rat	NMR	Li et al. (2015)
Chinese medicine including rhubarb, ethanol, and alpha- nephthylisothiolyanate	Pathogenic mechanism of yinhuang syndrome	MS	Tong et al. (2011)
Potentilla discolor	Treatment of type 2 diabetes in mouse	MS	Li et al. (2014)
Chaihuang-Yishen formula	Effect on diabetic nephropathy in rat	MS	Zhao et al. (2015)
Multi-component Chinese medicine (termed as SUB885C)	Effect on lipid biochemistry in mouse	MS	Weiet al. (2012)
Aristolochic acid	Induced nephrotoxicity in mouse	NMR	Tsai et al. (2013)

MS mass spectrometry, *NMR* nuclear magnetic resonance

orthogonal partial least squares analysis was used to analyze the metabolic patterns, and various compounds including, for example, phosphatidylethanolamine lipids and serotonin were linked in the pathogenesis of systemic lupus erythematosus. After treatment with jieduquyuziyin, the symptoms were alleviated, and metabolic differences were observed, following a suggestion that the herbal treatment could participate in the metabolism of unsaturated fatty acids, tryptophan, and phospholipids (Ding et al. 2014). Other animal studies with therapeutic approaches of various herbal medicines involving metabolomics include a focus on antidementia (Niu et al. 2015), anxiety (Liu et al. 2013), and acute pancreatitis (Li et al. 2015). An extreme approach to use metabolomics in an experimental setting was a study in which the pathogenic mechanism of Chinese medicine-induced yinhuang syndrome was examined in rats with a cocktail approach, and metabolomics data were collected on the organ level (Tong et al. 2011).

The potential role of herbal remedies in ameliorating or preventing type 2 diabetes has been widely investigated in various mouse models; these include the effect of *Potentilla discolor* in the treatment of type 2 diabetes mellitus in male C57BL/6 mice assessed with UPLC-Q-TOF-MS (Li et al. 2014), and the effect of Chaihuang-Yishen formula on rats with diabetic nephropathy by metabolomic and lipidomic analysis (Zhao et al. 2015b). A multi-component preparation used in Chinese medicine (termed as SUB885C) was administered to apolipoprotein E3 Leiden cholesteryl ester transfer protein (ApoE*3Leiden. CETP) mice followed by plasma and liver lipidomics analysis (Wei et al. 2012).

Metabolomics in Addressing the Effect of Herbal Treatment in Clinical Trials

The number of metabolomics analyses of controlled clinical trials with treatment by herbal medicines is still relatively low. Metabolomics approaches have been used when addressing the composition of herbal chemicals found in urine or plasma, and for evaluating various biomarker alterations related to endogenous metabolism in this context (Table 8.4). Patients'/subjects' responsiveness to herbal medicines and their efficacies have been followed by monitoring the patients'/subjects' endogenous metabolism including for example, carbohydrates, lipids, micronutrients, and vitamins in blood and/or urine samples. One goal of such research has been the identification of biomarker candidates to be used as an endpoint for pharmacodynamic responses. Clinical situations when such approaches have been used are, among others, depression (Tian et al. 2014), metabolic disorders including diabetes (van Wietmarschen et al. 2013), dysmennorrhea (Su et al. 2013), and hypertension (Feng et al. 2015). The duration of the study in the above-mentioned trials has varied between four and 12 weeks, and in some cases each subject acted as his or her own control during the trial. In the depression study (Tian et al. 2014), using the Hamilton depression scale scoring therapeutic cure, eight pharmacodynamic-like

Table 8.4 Metabolomics applications in human trials of herbal medicines

Substance	Study focus	Method	Reference
TCM formula Xiaoyaosan	Effect on depression	NMR	Tian et al. (2014)
Rehmannia six formula (R6)	Effect on clinical chemistry and metabolite profiles	MS	van Wietmarschen et al. (2013)
Shaofu Zhuyu formula concentrated-granule, SFZYFG	Effect on primary dysmenorrhea	MS	Su et al. (2013)
Qingrehuatan decoction	Effect on hypertension	NMR	Feng et al. (2015)

MS mass spectrometry, *NMR* nuclear magnetic resonance

metabolomics components increased or decreased as a response to the herbal therapy (urinary creatinine, taurine, 2-oxoglutarate and xanthurenic acid, citrate, lactate, alanine, and dimethylamine). The authors suggest that they can also be useful biomarkers for efficacy. However, in the metabolic disorders trial, clinical efficacy was mainly estimated according to the traditional Chinese medicine endpoints, but the relevant endpoints generally used in Western medicine (e.g., blood glucose, insulin, HbA1c, and triglycerides) were not used to establish clinical efficacy (van Wietmarschen et al. 2013). Metabolic endpoints were measured in serum where LDL-C, total cholesterol and phosphatidylcholine levels were decreased during the therapy. In the dysmenorrhea trial (Su et al. 2013), both plasma and urine were collected from patients as well as from healthy controls. Altogether, 19 metabolites in plasma and 16 metabolites in urine were up- or down-regulated, compared to the controls. These metabolic pathways represent sphingolipid metabolism, steroid hormone biosynthesis, and glycerophospholipid metabolism, and were considered to be pharmacodynamics efficacy signals. In the hypertension trial, serum samples were analyzed from 12 young hypertensive patients with phlegm-heat syndrome before and after four weeks on herbal therapy. Following the therapeutic response, altogether nine metabolite markers were increased, whereas five metabolites were decreased. (Feng et al. 2015).

One common feature for all the above-mentioned clinical examples is that the studies were not designed to cover present regulatory guidance (e.g., European Medicines Agency guidelines according to each therapeutic area). Rather, each study represents an experimental preliminary approach to identifying a "proof-of-concept" for the platform. However, based on these sporadic experiences, metabolomics approaches, together with indication-based true clinical hard endpoints, may finally serve patient monitoring, and after validation and qualification could establish clinical efficacy as "stand-alone."

Investigation of Adverse Effects of Herbal Medicines by Metabolomics

Herbal medicines are not always beneficial for the user. In the past, several cases were reported, in which the main reason for adverse, toxic, or even lethal responses were due to herbals (Teschke 2015). Based on these examples, in addition to other "omics" techniques, metabolomics data could be supportive when analyzing the mechanism(s) of adverse responses to the drug. With this goal in mind, various compositions of aristolochic acid in mice were studied together with histopathological end points. The main outcome of the study was that despite the origin or composition of aristolochic acid, all material studied proved to be nephrotoxic in the proximal tubular area. This study reinforced the earlier data about toxic characteristics of aristolochic acid; it is toxic itself, and toxicity is dependent on the dosage of "active substance" only (Tsai et al. 2013). The above-mentioned example is not a gold standard for studies aimed at evaluating potential adverse responses by metabolomics means, but at least it shows that when several techniques are prospectively used to identify warning signals, the targets can be identified in more detail.

Conclusions

Metabolomics has been used, and will increasingly be used, to identify the source of material, to trace potential contaminants, and, finally, to identify efficacy and safety variables both in vitro and in vivo. These last two uses are still too premature to be fully taken into account as clinically relevant end-points as a response to herbal medicine, or to any other therapy. However, together with extensive, well-defined clinical trials with clinical end-points, metabolomic biomarkers could well serve regulatory purposes as well. What can be expected in future studies on herbal medicines? There is no easy answer, but by using modern analytical approaches and "omics" in in vitro platforms, certain hazards can be identified. On the other hand, prospective metabolomics approaches in clinical trials with difficult clinical end-points will most likely serve well for investigations on both safety and efficacy.

References

AiHua Z, Hui S, Wang ZG, WenJun S et al (2010) Metabolomics: towards understanding traditional Chinese medicine. Planta Med 76:2026–2035

Allwood JW, Goodacre R (2010) An introduction to liquid chromatography-mass spectrometry instrumentation applied in plant metabolomic analyses. (Special Issue: metabolomics in plant and herbal medicine research). Phytochem Anal 21:33–47, many ref

Baker M (2011) Metabolomics: from small molecules to big ideas. Nat Meth 8:117–121

Blow N (2008) Metabolomics: biochemistry's new look. Nature 455:697–700

Bondia-Pons I, Savolainen O, Törrönen R, Martinez JA et al (2014) Metabolic profiling of Goji berry extracts for discrimination of geographical origin by non-targeted liquid chromatography coupled to quadrupole time-of-flight mass spectrometry. Food Res Int 63:132–138, Part B

Cao H, Zhang A, Zhang H, Sun H, Wang X (2015) The application of metabolomics in traditional Chinese medicine opens up a dialogue between Chinese and Western medicine. Phytother Res 29:159–166

Ding X, Hu J, Wen C, Ding Z et al (2014) Rapid resolution liquid chromatography coupled with quadrupole time-of-flight mass spectrometry-based metabolomics approach to study the effects of jieduquyuziyin prescription on systemic lupus erythematosus. PLoS One 9:e88223

EFSA CONTAM Panel (2011) Scientific opinion on pyrrolizidine alkaloids in food and feed. EFSA J 9:2406–2540

EMA/HMPC/893108/2011 Committee on Herbal Medicinal Products. Public statement on the use of herbal medicinal products containing toxic, unsaturated pyrrolizidine alkaloids (PAs). 2014, http://www.ema.europa.eu/docs/en_GB/document_library/Public_statement/2014/12/WC500179559.pdf

Farag MA, Porzel A, Wessjohann LA (2012) Comparative metabolite profiling and fingerprinting of medicinal licorice roots using a multiplex approach of GC-MS, LC-MS and 1D NMR techniques. Phytochemistry 76:60–72, 45 ref

Fashe MM, Juvonen RO, Petsalo A, Rahnasto-Rilla M et al (2014) Identification of a new reactive metabolite of pyrrolizidine alkaloid retrorsine: (3H-pyrrolizin-7-yl) methanol. Chem Res Toxicol 27:1950–1957

Fashe MM, Juvonen RO, Petsalo A, Vepsalainen J et al (2015a) In silico prediction of the site of oxidation by cytochrome P450 3A4 that leads to the formation of the toxic metabolites of pyrrolizidine alkaloids. Chem Res Toxicol 28:702–710

Fashe MM, Juvonen RO, Petsalo A, Rasanen J, Pasanen M (2015b) Species-specific differences in the *in vitro* metabolism of Lasiocarpine. Chem Res Toxicol 28(10):2034–2044

Feng X, Yang Z, Chu Y, Du B et al (2015) 1H nuclear magnetic resonance-based metabolomic study on efficacy of Qingrehuatan decoction against abundant phlegm-heat syndrome in young adults with essential hypertension. J Tradit Chin Med 35:28–35

Gao W, Yang H, Qi LW, Liu EH et al (2012) Unbiased metabolite profiling by liquid chromatography-quadrupole time-of-flight mass spectrometry and multivariate data analysis for herbal authentication: classification of seven Lonicera species flower buds. J Chromatogr A 1245:109–116

Hohmann N, Maus A, Carls A, Haefeli WE, Mikus G (2015) St. John's wort treatment in women bears risks beyond pharmacokinetic drug interactions. Arch Toxicol 90(4):1013–1015

Hou CC, Chen CH, Yang NS, Chen YP et al (2010) Comparative metabolomics approach coupled with cell- and gene-based assays for species classification and anti-inflammatory bioactivity validation of Echinacea plants. J Nutr Biochem 21:1045–1059, 54 ref

Kim EJ, Kwon J, Park SH, Park C et al (2011) Metabolite profiling of Angelica gigas from different geographical origins using 1H NMR and UPLC-MS analyses. J Agric Food Chem 59:8806–8815

Lan K, Xie G, Jia W (2013) Towards polypharmacokinetics: pharmacokinetics of multicomponent drugs and herbal medicines using a metabolomics approach. Evid Based Complement Alternat Med 2013:819147

Lee J, Jung Y, Shin JH, Kim HK et al (2014) Secondary metabolite profiling of Curcuma species grown at different locations using GC/TOF and UPLC/Q-TOF MS. Molecules 19:9535–9551

Li Y, Li JJ, Wen XD, Pan R et al (2014) Metabonomic analysis of the therapeutic effect of Potentilla discolor in the treatment of type 2 diabetes mellitus. Mol Biosyst 10:2898–2906

Li J, Zhu SF, Zhao XL, Liu YX et al (2015) Metabolomic profiles illuminate the efficacy of Chinese herbal Da-Cheng-Qi decoction on acute pancreatitis in rats. Pancreatology 15:337–343

Liu P, Wang P (2014) Application of metabolomics technology in the research of Chinese medicine. Chin J Integr Med 20:307–310

Liu X, Zhu W, Guan S, Feng R et al (2013) Metabolomic analysis of anti-hypoxia and anti-anxiety effects of Fu Fang Jin Jing oral liquid. PLoS One 8:e78281

Ludwig C, Viant MR (2010) Two-dimensional J-resolved NMR spectroscopy: review of a key methodology in the metabolomics toolbox. (Special Issue: Metabolomics in plant and herbal medicine research.). Phytochem Anal 21:22–32, many ref

Ma JY, Zhou X, Fu J, Hu T et al (2014) Metabolite profiling analysis of FR429, an ellagitannin purified from Polygonum capitatum, in rat and human liver microsomes, cytosol and rat primary hepatocytes in vitro. Chem Biol Interact 220:33–40

Modarai M, Yang M, Suter A, Kortenkamp A, Heinrich M (2010) Metabolomic profiling of liquid Echinacea medicinal products with in vitro inhibitory effects on cytochrome P450 3A4 (CYP3A4). Planta Med 76:378–385

Nicholson JK, Holmes E, Lindon JC, Wilson ID (2004) The challenges of modeling mammalian biocomplexity. Nat Biotechnol 22:1268–1274

Niu Y, Li F, Inada C, Tanaka K et al (2015) Chemical profiling with HPLC-FTMS of exogenous and endogenous chemicals susceptible to the administration of chotosan in an animal model of type 2 diabetes-induced dementia. J Pharm Biomed Anal 104:21–30

Park HW, In G, Kim JH, Cho BG et al (2014) Metabolomic approach for discrimination of processed ginseng genus (Panax ginseng and Panax quinquefolius) using UPLC-QTOF MS. J Ginseng Res 38:59–65

Patti GJ, Yanes O, Siuzdak G (2012) Innovation: Metabolomics: the apogee of the omics trilogy. Nat Rev Mol Cell Biol 13:263–269

Pellati F, Orlandini G, Benvenuti S (2012) Simultaneous metabolite fingerprinting of hydrophilic and lipophilic compounds in Echinacea pallida by high-performance liquid chromatography with diode array and electrospray ionization-mass spectrometry detection. J Chromatogr A 1242:43–58

Quan Y, Wang ZY, Xiong M, Xiao ZT, Zhang HY (2014) Dissecting traditional Chinese medicines by omics and bioinformatics. Nat Prod Commun 9:1391–1396

Rahimi R, Abdollahi M (2012) An update on the ability of St. John's wort to affect the metabolism of other drugs. Expert Opin Drug Metab Toxicol 8:691–708

Schripsema J (2010) Application of NMR in plant metabolomics: techniques, problems and prospects. (Special Issue: Metabolomics in plant and herbal medicine research). Phytochem Anal 21:14–21, many ref

Sheridan H, Krenn L, Jiang RW, Sutherland I et al (2012) The potential of metabolic fingerprinting as a tool for the modernisation of TCM preparations. (Special Issue: good practice in traditional Chinese medicine research in the post-genomic era). J Ethnopharmacol 140:482–491, many ref

Su S, Duan J, Wang P, Liu P et al (2013) Metabolomic study of biochemical changes in the plasma and urine of primary dysmenorrhea patients using UPLC-MS coupled with a pattern recognition approach. J Proteome Res 12:852–865

Teschke R, Eickhoff A (2015) Herbal hepatotoxicity in traditional and modern medicine: actual key issues and new encouraging steps. Front Pharmacol 6:72

Tian JS, Peng GJ, Gao XX, Zhou YZ et al (2014) Dynamic analysis of the endogenous metabolites in depressed patients treated with TCM formula Xiaoyaosan using urinary (1) H NMR-based metabolomics. J Ethnopharmacol 158(Pt A):1–10

Tong X, Sun H, Yan GL, Dong W et al (2011) Evaluation study on urine metabolomics in yinhuang rat model induced by triplet factors of rhubarb, ethanol, and alpha-nephthylisothiolyanate. Chin J Integr Med 17:369–375

Tsai DM, Kang JJ, Lee SS, Wang SY et al (2013) Metabolomic analysis of complex Chinese remedies: examples of induced nephrotoxicity in the mouse from a series of remedies containing aristolochic acid. Evid Based Comp Alternat Med 2013:263757

van Wietmarschen HA, van der Greef J, Schroen Y, Wang M (2013) Evaluation of symptom, clinical chemistry and metabolomics profiles during Rehmannia six formula (R6) treatment: an integrated and personalized data analysis approach. J Ethnopharmacol 150:851–859

Wang L, Chen C (2013) Emerging applications of metabolomics in studying chemopreventive phytochemicals. AAPS J 15:941–950

Wei H, Hu C, Wang M, van den Hoek AM et al (2012) Lipidomics reveals multiple pathway effects of a multi-components preparation on lipid biochemistry in ApoE*3Leiden. CETP mice. PLoS One 7:e30332

Xie G, Wang CZ, Yu C, Qiu Y et al (2015) Metabonomic profiling reveals cancer chemopreventive effects of American ginseng on colon carcinogenesis in Apc(Min/+) mice. J Proteome Res 14:3336–3347

Xin GZ, Qi LW, Shi ZQ, Li P et al (2011) Strategies for integral metabolism profile of multiple compounds in herbal medicines: pharmacokinetics, metabolites characterization and metabolic interactions. Curr Drug Metab 12:809–817

Yang M, Cheng C, Yang J, Guo DA (2012) Metabolite profiling and characterization for medicinal herbal remedies. Curr Drug Metab 13:535–557

Yang SO, Lee SW, Kim YO, Sohn SH et al (2013) HPLC-based metabolic profiling and quality control of leaves of different Panax species. J Ginseng Res 37:248–253

Yuliana ND, Jahangir M, Verpoorte R, Choi YH (2013) Metabolomics for the rapid dereplication of bioactive compounds from natural sources. Phytochem Rev 12:293–304, 54 ref

Zhang AH, Sun H, Wang X (2012) Recent highlights of metabolomics for traditional Chinese medicine. Pharmazie 67:667–675, many ref

Zhang Y, Li F, Huang F, Xie G et al (2014) Metabolomics analysis reveals variation in Schisandra chinensis cetabolites from different origins. J Sep Sci 37:731–737

Zhao H, Xu J, Ghebrezadik H, Hylands PJ (2015a) Metabolomic quality control of commercial Asian ginseng, and cultivated and wild American ginseng using (1) H NMR and multi-step PCA. J Pharm Biomed Anal 114:113–120

Zhao T, Zhang H, Zhang X, Zhao T et al (2015b) Metabolomic and lipidomic study of the protective effect of Chaihuang-Yishen formula on rats with diabetic nephropathy. J Ethnopharmacol 166:31–41

Chapter 9
Genotoxicity and Carcinogenicity of Herbal Products

Mélanie Poivre, Amandine Nachtergael, Valérian Bunel, Okusa Ndjolo Philippe, and Pierre Duez

Abstract In 2012, the World Health Organization (WHO) recorded 14 million new cases of cancer and 8.2 million cancer-related deaths. Remarkably, the WHO estimates that 30 % of cancer mortalities are due to lifestyle choices and environmental factors that can and should be avoided. In line with these recommendations, this chapter discusses the genotoxicity and carcinogenicity of herbal products. Although often perceived as innocuous by the general public, many herbs harbor phytochemicals that are either directly reactive towards DNA or likely to disturb cellular homeostasis, cell cycle, and/or genome maintenance mechanisms; this may translate into genotoxicity, carcinogenicity, or co-carcinogenicity. Genotoxicity refers to the deleterious effect of a chemical compound or a physical event on the genetic material; such genotoxic events are considered hallmarks of cancer risk. Nevertheless, much of the damage to the genetic material can be efficiently bypassed and/or repaired by the numerous genome maintenance mechanisms of the cell and may not lead to cancer. The long-term safety evaluation is probably better investigated through carcinogenicity, which denotes the capacity of a chemical substance or a mixture of chemical substances to induce cancer or increase its incidence. The major mechanisms of carcinogenicity are discussed along with biomarkers and approved regulatory guidelines. The recent development of innovative carcinogenicity testing strategies, especially based on functional genomics, are debated and evaluated for possible application to the precocious evaluation of herbal products'

M. Poivre • A. Nachtergael • O. Ndjolo Philippe • P. Duez (✉)
Unit of Therapeutic Chemistry and Pharmacognosy, Faculty of Medicine and Pharmacy, Research Institute for Health Sciences and Technology, University of Mons – UMONS, Mons, Belgium
e-mail: Pierre.DUEZ@umons.ac.be

V. Bunel
Unit of Therapeutic Chemistry and Pharmacognosy, Faculty of Medicine and Pharmacy, Research Institute for Health Sciences and Technology, University of Mons – UMONS, Mons, Belgium

Laboratory of Pharmacognosy, Bromatology and Human Nutrition, Faculty of Pharmacy, Université Libre de Bruxelles – ULB, Brussels, Belgium

O. Pelkonen et al. (eds.), *Toxicology of Herbal Products*,
DOI 10.1007/978-3-319-43806-1_9

long-term safety. Finally, this chapter provides some examples of proven or suspected carcinogenic herbal products reported in the current literature.

Keywords Herbal products • Medicinal plants • Natural products • Genotoxicity • Carcinogenicity

Abbreviations

2YRB	2-year rodent bioassay
AA	Aristolochic acid
AAN	Aristolochic acid nephropathy
BER	Base excision repair
CSC	Cancer stem cells
ECVAM	European Centre for the Validation of Alternative Methods
EMA	European Medicines Agency
FDA	Food and Drug Administration
HMP	Herbal medicinal products
HMPC	Committee on Herbal Medicinal Products
IARC	International Agency for Research on Cancer
ICH	International Conference on Harmonization
NER	Nucleotide excision repair
NOAEL	No observed adverse effect level
OECD	Organization for Economic Co-operation and Development
PA	Pyrrolizidine alkaloid
PFS	Plant food supplements
SAR	Structure-activity relationship
TCM	Traditional Chinese medicine
TFT	Trifluorothymidine
WHO	World Health Organization

Introduction

Reported cases of cancer and cancer-related deaths are increasing worldwide, partly due to increased longevity and higher diagnosis rate. In 2012, the World Health Organization (WHO) recorded 14 million new cases of cancer and 8.2 million cancer-related deaths. Remarkably, the WHO estimates that 30 % of cancer mortalities are due to lifestyle choices and environmental factors that can and should be avoided. Medicinal herbs are widely used throughout the world, both as primary healthcare solutions – mainly in developing countries – and as complementary or alternative medicines; their use has been continuously increasing for two decades in Western countries (Cheng and Leung 2012). The total market value of medicinal herbs accounts for about US $83 billion and is expected to reach US $107 billion by

2017 (Nutraceuticals 2012). This resurgence of interest in plant-based treatments seems to have various origins: it may come from patients' disappointment with standard treatments (in terms of efficacy and/or safety), from the rewarding feeling of active participation in the choice of therapeutic means, from the beliefs that the use of herbs is associated with a healthier lifestyle, and that herbal medicines, being "natural", are therefore harmless (Ekor 2014). Despite this positive perception of herbal treatments, their effectiveness and safety has most often not been evaluated per modern standards (Cheng and Leung 2012; Pelkonen et al. 2014); their quality, often unchecked, may be precarious, and cases of contamination, adulteration, toxicity, or poisoning are regularly detected (Vanherweghem et al. 1993; Liu et al. 2014). Until now, only a few quality toxicological studies have been carried out on the most widely used herbs; it is estimated that toxicological data are still missing for up to 90% of traditional Chinese herbal medicines (Cheng and Leung 2012), and the situation appears even worse for herbs used in developing countries, notably in African traditional medicine (Kahumba et al. 2015).

This chapter attempts to outline the issues of genotoxicity and carcinogenicity of herbal products. Indeed, many herbs harbor phytochemicals that are either directly reactive towards DNA, or are likely to disturb cellular homeostasis, cell cycle, and/or genome maintenance mechanisms, which may lead to genotoxicity, carcinogenicity, or co-carcinogenicity.

Genotoxicity

Genotoxicity describes the ability of chemical compounds and their metabolites to interact with DNA and/or the cellular machinery controlling the genome integrity (Butterworth 2006). Genotoxicants interact either directly with DNA or chromosomes to produce DNA damage such as adducts, strand breaks, chromosome breakages, etc., or indirectly, disturbing the genomic integrity through several mechanisms, notably by interaction (1) with proteins involved in DNA replication, transcription, or repair; (2) with components of mitotic spindle; or (3) with protein kinases in charge of cell cycle checkpoints (Magdolenova et al. 2014). Genotoxicants are usually classified according to their mutagenicity, through transformation of a DNA damage into a mutation, clastogenicity, through modification of chromosome structure, and aneugenicity, through changes in the number of chromosomes (loss or gain) (Muller et al. 2008; Botta 2013).

Mutagenicity

Mutagenesis, the process by which mutations appear, can arise spontaneously, without exposition to a mutagen (Smith 1992), or it can be induced by physical or chemical mutagens. The conversion of a DNA lesion into a permanent and heritable mutation requires DNA replication (Botta 2013), and so imbalances in the fidelity of undamaged and damaged DNA replication appear as major causes of mutagenesis (Sarasin 2003; Loeb and Harris 2008). Introduced base substitutions can lead to "missense" and "nonsense" mutations, whereas insertions or deletions can induce frameshift mutations, both leading to altered gene expression (Magdolenova et al. 2014). Moreover, some non-genotoxic agents are able to increase mutagenesis indirectly by acting on DNA repair mechanisms or by stimulating cell proliferation, which increases the replication frequency (Dixon and Kopras 2004). Inorganic arsenic is an example of non-genotoxic mutagens; although negative in most mutagenic activity tests, exposure to arsenite (arsenic oxoanion where arsenic has an oxidation state of +3) has been strongly associated with an increased risk of skin, bladder, lung, and liver cancers. The mutagenesis of arsenic lies in its ability to increase the mutagenic activity of carcinogenic agents, such as UV irradiations, by interfering with DNA repair mechanisms of base excision repair (BER) and nucleotide excision repair (NER), through several mechanisms not completely elucidated (Shen et al. 2013; Andrew et al. 2006; Hubaux et al. 2013; Rossman 2003).

Carcinogenicity

Carcinogenesis is a complex process that is subject to intensive research. The following sections give insight into the various mechanisms involved in the genesis of cancer and point out the major critical events in which carcinogenicity can arise.

Carcinogens are substances able to "induce tumor (benign or malignant), increase the incidence or reduce the delay time of a tumor after their penetration into the body through inhalation, injection, dermal contact or ingestion" (Mulware 2012). The Organisation for Economic Co-operation and Development (OECD) classifies carcinogens into two categories: genotoxic carcinogens that initiate carcinogenesis by direct interaction with DNA and that are easily characterized by genotoxicity assays; and non-genotoxic carcinogens causing structural and functional DNA alterations that result in altered gene expression or signal transduction (Mulware 2012) and that are generally negative in genotoxicity assays (OECD 2007).

Genotoxic Carcinogens and the Somatic Mutation Theory

In 1914, Theodor Boveri observed a link between genotoxicity, mutagenesis, and carcinogenesis. Accordingly, he concluded that "tumor growth is a consequence of incorrect chromosomal combination transmittable to daughter cells" (Balmain 2001). This discovery was the cornerstone of the somatic mutation theory that led to the development of *in vitro* screening assays for mutagenic compounds, notably the Ames test (Ames et al. 1973). In this theory, carcinogenesis is defined as a multistep process starting with initiation, in which genomic alterations occur through chemical, physical, or biological (pathogens) agents. Initiated cells with selective growth advantage install as transformed clones during the promotion step. Finally, progression is characterized by the transformation of preneoplastic lesions into clinically relevant cancer, with an increase of the metastatic potential and angiogenesis (Botta 2013; Loeb and Harris 2008; Monier 2008). This classical view of carcinogenesis is considered to be a simplification and cannot account for the various deregulated biological processes involved in cancer (Loeb and Harris 2008).

The Mutator Phenotype

In their paper "The Hallmarks of Cancer" published in 2000, D. Hanahan and R.A. Weinberg described the molecular, structural, and behavioral capacities of cancer cells (Hanahan and Weinberg 2000) as (1) sustaining proliferative signaling; (2) evading growth suppressors; (3) resisting cell death; (4) enabling replicative immortality; (5) inducing angiogenesis; and (6) activating invasion and metastasis. Recent advances in carcinogenesis led the authors to add two emerging characteristics to this list: (7) reprogramming of energy metabolism to support growth and continuous cell proliferation; and (8) evading immune destruction. According to the authors, these characteristics are underpinned by the genomic instability and inflammatory status of pre-malignant and malignant lesions (Hanahan and Weinberg 2011).

Mutator Phenotype and Genomic Instability

The discovery of some critical mutated genes in many cancers (Davies et al. 2002) has contributed to build the hypothesis that alterations in some specific genes are responsible for tumor initiation, maintenance, and progression (Quante and Wang 2008). A malignant transformation would require a number of independent mutations (Knudson 2001): (1) oncogene activation; (2) tumor suppressor gene inactivation; and (3) telomerase constitutive expression (e.g., hTERT) (Botta 2013; Dixon and Kopras 2004). Oncogenes are genes coding for growth factors (e.g., PDGF), tyrosine kinase surface receptors (e.g., EGFR, HER), anti-apoptotic proteins (e.g., BCL-2) (Martinez-Arribas et al. 2007), nuclear transcription factors (e.g., MYC), or signal transducing G-proteins (e.g., RAS) (Hesketh 1997), and are involved in signalization pathways that stimulate cell proliferation. These genes are mainly active during embryogenesis but can be activated during adulthood through mutation or chromosomal rearrangement. Tumor suppressor genes code for proteins associated with cell cycle arrest, apoptosis and DNA repair. These genes are classified either as "gate keepers" coding for proteins involved in the control and regulation of cell proliferation (e.g., P53, RB1, APC) or as "care takers" coding for proteins involved in the genome repair and stabilization (e.g., BRCA1, BRCA2, MSH2, MLH1) (Botta 2013; Dixon and Kopras 2004). Regions of repetitive DNA sequences at each end of a chromosome (telomeres) are synthesized by an enzyme called telomerase (hTERT) that prevents their degradation. The gradual reducing of telomeres during each cell division is a normal process for the cell, ultimately leading to apoptotic death. To counter this, many tumor cells constitutively express telomerases (Dixon and Kopras 2004).

In 1974 L.A. Loeb described for the first time the mutator phenotype hypothesis (Loeb et al. 1974). He calculated that the mutation rate of non-cancerous cells is insufficient to generate the large number of mutations found in cancerous cells. According to this hypothesis, mutations in specific genes governing genomic stability (oncogenes and tumor suppressor genes) lead to an enhanced genomic instability that substantially increases the mutation rate and justifies the multiple mutations observed in cancer cells (Loeb and Harris 2008; Loeb 2011). However, it still remains unclear whether genomic instability is a prerequisite or a consequence of cancer development, and arguments have developed on both sides (Marx 2002). The high prevalence of cancers among patients with genetic diseases linked to defects in genes responsible for genetic stability (Cleaver 2005), the existence of a mutator phenotype in DNA repair proteins deficient cells (Friedberg et al. 2002), and the demonstration of a mutation rate 200 times higher in tumor tissues (Bielas et al. 2006) are strong arguments in favor of such a mutator phenotype. On the other hand, it has been shown that, in highly proliferative tissues, the rate of spontaneous mutations is enough to allow the accumulation of mutations and provide a selective advantage required for clonal expansion (Sarasin 2003; Dixon and Kopras 2004; Wang et al. 2002). According to these arguments, the genomic instability would take place later in cancer development to contribute to its expansion in the body (Marx 2002).

Non-genotoxic Carcinogens

Non-genotoxic carcinogens exert various modes of action including (1) mitogen stimulation of growth through hormonal effects eventually mediated by a receptor (e.g., binding to estrogen receptor, disturbance of the synthesis or secretion of thyroid hormones by anti-thyroid substances); (2) promotion of tumors (modulation of DNA repair mechanisms and cell cycle control); (3) induction of a specific tissue toxicity and targeted inflammation, resulting in a regenerative hyperplasia; (4) immune suppression; (5) inhibition of intercellular communications through gap junctions, essential to cellular homeostasis; and (6) epigenetic modifications (Butterworth 2006; Hernandez et al. 2009). The great diversity in modes of action and tissue specificities of these agents, combined with their absence of genotoxicity, makes carcinogenicity prediction extremely challenging.

Stem Cells

The origin of cancer is also attributed to stem cells, a sub-population of cells able to divide and generate numerous copies identical to themselves (self-renewing), and differentiated cell lineages (Gonzalez and Bernad 2012). In the middle of the nineteenth century, two German pathologists, Julius Cohnheim and Rudolph Virchow, observed similarities between embryonic and cancerous tissues and hypothesized that tumors would arise by reactivation of sleeping embryonic rest tissue (Virchow

1855; Cohnheim 1875). This "embryonic rest hypothesis of cancer" then postulates that adult cancers develop from stem cells. In this theory, a tumor is seen as an aberrant and heterogeneous organ in which only a small portion of cancer cells, the cancer stem cells (CSC), is able to initiate tumor growth, proliferate extensively, and present a metastatic potential (Quante and Wang 2008). Recent technological developments allowed isolating this cancer stem cells sub-population from the other tumor cells using specific surface markers (Monier 2008). The CSCs are capable, just as normal stem cells, of self-renewal, differentiation, and asymmetric division to generate both a new identical stem cell and a progenitor cell with a limited lifetime but responsible for the proliferation (Monier 2008). The injection to immunodeficient mice of a small number of CSCs, but not of other tumor cells, effectively leads to tumor development (O'Brien et al. 2007). The origin of these cancer stem cells is still a matter of debate and seems to vary from one tumor type to another (Hanahan and Weinberg 2011). In some tumors, the CSC would originate from a tissue stem cell that has undergone a cancerous transformation. In others, CSC would drift from progenitor cells arising from the asymmetric division of a normal tissue stem cell (Monier 2008; Hanahan and Weinberg 2011; Quante and Wang 2008). The recent discovery of circulating progenitor cells with stem cells lineage specific properties raised questions about the existence of distinct stem cell populations for each tissue or the existence of a centralized stem cell source (Quante and Wang 2008; Shaked et al. 2006).

Influence of Epigenetics

It is nowadays acknowledged that the altered expression of oncogenes and tumor suppressor genes can also arise from epigenetic modifications; these involve DNA chemical modifications free of sequence alteration, such as nucleotide methylation, histone modification (acetylation, methylation and phosphorylation), chromatin remodeling, nucleosome positioning, and non-coding RNA modulation (e.g., microRNA) (Dixon and Kopras 2004; Vineis et al. 2010; Migheli and Migliore 2014). Genetic and epigenetic factors interact and influence themselves during carcinogenesis, and to date, no cancer has been detected with only a genetic or epigenetic background (Migheli and Migliore 2014; Burgio and Migliore 2015).

Influence of Tumor Microenvironment

Neglected for too long by the somatic mutation theory, the microenvironment of tumors plays a predominant role in carcinogenesis. This tumor microenvironment is defined as "the normal cells, molecules, and blood vessels that surround and feed a tumor cell. A tumor can change its microenvironment, and the microenvironment can affect how a tumor grows and spreads" (National Cancer Institute 2015). The information coming from the tumor environment induces dynamic mechanisms that yield phenotypic alterations, most probably through epigenetic modifications (Burgio and

Migliore 2015). During clonal expansion, the tumor microenvironment dictates selective conditions; a mutation on a specific gene can lead to clonal expansion if this mutation confers a selective advantage over normal cells towards the microenvironment (Sarasin 2003; Vineis et al. 2010; Wu and Starr 2014).

Cancer and Inflammation

An inflammatory state of pre-malignant and malignant lesions favor tumor progression through various mechanisms (Hanahan and Weinberg 2011). Indeed, it has been shown that chronic inflammatory situations such as viral infections (human papilloma virus, hepatitis-B, etc.), obesity, chronic gastric reflux, chronic colitis, and Crohn's disease are associated with cancer development (Coussens and Werb 2002). The underlying mechanisms involve oxygen and nitrogen reactive species, inflammatory cytokines, prostaglandins, and microRNA produced during inflammation. The chronic production of such mediators will cause DNA damage, alter gene expression, and provoke cellular proliferation changes (Loeb and Harris 2008; Quante and Wang 2008; Vineis et al. 2010).

Mechanism of Malignancy Is Still a Matter of Huge Debate

In January 2015, C. Tomasetti and B. Vogelstein investigated the significant variation in cancer risk between different types of tissues, showing an important correlation between the number of stem cell divisions in a particular tissue and the risk of cancer (Tomasetti and Vogelstein 2015). Based on this result, they showed that only a third of cancer risk variation from one tissue to another would be attributed to genetic predisposition and environmental factors. The majority of observed tissue-to-tissue variation would be due to "bad luck" arising from stochastic mutations during DNA replication in non-cancerous stem-cells. The bad luck hypothesis of cancer has been the subject of considerable criticism arguing, among others, that the authors have considered stem-cell division rates and extrinsic risk factors as entirely independent. In January 2016, Y.A. Hannun and his colleagues provided evidence that intrinsic risk factors contribute only modestly (less than 10–30 % lifetime risk) to the mechanism of malignancy (Wu et al. 2016). The exact contribution of external and internal factors in cancer development is still open to debate and could have implications in cancer prevention strategies. Primary prevention (e.g., lifestyle modification, HPV vaccines) would impact on the risk of cancers triggered by environmental factors but would not be effective on cancers for which the risk is attributable to "bad luck" or internal factors. For these cancers, secondary prevention (early detection of cancer) and chemoprevention (dietary agents modulating DNA replication and/or repair (Charles et al. 2012, 2014; Nachtergael et al. 2013)) would be the most effective strategies to decrease cancer mortality.

Genotoxicity and Carcinogenicity Assessment

There is currently no single validated test able to provide information on the three genotoxicity critical end-points, which are mutation induction, clastogenicity, and aneugenicity; a battery of tests is thus needed to determine the genotoxic and mutagenic potential of a compound. Moreover, due to the diversity of the end-points, genotoxicity and/or carcinogenicity cannot be assessed in a single assay (Maurici et al. 2004).

Genotoxicity Assessment

Genotoxicity assays are dedicated to the detection of compounds that can induce genetic damage by various mechanisms (ICH 2014). The major challenge in genotoxicity testing resides in the development of methods that can reliably and sensibly detect either such a vast array of damages, or a general cellular response to genotoxic insult. It is recognized that no single test can detect every genotoxicant, and therefore the concept of the battery of tests has been implemented in many regulatory guidelines (Billintona et al. 2008).

Methods for genotoxicity assessment include *in silico* and structure alert methods; as well as *in vitro* and *in vivo* methods.

In Silico and Structure Alert Methods

In silico methods

In silico methods aim at predicting biological activities of a molecule from its physicochemical properties (Combes 2012). These predictive methods generally rely on computational tools, mathematical calculation, and analysis of predicted or experimental data through computer-based models (Valerio 2009) that are generally classified as: (1) rule-based expert systems (e.g., DEREK), which estimate the presence of a DNA-reactive moiety in a given molecule (Greene 2002); (2) quantitative structure-activity relationship models, so-called "QSAR" models (e.g., TOPKAT) that use "electro-topological" descriptors rather than chemical structure to predict mutagenic reactivity with DNA; and (3) three-dimensional computational DNA-docking models to identify molecules that are capable of non-covalent DNA interaction.

- Benefits: *In silico* prediction systems have many advantages, such as their low-cost, rapidity, high reproducibility, low/no compound synthesis requirements, constant optimization, and potential to reduce or replace the use of animals (3R policy, aiming at replacing, reducing, and refining the use of animals (Fjodorova et al. 2010)).

- Limitations: The lack of factual toxicity data, inappropriate (simplistic) modeling of some endpoints, and poor domain applicability of models represent their main limitations. The application of *in silico* methods to complex mixtures such as herbal extracts is limited to the detection of known or new structural alerts for genotoxicity. However, they could help to elucidate which compounds are responsible for a proven effect (Valerio 2009; Ouedraogo et al. 2012).

Structure alert methods

Structural alerts, also called "toxicophores", are defined as molecules or moieties that are known to be associated with toxicity; their presence alerts the investigator to their potential toxicity. Well-characterized genotoxic compounds include (1) 1–2 unsaturated pyrrolizidine ester alkaloids from many *Boraginaceae*, *Asteraceae*, and *Fabaceae* (Fu et al. 2002; Chen et al. 2010; Xia et al. 2008); (2) aristolochic acids (AA), nitro-polyaromatic compounds notably responsible for terminal nephropathies observed upon intoxication by many *Aristolochia* species (Chan 2003; Mei et al. 2006); and (3) allylalkoxybenzenes, e.g., eugenol, methyleugenol, estragole, safrole (4-allyl-1,2-methylenedioxybenzene) or asarone, potentially genotoxic components from some essential oils. The notion of threshold for genotoxic insults is still a matter of serious debate; consequently, proved toxicophores should be proscribed from herbal medicines or at least severely limited (Ouedraogo et al. 2012).

In Vitro Methods

The term *in vitro* (Latin for "in the glass") refers to experiments carried out in a controlled environment, outside of a living organism. *In vitro* methods are based on the use of pro- or eukaryotic cells and tissue cultures (Brusick 1980). Increasingly, human cells are used since they better predict human toxicity (ECVAM 2015).

- Benefits: *In vitro* assays are relatively inexpensive, easy to conduct, and do not involve the use of animals. *In vitro* assays typically provide an initial indication of the genotoxicity of a chemical, and the results often guide eventual subsequent *in vivo* studies (OECD 2014).
- Limitations: A battery of tests is required to investigate the multiple aspects of genotoxicity. Oversensitivity and low specificity represent common problems, compared to *in vivo* situations. *In vitro* assays notably require supplementation with exogenous metabolic activation enzymes (e.g., S9 fraction of liver homogenate) in order to simulate mammalian metabolism. Moreover, *in vitro* testing on mammalian cells may use cell lines that are not relevant to predict genotoxic endpoints at target organs (Maurici et al. 2004; Ouedraogo et al. 2012; Brusick 1980). Effectively, the commonly used cell lines are often deficient in DNA repair, p53 function, or metabolic competency, and many derive from malignancies (Walmsley and Billinton 2011).

The emerging toxicogenomics area could lead to a better understanding of genotoxicity/mutagenicity processes and help in the development of more accurate *in vitro* models (Burgio and Migliore 2015; Hoet 2013). Nevertheless, to overcome the limitations of *in vitro* testing and fully replace the use of animals, refined and validated toxicokinetics and metabolism-competent model systems are sorely needed to accurately predict or mirror the *in vivo* situation.

In Vivo Methods

In vivo methods are generally recommended to complete the information gathered during *in vitro* investigations and/or to overcome their limitations (Ouedraogo et al. 2012). The term "*in vivo*" refers to experimentations based on a whole, living organism – as opposed to a partial or dead organism – and consist of either animal studies or clinical trials (ECVAM 2015). For evident ethical reasons, genotoxicity studies are only performed in animals.

- Benefits: *In vivo* studies include pharmacokinetic factors that are able to influence the outcomes of toxicity assessment, which allows better extrapolation of potential noxious effects to humans. The number of test animals, gender, dosage, time, and the use of suitable controls are important parameters to consider (Ouedraogo et al. 2012; Hartung 2011).
- Limitations: *In vivo* tests contradict the 3R philosophy, and entitle chronic studies with longer durations and costs. Moreover, the metabolism of drugs can vastly differ among mammals, so both negative and positive data may not be transferrable to humans. *In vivo* tests, such as the bone marrow micronucleus test, are relatively insensitive (Muller et al. 1999), so the established *in vitro* genotoxicity tests are still considered first-line tests, as they are sensitive enough to detect the great majority of genotoxins. Several tests (typically on bone marrow, blood, or liver) are deemed unable to provide additional useful information as compared to *in vitro* assays, especially for compounds with poor systemic absorption, e.g., radioimaging agents and aluminum-based antacids (ICH 2014; FDA 2012).

Guidelines on the Genotoxicity Assessment

The OECD, the Food and Drug Administration (FDA), the International Agency for Research on Cancer (IARC), the European Medicines Agency (EMA), and the European Centre for the Validation of Alternative Methods (ECVAM) are all organizations that investigate the validation of tests and provide a general framework, practical approaches, and rules for data interpretation (EMEA 2008a). Guidelines have been established by OECD, ICH, and EMA committees (ICH 2014; OECD 2014) to optimize genetic toxicology testing for the prediction of potential human risks. They also provide guidance on the interpretation of results, describing internationally acknowledged standards for follow-up testing and interpretation of positive *in vitro* and *in vivo* results, including the assessment of non-relevant findings (ICH 2014). The "Guideline on non-clinical documentation for herbal medicinal products in applications for marketing authorization (bibliographical and mixed applications) and in applications for simplified registration", implemented by the EMA Committee on Herbal Medicinal Products (HMPC), establishes a step-by-step procedure for assessing the genotoxicity of herbal medicinal products (HMPs) (EMEA 2008b).

Tests Approved for Genotoxicity Testing

Guidelines for the assessment of the genotoxicity of pharmaceuticals

The registration of pharmaceuticals requires a comprehensive assessment of their genotoxic potential. The recommended battery of tests is described in the ICH Harmonised Tripartite Guideline Report S2, "Guidance on genotoxicity testing and data interpretation for pharmaceuticals intended for human use". The recommended battery includes a bacterial reverse mutation test (Ames test), which effectively detects genetic changes and the majority of carcinogens genotoxic to rodents and humans, and mammalian *in vitro* and/or *in vivo* tests, mandatory or decided on obtained data (ICH 2014). Three *in vitro* mammalian assays are widely used, considered sufficiently validated, equally appropriate, and therefore interchangeable for the measurement of chromosomal damage when used together with other genotoxicity tests: (1) the *in vitro* metaphase chromosome aberration assay; (2) the *in vitro* micronucleus assay; and (3) the mouse lymphoma L5178Y cell thymidine kinase gene mutation assay. *In vivo* assays are included in the test battery to help identify false negatives (e.g., agents mutagenic *in vivo* but not *in vitro*), and to provide insights on the influence of pharmacokinetics.

Guidelines for the assessment of the genotoxicity of herbal medicinal products

HMPs present a number of characteristics that differentiate them from other medicinal products, explaining the need for specific guidance. Herbal products are complex mixtures containing a large number of constituents that are sometimes present in highly variable amounts. The complete composition of a preparation is often unknown, and thus structural alerts for toxicants can be unraveled; moreover, the composition may vary with many parameters (harvesting time, geographical origin, mode of preparation, contamination, adulteration) that could invalidate previously obtained genotoxicity data. Nevertheless, HMPs are framed by similar regulations as for other medicinal products for human use; as with other medicinal products, signals of adverse effects could arise occasionally through pharmacovigilance (EMEA 2008a).

The HMPC stepwise testing process for HMPs involves a battery of genotoxicity tests, as described in a decision tree (Fig. 9.1 The stepwise testing process of herbal medicinal products, adapted from EMA (FDA 2012)).

Step 1: The Ames test

The Ames test, a bacterial reverse gene mutation test, should be performed and interpreted in conformity with existing OECD and EU guidelines. Briefly, a set of mutated *Salmonella typhimurium* strains, each auxotroph for a specific amino acid, is incubated under selected pressure (low level of the specific amino acid), and in the presence of the studied substance/preparation, with or without a metabolic activation system. Mutations occurring in the non-functional gene will restore the capability of bacteria to synthesize the specific amino acid ("revertants"). The number of revertants correlates quite well with the mutagenic potential of a substance (ICH 2014; Ouedraogo et al. 2012; OECD 2014; Hoet 2013; EMEA 2008a).

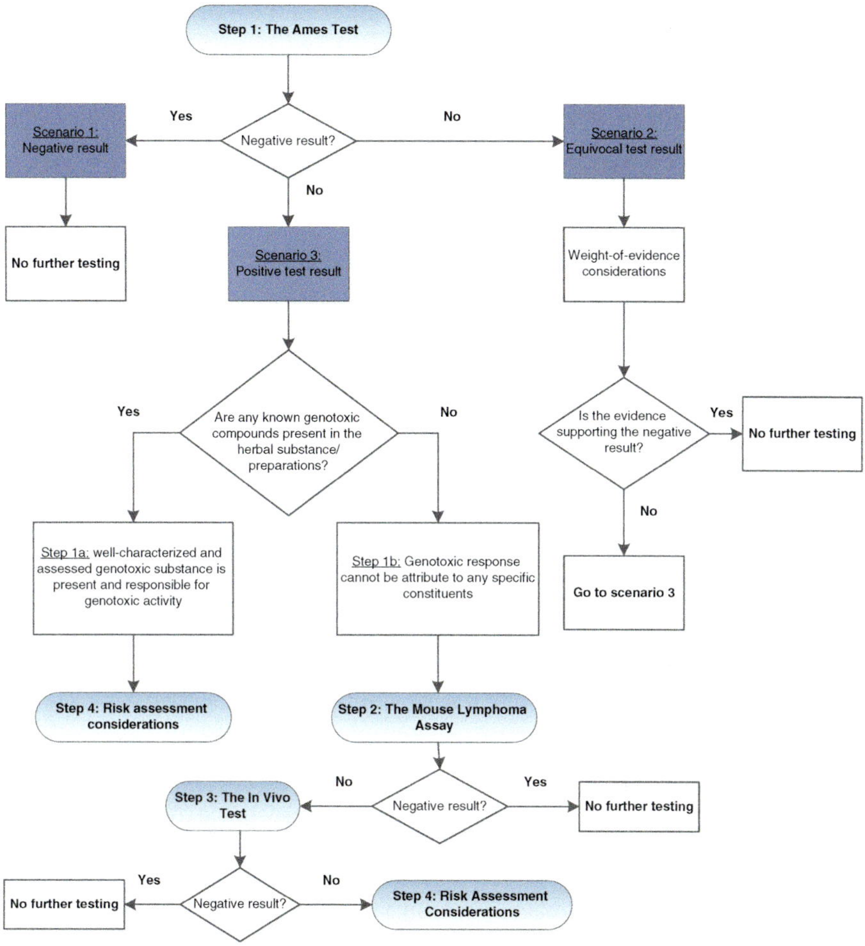

Fig. 9.1 The stepwise testing process of herbal medicinal products (Adapted from EMA (2008a))

- In the case of a negative result, no further genotoxicity testing is required.
- Equivocal test results require special considerations; a repetition of the experiment should generally be envisaged.
- In case of positive results:
  - the presence of acknowledged genotoxic compounds not known to be carcinogenic (e.g., quercetin) can tentatively explain the mutation.
  - the absence of such genotoxic compounds implies that the herbal product has to be studied in a Step 2 test.

Step 2: The mouse lymphoma assay or other mammalian cell assays

As for the Ames test, the mouse lymphoma assay should be performed and interpreted in conformity with existing OECD and EU guidelines. Briefly, L5178Y mouse lymphoma cells in culture are exposed to a compound or a preparation, and mutants in the thymidine kinase gene (mutation $TK^{+/-} \rightarrow TK^{-/-}$) are detected by their resistance to the cytotoxic pyrimidine analogue trifluorothymidine

(OECD 2014). This assay may confirm or refute positive findings in the Ames test. Moreover, it may give information on the ability of herbal products to cause chromosomal damage. If other mammalian cell assays are used for genotoxicity tests, their use has to be justified (OECD 2014; Hoet 2013).

- In the case of a negative result, no further testing is required.
- In the case of a positive result, the relevance of the finding should be thoroughly assessed, as it is known that the mouse lymphoma assay is associated with false positives. If the test is unequivocally positive (gene mutation or chromosomal damage), it is advisable to proceed to step 3. If the herbal preparation is known to contain a compound with chromosomal damaging properties, it may be advisable to perform the *in vitro* micronucleus test in mammalian cells in culture (EMEA 2008a).

Step 3: The rodent micronucleus test or other *in vivo* genotoxicity tests

The rodent micronucleus test should be performed and interpreted in conformity with the existing OECD and EU guidelines. Briefly, mice or rats are treated with the compound or preparation (in an appropriate vehicle and via appropriate route of administration). The proportion of micronuclei in bone marrow and/or peripheral blood cells can identify agents causing structural and numerical chromosome changes (OECD 2014; EMEA 2008a).

- In the case of a negative result, no further testing is required.
- In the case of a positive result, it is advisable to proceed to step 4.

Step 4: The risk assessment considerations

No single specific approach has been recommended for risk assessment (EMEA 2008a), since many points have to be taken into consideration. A risk assessment through the "Threshold of Toxicological Concern" approach is possible whenever a herbal preparation contains an identifiable genotoxic compound that presents a demonstrated threshold mechanism; permissible exposure levels – without appreciable risk of genotoxicity – can be established according to the usual "No Observable Effects Level" (NOEL) method. However, as herbal preparations are complex mixtures with partially unidentified components, it is quite possible that the compound(s) responsible for genotoxicity is/are still not identified at the end of the testing protocol. Thus, the usual procedure for toxicity testing and risk assessment of mixtures should consist in isolating and identifying various major constituents and testing them individually (EMEA 2008a) – which is a time-consuming, costly, and probably unrealistic approach for herbal medicines.

This HMPC stepwise testing process for herbal medicinal products effectively defines the Ames test as the primary endpoint which, if negative, accepts the drug as probably "non-genotoxic". This is not entirely satisfying, however (Ouedraogo et al. 2012), and has been greatly debated (EMEA 2008b); indeed (1) the Ames test does not detect every genotoxic insult; and (2) since some common compounds, including flavonoids, yield very positive Ames tests but are not carcinogens, they may effectively mask the genotoxic effect of real carcinogens.

Novel Approaches for Genotoxicity Testing

"Omics" studies involve a large number of measurements per endpoint to acquire comprehensive, integrated understanding of biology and to identify various factors simultaneously (e.g., genes, RNA, proteins and metabolites) rather than each of those individually. Toxicogenomics study the interactions between the structure and activity of the genome and the adverse biological effects of exogenous agents. The toxic effects of xenobiotics on biological systems are generally reflected at the cellular level by their impact on gene expression (transcriptomics), and on the production of proteins (proteomics) and small metabolites (metabonomics) (Ouedraogo et al. 2012; Borner et al. 2011). Genetic variation and expression signatures can be used to screen compounds for hazards, to assess cellular responses to various dosages, to classify toxicants on the basis of mechanisms of action, to monitor exposure of individuals to toxicants, and to predict individual variability in sensitivity to toxicants (NAP 2007). Toxicogenomics effectively allows understanding dose-response relationships, cross-species extrapolations, exposure quantification, underlying mechanisms of toxicity, and the basis of individual susceptibilities to particular compounds (NAP 2007). We recently reviewed novel approaches for genotoxicity testing, based on "omics" technologies, with their applications to herbal drugs, including their advantages and limitations (Ouedraogo et al. 2012).

Carcinogenicity Assessment

The assessment of carcinogenicity aims to identify a tumorigenic potential in animals and to evaluate the possible risk to humans (ICH 2015). Determining the carcinogenic potential is an important, complex, and imperfect exercise. The methods for such determinations are expensive and long, and they use many animals; moreover, the extrapolation of data from such studies to human risk is imprecise (Jacobson-Kram 2009).

Human Carcinogens Classification

Carcinogenic substances induce tumors (benign or malignant), increase their incidence or malignancy, or shorten the time for tumor occurrence. As discussed in section "Carcinogenicity", carcinogens are classified as either genotoxic or non-genotoxic, depending on their mode of action. They can also be classified as threshold-unlikely (DNA-reactive genotoxic compounds, for which NOAEL – no observed adverse effect level – cannot be estimated) or threshold-likely (non-DNA reactive genotoxins and non-genotoxic carcinogens, for which NOAEL can be estimated) (Hernandez et al. 2009). The IARC classifies human carcinogens into five groups depending on their carcinogenic potential to humans (Hernandez et al. 2009; IARC 2015):

- *Group 1*: Carcinogens to humans (117 agents). This group includes 61 chemicals, 9 viruses or pathogens (e.g., HIV), 19 exposure circumstances (e.g., chimney sweeping), and 16 mixtures (e.g., coal-tars) including as herbal products: areca nut, plants containing aristolochic acid, betel quid with and without tobacco, tobacco (smokeless, smoking, and passive smoking), and aflatoxins (from contamination by producing organisms).
- *Group 2A*: Probable carcinogens to humans (74 agents), including 50 chemicals, 2 viruses or pathogens, 7 exposure circumstances, and 7 mixtures, including as herbal products: emissions from high-temperature frying (applied to some herbal medicines processing) and hot beverages.
- *Group 2B*: Possible carcinogens to humans (287 agents), including 224 chemicals, 4 viruses or pathogens, 7 exposure circumstances, and 13 mixtures, including as herbal products: *Aloe vera*, whole leaf extract, bracken fern, coffee (urinary bladder), *Ginkgo biloba* extract, goldenseal root powder, kava extract, and pickled vegetables (traditional Asian), toxins derived from *Fusarium moniliforme* (from contamination by producing organisms). Moreover, monocrotaline and safrole are also both found in this group.
- *Group 3*: Non classifiable carcinogens (503 agents), including 496 chemicals, 8 exposure circumstances, and 11 mixtures, including as herbal products: Madder root (*Rubia tinctorum* L.) and mate (*Ilex paraguariensis* A.St.-Hil.). Retrorsine and eugenol are also both classified in this group.
- *Group 4*: Probably not carcinogenic to humans. This group contains a single agent: caprolactam.

Guidelines for Carcinogenicity Assessment of Pharmaceuticals

The strategy for testing the carcinogenic potential is developed according to the results of genetic and repeated-dose toxicology studies (EMA guidelines S2A and S2B), pharmacodynamics (selectivity, dose-response), and pharmacokinetics in animals and in humans (Guideline S1C), intended patient population, and clinical dosage regimen (guideline S1A) (ICH 2015). Nowadays, there is no carcinogenic assessment strategy recommended or required for HMPs. Guidelines for the carcinogenicity testing of pharmaceuticals could probably be extrapolated to them whenever deemed necessary; however, most of the time, the carcinogenic potential of an HMP is not evaluated and only genotoxicity data are used to "predict" carcinogenicity.

Factors to Consider for Carcinogenicity Testing of Pharmaceuticals

Duration and exposure

Carcinogenicity studies should be performed for any pharmaceutical for which expected clinical use is continuous for at least 6 months or repeated in an intermittent manner (e.g., for depression, anxiety, or allergy). Pharmaceuticals administered infrequently or for short durations of exposure (e.g., anesthetics and radiolabeled imaging agents) do not require carcinogenicity studies unless there is cause for concern (ICH 2015).

Cause for concern

Carcinogenicity assays may be recommended for some pharmaceuticals if there is concern about their carcinogenic potential, which includes: (1) previous demonstration of carcinogenic potential in the product class considered; (2) structure-activity relationship suggesting carcinogenic risk; (3) evidence of preneoplastic lesions in repeated-dose toxicity studies; and (4) long-term tissue retention of parent compound or metabolite(s), resulting in local tissue reactions or other pathophysiological responses (ICH 2015).

Genotoxicity

Unequivocally, genotoxic compounds (in the absence of other data) are presumed to be carcinogens and warrant long-term carcinogenicity studies (ICH 2015). In addition to their use as a screening tool, genotoxicity data constitute part of the weight of evidence when evaluating environmental chemicals and herbal medicines (HMPC strategy). In practice, environmental contaminants have not been regulated as carcinogens on the basis of positive genotoxicity results alone. Nonetheless, positive tests are generally indicative of chemicals capable of inducing cancer via a genotoxic or mutagenic activity (Guyton et al. 2009).

Indication and patient population

Pharmaceuticals developed to treat life-threatening or severely debilitating diseases do not always require carcinogenicity testing before market approval; this is particularly the case for anti-cancer agents. These time-consuming studies can be conducted post-approval in order to speed the availability of the product on the market (ICH 2015).

Route of exposure

If possible, the route of exposure in animals should be the same as the intended clinical routes. If similar metabolism and systemic exposure can be demonstrated by differing routes of administration, carcinogenicity studies should be conducted only by a single route (ICH 2015).

Extent of systemic exposure

Pharmaceuticals applied topically (e.g., dermal and ocular routes of administration) may need carcinogenicity studies. However, pharmaceuticals showing poor systemic exposure from topical routes in humans may not need oral administration studies. Moreover, for various salts, acids, or drug bases, evidence of no significant changes in pharmacokinetics, pharmacodynamics, or toxicity should be provided (ICH 2015).

Tests Approved for Carcinogenicity Assessment

The assessment of the carcinogenic potential of pharmaceuticals usually involves two rodent species (2-year rodent bioassay – mostly rat and mouse) (ICH 2015; Raghava et al. 2014). The species are selected according to data on pharmacology, repeated-dose toxicology, metabolism, toxicokinetics, and route of administration. In the absence of a clear advantage favoring a species, rat models are recommended (ICH 2015).

Further mechanistic studies are often useful for the interpretation of carcinogenicity data and can provide a perspective on their relevance in humans; these may investigate (1) cellular changes in relevant tissues, using morphological, histochemical, or functional criteria, e.g., dose-relationship for apoptosis, cell proliferation, liver foci, alteration or changes in intercellular communication; and (2) biochemical measurements, e.g., plasma hormone levels, growth factors, binding proteins (i.e., α2μ-globulin) and tissue enzyme activity (ICH 2015). In some cases, additional genotoxicity testing in appropriate models may be required (Butterworth 2006); this would be the case of compounds resulting in negative outcomes in the standard test battery, but which demonstrated effects a carcinogenicity test with no clear evidence for an epigenetic mechanism. This additional testing can include modified conditions for metabolic activation in *in vitro* tests or can include *in vivo* tests measuring genotoxic damage in target organs (ICH 2015).

Limitations of the 2YR Strategy

For each compound, the 2YR strategy requires more than 800 rodents with more than 40 histopathological tissue analyses for each of them. The cost of this approach can reach US $2.4 million per compound, depending on the route of administration, number of doses, and the chemical to evaluate. Thus, in addition to denying the 3R policy, the 2YR strategy is costly and time-consuming (Fjodorova et al. 2010; Raghava et al. 2014). As a result, only an estimated 1,500 of the 84,000 chemicals available for commercial use have been tested so far. Moreover, the relevance of animal models to human carcinogenicity risk has been seriously questioned (Raghava et al. 2014).

Novel Approaches for Carcinogenicity Testing

The short-term tests currently used to predict a chemical's ability to induce cancer are implemented based on scientific evidence that emerged in the 1970s, when links between DNA damage and mutation were described. Accordingly, these screening methodologies firstly aimed at identifying genotoxic agents under the premise that such agents would most likely pose cancer risks in humans (Guyton et al. 2009). Novel approaches aim at identifying all types of carcinogens, but with varying efficacy.

Structure-activity relationship (SAR)

As for genotoxicity, *in silico* and structure alert methods have been proposed, with the same advantages and limitations. The OECD principles for the validation of such models must be defined by a precise endpoint, an unambiguous algorithm, goodness-of-fit, robustness, predictivity, and applicability domain (Hernandez et al. 2009; Ouedraogo et al. 2012).

Replicative DNA synthesis (RDS)

In eukaryotic cells, the regulatory mechanisms for DNA replication are crucial to control the cell cycle (Rizwani and Chellappan 2009); as numerous (non-genotoxic) carcinogens are mitogenic inducers, an increase in cellular proliferation can be investigated by measuring the rate of replicative DNA synthesis upon exposure of cell cultures to tested agents. The major advantages of the RDS test are the *in vivo* response to the agent and the short duration of the assay. On the other hand, disadvantages are characterized by the requirement for high doses, false positives obtained because of regenerative cell proliferation (due to acute toxicity), and false negatives obtained if the studied organ is not the primary target for the agent. For these reasons, the RDS test should be performed in conjunction with other short-term assays (Hernandez et al. 2009).

In vitro cell transformation assay

These assays detect the carcinogenic potential of a chemical through morphological transformation of primary cultured cells. The main advantages of this assay are the use of a normal cell line, the low spontaneous transformation, the capacity for metabolic activation, the rapidity of phenotypic changes, and reproducibility. On the other hand, the major disadvantage is that the induction of cells' transformation is not yet fully understood. Thus, results of this assay need to be evaluated with caution. Other disadvantages include inter- and intra-laboratory variations due to subjectivity in the scoring of transformed cells, the requirement for regular preparation of primary cultures, variation in cloning efficiency, and transformation frequency due to the composition of culture serums and the use of an initiation-promotion protocol in order to enhance the transformation frequency (Hernandez et al. 2009; OECD 2014).

Toxicogenomics

There is evidence that suggests that gene expression profiles in model organisms or cells exposed to various compounds reflect underlying biological mechanisms of action and can be used in higher throughput assays to predict toxicity and, notably, carcinogenicity. Predicting the carcinogenicity of genotoxic and non-genotoxic compounds has been assessed from the expression profiles of exposed cell cultures, tissues, and animals, indicating that gene expression-based carcinogenicity prediction is possible (Hernandez et al. 2009; Raghava et al. 2014). Toxicogenomic methods have progressed to the extent that it may be possible to use them in acute or sub-chronic studies to predict carcinogenicity. Several research groups have recently identified cancer-relevant gene sets that can discriminate carcinogenic from non-carcinogenic compounds. Proof-of-concept studies using advances in toxicogenomic have provided an initial demonstration of the utility of these assays as predictive tools. However, further exploratory research, as well as validation efforts, are still needed (Guyton et al. 2009).

Genotoxic and Carcinogenic Herbal Products

Although genotoxicity data are being generated for a growing number of medicinal plants, relatively few herbs have been proven to be carcinogenic. Until now, carcinogenicity information is still lacking, and safety information mostly relies on genotoxicity testing; some carcinogens (notably indirect) are probably not detected, which presents a rather unsatisfying situation.

Pyrrolizidine Alkaloids

Pyrrolizidine alkaloids (PA), secondary metabolites found in 12 Angiosperm plant families, are produced for defense against herbivore insects. To date, more than 660 pyrrolizidine alkaloids and their N-oxide derivatives have been identified in over 6,000 plants grown virtually worldwide, including Africa, the West Indies, China, Jamaica, Canada, Europe, New Zealand, Australia, and the U.S (Xia et al. 2008; Wang et al. 2005a, b). The majority of these compounds are found in different genera from three botanical families: Boraginaceae, Compositae (Asteraceae) and Leguminosae (Fabaceae) (Wang et al. 2005a; Fu et al. 2002). The genus *Senecio* (Compositae) is particularly concerned. PAs-containing plants are probably the most common poisonous plants affecting livestock, wildlife, and humans, and PAs are among the first naturally occurring carcinogens identified in plants (Xia et al. 2008). People are exposed to PAs not only by the consumption of traditional medicines or herbal teas made from PA-containing plants, such as comfrey (*Symphytum officinale* L.) (Xia et al. 2008), but also by the consumption of contaminated human foodstuffs such as milk, honey, grains, herbal medicines, and dietary supplements (Wang et al. 2005b).

The classification of PAs is mostly based on the identity of the necine base, the presence or absence of a macrocyclic structure esterifying the alcohol groups, and the number of its members (e.g., 11-, 12- or 13-membered macrocycles), stereochemistry, and patterns of hydroxylation (Langel et al. 2010) (Fig. 9.2).

Pyrrolizidine alkaloids, particularly those from plants such as *Senecio, Crotalaria, Heliotropium* and *Amsinckia*, are highly toxic compounds, exhibiting acute toxicity, chronic toxicity and genotoxicity. Acute toxicity results in hepatic veno-occlusive disease, causing massive hepatotoxicity with hemorrhagic necrosis. Chronic poisoning takes place mainly in the liver, lungs, and blood vessels, and in some instances the kidneys, pancreas, gastrointestinal tract, bone marrow, and brain. Exposure over a longer period of time causes cell enlargement (megalocytosis),

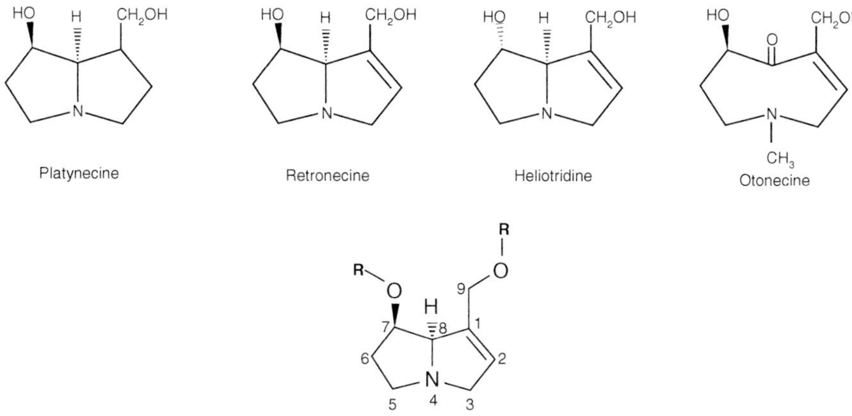

Fig. 9.2 Common necine bases of pyrrolizidine alkaloids (Fu et al. 2002)

veno-occlusion in liver and lungs, steatosis, nuclei enlargement with increasing nuclear chromatin, loss of metabolic function, inhibition of mitosis, proliferation of biliary tract epithelium, liver cirrhosis, nodular hyperplasia, and adenomas or carcinomas (Wang et al. 2005b; Langel et al. 2010; Mädge et al. 2015). PAs require metabolic activation to exert their genotoxicity and tumorigenicity. Upon ingestion, 1,2-unsaturated PAs (Fig. 9.2*:* Common necine bases of pyrrolizidine alkaloids) (Wang et al. 2005b) are oxidized by cytochromes P-450 to reactive pyrrolic bifunctional electrophiles that are potent DNA linkers, an event reputed critical in their toxicity and carcinogenesis. The platynecine-type pyrrolizidine alkaloids that do not harbor a 1–2 double bond have been found to be non-genotoxic. *In vivo* and/or *in vitro* metabolism of the tumorigenic retronecine-type (e.g., riddelliine, retrorsine, and monocrotaline), heliotridine-type (e.g., lasiocarpine) and otonecine-type (e.g., clivorine) pyrrolizidine alkaloids all generate a common set of 6,7-dihydro-7-hydroxy-1-hydroxymethyl-5H-pyrrolizine (so-called "DHP")-derived DNA adducts responsible for tumor induction (Xia et al. 2008), and for most of the genotoxicity of the parent pyrrolizidine alkaloids (Fu et al. 2002). There are three principal metabolic pathways, mainly in the liver (CYP3A and CPY2B6 isozymes) (Fu et al. 2002): (1) hydrolysis of the ester functional group to form the necine bases; (2) oxidation of the necine bases to the corresponding necine N-oxides (heliotridine type and retronecine type); and (3) hydroxylation at the C-3 or C-8 position of the necine bases to form 3- or 8-hydroxynecine derivatives followed by dehydration, to form the corresponding dehydropyrrolizidine (pyrrolic) derivatives. The third pathway is generally considered to be the metabolic activation responsible for intoxication, whereas N-oxidation and hydrolysis are considered to be detoxifying pathways; pyrrolic ester metabolites are very reactive and can bind to one or two molecules of glutathione to form glutathione conjugates for excretion, which is a further detoxification pathway (Fu et al. 2002). The genotoxicity of pyrrolizidine alkaloids includes DNA binding, DNA cross-linking, DNA-protein cross-linking, mutagenicity, and carcinogenicity (Mädge et al. 2015).

Volatile Alkenylbenzenes from Essential Oils

Safrole

Safrole (Fig. 9.3) is a phenylpropenic compound constituting up to 80% of the essential oil of sassafras (*Sassafras albidum* (Nutt.) Nees) root bark, a tree native to the northeast U.S. that is used for medicinal and culinary purposes, especially as a flavoring agent for beverages such as root beer (Segelman et al. 1976). It is found in other species, such as nutmeg and mace (*Myristica fragrans* Houtt.), *Ocotea pretiosa* Mez. (synonym of *O. odorifera* (Vell.) Rohwer), *O. cymbarum* Kunth, *Cinnamomum camphora* (L.) J. Presl (used for the production of camphor), in betel quid (leaves of *Piper betle* L.), and in areca nut (*Areca catechu* L.). Betel quid and areca nut are widely chewed in southeast Asian countries for their addictive psycho-stimulating effects; their regular consumption has, however, been linked to a 50-times increase in the prevalence of oral cancers (Thomas and MacLennan 1992; Chen et al. 1999). Safrole has also been shown to be a weak hepatocarcinogen (Homburger et al. 1965; Miller and Miller 1976; Wislocki et al. 1977). The carcinogenic effects of safrole have been recognized since 1960, when the U.S. FDA prohibited its use in food (U.S. Food and Drug Administration 1973). The IARC classifies betel quid and areca nut as acknowledged carcinogens to humans (IARC 2004).

The genotoxicity of safrole was further investigated on mammalian cells; it was found to induce chromosomal aberrations, gene mutations, and sister chromatid exchange (European Comission: Scientific Committee on Food 2002) and to trigger unscheduled DNA synthesis in cultured rat hepatocytes – but not in HeLa cells – and DNA damage (single-strand breaks) in cultured rat hepatocytes. The genotoxicity is not mediated through safrole itself, but rather from its activation into 1′-hydroxysafrole by cytochromes P450 2C9 and 2E1 (Ueng et al. 2004). This compound is subsequently sulfonated into an unstable sulfuric acid ester capable of forming adducts with DNA (Chung et al. 2008). Other oxidized metabolites, such as 1′-acetoxysafrole, safrole-2′,3′-oxide, 1′-acetoxysafrole, and 1′-oxosafrole are also suspected of being genotoxic (European Commission: Scientific Committee on Food 2002).

In vivo, safrole is able to induce chromosome aberrations, sister chromatid exchange, and DNA adducts in the hepatocytes of rats (Daimon et al. 1998). These DNA adducts, tentatively identified as N^2-(trans-isosafrol-3′-yl)2′-deoxyguanosine and N^2-(safrol-1′-yl)2′-deoxyguanosine, suggest that safrole is a genotoxic carcinogen in the liver. The presence of these adducts was effectively confirmed in the hepatic tissues from patients who developed hepatocellular carcinoma (Chung et al. 2008).

In another *in vivo* study, myristicin (methoxy-safrole, the major flavoring compound of nutmeg) along with safrole, were shown to induce hepatic DNA adducts in adult and fetal mice (Randerath et al. 1993). Transplacental passage of safrole's reactive metabolites was highlighted by the presence of DNA adducts in the livers of fetal mice whose mothers were exposed to safrole (Randerath et al. 1989). It is suggested that the dosages of safrole and myristicin ingested during the consumption of nutmeg are significantly lower than those that could be associated with psychogenic and toxic effects (Bruneton 2009).

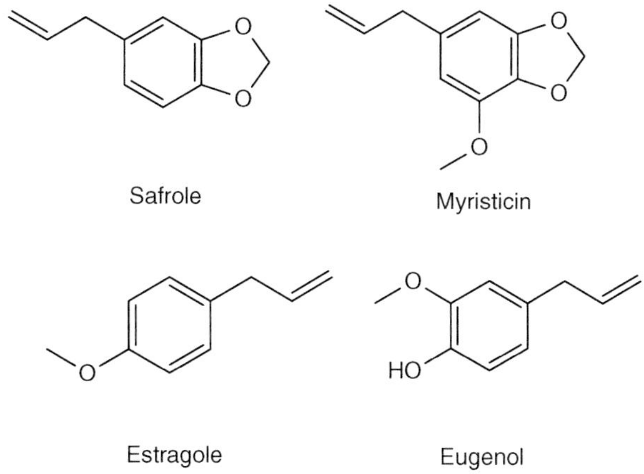

Fig. 9.3 Molecular structures of safrole, myristicin, estragole, and eugenol

Estragole

Estragole (Fig. 9.3) is present in many culinary herbs, including anise, star anise, basil, bay, tarragon, fennel, and marjoram. Widespread human exposure to estragole occurs through the consumption of these herbs and through the use of their essential oils as flavors and fragrances in numerous foods, cosmetics, and other consumer products (EFSA 2009). Previously recognized as safe and approved by the U.S. FDA for food use, estragole and its metabolites have been shown to be mutagenic in bacterial systems (Ames test) and to produce hepatomas in a susceptible strain of mice. The carcinogenicity of estragole proceeds through a genotoxic mechanism upon liver metabolism into 1′hydroxyestragole and several epoxide compounds; both estragole and its hydroxylated metabolite induce hepatic tumors in CD-1 or B6C3F1 mice either after dietary chronic exposure or after i.p. or s.c. injections, prior to or after weaning (males appear to be more susceptible than females) (Council of Europe 2005). Further strong supporting evidence of carcinogenicity comes from comparison with compounds structurally similar to estragole (e.g., safrole, methyleugenol), which produce liver tumors and tumors at other sites in rodents.

Eugenol

Eugenol (Fig.9.3) is a widely distributed component of essential oils. It is a major constituent of clove oil and is found in several spices including basil, cinnamon, and nutmeg (Zhou et al. 2013). It has been used since at least the nineteenth century, primarily as a flavoring agent in a variety of foods, pharmaceutical products, and as an analgesic and antiseptic in dental care.

Eugenol has been investigated for its carcinogenicity in mice and rats by oral administration of a diet containing various eugenol concentrations. At high dosages (diets containing 12,000 ppm of eugenol), it induced a significant increase in the incidence of liver tumors in female mice, whereas in males, the increase was significant only for those receiving the lower dosage (dietary level of 3,000 ppm of eugenol) (Carcinogenesis Studies of Eugenol 1983). Cytochrome P450-catalyzed metabolism has been suggested as a possible major bioactivation pathway *in vitro* (Munerato et al. 2005).

Other studies in mice by oral administration, skin application and intraperitoneal injection were inadequate for an evaluation of carcinogenicity, mainly due to the short duration of treatment. Thus there is only limited evidence of the carcinogenicity of eugenol in experimental animals. In the absence of epidemiological data, no evaluation could be made on the carcinogenicity of eugenol to humans. By contrast, methyleugenol – a derivative of eugenol also found in numerous dietary herbs – has shown clearer evidence of its hepatocarcinogenic activity in rodents (NTP 2000), which could be tentatively explained by the formation of DNA adducts (Williams et al. 2013).

Asarones

Asarones (α-, β- and γ-asarone; Fig. 9.4) are alkenylbenzenes isolated from a wide variety of herbs, including *Acorus calamus* L., *Acorus tatarinowii* Schott, *Asarum europaeum* L., *Asarum forbesii* Maxim., *Mosannona depressa* (Baill.) Chatrou, *Orthodon asaroniferum* Fujita, *Orthodon isomyristicineferum* Fujita[1] or *Piper lolot* DC (Zhou et al. 2013; Niir 2003; Chamorro et al. 1998).

α-asarone is an acknowledged hypolipidemic agent (Cassani-Galindo et al. 2005) that inhibits the hepatic 3-hydroxy-3-methylglutaryl coenzyme A reductase (Rodriguez-Paez et al. 2003); proposed for the prevention of atherosclerotic disease, the compound proved, however, to be genotoxic. The carcinogenic activity of β-asarone has been known since a 1967 toxicity study of *Acorus calamus* L. root oil in rodents. The herb contains up to 80 % of β-asarone and was formerly used as a flavoring agent in food and beverages (Taylor et al. 1967; Abel and Göggelmann 1986).

Numerous *in vitro* works also confirmed the genotoxicity of α-asarone. Exposure of murine connective tissue cells (L929 cell line) induced DNA fragmentation measured by the comet assay; also, in human lymphocytes, increased sister chromatid exchanges were observed (Morales-Ramírez et al. 1992). The Ames test does not denote any mutagenic effect of α-asarone (Marczewska et al. 2013) unless the compound is metabolically activated by preincubation with the S9 fraction (Cassani-Galindo et al. 2005), through cytochrome P450-mediated hydroxylation and sulfation. The subsequent loss of the sulfate moiety generates carbonium cations, which are able to react with DNA, thus triggering a genotoxic potential. Similarly, chromosomal aberrations were observed in human lymphocytes when β-asarone was metabolically activated (Abel and Göggelmann 1986). Due to its moderate capacity of increasing sister chromatid exchange, α-asarone has been shown to be mutagenic in both human lymphocytes *in vitro* and murine bone marrow cells *in vivo* (Kevekordes et al. 2001). The α-asarone reaches the gonads of male rats,

[1] These are not accepted botanical names; Fujita has classified *Orthodon* species according to their essential oils composition and, to the best of our knowledge, no relationship between botanical identification and essential oil classification has been established (Niir 2003).

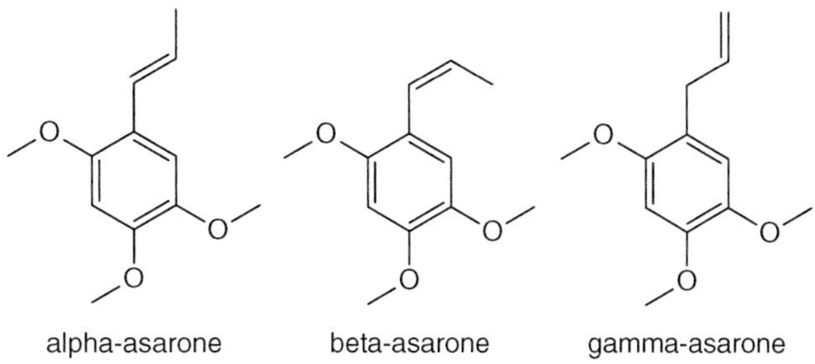

Fig. 9.4 Molecular structures of α-, β – and γ-asarones

affects the concentration and motility of spermatozoids and induces a teratogenic activity (Unger and Melzig 2012); the pregnancies developing after mating with exposed male rats resulted in an increased incidence of post-implantation loss and fetal malformations (Abel and Göggelmann 1986; Unger and Melzig 2012). An induction of micronuclei formation was observed in human hepatoma cells (Hep G2) and increased with exposure to α-asarone (Kevekordes et al. 2001). This result is, however, subject to controversy: using the same cell line, another team found that β-asarone – but not α-asarone – induced the formation of micronuclei (Unger and Melzig 2012). Furthermore, α-and β-asarones showed different cytotoxic profiles as revealed by a cellular proliferation (BrdU) assay; the more pronounced cytotoxicity of α-asarone was tentatively explained by an increased metabolism into the cytotoxic but non-genotoxic 2,4,5-trimethoxycinnamic acid (Hasheminejad and Caldwell 1994).

In vivo, asarones triggered unscheduled DNA synthesis in rat hepatocytes, suggesting genotoxicity and reinforcing the assumption of a hepatocarcinogenic potential (Hasheminejad and Caldwell 1994; Howes et al. 1990). The unscheduled DNA synthesis could be prevented when a cytochrome P450 inhibitor (cimetidine) was administered concomitantly.

Anthraquinones from **Rubia tinctorum** *L. and* **Morinda officinalis** *F.C. How*

Rubia tinctorum L. (madder) is a plant that grows in southern Europe, western Asia, and North Africa, and is cultivated elsewhere. Its roots, known as "madder roots", are used for dyeing (red coloring matter from roots), treating kidney and bladder stones, as a laxative, as a mild sedative and for menstrual and urinary disorders (IARC 2002). Anthraquinones are the main bioactive compounds found in *Rubia tinctorum* L.

The fresh roots of *Morinda officinalis* F.C. How have been used as a Chinese folk medicine for their tonic and analgesic properties. A number of compounds have been isolated from *M. officinalis,* including anthraquinones, terpenoids and scopoletin (IARC 2002; Zhang et al. 2010). Among the compounds found in these two herbs, 1-hydroxyanthraquinone and 1,3-dihydroxy-2-hydroxymethylanthraquinone (lucidin) are known to be potential carcinogens (Fig. 9.5).

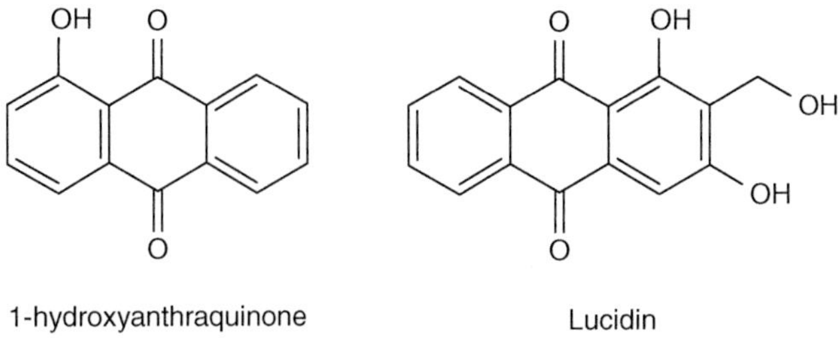

Fig. 9.5 Molecular structures of two genotoxic anthraquinones

1-hydroxyanthraquinone has been isolated from the roots of *Rubia cordifolia* L., *Morinda officinalis* F.C. How and *Damnacanthus indicus* C.F.Gaertn., from the heartwood of *Tabebuia avellanedae* (Mart. ex DC.) Mattos, and the herb of *Cassia occidentalis* L. In rats, 1-hydroxyanthraquinone has also been identified as a metabolite of alizarin primeveroside, found in *Rubia tinctorum*. Lucidin has been identified in plants from several genera, such as *Rubia, Coprosma, Morinda, Galium, Hymenodictyon* and *Commitheca*.

Madder root caused an increase in hepatocellular adenomas, and adenomas and carcinomas of the renal cortex in male and female rats in a single experiment. The 1-hydroxyanthraquinone is capable of inducing DNA repair synthesis in rat hepatocytes, suggesting a genotoxic potential (Kawai et al. 1986). Following oral administration, it induced adenocarcinomas of the large intestine, highlighting a carcinogenic activity (IARC 2002). Although no data have been obtained regarding the carcinogenicity of lucidin, the compound is highly suspected to portray similar properties.

Many medicinal plants used as laxatives (senna, cascara, frangula, rhubarb, and aloe) harbor anthraquinone glycosides as active principles (Bruneton 2009). Very little is known about their potential carcinogenicity, and these plants may probably be considered to be safe. Nevertheless, 1,8-dihydroxyanthraquinone (hydrolysis product of sennosides, the laxative ingredients of senna), formerly marketed as a laxative medicine (Dantron®), was withdrawn from the market in the United States in 1987 after it was shown to cause intestinal tumors *in vivo* (National Toxicology Program 2011).

Aristolochic Acids

Aristolochic acids (AA) are phenanthrene cyclic molecules found throughout herbs belonging to the *Aristolochia* and *Asarum* genera (Fig. 9.6). They are known for their nephrotoxicity as well as their genotoxic, mutagenic, and carcinogenic potential (Bruneton 2005; Barnes et al. 2007; Michl et al. 2014; Heinrich et al. 2009), and they have therefore been listed as poisonous plants and are prohibited in many countries (Zhou et al. 2013).

Species such as *Aristolochia clematitis* L. or *Aristolochia serpentaria* L. have been traditionally used in Europe as diuretic, emmenagogue, or oxytocic herbal medicines. Despite their toxic effects – which have only been confirmed in the 1960s by *in vivo* models (Bruneton 2005) – *Aristolochia* species have been included in numerous traditional medicines worldwide and are still frequently identified as

Fig. 9.6 Structures of AAI and AAII

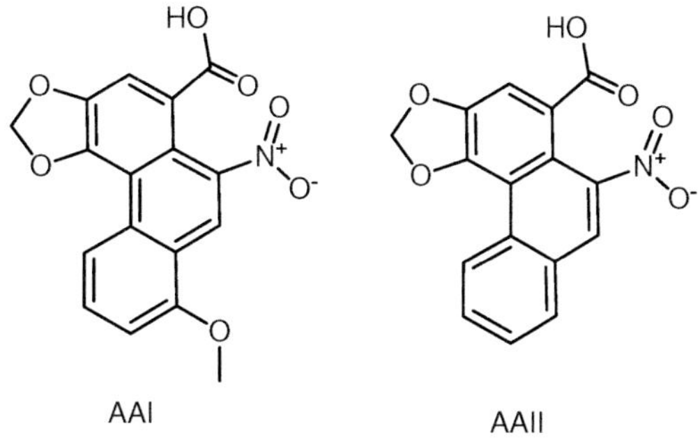

responsible for nephrotoxicity and/or carcinogenicity cases; indeed, in TCM *Aristolochia fangchi* Y.C.Wu ex L.D. Chow and S.M. Hwang has been used as a diuretic and for the treatment of rheumatism (Heinrich et al. 2009). The so-called "aristolochic acid nephropathy" (AAN) outbreak that took place in Belgium in the 1990s highlighted the nephrotoxic potential of *Aristolochia* species; a cohort of about 120 patients experienced a rapidly progressing renal interstitial fibrosis after administration of slimming capsules inadvertently containing *Aristolochia fangchi* in place of *Stephania tetrandra* S. Moore (Vanherweghem et al. 1993; Debelle et al. 2008). This regular intake of AA-containing herbs led to the complete loss of renal structure and function, requiring patients to undergo renal replacement therapy by dialysis or transplantation (Vanherweghem et al. 1993; Nortier et al. 2000).

During the follow-up of AAN patients, DNA adducts were identified in five kidney biopsies. Because of the involvement of these adducts in tumorogenesis, AAN patients appeared to be at risk of cancer development. This was confirmed in 1994, when cellular atypia was observed in three patients with AAN, throughout the urothelium of the kidneys removed during transplantation (Nortier et al. 2000; Nortier and Vanherweghem 2002). In 1997, from 39 patients who agreed to undergo prophylactic nephrectomy, 18 (46%) were positive for urothelial carcinoma, 19 had mild to moderate urothelial dysplasia, and 2 had normal urothelia. However, all renal samples were found positive for AA-DNA adducts, confirming (if still necessary) that the exposure to AA is responsible for the urothelial carcinoma onset.

Worldwide, 99 *Aristolochia* species have been identified as medicinal herbs used to treat a wide variety of ailments (Michl et al. 2013, 2014). In the Balkans, the consumption of contaminated wheat flour was identified as the cause of the so-called "Balkan endemic nephropathy". In 2007, the presence of AA-DNA adducts in renal biopsies confirmed the involvement of *Aristolochia clematitis* L. at the onset of the disease (De Broe 2012). In Maghreb, *Aristolochia baetica* L. and *Aristolochia debilis* Sieb and Zucc are still frequently used for the treatment of cancer, digestive tract disorders, and diabetes (Bellakhdar 1999; Yamani et al. 2015). In China, between 1964 and 1999, only five cases of AAN related to the consumption of *A. fangchi* or *Aristolochia manshuriensis* were reported (Li and Wang 2004). In 2008, the number of cases rose to 116 (Debelle et al. 2008), and it is expected that the incidence of the disease is still increasing, as Chinese herbalists still probably use aristolochia-containing remedies despite their prohibition.

In cytoplasm, AA undergo an enzymatic nitroreduction, leading to the formation of aristolactames (Debelle et al. 2008). These reactive metabolites are capable of forming DNA adducts: the positive charge of N-acylnitrenium ions can be delocalized and can react with amine functions of the puric bases adenine and guanine (Fig. 9.7).

AA-DNA adducts can persist for years after ingestion of *Aristolochia* and have thus been proposed as potential biomarkers of exposure (Nortier et al. 2013). These carcinogenic properties of AA are supported by the formation of DNA adducts and by the characteristic transverse mutation A→T in the p53 tumor suppressor gene (Gokmen et al. 2013). This is especially the case of dA-AAI adduct, which is found more frequently and is considered to be highly mutagenic (Nortier et al. 2013). An overexpression of P53 protein has been highlighted in AAN-associated urothelial cancers, suggesting that the p53 gene was mutated (Debelle et al. 2008), as was confirmed by the identification of a specific AAG to TAG mutation in codon 139 (Lys-Stop) of exon 5 of p53 gene. In a rodent model, the A→T mutation was also observed in codon 61 of the H-ras oncogene and may be responsible for tumorigenesis as well (Debelle et al. 2008).

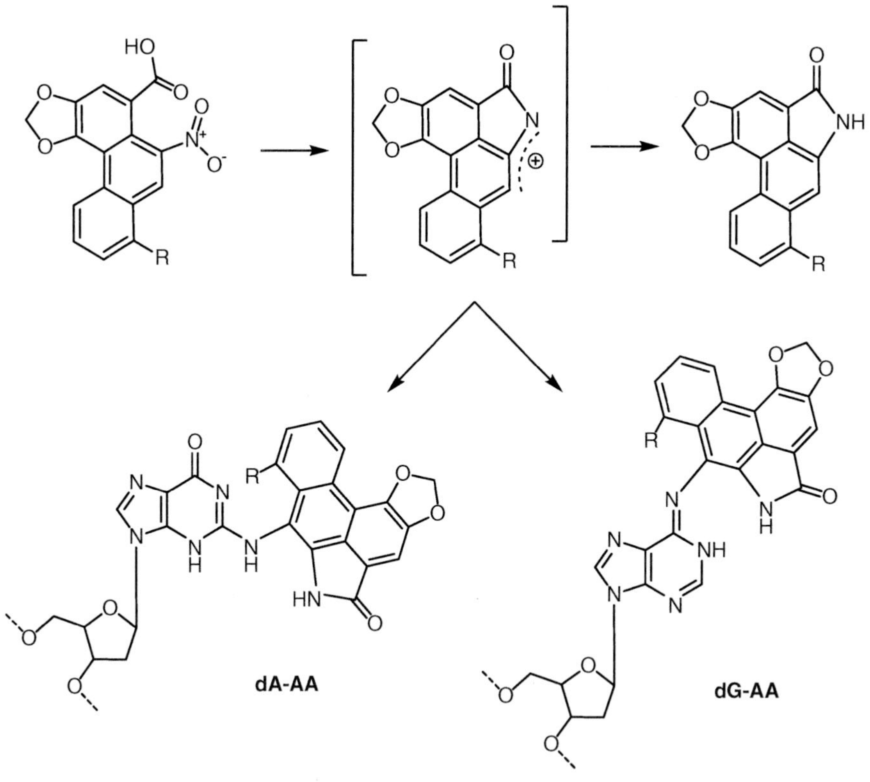

Fig. 9.7 Metabolic activation of AA (AAI: R = OCH_3/AAII: R = H) during which the nitroreduction leads to the corresponding aristolactams. The intermediate nitrenium cation can bind to DNA and form adducts, notably with adenine (dA-AA) and guanine (dG-AA) bases

Summing up

Table 9.1 provides an overview of the genotoxicity and carcinogenicity demonstrated for compounds described in this chapter. In practice, substances giving positive genetic toxicity data are considered to be carcinogenic until proven otherwise. However, it has become clear that many non-carcinogenic natural compounds (e.g., the flavonoid quercetin) produce misleading positive results in regulatory genotoxicity assays (Ames test). Given the wide variety of modes of action for carcinogenicity, the evaluation of natural products and herbal extracts still has to be carried out case by case, based on the weight-of-evidence approach. This method assesses the weight of all epidemiological and experimental data available, taking into account their strengths and weaknesses (Hernandez et al. 2009; Walmsley and Billinton 2011; Guyton et al. 2009; Berg et al. 2011).

Table 9.1 Overview of the genotoxicity and carcinogenity for the various compounds described in this chapter

Group	Compound	Main botanical families	Genotoxicity	Carcinogenicity	References
Pyrrolizidine alkaloids	Retrorsine, Heliotrine, Monocrotaline	Boraginaceae Compositae (Asteraceae) Leguminosae (Fabaceae)	Yes	Yes	Wang et al. (2005a), Fu et al. (2002)
Alkenylbenzenes	Safrole	Lauraceae Myristicaceae	Yes	Yes	Segelman et al. (1976), Thomas and MacLennan (1992), Chen et al. (1999)
	Myristicin	Myristicaceae	Yes	Yes	Barnes (2007)
	Estragole	Asteraceae	Yes	Yes	Barnes (2007)
	Eugenol	Myrtaceae	No	Equivocal	Barnes (2007)
	Asarone	Acoraceae	Yes	Equivocal	Ouedraogo et al. (2012), Niir (2003), Chamorro et al. (1998)
	Anthraquinones	Rubiaceae	Probable	Probable	IARC (2002), Kawai et al. (1986)
Nitrophenanthrene Carboxylic acid	Aristolochic acids I and II	Aristolochiaceae	Yes	Yes	Debelle et al. (2008), Nortier et al. (2013)

Conclusion

All effective drugs may produce adverse drug reactions, and herbal medicinal products are no exception (Liu et al. 2014); effectively, over the last decades, cases of poisoning due to herbal medicines have occurred in many countries (Zhou et al. 2013). The experience gained from traditional use is efficient enough to detect immediate or near-immediate relationships between administration and toxic effects, but is quite unlikely to detect medium- and long-term toxicities (Zhou et al. 2013). Notably, carcinogenicity and genotoxicity are not "obvious" adverse effects (such as gastrointestinal disorders or many autonomic nervous system modulations) but, as shown by the dramatic cases of *Aristolochia* poisoning, they are nonetheless dreadful for the patients' health. Such an apparent lack of toxicity of an herbal medicine can lead to a false sense of safety, to chronic use, and to reliance on its properties.

"Traditional" medicine often recommends the use of combinations of HMP, an additional major challenge in safety assessment. Consequently, toxicity can be caused not only by an individual HMP drug, but also by the interaction between two or more HMP drugs (Liu et al. 2014). Moreover, in China and other countries (mainly Asian), it is frequently recommended to use conventional drugs concurrently with traditional herbal medicines. Drug-herb interactions are then also possible, in addition to eventual food-drug interactions (Liu et al. 2014). We have recently published potentiated genotoxic effects measured for the association of *Magnolia* and *Aristolochia* species. Both plants were present in the weight-reducing capsules taken by Belgian women in the 1990s, which may possibly explain the rapid onset of Chinese herb nephropathies observed in the 1990s (Nachtergael et al. 2015).

Guidelines for genotoxicity or carcinogenicity assessment do not currently take such interactions into consideration. However, given the worldwide and constantly increasing use of herbal products, a better risk assessment certainly represents a very important point for the safety of patients.

References

Abel G, Göggelmann W (1986) Genotoxic activity of ß-asarone and commercial calamus drugs. Mutat Res/Environ Mutagen Relat Sub 164(4):287

Ames BN, Durston WE, Yamasaki E, Lee FD (1973) Carcinogens are mutagens: a simple test system combining liver homogenates for activation and bacteria for detection. Proc Natl Acad Sci USA 70(8):2281–2285

Andrew AS, Burgess JL, Meza MM, Demidenko E, Waugh MG, Hamilton JW et al (2006) Arsenic exposure is associated with decreased DNA repair *in vitro* and in individuals exposed to drinking water arsenic. Environ Health Perspect 114(8):1193–1198

Balmain A (2001) Cancer genetics: from Boveri and Mendel to microarrays. Nat Rev Cancer 1(1):77–82. doi:10.1038/35094086

Barnes J, Anderson L, Phillipson D (2007) Herbal medicines. 3rd edn. Press P, editor, London

Bellakhdar J (1999) In: Press I (ed) La pharmacopée marocaine traditionnelle. Ibis Press, Paris

van den Berg SJPL, Restani P, Boersma MG, Delmulle L, Rietjens IMCM (2011) Levels of genotoxic and carcinogenic ingredients in plant food supplements and associated risk assessment. Food Nutr Sci 02(09):989–1010
Bielas JH, Loeb KR, Rubin BP, True LD, Loeb LA (2006) Human cancers express a mutator phenotype. Proc Natl Acad Sci 103(48):18238–18242
Billintona N, Hastwellb P, Beerensc D, Birrella L, Ellisb P, Maskelld S et al (2008) Interlaboratory assessment of the GreenScreen HC GADD45a-GFP genotoxicity screening assay: an enabling study for independent validation as an alternative method. Mutat Res 653(1–2):23–33
Borner FU, Schutz H, Wiedemann P (2011) The fragility of omics risk and benefit perceptions. Toxicol Lett 201:249–257
Botta A (2013) Relations entre génotoxicité, mutagenèse et cancérogenèse. In: l'Environnement IPMdSd (ed) Journées Nationales de Santé au Travail dans le BTP. Service Hospitalo-universitaire de Médecine et Santé au Travail, Marseille, pp 9–13
Bruneton J (2005) Plantes toxiques: végétaux dangereux pour l'homme et les animaux. 3rd edn. Tec & Doc, editor. Lavoisier, Paris
Bruneton J (2009a) Pharmacognosie, phytochimie, plantes médicinales 4e Ed. Lavoisier S.A.S., editor, Cachan
Bruneton J (2009b) Pharmacognosie: phytochimie, plantes médicinales. 4th edn. Lavoisier, editor, Paris
Brusick D (1980) Principles of genetic toxicology. Plenum Press, New York
Burgio E, Migliore L (2015) Towards a systemic paradigm in carcinogenesis: linking epigenetics and genetics. Mol Biol Rep 42(4):777–790
Butterworth BE (2006) A classification framework and practical guidance for establishing a mode of action for chemical carcinogens. Regul Toxicol Pharmacol 45(1):9–23, Research Support, Non-U.S. Gov't
Carcinogenesis Studies of Eugenol (1983) (CAS No. 97-53-0) in F344/N rats and B6C3F1 mice (Feed Studies). Natl Toxicol Program Tech Rep Ser 223:1–159
Cassani-Galindo M, Madrigal-Bujaidar E, Chamorro G, Díaz F, Tamariz J, Espinosa-Aguirre J (2005) *In vitro* genotoxic evaluation of three α-asarone analogues. Toxicol Vitro 19(4):547–552
Chamorro G, Garduño L, Martínez E, Madrigal E, Tamariz J, Salazar M (1998) Dominant lethal study of α-asarone in male mice. Toxicol Lett 99(2):71–77
Chan S (2003) Determination of aristolochic acids in medicinal plant and herbal product by liquid chromatography–electrospray–ion trap mass spectrometry. Talanta 60(4):679–685
Charles C, Chemais M, Stevigny C, Dubois J, Nachergael A, Duez P (2012) Measurement of the influence of flavonoids on DNA repair kinetics using the comet assay. Food Chem 135(4):2974–2981
Charles C, Nachtergael A, Ouedraogo M, Belayew A, Duez P (2014) Effects of chemopreventive natural products on non-homologous end-joining DNA double-strand break repair. Mutat Res/Genet Toxicol Environ Mutagen 768:33–41
Chen CL, Chi CW, Chang KW, Liu TY (1999) Safrole-like DNA adducts in oral tissue from oral cancer patients with a betel quid chewing history. Carcinogenesis 20(12):2331–2334
Chen T, Mei N, Fu PP (2010) Genotoxicity of pyrrolizidine alkaloids. J Appl Toxicol 30(3):183–196
Cheng K-F, Leung P-C (2012) Safety in Chinese medicine research. Open J Saf Sci Technol 2(01):32
Chung Y-T, Chen C-L, Wu C-C, Chan S-A, Chi C-W, Liu T-Y (2008) Safrole-DNA adduct in hepatocellular carcinoma associated with betel quid chewing. Toxicol Lett 183(1–3):21–27
Cleaver JE (2005) Cancer in *xeroderma pigmentosum* and related disorders of DNA repair. Nat Rev Cancer 5(7):564–573. doi:10.1038/nrc1652
Cohnheim J (1875) Congenitales, quergestreiftes muskelsarkom der nieren. Virchows Arch 65(1):64–69
Combes R (2012) In silico methods for toxicity prediction. In: Balls M, Combes R, Bhogal N (eds) New technologies for genotoxicity testing. Springer US, New York city. pp 96–116

Council of Europe (2005) Estragole. In: Active principles (constituents of toxicological concern) contained in natural sources of flavouring. Council of Europe, Strasbourg

Coussens LM, Werb Z (2002) Inflammation and cancer. Nature 420(6917):860–867. doi:10.1038/nature01322

Daimon H, Sawada S, Asakura S, Sagami F (1998) *In vivo* genotoxicity and DNA adduct levels in the liver of rats treated with safrole. Carcinogenesis 19(1):141–146

Davies H, Bignell GR, Cox C, Stephens P, Edkins S, Clegg S et al (2002) Mutations of the BRAF gene in human cancer. Nature 417(6892):949–954. doi:10.1038/nature00766

De Broe ME (2012) Chinese herbs nephropathy and Balkan endemic nephropathy: toward a single entity, aristolochic acid nephropathy. Kidney Int 81(6):513–515

Debelle FD, Vanherweghem JL, Nortier JL (2008) Aristolochic acid nephropathy: a worldwide problem. Kidney Int 74(2):158–169

Dixon K, Kopras E (2004) Genetic alterations and DNA repair in human carcinogenesis. Semin Cancer Biol 14(6):441–448

ECVAM (2015) European union reference laboratory for alternatives to animal testing

EFSA (2009) Advice on the EFSA guidance document for the safety 505 assessment of botanicals and botanical preparations intended for use as food supplements, based on 506 real case studies

Ekor M (2014) The growing use of herbal medicines: issues relating to adverse reactions and challenges in monitoring safety. F Pharmacol [Review] 4(177):1–10

EMEA (2008) Guideline on the assessment of genotoxicity of herbal substances/preparations. London [06/02/2014]; 4–11/2. Available from: http://www.ema.europa.eu/docs/en_GB/document_library/Scientific_guideline/2009/09/WC500003569.pdf

EMEA (2008) Guideline on the assessment of genotoxicity of herbal substances/preparations. London [20/04/2015]; Available from: http://www.ema.europa.eu/docs/en_GB/document_library/Scientific_guideline/2009/09/WC500003569.pdf

European Commission: Scientific Committee on Food (2002) Opinion of the scientific committee on food on the safety of the presence of safrole (1-allyl-3,4-methylene dioxy benzene) in flavourings and other food ingredients with flavouring properties. European Commission, Brussels, Contract No.: SCF/CS/FLAV/FLAVOUR/6 ADD3

FDA (2012) Guidance for industry S2(R1) genotoxicity testing and data interpretation for pharmaceuticals intended for human use

Fjodorova N, Vračko M, Novič M, Roncaglioni A, Benfenati E (2010) New public QSAR model for carcinogenicity. Chem Central J 4(Suppl 1):S3

Friedberg EC, Wagner R, Radman M (2002) Specialized DNA polymerases, cellular survival, and the genesis of mutations. Science 296(5573):1627–1630

Fu P, Xia Q, Lin G, Chou M (2002) Genotoxic pyrrolizidine alkaloids – mechanisms leading to DNA adduct formation and tumorigenicity. Int J Mol Sci 3:948–964

Gokmen MR, Cosyns JP, Arlt VM, Stiborova M, Phillips DH, Schmeiser HH et al (2013) The epidemiology, diagnosis, and management of aristolochic acid nephropathy: a narrative review. Ann Intern Med 158(6):469–477

Gonzalez M, Bernad A (2012) Characteristics of adult stem cells. In: López-Larrea C, López-Vázquez A, Suárez-Álvarez B (eds) Stem cell transplantation. Springer-Verlag, New York, USA, pp 103–120

Greene N (2002) Computer systems for the prediction of toxicity: an update. Adv Drug Deliv Rev 54:417–431

Guyton KZ, Kyle AD, Aubrecht J, Cogliano VJ, Eastmond DA, Jackson M et al (2009) Improving prediction of chemical carcinogenicity by considering multiple mechanisms and applying toxicogenomic approaches. Mutat Res/Rev Mutat Res 681(2–3):230–240

Hanahan D, Weinberg RA (2000) The hallmarks of cancer. Cell 100(1):57–70

Hanahan D, Weinberg RA (2011) Hallmarks of cancer: the next generation. Cell 144(5):646–674

Hartung T (2011) From alternative methods to a new toxicology. Eur J Pharm Biopharm 77:338–349

Hasheminejad G, Caldwell J (1994) Genotoxicity of the alkenylbenzenes α- and β-asarone, myristicin and elemicin as determined by the UDS assay in cultured rat hepatocytes. Food Chem Toxicol 32(3):223–231

Heinrich M, Chan J, Wanke S, Neinhuis C, Simmonds MS (2009) Local uses of *Aristolochia* species and content of nephrotoxic aristolochic acid 1 and 2 – A global assessment based on bibliographic sources. J Ethnopharmacol 125(1):108–144

Hernandez LG, van Steeg H, Luijten M, van Benthem J (2009) Mechanisms of non-genotoxic carcinogens and importance of a weight of evidence approach. Mutat Res [Review] 682(2–3):94–109

Hesketh R (1997) The oncogene & tumour suppressor gene factsbook. In: Hesketh R (ed) Academic press, San Diego

Hoet P, Godderis L (2013) Genotoxicity and mutagenicity testing: a brief overview of the main tests, pitfalls and regulatory framework. In: BEMS- Belgian Environmental Mutagen Society B (ed) Toxicology as the scientific basis for management of chemical risk. BEMS, Elewijt

Homburger F, Bogdonoff PD, Kelley TF (1965) Influence of diet on chronic oral toxicity of Safrole and butter yellow in rats. Proc Soc Exp Biol Med 119(4):1106–1110

Howes AJ, Chan VSW, Caldwell J (1990) Structure-specificity of the genotoxicity of some naturally occurring alkenylbenzenes determined by the unscheduled DNA synthesis assay in rat hepatocytes. Food Chem Toxicol 28(8):537–542

Hubaux R, Becker-Santos DD, Enfield KSS, Rowbotham D, Lam S, Lam WL et al (2013) Molecular features in arsenic-induced lung tumors. Mol Cancer 12:20

IARC (2002) *Rubia tinctorium*, *Morinda officinalis* and anthraquinones. IARC Monogr Eval Carcinog Risks Hum 82:129–151, Some Traditional Herbal Medicines, Some Mycotoxins, Naphthalene and Styrene

IARC (2004) Betel-quid and areca-nut chewing and some areca-nut derived nitrosamines. IARC Monogr Eval Carcinog Risks Hum 85:1–334

IARC (2015) International agency for research on cancer: agents classified by the IARC monographs. Available from: http://monographs.iarc.fr/ENG/Classification/

ICH (2014) Guidance for industry S2B genotoxicity: a standard battery for genotoxicity testing of pharmaceuticals. [03/02/2014]. Available from: http://www.ich.org/products/guidelines

ICH (2015) Guideline on the need for carcinogenicity studies of pharmaceuticals S1A S1A. Available from: http://www.ich.org/products/guidelines

ICH (2015) International conference on harmonisation: testing for carcinogenicity of pharmaceuticals S1B. Available from: http://www.ich.org/products/guidelines

Jacobson-Kram D (2009) Cancer risk assessment approaches at the FDA/CDER: is the era of the 2-year bioassay drawing to a close? Toxicol Pathol 38(1):169–170

Kahumba JRT, Okusa PN, Bakari A, Bizumukama L, Kalonji J-B, Kiendrebeogo M, Rabemenantsoa C, El Jaziri M, Williamson EM, Duez P (2015) Traditional African medicine: from ancestral knowledge to a modern integrated future? Science 350(6262):S61–S63

Kawai K, Mori H, Sugie S, Yoshimi N, Inoue T, Nakamaru T et al (1986) Genotoxicity in the hepatocyte/DNA repair test and toxicity to liver mitochondria of 1-hydroxyanthraquinone and several dihydroxyanthraquinones. Cell Biol Toxicol 2(4):457–467

Kevekordes S, Spielberger J, Burghaus C, Birkenkamp P, Zietz B, Paufler P et al (2001) Micronucleus formation in human lymphocytes and in the metabolically competent human hepatoma cell line Hep-G2: results with 15 naturally occurring substances. Anticancer Res 21(1A):461–469

Knudson AG (2001) Two genetic hits (more or less) to cancer. Nat Rev Cancer 1(2):157–162. doi:10.1038/35101031

Langel D, Ober D, Pelser PB (2010) The evolution of pyrrolizidine alkaloid biosynthesis and diversity in the Senecioneae. Phytochem Rev 10(1):3–74

Li X, Wang H (2004) Aristolochic acid nephropathy: what we know and what we have to do. Nephrology (Carlton) 9(3):109–111

Liu X, Wang Q, Song G, Zhang G, Ye Z, Williamson EM (2014) The classification and application of toxic Chinese materia medica. Phytother Res 28(3):334–347

Loeb LA (2011) Human cancers express mutator phenotypes: origin, consequences and targeting. Nat Rev Cancer 11(6):450–457. doi:10.1038/nrc3063

Loeb LA, Harris CC (2008) Advances in chemical carcinogenesis: a historical review and prospective. Cancer Res 68(17):6863–6872

Loeb LA, Springgate CF, Battula N (1974) Errors in DNA replication as a basis of malignant changes. Cancer Res 34(9):2311–2321

Mädge I, Cramer L, Rahaus I, Jerz G, Winterhalter P, Beuerle T (2015) Pyrrolizidine alkaloids in herbal teas for infants, pregnant or lactating women. Food Chem 187:491–498

Magdolenova Z, Collins A, Kumar A, Dhawan A, Stone V, Dusinska M (2014) Mechanisms of genotoxicity. A review of *in vitro* and *in vivo* studies with engineered nanoparticles. Nanotoxicology 8(3):233–278

Marczewska J, Drozd E, Anuszewska E, Chilmonczyk Z, Łozowicka B (2013) Assessment of the genotoxic activity of alpha-asarone and its derivatives in the comet assay. Acta Pol Pharm 70(2):349–354

Martinez-Arribas F, Alvarez T, Del Val G, Martin-Garabato E, Nunez-Villar M-J, Lucas R et al (2007) Bcl-2 expression in breast cancer: a comparative study at the mRNA and protein level. Anticancer Res 27(1A):219–222

Marx J (2002) Debate surges over the origins of genomic defects in cancer. Science 29(5581): 544–546

Maurici D, Aardema M, Corvi R, Kleber M, Krul C, Laurent C et al (2004) Timetables for the phasing-out of animal testing in the framework of the 7th amendment to the cosmetics directive: genotoxicity and mutagenicity. Commission of the European Communities, Brussels

Mei N, Arlt VM, Phillips DH, Heflich RH, Chen T (2006) DNA adduct formation and mutation induction by aristolochic acid in rat kidney and liver. Mutat Res/Fundam Mol Mech Mutagen 602(1–2):83–91

Michl J, Jennings HM, Kite GC, Ingrouille MJ, Simmonds MSJ, Heinrich M (2013) Is aristolochic acid nephropathy a widespread problem in developing countries? A case study of *Aristolochia indica* L. in Bangladesh using an ethnobotanical–phytochemical approach. J Ethnopharmacol 149(1):235–244

Michl J, Ingrouille MJ, Simmonds MS, Heinrich M (2014) Naturally occurring aristolochic acid analogues and their toxicities. Nat Prod Rep 31(5):676–693

Migheli F, Migliore L (2014) Epigenetic perturbations in the context of the multi-hit hypothesis of carcinogenesis. In: Maulik N, Karagiannis T (eds) Molecular mechanisms and physiology of disease. Springer, New York, pp 383–399

Miller JA, Miller EC (1976) Carcinogens occurring naturally in foods. Fed Proc 35(6):1316–1321

Monier R (2008) News in carcinogenesis. C R Biol 331(2):101–103

Morales-Ramírez P, Madrigal-Bujaidar E, Mercader-Martínez J, Cassini M, González G, Chamorro-Cevallos G et al (1992) Sister-chromatid exchange induction produced by *in vivo* and *in vitro* exposure to alpha-asarone. Mutat Res/Genet Toxicol Environ Mutagen 279(4):269–273

Muller L, Kikuchi Y, Probst G, Schechtman L, Shimada H, Sofuni T et al (1999) ICH-Harmonised guidances on genotoxicity testing of pharmaceuticals: evolution, reasoning and impact. Mutat Res 436:195–225

Muller J, Decordier I, Hoet PH, Lombaert N, Thomassen L, Huaux F et al (2008) Clastogenic and aneugenic effects of multi-wall carbon nanotubes in epithelial cells. Carcinogenesis 29(2):427–433

Mulware SJ (2012) Trace elements and carcinogenicity: a subject in review. 3 Biotech 3(2):85–96

Munerato MC, Sinigaglia M, Reguly ML, de Andrade HH (2005) Genotoxic effects of eugenol, isoeugenol and safrole in the wing spot test of *Drosophila melanogaster*. Mutat Res 582(1–2): 87–94

Nachtergael A, Charles C, Spanoghe M, Gadenne M, Belayew A, Duez P (2013) Measurement of translesion synthesis by fluorescent capillary electrophoresis: 7, 8-Dihydro-8-oxodeoxyguanosine bypass modulation by natural products. Anal Biochem 440(1):23–31

Nachtergael A, Poivre M, Belayew A, Duez P (2015) *In vitro* genotoxicity tests point to an unexpected and harmful effect of a Magnolia and Aristolochia association. J Ethnopharmacol 174(C):178–186

NAP (2007) Applications of toxicogenomic technologies to predictive toxicology and risk assessment. Press NA, editor, Washington

National Cancer Institute NIH (2015) Tumor microenvironment national institute of health; [22/04/2015]. Available from: http://www.cancer.gov/dictionary?cdrid=561725

National Toxicology Program (2011) Danthron. Rep Carcinog 12:128–129

Niir (2003) In: Research NIaN (ed) The complete technology book of essential oils (aromatic chemicals), 1st edn. Asia Pacific Business Press, Delhi

Nortier JL, Vanherweghem JL (2002) Renal interstitial fibrosis and urothelial carcinoma associated with the use of a Chinese herb (*Aristolochia fangchi*). Toxicology 181–182:577–580

Nortier JL, Martinez MC, Schmeiser HH, Arlt VM, Bieler CA, Petein M et al (2000) Urothelial carcinoma associated with the use of a Chinese herb (*Aristolochia fangchi*). N Engl J Med 342(23):1686–1692

Nortier J, Pozdzik A, Roumeguere T, Vanherweghem J-L (2013) Néphropathie aux acides aristolochiques ("néphropathie aux herbes chinoises"). EMC – Néphrologie 10(2):1–14

NTP (2000) NTP toxicology and carcinogenesis studies of methyleugenol (CAS NO. 93-15-2) in F344/N rats and B6C3F1 mice (gavage studies). Natl Toxicol Program Tech Rep Ser 491:1–412

Nutraceuticals World (2012) Global herbal supplement market to reach $107 billion by 2017 [29/04/2015]. Available from: http://www.nutraceuticalsworld.com/contents/view_breaking-news/2012-03-07/global-herbal-supplement-market-to-reach-107-billion-by-2017

O'Brien CA, Pollett A, Gallinger S, Dick JE (2007) A human colon cancer cell capable of initiating tumour growth in immunodeficient mice. Nature 445(7123):106–110. doi:10.1038/nature05372

OECD (2007) Detailed review paper on cell transformation assays for detection of chemical carcinogens. In: Environment O (ed) Health and Safety Publications

OECD (2014) OECD guidelines for testing of chemicals [03/02/2014]. Available from: http://www.oecd-ilibrary.org/fr/environment/lignes-directrices-de-l-ocde-pour-les-essais-de-produits-chimiques-section-4-effets-sur-la-sante_20745842

Ouedraogo M, Baudoux T, Stévigny C, Nortier J, Colet J-M, Efferth T et al (2012) Review of current and "omics" methods for assessing the toxicity (genotoxicity, teratogenicity and nephrotoxicity) of herbal medicines and mushrooms. J Ethnopharmacol 140(3):492–512

Pelkonen O, Xu Q, Fan T-P (2014) Why is research on herbal medicinal products important and how can we improve its quality? J Tradit Complement Med 4(1):1–7

Quante M, Wang TC (2008) Inflammation and stem cells in gastrointestinal carcinogenesis. Physiology (Bethesda) 23:350–359

Raghava GPS, Gusenleitner D, Auerbach SS, Melia T, Gómez HF, Sherr DH et al (2014) Genomic models of short-term exposure accurately predict long-term chemical carcinogenicity and identify putative mechanisms of action. PLoS One 9(7):e102579

Randerath K, Liehr JG, Gladek A, Randerath E (1989) Use of the 32P-postlabelling assay to study transplacental carcinogens and transplacental carcinogenesis. IARC Sci Publ 96:189–205

Randerath K, Putman KL, Randerath E (1993) Flavor constituents in cola drinks induce hepatic DNA adducts in adult and fetal mice. Biochem Biophys Res Commun 192(1):61–68

Rizwani W, Chellappan S (2009) *In vitro* replication assay with Mammalian cell extracts. In: Chellappan S (ed) Chromatin protocols. Humana Press, New York city. pp 203–216

Rodriguez-Paez L, Juárez-Sanchez M, Antúnez-Solís J, Baeza I, Wong C (2003) Alpha-asarone inhibits HMG-CoA reductase, lowers serum LDL-cholesterol levels and reduces biliary CSI in hypercholesterolemic rats. Phytomedicine 10(5):397–404

Rossman TG (2003) Mechanism of arsenic carcinogenesis: an integrated approach. Mutat Res/Fundam Mol Mech Mutagen 533(1):37–65
Sarasin A (2003) An overview of the mechanisms of mutagenesis and carcinogenesis. Mutat Res/Rev Mutat Res 544(2–3):99–106
Segelman AB, Segelman FP, Karliner J, Sofia RD (1976) Sassafras and herb tea. Potential health hazards. JAMA 236(5):477
Shaked Y, Ciarrocchi A, Franco M, Lee CR, Man S, Cheung AM et al (2006) Therapy-induced acute recruitment of circulating endothelial progenitor cells to tumors. Science 313(5794): 1785–1787
Shen S, Wang C, Weinfeld M, Le XC (2013) Inhibition of nucleotide excision repair by arsenic. Chin Sci Bull 58(2):214–221
Smith K (1992) Spontaneous mutagenesis: experimental, genetic and other factors. Mutat Res 277(2):139–162
Taylor JM, Jones WI, Hagan EC, Gross MA, Davis DA, Cook EL (1967) Toxicity of oil of calamus (Jammu variety). Toxicol Appl Pharmacol 10(2):378–411
Thomas SJ, MacLennan R (1992) Slaked lime and betel nut cancer in Papua New Guinea. Lancet 340(8819):577–578
Tomasetti C, Vogelstein B (2015) Variation in cancer risk among tissues can be explained by the number of stem cell divisions. Science 347(6217):78–81
U.S. Food and Drug Administration (1973) Database of select committee on GRAS substances (SCOGS) reviews: nutmeg and mace. FDA, Silver Spring
Ueng YF, Hsieh CH, Don MJ, Chi CW, Ho LK (2004) Identification of the main human cytochrome P450 enzymes involved in safrole 1′-hydroxylation. Chem Res Toxicol 17(8):1151–1156
Unger P, Melzig MF (2012) Comparative study of the cytotoxicity and genotoxicity of alpha- and beta-asarone. Sci Pharm 80(3):663–668
Valerio LG Jr (2009) *In-silico* toxicology for the pharmaceutical sciences. Toxicol Appl Pharmacol 241:356–370
Vanherweghem JL, Tielemans C, Abramowicz D, Depierreux M, Vanhaelen-Fastre R, Vanhaelen M et al (1993) Rapidly progressive interstitial renal fibrosis in young women: association with slimming regimen including Chinese herbs. Lancet 341(8842):387–391
Vineis P, Schatzkin A, Potter JD (2010) Models of carcinogenesis: an overview. Carcinogenesis 31(10):1703–1709
Virchow R (1855) Editorial Archive fuer pathologische. Anatomie und Physiologie fuer klinische Medizin 8:23–54
Walmsley RM, Billinton N (2011) How accurate is in vitro prediction of carcinogenicity? Br J Pharmacol 162(6):1250–1258
Wang T-L, Rago C, Silliman N, Ptak J, Markowitz S, Willson JKV et al (2002) Prevalence of somatic alterations in the colorectal cancer cell genome. Proc Natl Acad Sci 99(5): 3076–3080
Wang YP, Yan J, Beger RD, Fu PP, Chou MW (2005a) Metabolic activation of the tumorigenic pyrrolizidine alkaloid, monocrotaline, leading to DNA adduct formation in vivo. Cancer Lett 226(1):27–35 [Research Support, Non-U.S. Gov't Research Support, U.S. Gov't, Non-P.H.S.]
Wang Y-P, Yan J, Fu PP, Chou MW (2005b) Human liver microsomal reduction of pyrrolizidine alkaloid N-oxides to form the corresponding carcinogenic parent alkaloid. Toxicol Lett 155(3):411–420
Williams GM, Iatropoulos MJ, Jeffrey AM, Duan JD (2013) Methyleugenol hepatocellular cancer initiating effects in rat liver. Food Chem Toxicol 53:187–196
Wislocki PG, Miller EC, Miller JA, McCoy EC, Rosenkranz HS (1977) Carcinogenic and mutagenic activities of safrole, 1′-hydroxysafrole, and some known or possible metabolites. Cancer Res 37(6):1883–1891
Wu J, Starr S (2014) Low-fidelity compensatory backup alternative DNA repair pathways may unify current carcinogenesis theories. Future Oncol 10(7):1239–1253

Wu S, Powers S, Zhu W, Hannun YA (2016) Substantial contribution of extrinsic risk factors to cancer development. Nature 529(7584):43–47

Xia Q, Yan J, Chou MW, Fu PP (2008) Formation of DHP-derived DNA adducts from metabolic activation of the prototype heliotridine-type pyrrolizidine alkaloid, heliotrine. Toxicol Lett 178(2):77–82, Research Support, U.S. Gov't, Non-P.H.S

Yamani A, Bunel V, Antoine M-H, Husson C, Stévigny C, Duez P et al (2015) Substitution between *Aristolochia* and *Bryonia* genus in North-Eastern Morocco: toxicological implications. J Ethnopharmacol 166:250–260

Zhang HL, Zhang QW, Zhang XQ, Ye WC, Wang YT (2010) Chemical constituents from the roots of *Morinda officinalis*. Chin J Nat Med 8(10):1–4

Zhou J, Ouedraogo M, Qu F, Duez P (2013) Potential genotoxicity of traditional Chinese medicinal plants and phytochemicals: an overview. Phytother Res 27(12):1745–1755

Chapter 10
Teratogenicity and Developmental Toxicity of Herbal Products

Ean-Jeong Seo and Thomas Efferth

Abstract Developmental toxicology and research in teratogenicity focus on xenobiotic substances that damage embryos and fetuses and lead to death, growth retardation, and/or malformation of offspring. While considerable information has been acquired about synthetic drugs and environmental xenobiotics, much less is known about the teratogenicity of herbal products. In this chapter we report on some major topics of developmental toxicity and teratogenicity, and discuss the safety of a few selected medicinal herbs in this context, i.e., *Artemisia annua* L., *Caulophyllum thalictroides* (L.) Michx., *Echinacea spec*,. *Glycyrrhiza spec*., herbs derived from Chinese medicine, *Hypericum perforatum* L., *Panax ginseng* C.A.Mey., *Valeriana officinalis* L., *and Zingiber officinale* Roscoe. Due to insufficient information on most herbal products, their safe use during pregnancy remains uncertain. Toxicological research for the professional development of safe herbal products should include

- Measures of quality control (botanical, organoleptic, chemical, and molecular biological plant identification as well as standardization of cultivation, harvesting, and production processes and chemical constitution)
- Identification of both pharmacological and toxicological modes of actions by means of state-of-the-art methods
- Testing for the toxicity of herbal products *in silico, in vitro*, and in animal experiments.

Keywords Animal experiments • Birth defects • Dietary supplements • Mode-of-action (MOA) analysis • Over-the-counter (OTC) products • Quality control

E.-J. Seo • T. Efferth (✉)
Department of Pharmaceutical Biology, Institute of Pharmacy and Biochemistry, Johannes Gutenberg University, Staudinger Weg 5, 55128 Mainz, Germany
e-mail: efferth@uni-mainz.de

O. Pelkonen et al. (eds.), *Toxicology of Herbal Products*,
DOI 10.1007/978-3-319-43806-1_10

Introduction

All drugs – whether synthetic or natural – produce both wanted and unwanted effects in the human body. Western academic medicine can no longer state that herbal medicines and botanical supplements act only because of "placebo effects." Over the last years, mounting evidence has shown the opposite. Indeed, the activity of many herbal products can be demonstrated with techniques and methods accepted and applied by Western medicine and natural sciences (Efferth et al 2007; Konkimalla and Efferth 2008). If we accept the fact that many herbal supplements are active drugs that efficiently treat patients, we also have to accept that fact that they may eventually produce unwanted toxicity (Allard et al 2013; Efferth and Kaina 2011). Even if this occurs only in relatively rare cases (at least in the current situation of notoriously insufficient case reporting and causality assessment (Shaw et al 2012; Zhang et al. 2012), it is mandatory to take the utmost care of safety issues related to herbal products. Among the various toxicities (hepatotoxicity, nephrotoxicity, genotoxicity and carcinogenesis, hematotoxicity, allergy and skin toxicity, gastrointestinal toxicity, etc.) that have to be taken into account for herbal products, developmental toxicity and teratogenicity deserve special attention (Ouedraogo et al. 2012) (Fig. 10.1).

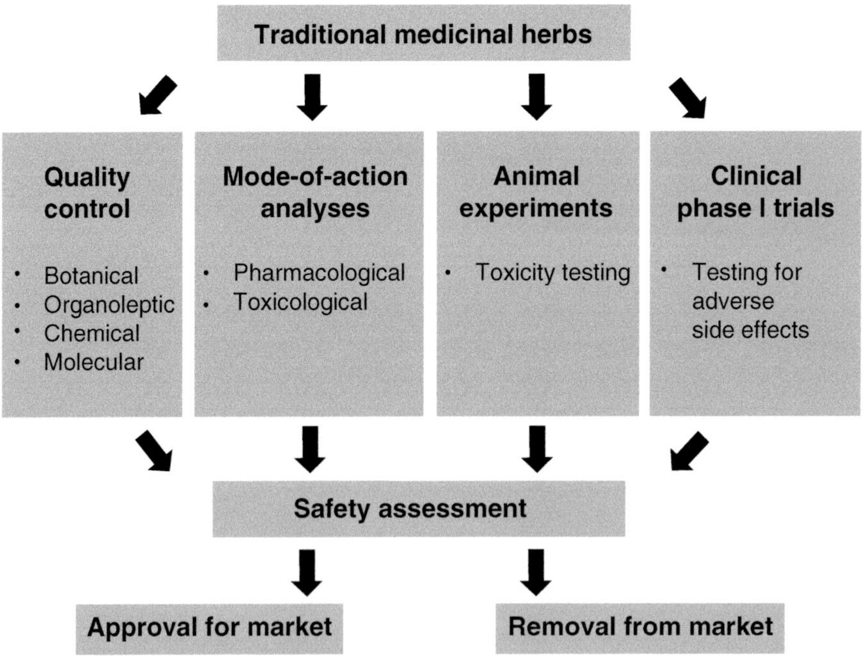

Fig. 10.1 Safety assessment of prevent developmental toxicity and teratogenicity of herbal products

Developmental Toxicity

Developmental toxicity, in its broadest sense, comprises any alteration that affects the development of offspring and is caused by exogenous factors (drugs, environmental toxins, diet, alcohol, etc.). The terms embryo- and fetotoxicity focus on substances that cross the placental barrier and directly affect embryos and fetuses, or both maternal and placental circulation, causing death, growth retardation, or malformation of embryos and fetuses. The in utero vulnerability of the developing embryo or fetus leading to malformation is also known as "teratogenesis." The Food and Drug Administration (FDA) has defined five classes of drugs according to their fetal risk (in increasing order):

- Category A: No risk
- Category B: Antibacterial drugs (e.g., penicillin, erythromycin), CNS drugs (e.g., acetaminophen)
- Category C: Antibacterial drugs (e.g., aminoglycosides, chloroquine), cardiovascular drugs (e.g., beta-blockers)
- Category D: Antibacterial drugs (e.g., tetracycline, gentamycin), cardiovascular drugs (e.g., ACE inhibitors)
- Category X: Highest risk (e.g., thalidomide)

Herbal products have not been intensively studied for their teratogenicity and developmental toxicity. Hence, clear-cut assignments according to this classification system are still not possible. There is an urgent need for research to allow proper assessment of herbal products for their teratogenic and developmental risks.

Use of Herbal Products in General and During Pregnancy

Herbal products are frequently regarded as "safe" by the general public; green medicine is often considered to be effective, but without side effects (Marcus and Snodgrass 2005). Herbal products became popular as evidenced by the sale of an estimated of \$4 billion worth in the United States, and \$6.7 billion in Europe (Gruenwald 2000). The U.S. sales for St. John's wort alone was \$48 million in 1997 and \$140 million in 1998; similarly, the sales of *Ginkgo biloba* increased from \$90 million in 1997 to \$150 million in 1998 (Blumenthal 1999; Landes 1998). Sales of herbal dietary supplements in the United States increased by 7.9 % in 2013, reaching \$6 billion (Lindstrom et al. 2014). One of the many explanations is that the general public may be dissatisfied with Western academic medicine. Other reasons may be the easy availability of dietary supplements as over-the-counter products, and effective promotion in media and marketing campaigns (Astin 1998; Beaubrun and Gray 2000; Ernst 2002a). Although many medicinal plants may indeed be safe with few or no side effects, a generalization such as this is dangerous since numerous plants exert significant toxicities or are even poisonous.

Nevertheless, an estimated 30–50 % of pregnant women (and more than 70 % in Japan) use herbal products to maintain good health and reduce the need for Western medication (Nordeng and Havnen 2004; Ong et al. 1983; Westfall 2001; Yu et al. 2004; Mantani et al. 2002). *Echinacea*, iron-rich herbs, ginger, chamomile, cranberry, and aloe are the most commonly used herbal drugs in pregnancy among Norwegian women (Nordeng and Havnen 2004). In the U.S., *echinacea*, *ginkgo biloba*, ginseng, or St. John's wort were taken by nearly 15 % of women (Yu et al. 2004). Tonic herbs – for instance, raspberry leaf (*Rubus idaeus* L.), partridge berry (*Mitchella repens* L.), and stinging nettle (*Urtica dioica* L.) – are defined as neither toxic nor strongly medicinal and therefore safe for regular consumption by pregnant women in North America (Westfall 2001). Indeed, many synthetic drugs are not recommended for use during pregnancy because of the toxic or teratogenic effects found in animal experiments. By contrast, herbal remedies were commonly used for centuries and, in most cases, there is no sufficient scientific documentation on their developmental toxicity (Wang et al. 2013). For the general public, this long-standing use and the absence of documentation are often referred to as "the absence of adverse effects".

Obstacles to the Safe Use of Herbal Products

Commercial trade with herbal products is not strictly regulated, although some countries issue legal regulations (Costa et al. 2004; Efferth and Greten 2012a, b; Eichhorn et al. 2011; Schulz et al. 1998). In the U.S., herbal preparations are regulated as dietary supplements under the Dietary Supplement Health and Education Act, which does not include evidence of efficacy or extensive proof of safety, although health claims are restricted (Hathcock 2001). Herbal products may have toxic effects per se, but also due to several issues, such as contamination or intended adulteration with foreign plants, intended adulteration with synthetic drugs, contamination with pesticides, heavy metals, radioactivity, microorganisms and fungal poisons (mycotoxins) (Ernst 2002a; Izzo and Ernst 2001; Klepser and Klepser 1999). Issues on efficacy and safety may derive from insufficient botanical identification, chemical standardization, variations of raw plant materials (due to genetic factors, climate, soil, growing conditions, etc.), methods of preparation, solvent used in the extraction process (Schulz et al. 1998), insufficient quality control during harvesting and processing (Efferth and Greten 2012b), insufficient legal frameworks in countries importing and distributing herbal products and so on.

Testing for Developmental Toxicity

Most teratogenicity studies are performed on animals (Moallem and Niapour 2008; Sairafianpour et al. 2001), since scientific evidence of congenital malformation identified in animals can be transferred to the situation in human beings (Barcz et al. 2007; Gallo et al. 2000). As a matter of course, teratogenic studies in human subjects are unethical and prohibited, something that further emphasizes the importance of animal experimentation. An entire panel of parameters are taken into account to estimate the developmental toxicity of compounds, e.g., maternal side effects, including weight loss, litter reduction, implantation failure, fetal resorption, and stillbirths, as well as perinatal effects on growth restriction, developmental delay, congenital malformations, and postnatal mortality.

Redox equilibrium plays an important role in teratogenesis as, during certain embryogenic phases, the embryo is more susceptible to oxidative stress and excessive ROS formation (Hansen 2006). Also, a series of drugs and environmental chemicals often leads to the formation of reactive oxygen species (ROS) (Harvey et al. 2002; Wells et al. 2005), a mechanism possibly involved in their teratogenic effects. Many but not all xenobiotic compounds, drugs, and environmental chemicals that potentially induce embryonic oxidative stress are eliminated or metabolized from the maternal body before reaching the embryo. On the other hand, a number of xenobiotics, such as phenytoin and benzo[a]pyrene, which are relatively non-toxic, can be enzymatically bioactivated within the embryo to highly toxic intermediates (Pelkonen et al. 1971; Juchau et al. 1992; Fantel 1996; Wells and Winn 1996; Wells et al. 1997; Hakkola et al. 1998). If not detoxified, xenobiotic electrophilic reactive intermediates can bind covalently to embryonic cellular macromolecules (e.g., proteins, DNA), while xenobiotic free radical reactive intermediates can react directly or indirectly with molecular oxygen to initiate the formation of ROS (Wells et al. 2005). These ROS may oxidatively damage cellular macromolecules such as lipids, proteins, RNA, and DNA (Wells et al. 2005); such macromolecular damage can interfere with embryonic and/or fetal development (Wells et al. 2005), which may be adversely affected by the reversible reaction of ROS with transduction proteins, thereby altering embryonic or fetal signal transduction pathways (Wells et al. 2005).

The testing for potential embryotoxicity is generally performed on pregnant mice; this is still the gold standard, although it is time-consuming and expensive. Therefore, more rapid and cost-effective screening methods are required. Many alternative in vitro methods have been described, using primary cells and permanent cell lines, and cultures of primary embryonic cells, as well as cultures of non-mammalian organisms (zebrafish), tissues and mammalian embryos (Brown et al. 1995; Daston 2011; Sogorb et al. 2014; Spielmann 1998; Moreno et al. 2012). Although these in vitro test systems do not often reflect the developmental processes from early embryonic stages up until terminally differentiated organs or individual organisms, they are useful for rapidly pinpointing possible concerns and for mechanistic studies.

Herbal Products

Herbal Products for Pregnancy

Chinese herbal medicines. A recent review by Wang et al. (2013) focuses on pregnancy outcomes, and embryonic and fetal development in maternal exposure to Chinese herbal products (Wang et al. 2013). Traditional Chinese medicine (TCM) has been used for many centuries to relieve adverse symptoms during pregnancy (Flaws 2005). Chinese herbal medicines are generally considered to be "efficient and safe" for preventing miscarriage and pre-term labor, and for managing low back pain, low fetal weight, placenta previa, uterine fibroids, and other obstetric problems (Cunningham et al. 2005; Fu and Fu 1978; Li 2011; Li et al. 2014; Li et al. 2012a, b, c; Liu 2002). On a global scale, many pregnant women consume CHMs, especially in Japan (78.7 %) (Mantani et al. 2002), Australia (36 %) (Forster et al. 2006), China (32 %) (Wang et al. 1995), and Taiwan (24 %) (Chuang et al. 2007).

More than 2000 TCM recipes are in use, of which more than 300 are applied against problems during pregnancy. Wang et al. (2013) reported that 30 of them should not be consumed during pregnancy and, among them, Largehead *Atractylodes* Rhizome (*Atractylodes macrocephala* Koidz.) has been regarded as the most harmful one. A meta-analysis of 37 clinical trials showed the effectiveness of TCM products compared to conventional drugs concerning miscarriage (Li et al. 2012a). However, none of these studies examined short- or long-term adverse effects on the mother and fetus. Two other controlled trials did not show an increased risk of adverse pregnancy outcomes (Li et al. 2006; Zhou 2006). However, some observational studies reported pre- labor rupture of membranes, prematurity, and associated neonatal mortality (Cui 1998; He 1997; Luo et al. 2007; Zhou 1997). Even congenital malformations have been reported (Chou 2002). A recent preclinical study to review the safety of commonly used Chinese medicines during pregnancy showed that the adverse pregnancy outcomes in mice at clinical doses were very common (Wang et al. 2012). The 20 most common Chinese medicines for pregnancy were selected, and the crude extract was administered to pregnant mice at clinical doses (Wang et al. 2012). Adverse pregnancy outcomes were commonly observed after maternal exposure to the herbal medicines, particularly during early pregnancy (Wang et al. 2012). Major events, including maternal and prenatal mortality, were recorded (Wang et al. 2012), and maternal weight gain, embryo growth, and postnatal weight gain were significantly decreased (Wang et al. 2012). Some reproductive toxicity, including fetal resorption, growth restriction, and congenital malformations were also found (Wang et al. 2012). The results suggest that Chinese medicines can be harmful during pregnancy (Wang et al. 2012). Adverse effects of Chinese medicines on pregrancy were reported, and embryonic and fetal development require further investigations (Wang et al. 2013). More systematic investigations of the safety implications of the use of Chinese medicines in animals should be carried out, and more studies and clinical trials in humans with a larger sample size are also necessary.

Ginger (*Zingiber officinale*) As most pregnant women suffer from nausea and vomiting, they often take herbal products to manage these symptoms (Woolhouse 2006). A popular remedy for nausea and vomiting in pregnancy is ginger. Its efficacy was assessed by evaluating double-blind, randomized controlled trials (RCTs) and safety by uncontrolled case reports, observational studies, and RCTs (Jewell and Young 2003). Six RCTs met the selection criteria with a total of 675 participants. Four of six randomized clinical trials ($n = 246$ participants) reported better reduction of nausea and vomiting in early pregnancy with the use of ginger compared to the control drug (vitamin B6). The other two trials ($n=429$ participants) observed equivalent efficacy. Both case reports and clinical trials did not reveal adverse effects on pregnancy outcomes. Further controlled studies should confirm these preliminary data on the safety of using ginger (Jewell and Young 2003).

The Committee on Herbal Medicinal Products (HMPC 2011) reported that a moderate amount of data on pregnant women ($n=490$) indicates no malformative or feto/neonatal toxicity from ginger root (HMPC 2011). However, safety during pregnancy and lactation has not been established, and in the absence of sufficient data, its use during pregnancy and lactation is not recommended (HMPC 2011).

Herbal Products for Other Illnesses and Ailments

Sweet wormwood (*Artemisia annua* L.) *Artemisinin* is a sesquiterpene from *Artemisia annua* L., which was used in China against fever and chills. Because of its strong and rapid actions against *Plasmodia*, derivatives of artesunate are nowadays used worldwide to treat malaria. Artemisinin-based combination therapies are currently indispensable for malaria management (Efferth 2010; Shayo et al. 2015). Although artemisinin and its clinically used derivatives, artemether and artesunate, are generally considered to be safe, their potential embryotoxicity and teratogenicity have been pointed out in animal models (Amorim et al. 2013; Clark 2009; Efferth and Kaina 2010).

Artemisinin-type drugs caused bone malformations, cardiac ventricular septal defects, and embryonic death in rats and rabbits (Clark et al. 2004, 2008b). Higher artesunate doses in later gestation periods induced embryotoxicity (Clark et al. 2008a; Li et al. 2008b). The primary underlying mechanism may be the depletion of reticulocytes (erythropoietic precursor cells), which leads to severe anemia (White et al. 2006). These drugs selectively target proerythroblasts and basophilic erythroblasts apoptosis (Finaurini et al. 2012). Their common active metabolite, dihydroartemisinin, was the most potent compound among those tested. Although reports on embryotoxicity and teratogenicity in human patients are missing, the World Health Organization (WHO) recommends not using artemisinin-type drugs during the first trimester of pregnancy (World Health Organization 2007).

Artemisia absinthium, a species of *Artemisia*, has been used to treat loss of appetite, indigestion, biliary disorders, and other gastrointestinal problems (HMPC 2009a). Most sources recommend contraindication in pregnancy and lactation due to the uterus's contractions (HMPC 2009b). Tests on reproductive toxicity have only been performed with a dry ethanolic extract of *A. absinthium* orally administered to pregnant rats. Results showed significantly reduced sites of implantations and a reduction in the number of born pups per rat (HMPC 2009a). Safety during pregnancy and lactation has not been established. Because of the amounts of thujone in the preparations covered by the assessment report, their use should be avoided during pregnancy and lactation (HMPC 2009b).

Other sesquiterpene-containing plants are also known to cause reproductive toxicity in animals. Bitterweed (*Hymenoxys odorata* DC.) and sneezeweeds (*Helenium spp.*) affected neonatal survival (James et al. 1992). Dehydroleucodine was highly toxic towards *Bufo arenarum* embryos during the first stages of development (Moreno et al. 2012).

Blue Cohosh (*Caulophyllum thalictroides* (L.) Michx.) The roots and rhizomes of blue cohosh, which is a traditional medicine used in its tea form by native tribes in the northeastern U.S.,to relieve menstrual cramps and pain during childbirth. The plant was also called "papoose root" or "squaw root.". Dugoua et al. (2008) reviewed the literature on the safety and efficacy of this plant during pregnancy and lactation (Dugoua et al. 2008). Abortifacient, emmenagogue, and uterine stimulant effects of blue cohosh have been reported (Farnsworth et al. 1975). A 21-year-old female developed tachycardia, diaphoresis, abdominal pain, vomiting, muscle weakness, and fasciculations after using blue cohosh in an attempt to induce abortion (Rao and Hoffman 2002). An active ingredient isolated from blue cohosh, caulosaponin, increases blood flow to the uterus and causes contractions. Because caulosaponin causes birth defects, it is a suspected teratogen (Small and Catling 1999). The alkaloid anagyrine in the roots causes a congenital deformity known as "crooked calf disease" in bovine stock (Keeler 1984). A case report described a comparable human congenital malformation (marked anemia, skeletal dysplasia, and vascular anomaly) in an infant, which might have been caused by maternal consumption of anagyrine-contaminated goat milk in early pregnancy (Ortega and Lazerson 1987). Another alkaloid in blue cohosh is methylcytisine, which was teratogenic in rats (Ganzera et al. 2003; Kennelly et al. 1999; Rao and Hoffman 2002). The alkaloid taspine, also a constituent of blue cohosh, was highly embryotoxic in rats (Kennelly et al. 1999; Rao and Hoffman 2002). Blue cohosh was listed in a 1995 Health Canada document as an herb that is unacceptable as a non-prescription drug product for oral use (Small and Catling 1999). Based on these data, blue cohosh is an abortifacient and should be avoided in pregnancy.

***Echinacea* species (*E. angustifolia* DC., *E. purpurea* (L.) Moench, *E. pallida* (Nutt.) Nutt.)** *Echinacea* belongs to the most commonly used medicinal herbs among Native North Americans for diverse ailments such as wounds, insect bites, infections, toothache, joint pain, and as antidote for rattlesnake bites (Perri et al. 2006). In the twentieth century, it was a remedy against cold and flu and was frequently applied as an anti-infective until the advent of modern antibiotics.

The use of *Echinacea* during the first trimester of pregnancy was not associated with an increased risk of malformations in 112 pregnant women taking *Echinacea* in the first trimester (Gallo et al. 2000). The various brands of this phytomedicine cover two species of *Eschinacea* – *E. angustifolia* DC. and *E. purpurea* (L.) Moench (Gallo et al. 2000). The German Commission E considers oral *Echinacea* to be safe when used during pregnancy, if applied in recommended doses (Blumenthal et al. 1998). Also according to the European Medicines Agency, limited data, i.e., several hundred exposed pregnancies to Echinacea, indicate no adverse effects of *Echinacea* on pregnancy or on the health of the fetus/newborn child (HMPC 2015). However, because there are insufficient data, its use in pregnancy and lactation is not recommended unless advised by a doctor (HMPC 2015).

Ginseng (*Panax ginseng* C.A. Meyer, *Panax quinquefolium* L.) Ginseng of the genus *Panax* represents one of the most popular herbs in Eastern as well as Western countries (Barnes et al. 2004; Helms 2004). In China, Japan, and Korea the Asian ginseng (*Panax ginseng* C. A. Meyer) has been used for centuries, while the Native Americans in America and Canada used American ginseng (*Panax quinquefolium L.*) (Borchers et al. 2000; Helms 2004). Commercial harvesting of wild American gingeng plants occured since 1700s (Schorger AW 1969). The bioactive constituents are probably the ginsenosides, which exert vasorelaxative, antioxidative, anti-inflammatory, and chemopreventive effects. Ginseng is definitely a commercial best-seller (Blumenthal 2005; Kaufman et al. 2002).

A previous update on the molecular mechanisms and medical applications of ginseng focused on safety issues (Lu et al. 2009). There are not many data on the safe use of ginseng during pregnancy. Chan et al. reported that ginsenosides Rb1 and Re (but not Rc) were embryotoxic at a concentration above 50 μg/mL (Chan et al. 2004). Toxicity was related to developmental delay rather than teratogenicity. Liu et al. have shown that ginsenoside Rb1 at 50 μg/ml significantly reduced embryonic crown-rump length, head length, and somite number compared to the control group (Liu et al. 2005). Data on human subjects do not exist; however, the cautious use of ginseng products may be recommended for pregnant women. HMPC reported that safety of ginseng during pregnancy and lactation has not been established, and due to the absence of sufficient data, its use during pregnancy and lactation is not recommended (HMPC 2012).

Licorice (*Glycyrrhiza glabra* L., *G. uralensis* Fisch., *G. pallidiflora* Maxim.) Licorice is a medicinal plant known in the East and the West alike (Isbrucker and Burdock 2006). Its traditional uses include the treatment of fever, liver ailments, dyspepsia, gastric ulcers, sore throats, asthma, bronchitis, Addison's disease, and rheumatoid arthritis. Moreover, it has been used as a laxative, antitussive, expectorant, etc. (Anon 2005; Schulz et al. 1998; Wang et al. 2000).

The teratogenic evaluation of ammoniated glycyrrhizin in mice, rats, hamsters and rabbits did not reveal any effect on nidation or on maternal or fetal survival in any of the species (Food and Drug Research Labs 1971). The animals were orally given with 27–1,000 mg/kg/day ammonium glycyrrhizin on day 6 of gestation for 5–13 days. More than a decade later, Itami et al. (1985) investigated the potential teratogenicity of disodium glycyrrhizin in pregnant rats, who received disodium glycyrrhizin in their diets (80–2,000 mg/kg) during days 0–20 of gestation. There were no significant toxic effects of glycyrrhizin administration whatsoever, except for one fetus with dilated renal pelvis, who was treated with 80 mg/kg. No other malformations or anomalies were observed. The authors concluded that disodium glycyrrhizin is not teratogenic in rats.

Another study focused on ammoniated glycyrrhizin in pregnant rats (Mantovani et al. 1988). Doses of 10–250 mg/100 mL drinking water were applied for 20 days after conception. The authors found significantly increased embryotoxicity and skeletal anomalies in the highest treatment groups; these anomalies included misaligned, asymmetric, and bipartic sternebrae and hemisternebrae. Soft-tissue anomalies were found in the kidney, and some fetuses revealed external hemorrhages. The authors concluded that ammoniated glycyrrhizin exhibited minor embryotoxicity in fetuses, but no toxicity in mother animals.

Hundertmark and colleagues investigated rat fetal lung development (Hundertmark et al. 2002). Pregnant rats were fed 10–1,000 mg/kg glycyrrhetic acid daily, commencing on the 13th day of gestation. Fetuses were examined on days 17, 19, and 21 of gestation, as well as on the first post-partum day. The authors used 11β-hydroxysteroid dehydrogenase (11-HSD) as a parameter to evaluate lung toxicity, since this marker is involved in pulmonary surfactant synthesis during development. Fetal lung 11-HSD activity was moderately, but significantly, reduced. Fetal lung surfactant protein A mRNA levels decreased at the highest dose. Histological examination of fetal lungs revealed reduced lamellar body contents and reduced numbers of alveolar lamellar body and surfactant clusters. Nevertheless, the rates of malformation and fetal deaths did not increase with glycyrrhetic acid exposure.

In rodent dominant lethal tests, male rats were fed with 4–40,000 ppm glycyrrhizin for 10 weeks prior to mating (SRI 1977). These concentrations correspond to approximately 500 tor 5,000 mg/kg/day of glycyrrhizin. At the highest dose, significantly increased numbers of dead implants and dead-per-total implants have been observed, indicating that glycyrrhizin was mutagenic at 40,000 ppm.

If male rats were fed with 400–4,000 mg/kg glycyrrhizin for 10 weeks prior to mating with females, increased the numbers of dead implants per pregnant female (Sheu et al. 1986). A dominant lethal effect in male mice was not detected. Furthermore, glycyrrhizin did not induce heritable chromosomal defects.

The weak or negligible teratogenic effects recorded in these two studies were confirmed by other authors, who administered glycyrrhizin salts (1,000 mg/kg/day) maternally to mice, rats, hamsters, or rabbits during gestation (Food and Drug Research Labs 1971; Itami et al. 1985; Mantovani et al. 1988). However, somewhat lower doses (400–500 mg glycyrrhizin/kg/day) provoked mutagenicity in offspring (Sheu et al. 1986; SRI 1977).

Taken together, safety can be assumed for purified glycyrrhizin at doses between 15 and 229 mg/kg/day. If interspecies differences by a factor of 10 would exist, this dose range would compare well with the acceptable daily intake proposed by van Gelderen et al. (2000) of 0.2 mg/kg/day, but it is less than the estimated U.S. consumption of 0.027–3.6 mg/kg/day (van Gelderen et al. 2000). 1.2–1.5 g of dry extracts of this plant can be taken daily by adults (HMPC 2010). However, the use in children and adolescents under 18 is not recommended (HMPC 2010). A study has shown that 18β-glycyrrhetinic acid crosses through the placental barrier and can be detected in the rat fetuses (HMPC 2010). Following feeding of dams with 100 mg 18β-glycyrrhetinic acid/kg/day commencing on the 13th day of gestation, the maternal plasma 18β-glycyrrhetinic acid concentrations were approximately 100 μg/ml on the 17th, 19th and 21st days of gestation, whereas the fetal concentrations were 5, 18, and 32 μ/ml, respectively (HMPC 2010).

In developmental toxicity studies, glycyrrhizin (ammonium salt) exhibited some embryotoxicity to the developing rat fetus, but the fetal effects were considered to be minor (HMPC 2010). These effects were shown at the dosages of 100 and 250 mg/kg of ammonium glycyrrhizin from 7th to 20th day of pregnancy (soft-tissue abnormalities, mostly renal, and external hemorrhages), and at the dosage of 1,000 mg/kg of 18 β-glycyrrhetinic acid from the 13th day of gestation (significant reduction in lamellar body content of lungs and reduced number of alveolar lamellar body and surfactant clusters, but no apparent increase in malformation or fetal death rate) (HMPC 2010). Another study suggested that 100 mg/kg of licorice extract repeated for 7 days may also aggravate body weight loss and malformations of fetuses that are induced by intrauterine exposure to cyclophosphamide (HMPC 2010). Safety during pregnancy and lactation has not been established, and the use during pregnancy and lactation is not recommended because of lack of sufficient data (HMPC 2010).

St John's wort (*Hypericum perforatum* L.) St. John's wort, known as a medicinal plant for 2,000 years (Schulz et al. 1998), is one of the most popular herbal remedies in Europe and the United States (Beaubrun and Gray 2000; Di Carlo et al. 2001). The antidepressant effects of flower extracts have been verified in a large number of studies (Barnes et al. 2001; Linde and Mulrow 2000; Linde et al. 1996); major pharmacokinetic interactions with conventional drugs are also well established. St. John's wort contains anthraquinones (e.g., hypericins), flavonoids, prenylated phloroglucinols (e.g., hyperforins), tannins, phenols, and other constituents (Barnes et al. 2001). Hypericin and hyperforin derivatives are regarded as main active phytochemicals (Laakmann et al. 1998).

In light of the frequent use of this plant, information on possible developmental toxicity is important. Rayburn et al. investigated a standardized *Hypericum* preparation (0.3 % hypericin, 900 mg/day) applied to mice for 2 weeks before mating and throughout gestation (Rayburn et al. 2000, Rayburn et al. 2001a, b). Only a few effects were found, including lower body weights in male mice at birth, fewer male pups who successfully performed the negative geotaxis task (a test for vestibular and postural reflexes requiring motor coordination), and transient hyperactivity on postnatal day 21. Cada and colleagues fed mice from gestational day 3 to postnatal day 21 with 5 to 15-fold higher extract doses than recommended in human beings (Cada et al. 2001). Significantly reduced body weights were observed.

The evidence of toxic effects of St. John's wort is limited, but the mild effects on developing fetuses indicate that its extracts should not be used during pregnancy and nursing (Grush et al. 1998; Klier et al. 2002). Studies on acute toxicity and repeated dose toxicity did not show signs of toxic effect (HMPC 2009a). Animal studies have shown equivocal results, and the potential risk for humans is unknown; therefore, the use during pregnancy is not recommended (HMPC 2009b).

Valerian (*Valeriana officinalis* L.) Valerian is a well-known herb in traditional European phytomedicine (Schulz et al. 1998). Valerian acts as a mild sedative and favors sleep induction (Houghton 1999; Stevinson and Ernst 2000). The plant contains monoterpenes (e.g., borneol), sesquiterpenes (e.g., valerianic acid), and valepotriates (Houghton 1999). As valepotriates are recognized mutagens, current extracts in use harbor these compounds only in trace levels.

Developmental anomalies have been investigated in one study and gestational exposure to valepotriates did not show any effects in rats (Tufik et al. 1994). Although no adverse effects were found in animals, these data have not yet been confirmed in other studies; therefore, the use of valerian is contraindicated during pregnancy and lactation (Wong et al. 1998). Extracts with ethanol and the essential oil of valerian root have shown low toxicity in rodents; however, due to lack of data, its use during pregnancy and lactation is not recommended (HMPC 2006).

Conclusions and Perspectives

As a matter of course, Western drugs and herbal products should comply with the same scientific and pharmacological principles (Efferth and Greten 2012a; Wang et al. 2013). This is important to point out, because the therapeutic practices may differ considerably, as in the case of TCM or specific traditional medicines applied by practitioners and healers. Phytochemicals should be considered to be complex chemical entities in a similar manner as synthetic drugs, which may cause desired pharmacological as well as adverse toxic effects. Therefore, there is an obvious need for more research to allow non-ambiguous statements on the safety or risks of herbal products during pregnancy. The current practice is that herbal products with insufficient information on their safe use are not recommended during pregnancy (at least not during the first trimester, when organogenesis takes place). In line with existing and current practices and regulations of pharmaceuticals, scientific and professional herbal drug development should include several steps:

Quality control High quality of herbal extracts, including raw material, its processing and the final production of phytotherapeutics, should be guaranteed for the sake of efficient and safe use of herbal products (Efferth and Greten 2012b). Herbs should be botanically identified based on macroscopic features, chemical constitution, and advanced fingerprinting techniques (see Chapter 2). The macroscopic and microscopic authentication should be performed by experienced pharmacists according to existing pharmacopoeias. Chemical analyses (thin-layer chromatography, high-performance liquid chromatography, and gas chromatography, eventually coupled with mass spectrometry) represent an indispensable condition not only for plant identification, but also for standardization of herbal products. Recently, molecular biological techniques have been developed; these serve as a valuable asset for the identification of herbal products. The genetic code of plant species is effectively suited to authenticate them on a molecular level; polymerase chain reaction (PCR), direct amplification of length polymorphism (DALP), amplified fragment length polymorphisms, and related techniques have been successfully applied in this context. Such methods, highly valuable, however, yield no indication on the chemical profiles of authenticated herbs and cannot substitute analytical profiling.

Pharmacological mode of action (MOA) analyses Pharmacological and toxicological mechanisms should be analyzed with state-of-the-art methods as with any other synthetic drug (Youns et al. 2010). In addition to MOA analyses, drug interactions among phytochemicals and between phytochemicals and conventional drugs play an important role (see Chapter 5). Especially in the context of complex herbal mixtures, many different compounds may play together, not only to produce wishful synergistic effects important for therapeutic efficacy, but also to potentiate toxicities.

Safety investigation by animal experiments and clinical trials A *sine qua non* condition is testing for toxicity in animal experiments (Rutteman et al. 2013). Herbal products with signs of teratogenicity and developmental toxicity have to be removed from the market. Legal regulations should provide the framework that this is not only true for herbal products approved according to the FDA and European Medicines Agency (EMA) rules for drug development, but also for over-the-counter products and dietary supplements (Efferth and Greten 2012a; Eichhorn et al. 2011). Drugs tested to be safe in animal experiments may further be investigated in randomized, placebo-controlled clinical trials (Jansen et al. 2011; Krishna et al. 2015). For instance, artenimol-R, which has an excellent safety profile when used as an antimalarial drug (Adjuik et al. 2004; Staedke et al. 2008), underwent a pilot clinical pharmacological study in 10 patients with stage III and IV cervical carcinoma (Jansen et al. 2011). Artenimol-R treatment led to a rapid improvement of the clinical symptoms as defined by vaginal discharge and pain (Jansen et al. 2011). Artesunate, which is extracted from *Artemisia annua* L. and is a widely used antimalarial that can be administered by oral, rectal or parenteral routes (Gomes et al. 2009; Kremsner and Krishna 2004; Kremsner et al. 2012; Nealon et al. 2002; Hien et al. 1994, 1992; Jiang et al. 1982), was investigated by a single-center, double-blind, placebo-controlled trial with balanced randomization of colorectal cancer patients (1:1) (Krishna et al. 2015). This trial was approved by the Wandsworth Ethics Committee (Wandsworth UK, Ref: 08/H0803/3) and was registered (ISCRTN05203252) (Krishna et al. 2015). The artesunate treatment affects reducing Ki67 and increasing CD31 expression, indicating that artesunate has anti-proliferative properties in colorectal cancer (Krishna et al. 2015).

During the regular drug development process, special emphasis is on adverse effects (especially phase I trials). In this context, it is important that for herbal products, the right extract, dosages, and treatment duration according to the traditional practices be investigated. The saying of Paracelsus, "All things are poisons, for there is nothing without poisonous qualities. It is only the dose which makes a thing poison" is true for herbal products as well. There is no doubt that any herb, but not all herbs, might provoke side effects, if only the dose is high enough. Toxicity must be assessed at therapeutically relevant doses. Data obtained with concentrations beyond therapeutically relevant ranges may not be helpful to encourage the effective and safe use of herbal products (Wu et al. 2008).

Conflict of Interest The authors declare that there is no conflict of interest.

References

Adjuik M, Babiker A, Garner P, Olliaro P, Taylor W, White N (2004) Artesunate combinations for treatment of malaria: meta-analysis. Lancet 363:9–17

Allard T, Wenner T, Greten HJ, Efferth T (2013) Mechanisms of herb-induced nephrotoxicity. Curr Med Chem 20:2812–2819

Amorim MH, Gil da Costa RM, Lopes C, Bastos MM (2013) Sesquiterpene lactones: adverse health effects and toxicity mechanisms. Crit Rev Toxicol 43:559–579

Anon (2005) Glycyrrhiza glabra. Monograph. Altern Med Rev 10:230–237

Astin JA (1998) Why patients use alternative medicine: results of a national study. JAMA 279:1548–1553

Barcz E, Sommer E, Nartowska J, Balan B, Chorostowska-Wynimko J, Skopinska-Rozewska E (2007) Influence of Echinacea purpurea intake during pregnancy on fetal growth and tissue angiogenic activity. Folia Histochem Cytobiol 45(Suppl 1):S35–S39

Barnes J, Anderson LA, Phillipson JD (2001) St John's wort (Hypericum perforatum L.): a review of its chemistry, pharmacology and clinical properties. J Pharm Pharmacol 53:583–600

Barnes PM, Powell-Griner E, McFann K, Nahin RL (2004) Complementary and alternative medicine use among adults: United States, 2002. Adv Data 343:1–19

Beaubrun G, Gray GE (2000) A review of herbal medicines for psychiatric disorders. Psychiatr Serv 51:1130–1134

Blumenthal M (1999) Herb market levels after five years of boom. HerbalGram 47:64–65

Blumenthal M (2005) Herb sales down 7.4 percent in mainstream market; garlic is top-selling herb; herb combinations see increase. HerbalGram:63

Blumenthal M, Busse WR, Goldberg A, Gruenwald J, Hall T, Riggins CW, Rister RS, Klein S, Rister RS (1998) The Complete German Commission E Monographs. In: Council., A.B (ed) Therapeutic guide to herbal medicines. Integrative Medicine Communication, Boston/Austin

Borchers AT, Keen CL, Stern JS, Gershwin ME (2000) Inflammation and Native American medicine: the role of botanicals. Am J Clin Nutr 72:339–347

Brown NA, Spielmann H, Bechter R, Flint OP, Freeman SJ, Jelinek RJ, Koch E, Nau H, Newall DR, Palmer AK, Renault JY, Repetto MF, Vogel R, Wiger R (1995) Screening chemicals for reproductive toxicity: the current alternatives. The report and recommendations of an ECVAM/ETS workshop (ECVAM workshop 12). Altern Lab Anim 23:868–882

Cada AM, Hansen DK, LaBorde JB, Ferguson SA (2001) Minimal effects from developmental exposure to St. John's wort (*Hypericum perforatum*) in Sprague-Dawley rats. Nutr Neurosci 4:135–141

Chan LY, Chiu PY, Lau TK (2004) Embryotoxicity study of ginsenoside Rc and Re in in vitro rat whole embryo culture. Reprod Toxicol 19:131–134

Chou HG (2002) Clinical observation on 112 cases of abortion treated by supplemented shoutai wuzi soup. J Youjiang Med Coll National 2:301–302

Chuang CH, Hsieh WS, Guo YL, Tsai YJ, Chang PJ, Lin SJ, Chen PC (2007) Chinese herbal medicines used in pregnancy: a population-based survey in Taiwan. Pharmacoepidemiol Drug Saf 16:464–468

Clark RL (2009) Embryotoxicity of the artemisinin antimalarials and potential consequences for use in women in the first trimester. Reprod Toxicol 28:285–296

Clark RL, White TE, Clode AC, Gaunt I, Winstanley P, Ward SA (2004) Developmental toxicity of artesunate and an artesunate combination in the rat and rabbit. Birth Defects Res B Dev Reprod Toxicol 71:380–394

Clark RL, Arima A, Makori N, Nakata Y, Bernard F, Gristwood W, Harrell A, White TE, Wier PJ (2008a) Artesunate: developmental toxicity and toxicokinetics in monkeys. Birth Defects Res B Dev Reprod Toxicol 83:418–434

Clark RL, Lerman SA, Cox EM, Gristwood WE, White TE (2008b) Developmental toxicity of artesunate in the rat: comparison to other artemisinins, comparison of embryotoxicity and kinetics by oral and intravenous routes, and relationship to maternal reticulocyte count. Birth Defects Res B Dev Reprod Toxicol 83:397–406

Costa LG, Steardo L, Cuomo V (2004) Structural effects and neurofunctional sequelae of developmental exposure to psychotherapeutic drugs: experimental and clinical aspects. Pharmacol Rev 56:103–147

Cui SH (1998) Treatment on 47 cases of abortion. Shanxi J Tradit Chin Med 19:242

Cunningham F, Leveno KL, Bloom SL, Hauth J, Gilstrap L, Wenstrom K (2005) Williams obstetrics, 22nd edn. McGraw-Hill Professional Publishing, New York

Daston GP (2011) Laboratory models and their role in assessing teratogenesis. Am J Med Genet C Semin Med Genet 157C:183–187

Di Carlo G, Borrelli F, Ernst E, Izzo AA (2001) St John's wort: Prozac from the plant kingdom. Trends Pharmacol Sci 22:292–297

Dugoua JJ, Perri D, Seely D, Mills E, Koren G (2008) Safety and efficacy of blue cohosh (Caulophyllum thalictroides) during pregnancy and lactation. Can J Clin Pharmacol 15:e66–e73

Efferth T (2010) Personalized cancer medicine: from molecular diagnostics to targeted therapy with natural products. Planta Med 76:1143–1154

Efferth T, Greten HJ (2012a) The European directive on traditional herbal medicinal products: friend or foe for plant-based therapies? Zhong Xi Yi Jie He Xue Bao 10:357–361

Efferth T, Greten HJ (2012b) Quality control for medicinal plants. Med Aromat Plants 1:1–3

Efferth T, Kaina B (2010) Toxicity of the antimalarial artemisinin and its dervatives. Crit Rev Toxicol 40:405–421

Efferth T, Kaina B (2011) Toxicities by herbal medicines with emphasis to traditional Chinese medicine. Curr Drug Metab 12:989–996

Efferth T, Li PC, Konkimalla VS, Kaina B (2007) From traditional Chinese medicine to rational cancer therapy. Trends Mol Med 13:353–361

Eichhorn T, Greten HJ, Efferth T (2011) Self-medication with nutritional supplements and over-the-counter products. Nat Prod Bioprospecting 1:62–70

Ernst E (2002) The risk-benefit profile of commonly used herbal therapies: Ginkgo, St. John's Wort, Ginseng, Echinacea, Saw Palmetto, and Kava. Ann Intern Med 136:42–53

Fantel AG (1996) Reactive oxygen species in developmental toxicity: review and hypothesis. Teratology 53:196–217

Farnsworth NR, Bingel AS, Cordell GA, Crane FA, Fong HH (1975) Potential value of plants as sources of new antifertility agents I. J Pharm Sci 64:535–598

Finaurini S, Basilico N, Corbett Y, D'Alessandro S, Parapini S, Olliaro P, Haynes RK, Taramelli D (2012) Dihydroartemisinin inhibits the human erythroid cell differentiation by altering the cell cycle. Toxicology 300:57–66

Flaws B (2005) Chinese medical obstetrics. Blue Poppy Press, Boulder

Food and Drug Research Labs F (1971) Teratogenic evaluation of FDA 71-1 (ammonium glycyrrhizinate). In: Report, U.N. (Ed.)

Forster DA, Denning A, Wills G, Bolger M, McCarthy E (2006) Herbal medicine use during pregnancy in a group of Australian women. BMC Pregnancy Childbirth 6:21

Fu S, Fu QZNK (1978). Shanghai People's Publications House, Shanghai

Gallo M, Sarkar M, Au W, Pietrzak K, Comas B, Smith M, Jaeger TV, Einarson A, Koren G (2000) Pregnancy outcome following gestational exposure to echinacea: a prospective controlled study. Arch Intern Med 160:3141–3143

Ganzera M, Dharmaratne HR, Nanayakkara NP, Khan IA (2003) Determination of saponins and alkaloids in Caulophyllum thalictroides (blue cohosh) by high-performance liquid chromatography and evaporative light scattering detection. Phytochem Anal 14:1–7

Gomes MF, Faiz MA, Gyapong JO et al (2009) Pre-referral rectal artesunate to prevent death and disability in severe malaria: a placebo-controlled trial. Lancet 373:557–566

Gruenwald J (2000) The supplement markets in the US and Europe. Neutraceuticals World:36–7

Grush LR, Nierenberg A, Keefe B, Cohen LS (1998) St John's wort during pregnancy. JAMA 280:1566

Hakkola J, Pelkonen O, Pasanen M, Raunio H (1998) Xenobiotic-metabolizing cytochrome P450 enzymes in the human feto-placental unit: role in intrauterine toxicity. Crit Rev Toxicol 28(1):35–72

Hansen JM (2006) Oxidative stress as a mechanism of teratogenesis. Birth Defects Res C Embryo Today 78:293–307

Harvey AJ, Kind KL, Thompson JG (2002) REDOX regulation of early embryo development. Reproduction 123:479–486

Hathcock J (2001) Dietary supplements: how they are used and regulated. J Nutr 131:1114S–1117S

He YP (1997) Shou Tai Pill for threatened miscarriage. Guangzhou J Chin Med Coll 14:84–6

Helms S (2004) Cancer prevention and therapeutics: Panax ginseng. Altern Med Rev 9:259–274

Hien TT, Arnold K, Vinh H et al (1992) Comparison of artemisinin suppositories with intravenous artesunate and intravenous quinine in the treatment of cerebral malaria. Trans R Soc Trop Med Hyg 86:582–583

Hien TT, Arnold K, Hung NG et al (1994) Single dose artemisinin-mefloquine treatment for acute uncomplicated malaria. Trans R Soc Trop Med Hyg 88:688–691

HMPC (2006) Community herbal monograph on *Valeriana officinalis* L., radix

HMPC (2009) Community herbal monograph on Hypericum perforatum L., Herba

HMPC (2010) Community herbal monograph on *Glycyrrhiza glabra* L. and/or *Glycyrrhiza inflata* Bat. and/or *Glycyrrhiza uralensis* Fisch., radix, 2010

HMPC (2011) Community herbal monograph on *Zingiber officinale* Roscoe, rhizoma

HMPC (2012) Community herbal monograph on *Panax ginseng* C.A. Meyer, radix

HMPC (2015) European union herbal monograph on *Echinacea purpurea* (L.) Moench, herba recens

HMPC (Committee on Herbal Medicinal Products) (2009) Assessment report on *Artemisia absinthium* L., Herba

Houghton PJ (1999) The scientific basis for the reputed activity of Valerian. J Pharm Pharmacol 51:505–512

Hundertmark S, Dill A, Buhler H, Stevens P, Looman K, Ragosch V, Seckl JR, Lipka C (2002) 11beta-hydroxysteroid dehydrogenase type 1: a new regulator of fetal lung maturation. Horm Metab Res 34:537–544

Isbrucker RA, Burdock GA (2006) Risk and safety assessment on the consumption of Licorice root (Glycyrrhiza sp.), its extract and powder as a food ingredient, with emphasis on the pharmacology and toxicology of glycyrrhizin. Regul Toxicol Pharmacol 46:167–192

Itami T, Ema M, Kanoh S (1985) Effect of disodium glycyrrhizinate on pregnant rats and their offspring. J Food Hyg Soc Jpn 26:460–464

Izzo AA, Ernst E (2001) Interactions between herbal medicines and prescribed drugs: a systematic review. Drugs 61:2163–2175

James LF, Panter KE, Nielsen DB, Molyneux RJ (1992) The effect of natural toxins on reproduction in livestock. J Anim Sci 70:1573–1579

Jansen FH, Adoubi I, J C CK, DE Cnodder T, Jansen N, Tschulakow A, Efferth T (2011) First study of oral Artenimol-R in advanced cervical cancer: clinical benefit, tolerability and tumor markers. Anticancer Res 31:4417–22

Jewell D, Young G (2003) Interventions for nausea and vomiting in early pregnancy. Cochrane Database Syst Rev 9:CD000145

Jiang JB, Li GQ, Guo XB, Kong YC, Arnold K (1982) Antimalarial activity of mefloquine and qinghaosu. Lancet ii:285–288

Juchau MR, Boutelet-Bochan H, Huang Y (1998) Cytochrome P450-dependent biotransformation of xenobiotics in human and rodentembryonic tissues. Drug Metab Rev 30:541–568

Juchau MR, Lee QP, Fantel AG (1992) Xenobiotic biotransformation/bioactivation in organogenesis-stage conceptual tissues: implications for embryotoxicity and teratogenesis. Drug Metab Rev 24:195–238

Kaufman DW, Kelly JP, Rosenberg L, Anderson TE, Mitchell AA (2002) Recent patterns of medication use in the ambulatory adult population of the United States: the Slone survey. JAMA 287:337–344

Keeler RF (1984) Teratogens in plants. J Anim Sci 58:1029–1039

Kennelly EJ, Flynn TJ, Mazzola EP, Roach JA, McCloud TG, Danford DE, Betz JM (1999) Detecting potential teratogenic alkaloids from blue cohosh rhizomes using an in vitro rat embryo culture. J Nat Prod 62:1385–1389

Klepser TB, Klepser ME (1999) Unsafe and potentially safe herbal therapies. Am J Health Syst Pharm 56:125–138; quiz 139–41

Klier CM, Schafer MR, Schmid-Siegel B, Lenz G, Mannel M (2002) St. John's wort (Hypericum perforatum) – is it safe during breastfeeding? Pharmacopsychiatry 35:29–30

Konkimalla VB, Efferth T (2008) Evidence-based Chinese medicine for cancer therapy. J Ethnopharmacol 116:207–210

Kremsner PG, Krishna S (2004) Antimalarial combinations. Lancet 364:285–294

Kremsner PG, Taylor T, Issifou S et al (2012) A simplified intravenous artesunate regimen for severe malaria. J Infect Dis 205:312–319

Krishna S, Ganapathi S, Ster IC, Saeed MEM, Cowan M, Finlayson C, Kovacsevics H, Jansen H, Kremsner PG, Efferth T, Kumar D (2015) A randomised, double blind, placebo-controlled pilot study of oral artesunate therapy for colorectal cancer. EbioMedicine 2:82–90

Laakmann G, Schule C, Baghai T, Kieser M (1998) St. John's wort in mild to moderate depression: the relevance of hyperforin for the clinical efficacy. Pharmacopsychiatry 31(Suppl 1):54–59

Landes P (1998) Market report. HerbalGram 42:64–65

Li L (2011) Efficacy and safety of Chinese medicines for pregnancy. The Chinese University of Hongkong, Hongkong

Li L, Zhou ZX, Liu JY (2006) Clinical studies of Bu Shen Gu Tai decoction for threatened miscarriage. Guangxi Chin Med 29:17–18

Li Q, Si Y, Smith KS, Zeng Q, Weina PJ (2008) Embryotoxicity of artesunate in animal species related to drug tissue distribution and toxicokinetic profiles. Birth Defects Res B Dev Reprod Toxicol 83:435–445

Li L, Dou LX, Neilson JP, Leung PC, Wang CC (2012a) Adverse outcomes of Chinese medicines used for threatened miscarriage: a systematic review and meta-analysis. Hum Reprod Update 18:504–524

Li L, Leung PC, Wang CC (2012b) Efficacy and safety of Chinese medicines for threatened miscarriage. In: Tranquilli AL (ed) Miscarriages: causes, symptoms and prevention. Nova Science, New York, pp 155–186

Li L, Dou L, Leung PC, Wang CC (2012a) Chinese herbal medicines for threatened miscarriage. Cochrane Database Syst Rev (5):CD008510

Li L, Leung PC, Chung TK, Wang CC (2014) Systematic review of Chinese medicine for miscarriage during early pregnancy. Evid Based Complement Alternat Med 2014:753856

Linde K, Mulrow CD (2000) St John's wort for depression. Cochrane Database Syst Rev 2:CD000448

Linde K, Ramirez G, Mulrow CD, Pauls A, Weidenhammer W, Melchart D (1996) St John's wort for depression – an overview and meta-analysis of randomised clinical trials. BMJ 313: 253–258

Lindstrom A, Ooyen C, Lynch ME, Blumenthal M, Kawa K (2014) Sales of herbal dietary supplements increase by 7.9% in 2013, marking a decade of rising sales: turmeric supplements climb to top ranking in natural channel. HerbalGram 103:52–56

Liu RF (2002) Clinical observation of integrated traditional and western medicine. Chin J Integr Tradit West Med 22:68–69

Liu P, Xu Y, Yin H, Wang J, Chen K, Li Y (2005) Developmental toxicity research of ginsenoside Rb1 using a whole mouse embryo culture model. Birth Defects Res B Dev Reprod Toxicol 74:207–209

Lu JM, Yao Q, Chen C (2009) Ginseng compounds: an update on their molecular mechanisms and medical applications. Curr Vasc Pharmacol 7:293–302

Luo HT, Zhao SP, Juan HX (2007) 58 cases of ultrasound observations on treatened abortions with PPS. J Shanxi Coll Tradit Chin Med 30:26

Mantani N, Kasahara Y, Kamata T, Sekiya N, Shimada Y, Usuda K, Sakakibara I, Hattori N, Terasawa K (2002) Effect of Seihai-to, a Kampo medicine, in relapsing aspiration pneumonia – an open-label pilot study. Phytomedicine 9:195–201

Mantovani A, Ricciardi C, Stazi AV, Macri C, Piccioni A, Badellino E, De Vincenzi M, Caiola S, Patriarca M (1988) Teratogenicity study of ammonium glycyrrhizinate in the Sprague-Dawley rat. Food Chem Toxicol 26:435–440

Marcus DM, Snodgrass WR (2005) Do no harm: avoidance of herbal medicines during pregnancy. Obstet Gynecol 105:1119–1122

Moallem SA, Niapour M (2008) Study of embryotoxicity of Perovskia abrotanoides, an adulterant in folk-medicine, during organogenesis in mice. J Ethnopharmacol 117:108–114

Moreno LE, Juarez AO, Pelzer LE (2012) Lethal effect of dehydroleucodine (DhL) on amphibian Bufo arenarum embryos. Food Chem Toxicol 50:672–674

Nealon C, Dzeing A, Muller-Romer U et al (2002) Intramuscular bioavailability and clinical efficacy of artesunate in Gabonese children with severe malaria. Antimicrob Agents Chemother 46:3933–3939

Nordeng H, Havnen GC (2004) Use of herbal drugs in pregnancy: a survey among 400 Norwegian women. Pharmacoepidemiol Drug Saf 13:371–380

Ong LL, Schardein JL, Petrere JA, Sakowski R, Jordan H, Humphrey RR, Fitzgerald JE, de la Iglesia FA (1983) Teratogenesis of calcium valproate in rats. Fundam Appl Toxicol 3:121–126

Ortega JA, Lazerson J (1987) Anagyrine-induced red cell aplasia, vascular anomaly, and skeletal dysplasia. J Pediatr 111:87–89

Ouedraogo M, Baudoux T, Stevigny C, Nortier J, Colet JM, Efferth T, Qu F, Zhou J, Chan K, Shaw D, Pelkonen O, Duez P (2012) Review of current and "omics" methods for assessing the toxicity (genotoxicity, teratogenicity and nephrotoxicity) of herbal medicines and mushrooms. J Ethnopharmacol 140:492–512

Pelkonen O, Arvela P, Kärki NT (1971) 3,4-Benzpyrene and N-methylaniline metabolizing enzymes in the immature human foetus and placenta. Acta Pharmacol Toxicol (Copenh) 30(5):385–395

Perri D, Dugoua JJ, Mills E, Koren G (2006) Safety and efficacy of echinacea (Echinacea angustafolia, e. purpurea and e. pallida) during pregnancy and lactation. Can J Clin Pharmacol 13:e262–e267

Rao RB, Hoffman RS (2002) Nicotinic toxicity from tincture of blue cohosh (Caulophyllum thalictroides) used as an abortifacient. Vet Hum Toxicol 44:221–222

Rayburn WF, Christensen HD, Gonzalez CL (2000) Effect of antenatal exposure to Saint John's wort (Hypericum) on neurobehavior of developing mice. Am J Obstet Gynecol 183:1225–1231

Rayburn WF, Gonzalez CL, Christensen HD, Harkins TL, Kupiec TC (2001a) Impact of hypericum (St. John's wort) given prenatally on cognition of mice offspring. Neurotoxicol Teratol 23:629–637

Rayburn WF, Gonzalez CL, Christensen HD, Stewart JD (2001b) Effect of prenatally administered hypericum (St John's wort) on growth and physical maturation of mouse offspring. Am J Obstet Gynecol 184:191–195

Rutteman GR, Erich SA, Mol JA, Spee B, Grinwis GC, Fleckenstein L, London CA, Efferth T (2013) Safety and efficacy field study of artesunate for dogs with non-resectable tumours. Anticancer Res 33:1819–1827

Sairafianpour M, Christensen J, Staerk D, Budnik BA, Kharazmi A, Bagherzadeh K, Jaroszewski JW (2001) Leishmanicidal, antiplasmodial, and cytotoxic activity of novel diterpenoid 1,2-quinones from Perovskia abrotanoides: new source of tanshinones. J Nat Prod 64:1398–1403

Schorger AW (1969) Ginseng: A pioneer resource. Trans Wis Acad Sci Arts Letts 57:65–74

Schulz V, Haensel R, Tyler VE (1998) Rational phytotherapy. A physicians's guide to herbal medicine. Springer, Berlin, pp 160–187

Shaw D, Ladds G, Duez P, Williamson E, Chan K (2012) Pharmacovigilance of herbal medicine. J Ethnopharmacol 140:513–518

Shayo A, Buza J, Ishengoma DS (2015) Monitoring of efficacy and safety of artemisinin-based anti-malarials for treatment of uncomplicated malaria: a review of evidence of implementation of anti-malarial therapeutic efficacy trials in Tanzania. Malar J 14:135

Sheu CW, Cain KT, Rushbrook CJ, Jorgenson TA, Generoso WM (1986) Tests for mutagenic effects of ammoniated glycyrrhizin, butylated hydroxytoluene, and gum Arabic in rodent germ cells. Environ Mutagen 8:357–367

Small E, Catling PM (1999) Canadian medicinal crops. NRC Research Press, Ottawa, pp 35–39

Sogorb MA, Pamies D, de Lapuente J, Estevan C, Estevez J, Vilanova E (2014) An integrated approach for detecting embryotoxicity and developmental toxicity of environmental contaminants using in vitro alternative methods. Toxicol Lett 230:356–367

Spielmann H (1998) Reproduction and development. Environ Health Perspect 106(Suppl 2):571–576

SRI, S.R.I. (1977) Study of the mutagenic effects of ammoniated glycyrrhizin (71-1) by the dominant lethal test in rats. US NTIS

Staedke SG, Jagannathan P, Yeka A, Bukirwa H, Banek K, Maiteki-Sebuguzi C, Clark TD, Nzarubara B, Njama-Meya D, Mpimbaza A, Rosenthal PJ, Kamya MR, Wabwire-Mangen F, Dorsey G, Talisuna AO (2008) Monitoring antimalarial safety and tolerability in clinical trials: a case study from Uganda. Malar J 7:107–117

Stevinson C, Ernst E (2000) Valerian for insomnia: a systematic review of randomized clinical trials. Sleep Med 1:91–99

Tufik S, Fujita K, Seabra Mde L, Lobo LL (1994) Effects of a prolonged administration of valepotriates in rats on the mothers and their offspring. J Ethnopharmacol 41:39–44

van Gelderen CE, Bijlsma JA, van Dokkum W, Savelkoul TJ (2000) Glycyrrhizic acid: the assessment of a no effect level. Hum Exp Toxicol 19:434–439

Wang XJ, Shi SZ, Lin GW, Wang Q (1995) Investigation on the application of drugs in 416 pregnant women in Shanghai. Chin J Pharmacoepidemiol 4:167–170

Wang ZY, Athar M, Bickers DR (2000) Licorice in foods and herbal drugs: chemistry, pharmacology, toxicology and uses. In: Mazza G, Oomah BD (eds) Herbs, botanicals and teas. Technomic Publishing Co. Inc., Lancaster, pp 321–353

Wang CC, Li L, Tang LY, Leung PC (2012) Safety evaluation of commonly used Chinese herbal medicines during pregnancy in mice. Hum Reprod 27:2448–2456

Wang CC, Li L, San Lau CB, Leung PC, Fung KP (2013) Pregnancy outcomes, embryonic and fetal development in maternal exposure to Chinese medicines. Birth Defects Res C Embryo Today 99:275–291

Wells PG, Winn LM (1996) Biochemical toxicology of chemical teratogenesis. Crit Rev Biochem Mol Biol 31:1–40

Wells PG, Kim PM, Nicol CJ, Parman T, Winn LM (1997) Chapter 17. Reactive intermediates. In: Kavlock RJ, Daston GP (eds) Drug toxicity in embryonic development. Springer, Heidelberg, pp 453–518

Wells PG, Bhuller Y, Chen CS, Jeng W, Kasapinovic S, Kennedy JC, Kim PM, Laposa RR, McCallum GP, Nicol CJ, Parman T, Wiley MJ, Wong AW (2005) Molecular and biochemical mechanisms in teratogenesis involving reactive oxygen species. Toxicol Appl Pharmacol 207:354–366

Westfall RE (2001) Herbal medicine in pregnancy and childbirth. Adv Ther 18:47–55

White TE, Bushdid PB, Ritter S, Laffan SB, Clark RL (2006) Artesunate-induced depletion of embryonic erythroblasts precedes embryolethality and teratogenicity in vivo. Birth Defects Res B Dev Reprod Toxicol 77:413–429

Wong AH, Smith M, Boon HS (1998) Herbal remedies in psychiatric practice. Arch Gen Psychiatry 55:1033–1044

Woolhouse M (2006) Complementary medicine for pregnancy complications. Aust Fam Physician 35:695

WorldHealthOrganization W (2007) Assessment of the safety of artemisinin compounds in pregnancy. WHO, Geneva

Wu KM, Ghantous H, Birnkrant DB (2008) Current regulatory toxicology perspectives on the development of herbal medicines to prescription drug products in the United States. Food Chem Toxicol 46:2606–2610

Youns M, Hoheisel JD, Efferth T (2010) Toxicogenomics for the prediction of toxicity related to herbs from traditional Chinese medicine. Planta Med 76:2019–2025

Yu SM, Ghandour RM, Huang ZJ (2004) Herbal supplement use among US women, 2000. J Am Med Womens Assoc 59:17–24

Zhang L, Yan J, Liu X, Ye Z, Yang X, Chan K, Shaw D, Duez P (2012) Pharmacovigilance practice and risk control of TCM in China. J Ethnopharmacol 140:519–525

Zhou Y (1997) Clinical observation on 305 cases of threatened abortion treated by integrated traditional Chinese and western medicine. Pract J Integrating Chin Mod Med 10:1614

Zhou Y (2006) Clinical observation of Shoutai Pill and Shixiao Powder for the treatment of threatened abortion with kidney deficiency and blood stasis. J Guangzhou Univ Tradit Chin Med 23:25–29

Chapter 11
Sensitization and Allergies of Herbal Products

Jacqueline Wiesner

Abstract The most common example for an allergy to herbal material is rhinoconjunctivitis, better known as "hay fever." It is estimated that between 10 and 40 % of the world's population suffers from allergic rhinoconjunctivitis (Bachert et al., Allergy 65 (Suppl 93): 1–13, 2010). But food allergies, which are estimated to affect 1–10 % of the world's population, including allergies to plant-derived materials such as wheat, soy, peanuts and tree nuts, are also prevalent (Quake and Nadeau, Semin Cell Dev Biol, 43: 125–130, 2015). While most of the cases will also refer to animal proteins (milk, egg, fish), separate numbers for plant food allergy are not available. Also, medicinal products containing herbal substances/preparations may provoke allergies. Not only can the processed forms trigger allergies, the starting material (plants) may provoke allergic reactions as well, either in individuals involved in harvesting or processing, or in persons concocting preparations (Sticher et al., Hänsel/Sticher – Pharmakognosier Phytopharmazie. Wissenschaftliche Verlagsgesellschaft mbH, Stuttgart, 2015), such as pharmacists or nurses; this has in particular been reported for *Psyllium*.

Beside these facts, it should not be forgotten that a "food/plant allergy" is inferred by patients or consumers, whereas signs of allergy have effectively been triggered by food additives (i.e., artificial coloring), excipients, fungal spores, or contaminants. Therefore, in most cases, it is not, or at least hardly ever, possible to pinpoint the triggering agent and to confirm or exclude herbal preparations as allergic agents. This chapter will focus mainly on herbal preparations found in food supplements or medicinal products; however, there are flowing transitions to food, cosmetics, and environmental herbal products such as pollen.

Keywords Adverse effects • Allergy • Cosmetics • Food • Food supplements • Herbal medicinal products • Labelling • Plant allergens • Test systems

---

Disclaimer: The views presented in this chapter represent the personal views of the author and should not be cited or understood as the position of the Federal Institute for Drugs and Medical Devices (BfArM) or the European Medicines Agency (EMA).

J. Wiesner
Federal Institute for Drugs and Medical Devices (BfArM),
Kurt-Georg-Kiesinger-Allee 3, 53175 Bonn, Germany
e-mail: jacqueline.wiesner@bfarm.de

O. Pelkonen et al. (eds.), *Toxicology of Herbal Products*,
DOI 10.1007/978-3-319-43806-1_11

Sensitization and Allergy: A Brief Introduction

It is not the objective of this chapter to describe the immunological processes of sensitization and allergy in detail (textbooks on immunology are more appropriate for this purpose); however, the effort should be made here to at least grasp the basic principles.

Today, allergy is defined as any exaggerated immune response to a foreign antigen, regardless of the mechanism of response, with the antigen being harmless for most people. Therefore sensitization is seen as the induction of allergic responses, and it can be long-lasting due to the immunologic memory. However, it is useful to note that sensitization will not always lead to symptoms or clinical disease (Rosenstreich et al. 2016). The substances that provoke allergic reaction (antigens) are called "allergens," and in most cases they are small (5–100 kDa), water-soluble proteins, often with carbohydrate side chains. However, in addition, other smaller molecules, pure carbohydrates or hydrophobic proteins, might act as allergens (Scheurer et al. 2015). But also "haptens" might provoke allergic reactions; these are defined as small organic compounds (also reactive secondary metabolites, such as quinones, aldehydes, sesquiterpenic lactones, etc.) that are susceptible to electrophilic additions to proteins (so-called "haptenization") and therefore generating allergens (Dudeck et al. 2011).

Different Types of Hypersensitivity Reactions

Gell and Coombs (1963) proposed a subdivision of hypersensitivity reactions that classifies those reactions based on the underlying immune mechanisms (Fig. 11.1). It is still used today as a general basis although with current information, additions/corrections are often made by various authors.

In the context of herbal material, all types of hypersensibility are possible. Usually most plant allergy reactions belong to type I (approximately 48 %), followed by type IV (approximately 18 %), and types III and II (10 % and 6 %, respectively) (Żukiewicz-Sobczak et al. 2013). Type I and type IV reactions seem significant and will therefore be highlighted, even though none of the reactions may occur in isolation; rather, mixed responses are conceivable (Descotes and Choquet-Kastylevsky 2001).

Type I Reaction

Type I hypersensitivity is the form of hypersensitivity that is often simply called "allergy." The reaction occurs "immediately," which means within seconds or minutes, and, in most cases, IgE antibodies are responsible for such allergic reactions. With "atopy" or "atopic," the predisposition to become IgE-sensitized to allergens

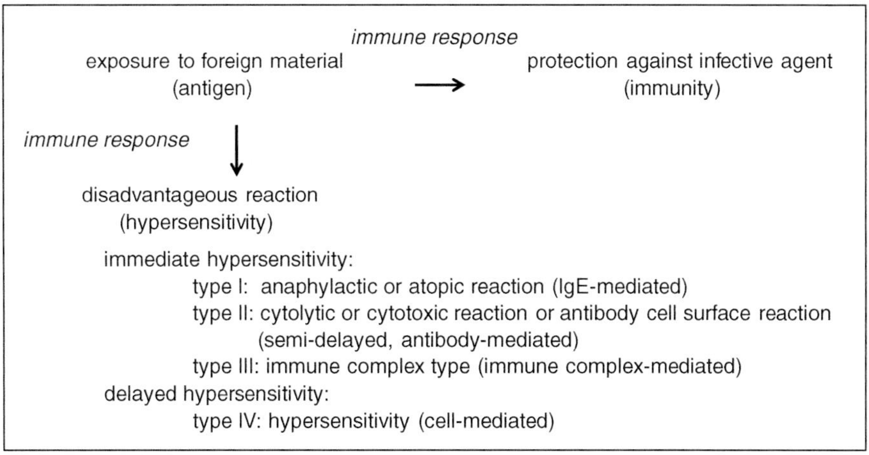

Fig. 11.1 Hypersensitivity reactions to foreign material according to the Gell and Coombs classification (Reprinted from *Toxicology*, Vol. 158, nos. 1-2, Jacques Descotes and Geneviève Choquet-Kastylevsky, "Gell and Coombs's Classification: Is It Still Valid?," Fig. 1, Copyright 2001, with permission from Elsevier)

is described. In predisposed people, the first contact with an allergen will lead to sensitization, and those individuals can develop typical symptoms of allergy. Unfortunately, there is no adequate understanding as to why allergens promote allergic responses or what the mechanisms behind the complex cascade are.

The prerequisite for developing an allergy seems to be the ability of allergens to penetrate mucosal tissue. It is presumed that a genetic basis underlies the development of allergies, but epigenetic and environmental factors are also probably involved (De Swert 1999; Sabounchi et al. 2015).

It seems to be a given that sensitization begins with antigen-presenting cells that present the processed antigen to naïve T-helper cells (Th-cells), via a major histocompatibility class II (MHC-II) complex and co-stimulating and soluble factors. After stimulation by these three signals, the naïve Th-cells will polarize into Th2-cells in atopic patients, while in non-atopic patients there will be no such change due to poorly understood genetic and environmental factors (Grammatikos 2008; van Ree et al. 2014). The activation of Th2-cells and their interaction with B-cells will lead to isotype switching of B-cells to IgE-producing cells. While the majority of activated B-cells are short-lived, there are also long-lived resident plasma cells that might survive up to several months or even an entire lifetime, mainly in the spleen, bone marrow, and inflamed tissue. IgE antibodies secreted from activated B-cells bind to specific high-affinity Fc receptors on the surface of mast cells and basophils. The IgE-coated cells, at this stage, are sensitized to the allergen, since the coating makes those cells sensitive to activation by subsequent encounters with that antigen. IgE antibodies also stimulate the activation and proliferation of mast cells and eosinophils.

In sensitized subjects, at later exposure, the allergen can bind to the IgE molecules held on the surface of the mast cells or basophils. Those cells are activated by the cross-linking of IgE and high-affinity Fc receptors (FcεRI) if the allergen binds to two or more IgE antibodies on the cell. The cross-linking of IgE and FcεRI triggers biochemical signal cascades that lead to the rapid release of granule content (degranulation) (e.g.,, histamine, serine proteases, carboxypeptidase A, proteoglycans, sulfatases) and synthesis and secretion of lipid mediators (e.g., prostaglandins, thromboxanes, leukotrienes) and cytokines (e.g., TNF-α, IL-1α, IL-4, IL-6, IL-13) (Paul 2003).

There are various categories of allergic disorders, describing the anatomical site where the disease is seen: atopic dermatitis, atopic rhinitis, atopic asthma, (food) allergy, and anaphylaxis (see Table 11.1). The effector cascade for all these forms will be more or less the same; however, the route and dosage of allergen exposure, and the site of initial sensitization, etc., might be different. Some patients will experience even two or more forms.

For food allergens – and here herbal preparations used in medicinal products or food supplements will be covered as well – two types are described. Class I food allergens will cause allergic reactions via primary sensitization (ingestion via the gut). They are called "classical," "true," or "complete" food allergens. Sensitization with class I food allergens is often associated with severe (sometimes anaphylactic) reactions. Class II food allergens produce sensitization via various routes (mainly inhalation). A reaction represents cross-reactivity of IgE antibodies with food proteins from the same protein family as the primary allergen; mostly mild to moderate reactions are seen (Lorenz et al. 2015).

Type IV Reaction

Type IV reactions are called "delayed hypersensitivity," since clinical symptoms peak 48–72 h after contact with the allergen in sensitized persons. Type IV reactions can be subdivided either by time of onset, clinical manifestation and cells involved, or by effector cells and mediators involved, leading to three or four sub-categories, respectively. Antigen-specific Th1 and cytotoxic T-lymphocytes, but also Th2- and T-cells, are mediators of such delayed hypersensitivity reactions.

Allergens for type IV reactions might be pathogens such as bacteria, fungi, or viruses, but also proteins and low-molecular-weight-chemicals. Such smaller molecules may act as haptens, which mean that they become allergens only after conjugation with proteins (prohaptens). But there are also substances that act as prehaptens, meaning that the components act as haptens only after external activation. An example of this is linalool, a substance occurring in the essential oil of many plants, which acts as an allergen after autoxidation. For other substances, such as geraniol (also in essential oils of plants), both activation ways are known (Peiser et al. 2012). Such substances (essential oils) are often used in flavoring agents, which might contribute to the allergenicity of the finished product.

Table 11.1 Manifestations of type I hypersensitivity

Organ system	Clinical features	Remarks
Eyes	Allergic conjunctivitis	Often associated with rhinitis, but not always
		As reaction to food mainly in pollen-sensitized individuals, but less frequently than asthma
Respiratory tract	Allergic rhinitis; allergic sinusitis	As reaction to food, less frequently than asthma
	Cough; stridor; asthma	As manifestation of a food-allergic reaction, sometimes the dominating symptom, but often associated with eczema, urticaria or gastrointestinal symptoms
		Deaths from anaphylactic reactions more often caused by respiratory problems than by circulatory failure
Gastrointestinal tract	Oral allergy syndrome (OAS); nausea/vomiting; gastro-oesophageal reflux disease; abdominal pain; diarrhea; enteropathies; infantile colic; constipation; failure to thrive	While OAS is IgE-mediated, all other forms are mainly mixed forms
		OAS can be restricted to the mouth/pharynx but may also involve several organs even reaching anaphylaxis
		Some of the conditions mainly occur in childhood and often cows' milk seems to be responsible
Skin	Atopic dermatitis; pruritus; angioedema; urticaria; erythema	Atopic dermatitis usually occurs in early infancy and persists sometimes in adulthood; children with atopic dermatitis often develop allergic rhinitis and asthma later
		Urticaria (synonyms are hives or nettle rush) due to food ingestion generally occurs within hours and fades within 3 h
Generalized (systemic)	Anaphylaxis	Involves cardiovascular symptoms (e.g., tachycardia, hypotension, cardiovascular collapse); respiratory involvement (e.g., bronchospasm, dyspnea, wheezing); cutaneous symptoms (e.g., urticaria, erythema, pruritus); edema of the pharynx (inducing difficulties in talking, breathing, swallowing); gastrointestinal symptoms (e.g., abdominal pain), singly or in combination
		In fatal cases, initial symptoms develop within 3-30 min and severe respiratory symptoms between 20 and 150 min of exposure

According to EFSA (2014)

The first step in the initiation of such a reaction is the internalization of the antigen by antigen-presenting cells (e.g., Langerhans cells and dermal dendritic cells), the presentation of the antigen together with MHC-class II molecules, and the secretion of interleukins. Naïve CD4-T-cells will therefore be activated, and memory cells will be formed. This phase will last 1–2 weeks. Also, for type IV reactions, microbial triggers and reactive oxygen species (ROS) may possibly be involved in the development of sensitization (Martin 2015).

The subsequent exposure of the antigen-activated cells will lead to the secretion of several cytokines. These will induce blood monocytes to adhere to vascular endothelial cells and to migrate to surrounding tissues. Lytic enzymes are excreted by macrophages and will lead to nonspecific destruction of host cells and tissue damage; the growth factor produced by macrophages will stimulate the proliferation and differentiation of fibroblasts that lead to the formation of fibrotic tissues.

Type IV reactions are generally confined to the contact site, but generalized reactions may occur (Baldo and Pham 2013). Furthermore – at least on the basis of animal studies – it is known that cutaneous sensitization may predispose to intestinal allergy (Ashley et al. 2015). On the other hand, there are known cases in which an allergic dermatitis reaction may develop after systemic exposure to a hapten that reaches the skin through hematogenous transport. While this condition has traditionally been described following topical exposure, it can also be observed without previous cutaneous sensitization to the hapten (Thyssen and Maibach 2008).

It is estimated that 15–20 % of the general population suffer from contact dermatitis to at least one allergen. Acquired risk factors, such as inflammatory skin diseases and hereditary risk factors such as genes, age, gender, and ethnicity, have been described (Peiser et al. 2012).

A special case of type IV reaction is photoallergy. Contact (mostly dermal) with photoallergens leads to allergic reactions, whereby photoallergens are often haptens that form reactive species under UV radiation; they then covalently bind to proteins (human serum albumins) to develop into full allergens. Other classes of photoallergens may be activated by radiant energy and transform chemically when returning to a resting state; the released energy then promotes conjugation of this new chemical entity to a carrier protein, forming a completely new antigen (Stein and Scheinfeld 2007).

Finally, there are terms used in (lay) literature that sometimes cause confusion. These terms should not be confused with hypersensitivity reactions, which always require the involvement of the immune system:

Pseudoallergy/anaphylactoid intolerance:

Terms such as "pseudoallergy," "anaphylactoid reactions," and "intolerance" describe non-immune responses that do not require a sensitization step; their definitions are not consistent among authors. Symptoms may occur mainly because of absent or defective enzymes (enzymopathy), activation of complement, unstable cell membranes of mast cells, or basophile granulocytes or metabolic disorders of the arachidonic acid pathway. Such symptoms often resemble type I-hypersensitivity reactions, but without proof of IgE antibodies involvement. Allergy-like symptoms can also mirror pharmacological effects of herbal material, for instance by its histamine or vasoactive amines content, which might lead to symptoms comparable to an allergic reaction (e.g.,, rash or abdominal pain). It is sometimes claimed that such reactions are as frequent as true IgE-mediated reactions (Pichler 2007).

Irritation:

According to the U.S. Occupational Safety and Health Administration (OSHA 1994), "…Irritants are noncorrosive substances that cause a temporary inflammation on direct contact with the skin, eyes, nose, or respiratory system by a chemical action at the point of contact," It is questionable whether the irritant effects (of contact allergens) may trigger the development of type IV hypersensitivity (Martin 2015).

Phototoxicity:

Phototoxicity may arise from systemically administered agents or from direct contact, and a sufficient dose will be needed. Such compounds will cause harm, either by the formation of free radicals or by the formation of stable phototoxic products after absorption of photons (energy). Erythema that is limited to sun-exposed skin will appear (Stein and Scheinfeld 2007).

Non-clinical Test Methods

Several test methods and international guidance documents cover medicinal products, food/food supplements, and cosmetics. As for most other toxicological methods, these tests describe methods that have been developed for single substances rather than for mixtures of substances (extracts) that may vary in composition. The tests mentioned here do not aim for completeness; rather, they are proposed as examples. It is also acknowledged that recent guidelines for assessing the genetically modified food and feed have been published (EFSA 2011) but they are not discussed here.

Systemic Hypersensitivity

Usually it is expected to find signs of hypersensitivity in chronic toxicity studies (e.g., microscopic findings of lymphoid tissue, hematology, lymphocyte subsets, or pathology at administration site). Concerning specific methods, standard hypersensitivity tests that have proven useful for detecting contact sensitization (see the section on ____) are not very useful for the identification of systemic sensitizers (Hastings 2001; Bala et al. 2005). It is quite likely that methods that exist for well-defined and known proteins (such as IgE-binding studies with human sera from individuals known to be allergic to the identified allergen source, which notably require standardization of test materials and stability to in vitro pepsin digestion) will not be applicable for multi-component mixtures, especially if no information about the triggering structure(s) is available.

However, three in vivo methods are described to detect substance-induced specific antibodies (FDA 2002):

- Passive cutaneous anaphylaxis assay (PCA-assay), which represents a localized cutaneous allergic response as a consequence of allergen-induced vascular permeability and plasma extravazation;
- Active cutaneous anaphylaxis test (ACA-assay), which is performed similarly to PCA, but without dye; instead, ear swelling, skin lesions, or severity of symptoms of anaphylactic reactions such as respiratory distress, increased respiratory rate, dyspnea, cyanosis, and mortality can be measured;
- Active systemic anaphylaxis assay (ASA-assay), which represents a generalized allergic reaction that manifests as hypotension, bronchoconstriction, or hypothermia.

Usually the tests are performed on guinea pigs, but since not only IgE-triggered reactions occur in guinea pigs but IgG-triggered reactions as well, all three tests are considered limited for safety assessment and consequently are not recommended for routine testing. Furthermore, the testing of small molecular weight compounds remains challenging, especially if biotransformation is important for the production of potential haptens (FDA 2002; Luebke et al. 2006; Gad 2009).

Over the last few years, a number of experimental mouse models of oral antigen-induced anaphylaxis have been described (Hogan et al. 2012), but (as with guinea pigs) questions about the human relevance of these studies remain, due to the differences in anaphylaxis between both animal species and humans (Verdier et al. 1994; Finkelman 2007). *In vitro* (ex-vivo) models have also been proposed to evaluate the allergic potential of orally administered compounds (Berin and Mayer 2009). Neither approach (mouse or ex-vivo models) has been developed into a guideline so far.

Respiratory Hypersensitivity

These mainly consist of adaptions of assays, such as the Local Lymph Node Assay (LLNA) for the detection of type IV hypersensitivity but with exposure via inhalation (FDA 2002; Derelanko and Auletta 2014), or the ACA-assay, performing the sensitization via the respiratory tract (Muller and Healy 1973).

Allergic Contact Dermatitis

From the numerous assays to detect a dermal sensitizing potential, the most common methods are Local Lymph Node Assays (LLNA) (OECD 2010a, b, c) in mice and guinea pigs, notably the Guinea Pig Maximization Test (GPMT – adjuvant test in which the acquisition of sensitization is potentiated by the use of Freund's Complete Adjuvant) and the Buehler test (OECD 1992). The Organisation for Economic Co-operation and Development (OECD) Guidelines (OECD 2010b, c) propose non-radioactive modifications of the LLNA. Other techniques, such as the mouse ear swelling test (MEST) (Gad et al. 1986; Auttachoat et al. 2011) or the Draize test (Draize et al. 1944), have also been described. While the goal of the LLNA is the afferent phase of the hypersensitivity (initial exposure through clonal expansion and release of memory cells), all other tests on guinea pigs and mice describe the efferent phase (local recognition of the antigen by the memory cells, release of lymphokines and activity of the inflammatory mediators) (Hayes and Kruger 2014).

The mouse ear swelling test (MEST) and variations on it were developed in the early 1980s (Gad et al. 1986), but it was shown to be unreliable for detecting weak to moderate sensitizers (Cornacoff et al. 1988). A modified test procedure has been described (Auttachoat et al. 2011) that increases the explanatory power. Even if no standardized guideline is available, the MEST is currently listed as an accepted test system under OECD guidance (OECD 1992).

The OECD published a guideline on acute eye irritation/corrosion (OECD 2012) that is based on the original idea behind the Draize test. However, it is clearly suggested that before performing this test, a weight-of-evidence analysis be performed on the existing data and that validated and accepted in vitro tests (e.g., OECD 2013, 2015a) be preferred; furthermore, the Draize test can also be performed on animal trunks (Hayes and Kruger 2014).

The four biological key events accounting for a skin sensitization process are well known and have been summarized in OECD (2014b): (1) the covalent binding of the chemical to skin proteins (haptenation); (2) the release of pro-inflammatory cytokines and the induction of cyto-protective pathways in keratinocytes; (3) the maturation and mobilization of dendritic cells; and (4) the antigen presentation to naïve T-cells and proliferation of memory T-cells. Over the past few years, efforts have been made to develop alternative (non-animal) methods to address these key elements. Until now, two in chemico/*in vitro* methods have been integrated into the evaluation of skin sensitization: the Direct Peptide Reactivity Assay (DPRA), which addresses reactivity towards peptides (key event 1) (OECD 2015b), and the ARE-Nrf2 Luciferase Test Method, which addresses the keratinocyte induction of cyto-protective gene pathways (key event 2) (OECD 2015c). Furthermore, a draft proposal for the human Cell Line Activation Test (h-CLAT), which measures the activation of dendritic cells (key event 3) was published by the OECD (2014a). It is acknowledged that only the combination of information from such alternative test methods could replace animal testing in future. Further mechanism-based methods are being developed and will likely contribute to risk assessment.

Therefore it seems that – especially in the field of allergic contact dermatitis – there are advanced efforts under way to replace animal testing. This was notably seen in the 7th Amendment to the Cosmetic Guideline (Guideline 2003/15/EC) (EU 2003) that banned animal tests for cosmetic finished products (implemented in 2004) and for cosmetic ingredients (implemented in 2009) throughout the European Union.

Photoallergenicity

While there is no valid assay that predicts photoallergenicity, some in vitro (e.g., Lovell 1993; Karschuk et al. 2010) and in vivo models (e.g., Scholes et al. 1991; Ulrich et al. 1998; Descotes 2004) have been developed for this purpose. For herbal preparations containing compounds with a molar extinction coefficient value greater than 1,000 $l\ mol^{-1}\ cm^{-1}$ at any wavelength between 290 and 700 nm, a photoallergy assessment would be crucial. However, as the predictability of nonclinical photoallergy tests is not known, clinical testing, using the to-be-marketed formulation and conducted during phase 3 of the clinical trials (ICH S10 2013), would be needed.

Pseudoallergy

Even though predicting the potential to induce pseudoallergic reactions are limited in animal models, biochemical markers of an anaphylactoid reaction can be observed in non-clinical toxicology studies (e.g., detection of serum anaphylactic complement products in animals showing signs of anaphylaxis; measurement of histamine plasma levels) (Descotes 2004). Mostly, it is pointed out that in vitro assays using human cells or peripheral blood may be more valuable; histamine release, basophil degranulation, or complement activation can be easily tested using increasing concentrations of the test article (FDA 2002; Descotes 2004).

Herbal Preparations and Herbal Substances with Sensitizing/Allergic Reactions

Allergies After Oral Intake of Herbal Preparations

In theory, all foods/plants can cause Allergic Reactions, but in reality only a small part is responsible for allergies to food or plants. However, in some publications, soy, some fruits (especially cherries, peaches, plums, and apricots), as well as oleaginous fruits (nuts, seeds), and peanuts are most often associated with allergic reactions (Żukiewicz-Sobczak et al. 2013), although in most cases there is no proof of the basic involvement of these materials in the allergic reaction.

While there are plants/plant parts that are used as food supplements (regulated as foods), cosmetics (regulated as cosmetics) as well as herbal medicinal products (regulated as medicinal products), there are also cases where plant/plant parts are only or mainly used in one category. Especially in the field of food supplements/cosmetics, new (at least to Europe) ingredients can be easily used, so that the number of plant preparations used in all fields cannot be calculated to a reasonable amount.

Reliable information on the allergenicity of herbal medicinal products and herbal food supplements is scarce and available only for some major preparations or components thereof. It might be debated whether excipients, which are also often found within such industrial products, are involved in such allergenic reactions. The literature indicates that IgE-related reactions to excipients (such as coloring agents or benzoate derivatives) are rare, and most reactions are described as non-IgE-mediated histamine release. For other excipients, such as soy, guar, tragacanth, and gum arabic allergic reactions are conceivable. While such excipients are often used in smaller amounts, they may be present in many food supplements/industrially prepared food, so that daily exposure may be relevant.

Furthermore, it also has to be kept in mind that plant material with a natural content of allergens, such as nickel or salicylic acid, might lead to (pseudo)allergic reactions due to the presence of this component (de Medeiros et al. 2008; Baenkler 2008).

The pollen-food-allergy syndrome is a situation in which food allergy develops in relation to inhalant allergens. The incidence is highest in patients with pollen allergy, and the symptoms occur mainly after ingestion of raw herbal materials such as fruits, nuts, vegetables, and spices. Even for processed dosage forms, this may be applicable as raw material (such as herbal powders) may have been used for production. For instance, such association with aeroallergens have been described for fennel, soybean, caraway seeds, aniseed, or dandelion (Price et al. 2015).

Both food supplements and herbal medicinal products might contain fragrances or herbal components that are used also in fragrances, implying a problem common with the fragrance field. In the general population, fragrance allergy is among the most frequently detected allergies and has a prevalence ranging from 1.0 to 4.2 % (Carlsen et al. 2007). In the "Opinion on Fragrance Allergens in Cosmetic Products" (SCCS 2012), a number of established contact allergens in humans were published; indeed, several natural compounds known as "fragrance allergens" are most frequently reported and well recognized consumer allergens: amyl cinnamal, amyl cinnamyl alcohol, benzyl alcohol, benzyl salicylate, cinnamal, cinnamyl alcohol, citral, coumarin, eugenol, geraniol, hydroxycitronnellal, and isoeugenol. Furthermore, substances that are less frequently reported and thus less documented were listed; these include anisyl alcohol, benzyl benzoate, benzyl cinnamate, citronellol, farnesol, hexyl cinnamaldehyde, d-limonene or linalool. It should be pointed out that for the oxidized forms of limonene and linalool, a significant rate of allergies (approximately 5 % of the patients tested) could be shown (Audrain et al. 2014). Additionally, extracts are also mentioned in the SCCS paper (2012) (Table 11.2). Even if it were possible to completely avoid these fragrances (which appears to be virtually impossible), the main problem will be that the same substances/extracts might be taken orally via food, herbal supplements, or herbal medicinal products. Other plants, such as *Ocimum basilicum* or *Pimenta racemosa,* are also mentioned because of their content of established human allergens, although publications regarding human data are lacking.

Table 11.2 Natural extracts, which are established as contact allergens in humans or for which at least positive human data exist, but which are, however, not sufficient to be categorized as established contact allergens in humans

Extract/preparation	Category
Acorus calamus (root oil)	Positive results
Cananga odorata and *Ylang-ylang* (oil)	Established contact allergen in humans
Cedrus atlantica (bark oil)	Established contact allergen in humans
Cedrus deodara (wood oil)	Positive results
Cinnamomum cassia (leaf oil)	Established contact allergen in humans
Cinnamomum zeylanicum (bark oil)	Established contact allergen in humans
Citrus aurantium amara (flower/peel oil)	Established contact allergen in humans
Citrus aurantium amara (leaf oil)	Positive results
Citrus bergamia (peel oil expressed)	Established contact allergen in humans
Citrus limonum (peel oil expressed)	Established contact allergen in humans
Citrus sinensis (peel oil expressed)	Established contact allergen in humans
Citrus tangerina	Positive results
Cymbopogon citratus/schoenanthus (oils)	Established contact allergen in humans
Cymbopogon nardus winterianus (herb oil)	Positive results
Eucalyptus ssp. (leaf oil)	Established contact allergen in humans
Eugenia caryophyllus (leaf/flower oil)	Established contact allergen in humans
Evernia furfuracea (extract)	Established contact allergen in humans
Evernia prunastri (extract)	Established contact allergen in humans
Illicium verum (fruit oil)	Positive results
Jasminum grandiflorum/officinale	Established contact allergen in humans
Juniperus virginiana	Established contact allergen in humans
Laurus nobilis	Established contact allergen in humans
Lavandula hybrida	Established contact allergen in humans
Lavandula officinalis	Established contact allergen in humans
Lavandula spica	Positive results
Litsea cubeba	Positive results
Mentha piperita	Established contact allergen in humans
Mentha spicata	Established contact allergen in humans
Myroxylon pereirae (balsam of Peru)	Established contact allergen in humans
Narcissus ssp.	Established contact allergen in humans
Pelargonium graveolens	Established contact allergen in humans
Pelargonium roseum	Positive results
Pinus mugo/pumila	Established contact allergen in humans
Pogostemon cablin	Established contact allergen in humans
Rosa ssp. (flower oil)	Established contact allergen in humans
Rosmarinus officinalis	Positive results
Santalum album	Established contact allergen in humans
Salvia ssp.	Positive results
Tagetes patula	Positive results
Thymus ssp.	Positive results
Turpentine (oil)	Established contact allergen in humans
Verbena absolute	Established contact allergen in humans
Vetiveria zizanoides	Positive results

According to SCCS (2012)

Herbal Food Supplements

According to the Commission to the Council and the European Parliament (COM 2008) "Food supplements containing substances other than vitamins or minerals are foodstuffs within the meaning of Article 2 of Regulation (EC) No 178/2002 of the European Parliament and of the Council, which states that "foodstuff" (or "food") means any substance or product, whether processed, partially processed or unprocessed, intended to be, or reasonably expected to be ingested by humans;" therefore such food supplements containing herbal preparations are regarded as "food."

In the field of food/food supplements, in 2014 the European Food Safety Agency (EFSA) published a document that deals with the most common allergens found in food. Table 11.3 summarizes the information given concerning plant-derived allergens; it should be pointed out that the EFSA document refers only to immune-mediated adverse reactions and covers only the most important food allergens. The minimum dosages taken from EFSA (2014) should be interpreted carefully, since (1) in the underlying literature, it is often not stated whether the values refer to discrete or cumulative doses; and (2) in some studies, the allergenic food was not administered in the form in which it is usually eaten (e.g., freeze-dried). So EFSA declares that these values do not represent a scientific based NOAEL, nor might they be taken to recommend an acceptable level of intake for individuals.

EFSA (2014) refers to a publication of Hompes et al. (2011), which maintains that most cases of anaphylactic reaction (defined as severe systemic allergic reactions with concomitant pulmonary and/or cardiovascular symptoms) registered between 2006 and 2009 in the anaphylaxis registry of German-speaking countries in children and adolescents, were traced to legumes – in particular peanuts, followed by tree nuts. Most important plant allergens belong to one of four main families on

Table 11.3 The most common allergens derived from plants found in food/food supplements according to the EFSA (2014) (it should be mentioned that the minimal dosages for sensitized persons might be much lower, since persons who are known to have severe reactions are mostly excluded from challenge studies)

Herbal substance/ plant material	Plant family	Min. dosages triggering allergic reactions	Cross-reactivity
Apium graveolens (celery)	Apiaceae (Umbelliferae)	~0.7 g (raw celery root) ~0.16 g (celery spice)	Parsley, peach, olive, timothy grass, bermuda grass, sunflower, soy, peanut, pear, cherry Pollen-allergy: Birch-mugwort-celery syndrome Celery-carrot-mugwort-spice syndrome
Arachis hypogea (peanut)	Fabacea	~0.1 mg (protein)	Extensive serological cross-reactivity with members of the legume family Tree nut

(continued)

Table 11.3 (continued)

Herbal substance/ plant material	Plant family	Min. dosages triggering allergic reactions	Cross-reactivity
Brassica junca (brown/oriental mustard) *Brassica nigra* (black mustard) *Sinapis alba* (white/ yellow mustard) (Or mixtures out of them)	Brassicaceaea	~1 mg (protein)	*Brassica napus* (rapeseed), turnip rape; *Brassica rapa* subsp. *oleifera* (turnip rape), *Brassica napus* subsp. *oleifera* (oilseed rape) Almond, walnut, pistachio, hazelnut, tree nut, peanut, fruits of the Rosaceae family Pollen allergy: Celery-mugwort-birch-spice syndrome Mugwort-mustard allergy syndrome
Glycine max (soy, soybean)	Fabacea	~0.2 mg (protein) (To be taken into account also for residual proteins in lecithin and soybean oil)	Members of the legume family (peanut, green pea, lima bean, string bean), wheat flour, casein Pollen-allergy: Birch pollen allergy
Lupin species (lupin)	Leguminosaea	~50 mg (protein) Subjective symptoms from ~0.5 mg (lupin flour)	Members of the legume family (peanut, soybean, lentils, beans, chickpeas, peas)
Nuts Such as hazelnut, Brazil nut, walnut, almond, cashew, macadamia, pecan, chestnut	Several e.g., Betulaceae, Juglandaceae, Rosaceae, Anacardiaceae	<1 mg (protein)	If allergy to a single nut is demonstrated, the nuts of the entire nut group should be avoided (often associated with botanical family but also cross-reactivity among nuts not showing taxonomic relationship is reported) Hazelnut: birch (pollen) Chestnut: latex Nuts: peach
Sesamum indicum (sesame)	Pedialaceae	6 mg (seed) 1 ml (oil) Few mg (protein)	Peanut, hazelnut, egg, walnut, almond, tree nut
Wheat and other cereals (e.g., barley, rye, oats) Gluten and similar cereal storage proteins (not IgE- but IgA-mediated) Non-gluten proteins	Mostly Gramineae	Gluten intake of <50 mg/day is considered safe for most patients with celiac disease Children: ~2.6 mg (wheat protein) Adults: ~100 mg wheat flour	Within members of the Graminaea family Grass pollen

the basis of sequence homology, conserved 3-D structures, and function: the prolamin, cupin, profilins, and Bet v 1 superfamilies (Radauer and Breitender 2007; Wang and Sampson 2011; EFSA 2014; Lorenz et al. 2015).

The largest number of plant food allergens contain the prolamin superfamily: 2S seed storage albumins; cereal seed storage proteins; cereal α-amylase/trypsin inhibitors; and non-specific lipid transfer proteins (nsLTPs). Most of the proteins have a defensive/protective role against pathogens, or they are needed to provide proteins to the developing seed. While major allergens in tree nuts, sesame, and mustard seeds belong to the 2S seed storage albumins, allergens present in wheat, barley, rice, and maize belong to the α-amylase/trypsin inhibitors family. Lipid transfer proteins are frequent and potentially severe allergens; they are responsible for most of the severe allergic reactions to fruits from the Rosaceae family (EFSA 2014), but also for allergies to vegetables such as asparagus, cabbage, and lettuce (James et al. 2012).

The proteins of the cupin superfamily are the cause of most allergic reactions to legumes and nuts, while profilins are cytosolic proteins, which are exclusively found in flowering plants, such as peanut, apple, and celery, and which account for a strong serological cross-reactivity with other plant foods, pollens and *Hevea* latex, which may be of clinical significance (EFSA 2014). Eight families are counted within the Bet v 1 superfamily, among which are the "pathogenesis-related proteins 10," the major latex proteins. These allergens are homologous to the major birch pollen allergen and are present in fruits of the Rosaceae family (e.g., apple, cherry, apricot, and pear) and Apiaceae vegetables (e.g., celery and carrot) (EFSA 2014). Bet v 1 is reported to act as an inhalant allergen and, only after sensitization, individuals develop allergies to a variety of fresh fruits, vegetables, nuts, and seeds.

Taking into account the information from EFSA (2014) (see Table 11.2), consequently within the Annex II of the Regulation (EU) No 1169/2011 of the European Parliament and of the Council of 25 October 2011, the following plant-derived substances or products may cause allergies or intolerances, which are mandatory to label to protect vulnerable consumers from inadvertent consumption:

- Cereals containing gluten – namely, wheat, rye, barley, oats, spelt, kamut, or their hybridized strains, and products thereof (with some exceptions, e.g., wheat-based glucose syrups including dextrose or cereals used for making alcoholic distillates including ethyl alcohol of agricultural origin);
- Peanuts and products thereof;
- Soybeans and products thereof (with some exceptions, e.g., fully refined soybean oil and fat or vegetable oil-derived phytosterols and phytosterol esters from soybean sources);
- Nuts, namely almonds, hazelnuts, walnuts, cashews, pecans, Brazil nuts, pistachios, macadamia, or Queensland nuts, and products thereof (exception: nuts used for making alcoholic distillates including ethyl alcohol of agricultural origin);
- Celery and products thereof;
- Mustard and products thereof;
- Sesame seeds and products thereof;
- Lupin and products thereof (EU 2011).

Food-associated, exercise-induced anaphylaxis is a particular case of food-induced anaphylaxis. It was reported for the first time only in 1979 but its incidence seems to be increasing over the past few decades. In affected individuals, the ingestion of causal food(s) followed by exercise leads to a rapid onset of anaphylaxis, while food and exercise are independently tolerated (James et al. 2012). It is assumed, although the pathogenesis is poorly understood, that during exercise the gut permeability increases so that larger amounts of allergenic proteins might reach the host's gut-associated immune system.

Some countries have implemented national measures, such as the Allergy Vigilance Network in France, to record severe adverse allergic effects (anaphylactic reactions) of food/food supplements or medicinal products. In the evaluation of the time frame of January 2001 until December 2004, the most important plant allergens in France were peanuts and other legumes (20 %), nuts (14 %), the latex group (7 %), wheat (6 %), and celery (5 %). From previous publications it was known that also rarer allergens, such as chamomile, boldo, caffeine ,and gum arabic, can cause severe anaphylaxis (Moneret-Vautrin et al. 2005). Such a network can also be used to conduct studies, for instance on sensitization prevalence; it was reported that among atopic patients, 11.8 % were sensitized against oilseed rape pollen, 26 % against maize pollen, 7.7 % against oilseed rape seeds, and 8.3 % against corn seeds (out of 5,372 subjects studied) (Moneret-Vautrin et al. 2012).

Besides the major allergens (see above), cases have been described that concern less frequently reported plant allergens, some of which should be mentioned here. A patient sensitized to grasses had allergic reactions to foods containing oregano or thyme; his skin prick test also revealed positive results for basil, lavender, hyssop, marjoram, peppermint, and sage, and in vitro testing of serum IgE revealed specific IgE levels for almost all these herbs. The authors concluded a cross-sensitivity involving plants of the Labiatae family (Benito et al. 1996). Armentia et al. (2014) report allergy to cannabis; the most important allergens seem to be lipid transfer proteins, which also provoked positive responses to lipid transfer proteins, mainly from tomato, mugwort and tobacco. Hence, cannabis lipid transfer proteins may act as primary sensitizers and can therefore be responsible for the induction of further food allergies.

It is worth pointing out that various EFSA panels and their cohorts are working on the assessment of preparations/substances, such as gums and food additives from natural sources. Such assessments will be published in the EFSA Journal and will include the known data on allergenicity, hypersensitivity, and intolerance, if available.

It should not be forgotten that "health food products" (often advertised as "all natural") might be adulterated by other substances, also including potent chemical medicines, such as phosphodiesterase type 5 inhibitors or synthetic corticosteroids (Ramsay et al. 2003; Lee et al. 2013). Such chemicals might be responsible for possible allergic reactions, although it will be attributed to the herbal preparations.

Herbal Medicinal Products

When discussing food/food supplements the question of allergy focuses mainly on proteins, but the situation is unclear for herbal medicinal products. Non-protein structures (secondary metabolites) are often discussed as sources for allergenicity, but proteins (possibly present as trace amounts in extracts) cannot be excluded as potential agents for adverse reactions.

The monographs of the Committee of Herbal Medicinal Products (HMPC) of the European Medicines Agency (EMA) certainly present the most condensed information on (traditional) herbal medicinal products. This committee is responsible for assessing the efficacy and safety of herbal substances/preparations marketed within the European Union (EU); data from the literature as well as pharmacovigilance data, are taken into account, and more than 150 monographs have been published. Table 11.4 lists all the undesirable effects associated with allergic reactions (including gastrointestinal disturbances). While the monographs typically cover several preparations (e.g., aqueous extracts and a high percentages of ethanol extracts), the undesirable effects are generally not split for the various extracts; it is possible, however, that chemical or thermal influences modify the composition in potential allergic structures and therefore influence the allergenicity of the various preparations.

Unfortunately, for different reasons, some assessments of the HMPC lead to "public statements" rather than monographs, which means that no recommendation could be given for the use of certain plants or plant parts. In such cases, the risk assessment did not always take place; therefore it might be that some allergenic plants are missing in the overview (such as *Adhatoda vasica*, *Withania somnifera*). For other herbal materials, for which allergenicity is probable (e.g., lecithin), an assessment has not yet been published. In addition, it is of course also questionable whether the plants for which no allergic potential was described, in fact exhibit very low allergenicity; if their usage is marginal compared to other plants, allergic reactions are not reported or correctly attributed. On the other hand, especially for effects related to the gastrointestinal tract, it cannot be taken for granted that these correspond to true allergic events. Also intolerances, irritations, or even the worsening of symptoms of the disease might be possible. For example, *Gentiana lutea* is traditionally taken in cases of mild dyspeptic/gastrointestinal disorders; the described undesirable effects called "gastrointestinal disorders" might reflect a "failure of therapy."

Not only patients taking herbal medicinal products develop allergic reactions. A study by Bernedo et al. (2008) in a sample of healthcare workers in geriatric care homes repeatedly exposed to *Plantago ovata* seed (ispaghula seeds) products revealed that about 9% suffered allergic reactions confirmed by allergy tests. Furthermore, pharmacovigilance data on allergic reactions (respiratory symptoms such as rhinitis and asthma) have been described in persons who inadvertently inhale a powder while preparing it for administration. Such cases have been reported in pharmaceutical industry workers who work with ispaghula seeds during their preparation (HMPC 2012).

Preparations/substances derived from plants are also used as excipients in herbal medicinal products. These might be either comminuted herbal substances – in herbal teas, for example – or herbal preparations, such as tinctures, to improve the flavor of the finished product. As required by medicinal products regulation, all the ingredients of the finished product must be indicated in the package leaflet, but a case-by-case decision is needed regarding eventual allergenicity warnings related to such excipients. In the Volume 3B of the Annex of the Notice to Applicants, several plant-derived preparations are mentioned, which in any case require special labeling due to their allergenic properties (Table 11.5) (EC 2003).

Contact Dermatitis Due to Herbal Preparations

In Tables 11.3, 11.5, and 11.6, examples of cases of contact allergy are shown. Mainly molecules with a molecular weight between 100 and 1,000 are considered to be responsible for such effects (Merfort 2002). It is estimated that 80 % of contact dermatitis cases are due to irritant events, while only 20 % correspond to allergenic reactions. Pharmaceutical products (or cosmetics) might be applied on diseased, inflamed, or dry skin. Therefore it can be anticipated that the barrier function of the skin might be disturbed and even weak allergens – either active principles or excipients/vehicle components – could thus induce sensitization.

Data of the HMPC mirror the data from other publications; mainly, Asteraceae (Compositae) are known for their allergic potential, followed by Primulaceae, Apiaceae, and a few other plant families (Aberer 2008). This is especially noteworthy since the Asteraceae family comprises some of the oldest and most valued medicinal plants, sometimes also used for their anti-inflammatory activity, such as *Calendula officinalis* (marigold). At least 15 species, including *Arnica montana* (arnica), *Chamomilla recutita* (German chamomile) or *Echinacea* sp. are associated with sensitization and/or allergies. On the basis of case reports and testing described in the literature, Paulsen (2002) remarked that only a few species, such as arnica, are associated with a high frequency of sensitization while for the majority of species, frequency is rare. Other authors claim a low incidence, even concerning contact allergy to arnica or chamomile (Merfort 2002; Aberer 2008), and discussions on the responsible allergens are ongoing. While sesquiterpene lactones are seen as very important allergens from Asteraceae, in addition, sensitization cases due to a coumarin, a sesquiterpene alcohol, epoxythymol-derivatives, polyacetylenes and thiophenes are known (Merfort 2002). Also vanillic acid, cinnamic acid, ferulic acid, caffeic acid and a variety of mono-caffeyl and di-caffeyl esters of quinic acid are discussed (Olennikov and Kashchenko 2013). As some of the plants are also taken orally, the question of allergy-triggering substances arises for this usage as well.

Table 11.4 Plant/plant extracts or preparations for which hypersensitivity reactions after oral intake are reported according to the monographs of the HMPC (positive assessment)

Herbal substance	Plant family	Undesirable effect noticed
Achillea millefolium (flos + herba)	Asteraceae (Compositae)	Hypersensitivity reactions of the skin
Aesculus hippocastanum (semen)	Sapindaceae	Itching and allergic reactions; gastrointestinal complaints
Aloe [various species], folium	Xanthorrhoeaceae	Hypersensitivity reactions; abdominal pain and spasms
Arctium lappa (radix)	Asteraceae (Compositae)	Anaphylactic reactions
Arctostaphylos uva-ursi (folium) (bearberry leaf)	Ericaceae	Nausea, vomiting, stomachache
Betula pendula and/or *Betula pubescens* as well as hybrids of both species (folium)	Betulaceae	Allergic reactions (itching, rash, urticaria, allergic rhinitis); gastrointestinal complaints (nausea, vomiting, diarrhea)
Cassia senna and *Cassia angustifolia* (fructus + folium)	Fabaceae	Hypersensitivity reactions (pruritus, urticaria, local or generalized exanthema); abdominal pain and spasms
Cimicifuga racemosa (rhizome)	Ranunculaceae	Skin reactions (urticaria, itching, exanthema), facial edema, peripheral edema; gastrointestinal symptoms (i.e., dyspeptic disorders, diarrhea)
Cinnamomum verum (corticis aetheroleum)	Lauraceae	Local irritation of the oral mucosa
Cucurbita pepo (semen)	Cucurbitaceae	Mild gastrointestinal complaints
Curcuma longa (rhizome)	Zingiberaceae	Mild symptoms of flatulence and gastric irritation
Curcuma xanthorrhiza (rhizome)	Zingiberaceae	Mild gastrointestinal symptoms such as dry mouth, flatulence, and gastric irritation
Cynara scolymus (folium)	Asteraceae (Compositae)	Slight diarrhea with abdominal spasm, epigastric complaints such as nausea and heartburn; allergic reactions
Echinacea angustifolia (radix)	Asteraceae (Compositae)	Hypersensitivity reactions (skin reactions)
Echinacea pallida (radix)	Asteraceae (Compositae)	Hypersensitivity reactions (skin reactions)
Echinacea purpurea (herba recens)	Asteraceae (Compositae)	Hypersensitive reactions in the form of rash, urticaria, itching, swelling of the face; cases of severe hypersensitivity reactions, such as Stevens-Johnson syndrome, angioedema of the skin, Quincke edema, bronchospasm with airway obstruction, asthma and anaphylactic shock

(continued)

Table 11.4 (continued)

Herbal substance	Plant family	Undesirable effect noticed
Echinacea purpurea (radix)	Asteraceae (Compositae)	Hypersensitivity reactions (skin reactions)
Equisetum arvense (herba)	Equisetaceae	Allergic reactions (e.g., rash); mild gastrointestinal complaints
Foeniculum vulgare subsp. vulgare var. vulgare (aetheroleum)	Apiaceae (Umbelliferae)	Allergic reactions affecting the skin or the respiratory system
Foeniculum vulgare subsp. vulgare var. dulce (fructus)	Apiaceae (Umbelliferae)	Allergic reactions affecting the skin or the respiratory system
Foeniculum vulgare subsp. vulgare var. vulgare (fructus)	Apiaceae (Umbelliferae)	Allergic reactions affecting the skin or the respiratory system
Gentiana lutea (radix)	Gentianaceae	Pruritus; gastrointestinal disorders
Ginkgo biloba (folium)	Ginkgoaceae	Hypersensitivity reactions (allergic shock); allergic skin reactions (erythema, edema, itching and rash); gastrointestinal disorders (diarrhea, abdominal pain, nausea, vomiting)
Harpagophytum procumbens and/or *Harpagophytum zeyheri* (radix)	Pedaliaceae	Allergic skin reactions; gastrointestinal disorders: (diarrhea, nausea, vomiting, abdominal pain)
Hedera helix (folium)	Araliaceae	Allergic reactions (urticaria, skin rash, couperoses, dyspnea); gastrointestinal reactions (nausea, vomiting, diarrhea)
Hypericum perforatum (herba)	Hypericaceae	Allergic skin reactions; fair-skinned individuals may react with intensified sunburn-like symptoms under intense sunlight; gastrointestinal disorders
Juniperus communis (aetheroleum)	Cupressaceae	Allergic skin reactions
Juniperus communis (pseudo-fructus)	Cupressaceae	Allergic skin reactions
Linum usitatissimum (semen)	Linaceae	Hypersensitivity including anaphylaxis-like reactions
Matricaria recutita (flos)	Asteraceae (Compositae)	Hypersensitivity reactions including severe allergic reaction (dyspnea, Quincke's disease, vascular collapse, anaphylactic shock)
Melilotus officinalis (herba)	Fabaceae	Allergic reactions; gastrointestinal disorders
Mentha x piperita (aetheroleum)	Lamiaceae (Labiatae)	Allergic reactions with headache, bradycardia, muscle tremor, ataxia, anaphylactic shock and erythematous skin rash; nausea and vomiting

Table 11.4 (continued)

Herbal substance	Plant family	Undesirable effect noticed
Oenothera biennis; *Oenothera lamarckiana* (oleum)	Onagraceae	Hypersensitive reactions such as exanthema and headache; gastrointestinal effects, indigestion, nausea, softening of the stool
Olea europaea (folium)	Oleaceae	Pollinosis in the form of rhinitis or bronchial asthma
Panax ginseng (radix)	Araliaceae	Hypersensitivity reactions (urticaria, itching); gastrointestinal disorders such as abdominal discomfort, nausea, vomiting, diarrhea, constipation
Pelargonium sidoides and/ or *Pelargonium reniforme* (radix)	Geraniaceae	Allergic reactions; mild gastrointestinal complaints (diarrhea, epigastric discomfort, nausea or vomiting, dysphagia)
Peumus boldus (folium)	Monimiaceae	Hypersensitivity (anaphylaxis)
Pimpinella anisum (aetheroleum)	Apiaceae (Umbelliferae)	Allergic reactions affecting the skin or the respiratory system
Pimpinella anisum (fructus)	Apiaceae (Umbelliferae)	Allergic reactions affecting the skin or the respiratory system
Plantago afra or *Plantago indica* (semen)	Plantaginaceae	Hypersensitivity reactions such as rhinitis, conjunctivitis, bronchospasm, and in some cases, anaphylaxis
		Cutaneous symptoms such as exanthema and/or pruritus [*also contact with skin or cases of inhalation of the powder*]
Plantago ovata (semen + seminis tegumentum)	Plantaginaceae	Hypersensitivity reactions such as rhinitis, conjunctivitis, bronchospasm, and in some cases, anaphylaxis
		Cutaneous symptoms such as exanthema and/or pruritus [also contact with skin or cases of inhalation of the powder]
Potentilla erecta (rhizome)	Rosaceae	mild gastrointestinal complaints such as nausea and vomiting
Primula veris and/or *Primula elatior* (flos)	Primulaceae	Allergic reactions
Quercus robur, *Quercus petraea*, *Quercus pubescens* (cortex)	Fagaceae	Allergic reactions
Rhamnus purshianus (cortex)	Rhamnaceae	Hypersensitivity reactions; abdominal pain and spasm and passage of liquid stools
Rhamnus frangula (cortex)	Rhamnaceae	Hypersensitivity reactions; abdominal pain and spasm and passage of liquid stools

(continued)

Table 11.4 (continued)

Herbal substance	Plant family	Undesirable effect noticed
Rheum palmatum or *Rheum officinale* or their hybrids, or a mixture of these two species and/or their hybrids (radix)	Polygonaceae	Hypersensitivity reactions; abdominal pain and spasm and passage of liquid stools
Rosmarinus officinalis (folium)	Lamiaceae (Labiatae)	Hypersensitivity (contact dermatitis and occupational asthma)
Rosmarinus officinalis (aetheroleum)	Lamiaceae (Labiatae)	Hypersensitivity (contact dermatitis and asthma)
Ruscus aculeatus (rhizome)	Asparagaceae	Nausea, gastrointestinal complaints, diarrhea, lymphocytic colitis
Salix [various species including *S. purpurea*, *S. daphnoides*, *S. fragilis*] (cortex)	Salicaceae	Allergic reactions such as rash, pruritis, urticaria, asthma, exanthema; gastrointestinal symptoms such as nausea, vomiting, abdominal pain, diarrhea, dyspepsia, heartburn
Serenoa repens (fructus)	Aracaceae	Gastrointestinal disorders (abdominal pain, nausea, vomiting, diarrhea); skin and subcutaneous tissue disorders (skin rash); nervous system disorders (headache); allergic or hypersensitivity reactions
Solidago virgaurea (herba)	Asteraceae (Compositae)	Hypersensitivity reactions; gastrointestinal disorders
Tanacetum parthenium (herba)	Asteraceae (Compositae)	Gastrointestinal disturbances
Taraxacum officinale (folium)	Asteraceae (Compositae)	Allergic reactions
Taraxacum officinale (radix cum herba)	Asteraceae (Compositae)	Allergic reactions; epigastric pain
Thymus vulgaris or *Thymus zygis* or a mixture of both species (aetheroleum)	Lamiaceae (Labiatae)	Hypersensitivity reactions
Thymus vulgaris and *Thymus zygis* or a mixture of both species (herba)	Lamiaceae (Labiatae)	Gastric disorders
Trigonella foenum-graecum (semen)	Fabaceae	Allergic reactions (facial angioedema, wheezing) or ingestion (asthma, allergic rhinitis); gastrointestinal disorders: flatulence, diarrhea
Urtica dioica or *Urtica urens*, their hybrids or mixtures (radix)	Urticaceae	Allergic reactions, i.e., pruritus, rash, urticaria; gastrointestinal complaints such as nausea, heartburn, feeling of fullness, flatulence, diarrhea

Table 11.4 (continued)

Herbal substance	Plant family	Undesirable effect noticed
Urtica dioica or *Urtica urens* or a mixtures of the two species (folium)	Urticaceae	Skin reactions (e.g., itching, exanthema, hives); mild gastrointestinal complaints (e.g., nausea, vomiting, diarrhea)
Urtica dioica or *Urtica urens*, their hybrids or mixtures (herba)	Urticaceae	Skin reactions (e.g., itching, exanthema, hives); mild gastrointestinal complaints (e.g., nausea, vomiting, diarrhea)
Valeriana officinalis (radix)	Caprifoliaceae	Gastrointestinal symptoms (e.g., nausea, abdominal cramps)
Vitex agnus-castus (fructus)	Lamiaceae (Labiatae)	Severe allergic reactions with facial swelling, dyspnea and swallowing difficulties; (allergic) skin reactions (rash and urticaria); gastrointestinal disorders (such as nausea, abdominal pain)
Vitis vinifera (folium)	Vitaceae	Contact allergy and/or hypersensitivity reactions of the skin (itching and erythema, urticaria); nausea, gastrointestinal complaints
Zingiber officinale (rhizome)	Zingiberaceae	Minor gastrointestinal complaints (stomach upset, eructation, dyspepsia, nausea)

Some plants commonly used in Chinese topical medicinal products show positive reactions in patch tests in patients. Examples of such plants are *Syzygium aromaticum* (flos), *Angelica pubescens* (radix), *Cinnamomum verum* (cortex), *Cnidium monnieri* (fructus), *Gentiana macrophylla* (radix) and *Eleutherococcus senticosus* (cortex radix). In most of these positive reactions, a concomitant allergy to colophonium was also found (Chen et al. 2003).

Strong contact sensitizers such as alk(en)yl catechols (urushiols), from the Anarcardiaceae plant family, e.g., poison ivy (*Toxicodendron rydbergii*), poison oak (*Toxicodendron toxicarium*), poison sumac (*Toxicodendron striatum*) and lacquer tree (*Toxicodendron vernicifluum*), and alk(en)yl resorcinols, which were identified in different plants, such as cashew nut (*Anacardium occidentale*), mango (*Mangifera indica*) or *Philodendron* ssp. (Christensen 2014) should not be used in products applied to skin. However, lacquer allergy (due to the usage of lacquer tree products) is a serious occupational skin disease of lacquerware workers, especially in East Asia (Christensen 2014). Similarly, this might be true for a couple of other substances such as primin, a benzoquinone found in some *Primula* spp., which might be important mainly for gardeners, florists or herbalists.

Table 11.5 Excipients and information for the package leaflet taken from the notice to applicants in Volume 3B concerning plant-derived products connected to allergies (EC 2003)

Name	Route of administration	Threshold	Information for the package leaflet
Arachis oil (peanut oil)	All	Zero	(Medicinal product) contains arachis oil (peanut oil). *If you are allergic to peanuts or soy, do not use this medicinal product*
Balsam of Peru	Topical	Zero	May cause skin reactions
Bergamot oil Bergapten	Topical	Zero	May increase sensitivity to UV light (natural and artificial sunlight). [*Does not apply when bergapten is shown to be absent from the oil!*]
Castor oil polyoxyl and hydrogenated castor oil polyoxyl hydrogenated	Parenteral		May cause severe allergic reactions
	Oral		May cause stomach upset and diarrhea
	Topical		May cause skin reactions
Latex natural rubber (latex)	All	Zero	The container of this medicinal product contains latex rubber. May cause severe allergic reactions
Sesame oil	All	Zero	May rarely cause severe allergic reactions
Soy oil (and hydrogenated soya oil)	All	Zero	(Medicinal product) contains soy oil. *If you are allergic to peanut or soy, do not use this medicinal product*
Wheat starch	Oral	Zero	Suitable for people with celiac disease. *Patients with wheat allergy (different from celiac disease) should not take this medicine.* [*Wheat starch may contain gluten, but only in trace amounts, and is therefore considered safe for people with celiac disease.* (*Gluten in wheat starch is limited by the test for total protein described in the PhEur monograph*)]

Not only direct contact may induce allergic contact dermatitis; also aerogenic contact dermatitis is described. Here, plant parts that are transferred by air (e.g., plant hairs, small fruits, or withered plant particles) may reach the skin and lead to responses.

Photoallergy

Cases of real photoallergy due to herbal material is rare; most cases will include phototoxic reactions. Very rare cases of photoallergy after exposition to furanocoumarins have been described (Hausen and Vieluf 1997), but such a sensitization possibility remains to be confirmed.

Table 11.6 Plant/plant extracts or preparations for which hypersensitivity reactions after cutaneous use have been reported according to the monographs of the HMPC (positive assessment)

Herbal substance	Plant family	Undesirable effect noticed
Achillea millefolium (flos + herba)	Asteraceae (Compositae)	Hypersensitivity reactions of the skin
Aesculus hippocastanum (semen)	Sapindaceae	Hypersensitivity reactions of the skin (itching and erythema)
Arnica montana (flos)	Asteraceae (Compositae)	Itching, redness of the skin and eczema
Avena sativa (fructus)	Poaceae	Skin reactions
Calendula officinalis (flos)	Asteraceae (Compositae)	Skin sensitization [*also oromucosal use*]
Capsicum annuum var. minimum and small fruited varieties of *Capsicum frutescens* (fructus)	Solanaceae	Skin hypersensitivity and allergic reactions (e.g., urticaria, blisters or vesiculation)
Commiphora molmol (gummi-resina)	Burseraceae	Allergic skin reactions [*also oromucosal use*]
Echinacea purpurea (herba recens)	Asteraceae (Compositae)	Hypersensitive reactions (local rash, contact dermatitis, eczema and angioedema of the lips)
Echinacea purpurea (radix)	Asteraceae (Compositae)	Hypersensitivity reactions (skin reactions) [oromucosal use]
Hamamelis virginiana (cortex + folium)	Hamamelidaceae	Allergic contact dermatitis [also oromucosal, rectal, anorectal use]
Hamamelis virginiana (folium et cortex aut ramunculus)	Hamamelidaceae	Allergic contact dermatitis conjunctivitis [ocular use]
Hypericum perforatum (herba)	Hypericaceae	Allergic skin reactions
Juniperus communis (aetheroleum)	Cupressaceae	Allergic skin reactions
Matricaria recutita (aetheroleum)	Asteraceae (Compositae)	Hypersensitivity reactions including severe allergic reaction (dyspnea, Quincke's disease, vascular collapse, anaphylactic shock) [bath additive]
Matricaria recutita (flos)	Asteraceae (Compositae)	Hypersensitivity reactions including severe allergic reaction (dyspnea, Quincke's disease, vascular collapse, anaphylactic shock) [inhalative; oromucosal]
Melaleuca alternifolia, *Melaleuca linariifolia*, *Melaleuca dissitiflora* and/ or other species of Melaleuca, (aetheroleum)	Myrtaceae	Adverse skin reactions, including smarting pain, mild pruritus, burning sensation, irritation, itching, stinging, erythema, edema (contact dermatitis) or other allergic reactions [also oromucosal use]
Melilotus officinalis (herba)	Fabaceae	Allergic reactions

(continued)

Table 11.6 (continued)

Herbal substance	Plant family	Undesirable effect noticed
Mentha x piperita (aetheroleum)	Lamiaceae (Labiatae)	Hypersensitivity reactions such as skin rash, contact dermatitis, and eye irritation [also transdermal use]
		Apnea, broncho- and laryngoconstriction [inhalative use]
		Contact sensitivity with intra-oral symptoms in association with burning mouth syndrome, recurrent oral ulceration or a lichenoid reaction [oromucosal use]
Quercus robur, Quercus petraea, Quercus pubescens (cortex)	Fagaceae	Allergic reactions [also oromucosal; anorectal]
Rosmarinus officinalis (folium)	Lamiaceae (Labiatae)	Hypersensitivity (contact dermatitis and occupational asthma) [bath additive]
Rosmarinus officinalis (aetheroleum)	Lamiaceae (Labiatae)	Hypersensitivity (contact dermatitis and asthma) [also as bath additive]
Syzygium aromaticum (floris aetheroleum)	Myrtaceae	Allergic reactions [oromucosal; dental]
Thymus vulgaris or *Thymus zygis* or a mixture of both species (aetheroleum)	Lamiaceae (Labiatae)	Hypersensitivity reactions and skin irritation [also as bath additive]
Trigonella foenum-graecum (semen)	Fabaceae	Allergic reactions (facial angioedema, wheezing) or ingestion (asthma, allergic rhinitis)
Vitis vinifera (folium)	Vitaceae	Contact allergy and/or hypersensitivity reactions of the skin (itching and erythema, urticaria)

Other ways in addition to cutaneous use – other than oral and cutaneous – are italics

Herbal Preparations Used in Inflammation and Allergic Reactions

Natural preparations able to regulate allergic responses, via various mechanisms, including inhibition of allergen diffusion into epithelial cells, are discussed, as are the suppression of Th2-related cytokine production, the inhibition of T-cell differentiation, and/or the inhibition of degranulation of mast cells. For each mechanism, non-clinical data exist for herbal preparations or their compounds. For instance, (1) an extract of *Scutellaria baicalensis* could be shown to inhibit ovalbumin permeation via Caco-2 cells monolayers; (2) *Trigonella foenum-graecum* is known to increase Th-1 response and decrease Th-2 response; and (3) compounds such as curcumin or resveratrol are able to suppress the Th-2 cell response (Shin and Shon 2015). Although clinical proof for such actions is still lacking, the fact that part of our diet can show antiallergic effects should not be ignored.

Some examples of in vitro or in vivo anti-allergenic activity of herbal preparations have been well documented:

Food Allergy Herbal Formula-2

A product called Food Allergy Herbal Formula-2 (FAHF-2) (an extract of nine herbs: *Prunus mume* fruit, *Zanthoxylum schinifolium* fruit skin, *Angelica sinensis* root, *Zingiber officinale* rhizome, *Cinnamomum cassia* twigs, *Phellodendron chinense* bark, *Coptis chinensis* rhizome, *Panax ginseng* root and *Ganoderma lucidum* fruiting body) was tested in a peanut allergic murine model. The protection against peanut-induced anaphylactic symptoms persisted for at least 6 months post-therapy following a single 7 week course of treatment. A reduction of Th2-cytokines and serum IgE-levels and an increase of IFN-γ and IgG2a could be seen. Also, a reduction in basophil and mast cell numbers and mast cell activation was demonstrated (Wang and Li 2012). After an acute, 1 week, randomized double-blind placebo-controlled, dose escalation phase I trial in subjects with peanut and/or tree nut, fish and shellfish allergies, an extended phase I clinical trial (open-label study) was performed for 6 months in 14 patients. During the course of the study, basophil activation and basophil and eosinophil numbers were evaluated. While no significant drug-associated differences in laboratory parameters, pulmonary function studies, or electrocardiographic findings before and after treatment were found, there was a significant reduction in basophil expression in response to *ex vivo* stimulation at month 6. There was also a trend towards a reduction of eosinophil and basophil numbers after treatment (Patil et al. 2011); however, although clinical safety could be proven, clinical data on efficacy are still lacking.

Petasites hybridus leaves

A CO_2 supercritical fluid extract of the leaves of *Petasites hybridus* (petasin chemovariety), given intranasally in mice, showed leukotriene-inhibiting properties that led to a reduced allergic airway inflammation (Brattström et al. 2010). It was suggested that petasin (a sesquiterpene) inhibits L-type Ca^{2+}-channels. The same extract has been studied in three placebo-controlled clinical studies showing that it may be effective for the relief of symptoms or improved peak nasal inspiratory flow (Guo et al. 2007). An oral product containing this extract was authorized in Switzerland for the treatment of allergic rhinitis (Zeller Medical 2012).

Urtica dioica

In vitro, an *Urtica dioica* extract inhibited several key inflammatory events (antagonist activity against the Histamine-1 receptor and inhibition of mast cell tryptase; inhibition of cyclooxygenase-1, cyclooxygenase-2 and hematopoietic prostaglandin D(2) synthase) that cause the symptoms of seasonal allergies (Roschek et al. 2009). These might justify, at least in part, the folk medicine use of *Urtica dioica* in such cases of seasonal rhinitis.

The use of herbal preparations in cases of cutaneous allergy

While a couple of herbal preparations are used cutaneously in cases of dry or inflamed skin, nothing is really known about the use of herbal preparations in cases of allergy. Of course it can be assumed that some cases of "inflamed skin" might also mirror allergic skin conditions, but this is only speculation.

Some non-clinical results concerning plants of traditional Chinese medicine have been published. In various models (in vivo, in vitro, or ex vivo), the ethanol extract of the radix of *Achyranthis bidentata*, methanol extract of *Schisandra chinensis* fruits, methanol extract of radix of *Sanguisorba officinalis*, ethanol extract of the defatted radix of *Scutellaria baicalensis*, ethanol extract of *Zizyphus jujube* fruits, extracts of *Rubia cordifolia* and *Dianthus superbus*, demonstrated anti-allergic activity (Jung et al. 2015; Lee et al. 2015; Jo et al. 2015; Li et al. 2014; Naik et al. 2013; Wang and Li 2012). The same applies for artesunate, a semisynthetic derivative of artemisinin, an active component of *Artemisia annua* (Cheng et al. 2013), or for Δ^9-tetrahydrocannabinol, an active constituent of *Cannabis sativa* (Gaffal et al. 2013).

Future Considerations

Allergies due to herbal preparations or substances isolated from plants (and later manufactured synthetically) are described, and it often seems that the main structures responsible for it are known – or at least suspected to be known. While it seems that the majority of allergies are associated with a manageable number of plant allergens, a few rare cases of allergies associated with different structures have also been reported.

It would be desirable to establish surveillance systems in the fields of food (supplements)/herbal medicinal products/cosmetics to improve information on potential allergens, since this is the basis of the prevention of allergies; this means, however, that for all product categories, an accurate diagnosis of IgE-mediated allergy should be performed by specialists who have been educated to interpret the results of such testing in an appropriate way and to use standardized allergen extracts.

For a long time, strict avoidance of specific antigen/herbal preparations was seen as the only possible way to avoid allergic reactions. A strict labelling of antigens/herbal preparations (considering the main/active ingredient, excipient, and possible contaminations) is a prerequisite for the patient. In food regulation, there is an ongoing discussion about the labelling of food/food supplements that *could* contain allergenic substances such as nuts, etc., as contamination (e.g., "May contain traces of …"). A regulatory standardized system is lacking in Europe, leading to inadequate labelling, and therefore to misinformation and/or an erroneous feeling of safety. Some countries, such as Australia and New Zealand, have undertaken to improve precautionary allergen labeling. The establishment of appropriate parameters (reference doses) is discussed, to label only the relevant level of allergens to avoid "overuse" of precautionary labeling, and to ensure the protection of most of the food-allergic population. But therefore, of course, manufacturers should quantitatively determine the degree of contamination – which may present a huge analytical problem as the level of detection of analytical methods can be too high, compared to the level of immunoreactivity in sensitized patients. The most sensitive methods, based on DNA (qPCR) or protein (immunoassays) detection, may also be impossible in cooked foods (heat denaturation). Furthermore, technical steps are desirable to prevent contamination as much as possible.

During the last years, immunotherapeutic strategies were developed to avoid serious adverse effects, such as standard subcutaneous immunotherapy or oral immunotherapy. Final conclusions are still lacking on the induction of short-term desensitization or even long-term tolerance by these treatments, and more meaningful clinical trials are needed. Furthermore, innovative drugs need to be developed, which may take into account some promising plant extracts or plant-derived compounds. There is some information on herbal preparations that might be useful in the treatment of allergies, but until now, mainly in vitro or animal data indicate that such activities and clinical data are missing in most cases.

Food, its preparation and the use of food supplements might change in the future, or may have changed already. There are bound to be changes not only in eating behaviors, but also in changes in production processes/technologies and the introduction of product innovations. Changes in allergenic properties or in hypersensibility patterns that arise from these altered conditions are not really known so far. Further research is definitely needed.

References

Aberer W (2008) Kontaktallergie und Arzneipflanzen. JDDG 6:15–24

Armentia A, Herrero M, Martín-Armentia B, Rihs HP, Postigo I, Martínez-Quesada J (2014) Molecular diagnosis in cannabis allergy. J Allergy Clin Immunol Pract 2(3):351–352

Ashley S, Dang T, Koplin J, Martino D, Prescott S (2015) Food for thought: progress in understanding the causes and mechanisms of food allergy. Curr Opin Allergy Clin Immunol 15(3):237–242

Audrain H, Kenward C, Lovell CR, Grenn C, Ormerod AD, Sansom J et al (2014) Allergy to oxidized limonene and linalool is frequent in the U.K. Br J Dermatol 171(2):292–297

Auttachoat W, Germolec DR, Smith MJ, White KL Jr, Guo TL (2011) Contact sensitizing potential of annatto extract and its two primary color components, cis-bixin and norbixin, in female BALB/c mice. Food Chem Toxicol 49(10):2638–2644

Bachert C, Kuna P, Zuberbier T (2010) Bilastine in allergic rhinoconjunctivitis and urticaria. Allergy 65(Suppl 93):1–13

Baenkler HW (2008) Salicylatintoleranz. Deut Ärzteblatt 105(8):137–142

Bala S, Weaver J, Hastings KL (2005) Clinical relevance of preclinical testing for allergic side effects. Toxicology 209(2):195–200

Baldo BA, Pham NH (2013) Drug allergy: clinical aspects, diagnosis, mechanisms, structure-activity relationships. Springer, New York

Benito M, Jorro G, Morales C, Peláez A, Fernández A (1996) Labiatae allergy: systemic reactions due to ingestion of oregano and thyme. Ann Allergy Asthma Immunol 76(5):416–418

Berin MC, Mayer L (2009) Immunophysiology of experimental food allergy. Mucosal Immunol 2(1):24–32

Bernedo N, García M, Gastaminza G, Fernández E, Bartolomé B, Algorta J et al (2008) Allergy to laxative compound (Plantago ovata seed) among health care professionals. J Investig Allergol Clin Immunol 18(3):181–189

Brattström A, Schapowal A, Maillet I, Schnyder B, Ryffel B, Moser R (2010) Petasites extract Ze 339 (PET) inhibits allergen-induced Th2 responses, airway inflammation and airway hyperreactivity in mice. Phytother Res 24(5):680–685

Carlsen BC, Menne T, Johansen JD (2007) 20 years of standard patch testing in an eczema population with focus on patients with multiple contact allergies. Contact Dermatitis 57:76–83

Chen HH, Sun CC, Tseng MP, Hsu CJ (2003) A patch test study of 27 crude drugs commonly used in Chinese topical medicaments. Contact Dermatitis 49(1):8–14

Cheng C, Ng DSW, Chan TK, Guan SP, Ho WE, Koh AHM et al (2013) Anti-allergic action of anti-malarial drug artesunate in experimental mast cell-mediated anaphylactic models. Allergy 68:195–203

Christensen LP (2014) Polyphenols and polyphenol-derived compounds and contact dermatitis. In: Watson RR, Preedy VR, Zibadi S (eds) Polyphenols in human health and disease – volume 1: polyphenols in disease and their mechanism of action. Elsevier, London

COM (2008) Report from the Commission to the Council and the European Parliament on the use of substances other than vitamins and minerals in food supplements (final). Available at: http://eur-lex.europa.eu/legal-content/EN/TXT/?uri=celex:52008DC0824

Cornacoff JB, House RV, Dean JH (1988) Comparison of a radioisotopic incorporation method and the mouse ear swelling test (MEST) for contact sensitivity to weak sensitizers. Fundam Appl Toxicol 10(1):40–44

de Medeiros LM, Fransway AF, Taylor JS, Wyman M, Janes J, Fowler JF Jr et al (2008) Complementary and alternative remedies: an additional source of potential systemic nickel exposure. Contact Dermatitis 58(2):97–100

De Swert LF (1999) Risk factors for allergy. Eur J Pediatr 158(2):89–94

Derelanko MJ, Auletta CS (eds) (2014) Handbook of toxicology, 3rd edn. CRC Press Taylor & Friends Group, Boca Raton

Descotes J (ed) (2004) Nonclinical evaluation of the potential for hypersensitivity. In: Immunotoxicology of drugs and chemicals: an experimental and cinical approach – volume 1: principles and methods of immunotoxicology. 3rd edn. Elsevier B.V., Amsterdam

Descotes J, Choquet-Kastylevsky G (2001) Gell and Coombs's classification: is it still valid? Toxicology 158(1–2):43–49

Draize JH, Woodard G, Calvery HO (1944) Methods for the study of irritation and toxicity of substances applied to the skin and mucous membranes. J Pharmacol Exp Ther 8:377–390

Dudeck A, Dudeck J, Scholten J, Petzold A, Surianarayanan S, Köhler A et al (2011) Mast cells are key promoters of contact allergy that mediate the adjuvant effects of haptens. Immunity 34(6):973–984

EC (2003) Notice to applicants. Volume 3B. Excipients in the label and package leaflet of medicinal products for human use. Available at: http://www.ema.europa.eu/docs/en_GB/document_library/Scientific_guideline/2009/09/WC500003412.pdf

EFSA (2011) Guidance for risk assessment of food and feed from genetically modified plants. EFSA J 9(5):2150

EFSA (2014) Scientific opinion on the evaluation of allergenic foods and food ingredients for labelling purposes. EFSA J 12(11):3894

EU (2003) Directive 2003/15/EC of the European Parliament and of the council of 27 February 2003 amending Council Directive 76/768/EEC on the approximation of the laws of the Member States relating to cosmetic products. Available at: http://ec.europa.eu/consumers/sectors/cosmetics/files/doc/200315/200315_en.pdf

EU (2011) Regulation (EU) No 1169/2011 of the European Parliament and of the council of 25 October 2011 on the provision of food information to consumers, amending Regulations (EC) No 1924/2006 and (EC) No 1925/2006 of the European Parliament and of the Council, and repealing Commission Directive 87/250/EEC, Council Directive 90/496/EEC, Commission Directive 1999/10/EC, Directive 2000/13/EC of the European Parliament and of the Council, Commission Directives 2002/67/EC and 2008/5/EC and Commission Regulation (EC) No 608/2004. Available at: http://eur.lex.europa.eu

FDA (2002) Immunotoxicology evaluation of investigational new drugs. Guidance for Industry. Food and Drug Administration Center for Drug Evaluation and Research (CDER). Available at:

http://www.fda.gov/downloads/drugs/guidancecomplianceregulatoryinformation/guidances/ucm079239.pdf
Finkelman FD (2007) Anaphylaxis: lessons from mouse models. J Allergy Clin Immunol 120(3):506–515
Gad SC (2009) Drug safety evaluation, 2nd edn. Wiley, Hoboken
Gad SC, Dunn BJ, Dobbs DW, Reilly C, Walsh RD (1986) Development and validation of an alternative dermal sensitization test: the mouse ear swelling test (MEST). Toxicol Appl Pharmacol 84(1):93–114
Gaffal E, Cron M, Glodde N, Tüting T (2013) Anti-inflammatory activity of topical THC in DNFB-mediated mouse allergic contact dermatitis independent of CB1 and CB2 receptors. Allergy 68(8):994–1000
Gell PGH, Coombs RRA (1963) The classification of allergic reactions underlying disease. In: Coombs RRA, Gells PGH (eds) Clinical aspects of immunology, 1st edn. Blackwell, Oxford
Grammatikos AP (2008) The genetic and environmental basis of atopic diseases. Ann Med 40:482–495
Guo R, Pittler MH, Ernst E (2007) Herbal medicines for the treatment of allergic rhinitis: a systematic review. Ann Allergy Asthma Immunol 99(6):483–495
Hastings KL (2001) Pre-clinical methods for detecting the hypersensitivity potential of pharmaceuticals: regulatory considerations. Toxicology 158(1–2):85–89
Hausen BM, Vieluf IK (1997) Allergiepflanzen – Pflanzenallergene, 2nd edn. Ecomed Verlagsgesellschaft mbH, Landberg
Hayes AW, Kruger CL (eds) (2014) Hayes' principles and methods of toxicology, 6th edn. CRC Press Taylor & Friends Group, Boca Raton
HMPC (2012) Assessment report on *Plantago ovata* Forskk., semen. EMA/HMPC/304360/2012. Available at: http://www.ema.europa.eu/docs/en_GB/document_library/Herbal-HMPC_assessment_report/2013/07/WC500146512.pdf
Hogan SP, Wang YH, Strait R, Finkelman FD (2012) Food-induced anaphylaxis: mast cells as modulators of anaphylactic severity. Semin Immunopathol 34(5):643–653
Hompes S, Köhli A, Nemat K, Scherer K, Lange L, Rueff F et al (2011) Provoking allergens and treatment of anaphylaxis in children and adolescents – data from the anaphylaxis registry of German-speaking countries. Pediatr Allergy Immunol 22(6):568–574
ICH S10 (2013) Photosafety evaluation of pharmaceuticals. International conference on harmonisation of technical requirements for registration of pharmaceuticals for human use. Available at: http://www.ich.org/products/guidelines/safety/article/safety-guidelines.html
James JM, Burks W, Eigenmann PA (2012) Food allergy. Elsevier Saunders, Philadelphia
Jo S, Ryu J, Kim H, Kim M, Ryu MH, Kim H et al (2015) Anti-inflammatory effects of *Sanguisorbae Radix* on contact dermatitis induced by dinitrofluorobenzene in mice. Chin J Integr Med [Epub ahead of print]
Jung SK, Choi DW, Kwon DA, Kim MJ, Seong KS, Shon DH (2015) Oral administration of achyranthis radix extract prevents TMA-induced allergic contact dermatitis by regulating Th2 cytokine and chemokine production in vivo. Molecules 20(12):21584–21596
Karschuk N, Tepe Y, Gerlach S, Pape W, Wenck H, Schmucker R et al (2010) A novel in vitro method for the detection and characterization of photosensitizers. PLoS One 5(12):e15221
Lee JH, Kim NS, Han KM, Kim SH, Cho S, Kim WS (2013) Monitoring by LC-MS/MS of 48 compounds of sildenafil, tadalafil, vardenafil and their analogues in illicit health food products in the Korean market advertised as enhancing male sexual performance. Food Addit Contam Part A Chem Anal Control Exp Risk Assess 30(11):1849–1857
Lee HJ, Jo S, Ryu J, Jeong HS, Lee G, Ryu MH et al (2015) Effects of *Schisandra chinensis* Turcz. fruit on contact dermatitis induced by dinitrofluorobenzene in mice. Mol Med Rep 12(2):2135–2139
Li X, Wang J, Shi H, Gao L, Wang X (2014) Inhibition of three novel *Radix Scutellariae* extracts on immediate hypersensitivity. Afr J Tradit Complement Altern Med 11(5):54–60
Lorenz AR, Scheurer S, Vieths S (2015) Food allergens: molecular and immunological aspects, allergen databases and cross-reactivity. Chem Immunol Allergy 101:18–29
Lovell WW (1993) A scheme for in vitro screening of substances for photoallergenic potential. Toxicol In Vitro 7(1):95–102

Luebke R, House RV, Kimber I (eds) (2006) Immunotoxicology and Immunopharmacology, 3rd edn. CRC Press Taylor & Friends Group, Boca Raton

Martin SF (2015) New concepts in cutaneous allergy. Contact Dermatitis 72(1):2–10

Merfort I (2002) Heil- und Nutzpflanzen mit Haut-Tücken. Pharmazeutische Zeitung online. Available at: http://www.pharmazeutische-zeitung.de/index.php?id=titel_03_2002

Moneret-Vautrin DA, Morisset M, Flabbee J, Beaudouin E, Kanny G (2005) Epidemiology of life-threatening and lethal anaphylaxis: a review. Allergy 60(4):443–451

Moneret-Vautrin DA, Peltre G, Gayraud J, Morisset M, Renaudin JM, Martin A (2012) Prevalence of sensitisation to oilseed rape and maize pollens in France: a multi-center study carried out by the Allergo-Vigilance Network. Eur Ann Allergy Clin Immunol 44(6):225–235

Muller HK, Healy DL (1973) Active cutaneous anaphylaxis in the guinea-pig. Immunological and inflammatory reactions. Immunology 24(6):1099–1112

Naik SR, Bhagat S, Shah PD, Tare AA, Ingawale D, Wadekar RR (2013) Evaluation of anti-allergic and anti-anaphylactic activity of ethanolic extract of *Zizyphus jujuba* fruits in rodents. Rev Bras Farmacognosia 23:811–818

OECD (1992) Test No. 406: Skin Sensitisation, OECD guidelines for the testing of chemicals, section 4. OECD Publishing, Paris, Available at: http://dx.doi.org/10.1787/9789264070660-en

OECD (2010a) Test No. 429: Skin Sensitisation: Local Lymph Node Assay, OECD guidelines for the testing of chemicals, section 4. OECD Publishing, Paris, Available at: http://dx.doi.org/10.1787/9789264071100-en

OECD (2010b) Test No. 442A: Skin Sensitization: Local Lymph Node Assay: DA, OECD guidelines for the testing of chemicals, section 4. OECD Publishing, Paris, Available at: http://dx.doi.org/10.1787/9789264090972-en

OECD (2010c) Test No. 442B: Skin Sensitization: Local Lymph Node Assay: BrdU-ELISA, OECD guidelines for the testing of chemicals, section 4. OECD Publishing, Paris, Available at: http://dx.doi.org/10.1787/9789264090996-en

OECD (2012) Test No. 405: Acute Eye Irritation/Corrosion, OECD guidelines for the testing of chemicals, section 4. OECD Publishing, Paris, Available at: http://dx.doi.org/10.1787/9789264185333-en

OECD (2013) Test No. 437: Bovine Corneal Opacity and Permeability Test Method for Identifying i) Chemicals Inducing Serious Eye Damage and ii) Chemicals Not Requiring Classification for Eye Irritation or Serious Eye Damage, OECD guidelines for the testing of chemicals, section 4, OECD Publishing, Paris. Available at: http://dx.doi.org/10.1787/9789264203846-en

OECD (2014a) Draft Proposal for a new Test Guideline: In Vitro Skin Sensitisation: human Cell Line Activation Test (h-CLAT). Available at: http://www.oecd.org/chemicalsafety/testing/Draft-new-Test-Guideline-Skin-Sensitisation-h-CLAT-July-2014.pdf

OECD (2014b) The Adverse Outcome Pathway for Skin Sensitisation Initiated by Covalent Binding to Proteins, OECD series on testing and assessment, No. 168. OECD Publishing, Paris. Available at: http://dx.doi.org/10.1787/9789264221444-en

OECD (2015a) Test No. 431: In vitro skin corrosion: reconstructed human epidermis (RHE) test method, OECD guidelines for the testing of chemicals, section 4. OECD Publishing, Paris, Available at: http://dx.doi.org/10.1787/9789264242753-en

OECD (2015b) Test No. 442C: In Chemico Skin Sensitisation: Direct Peptide Reactivity Assay (DPRA), OECD guidelines for the testing of chemicals, section 4. OECD Publishing, Paris, Available at: http://dx.doi.org/10.1787/9789264229709-en

OECD (2015c) Test No. 442D: In Vitro Skin Sensitisation: ARE-Nrf2 Luciferase Test Method, OECD guidelines for the testing of chemicals, section 4. OECD Publishing, Paris, Available at: http://dx.doi.org/10.1787/9789264229822-en

Olennikov DN, Kashchenko NI (2013) New isorhamnetin glucosides and other phenolic compounds from *Calendula officinalis*. Chem Nat Comp 49(5):833–840

OSHA (Occupational Safety and Health Administration) (1994) OSHA Hazard Communication Standard (HCS) in 29 CFR 19101200. OSHA, Washington, DC

Patil SP, Wang J, Song Y, Noone S, Yang N, Wallenstein S et al (2011) Clinical safety of Food Allergy Herbal Formula-2 (FAHF-2) and inhibitory effect on basophils from patients with food allergy: extended phase I study. J Allergy Clin Immunol 128(6):1259–1265

Paul W (ed) (2003) Fundamental immunology, 5th edn. Lippincott Williams & Wilkins, Philadelphia
Paulsen E (2002) Contact sensitization from compositae-containing herbal remedies and cosmetics. Contact Dermatitis 47:189–198
Peiser M, Tralau T, Heidler J, Api AM, Arts JH, Basketter DA et al (2012) Allergic contact dermatitis: epidemiology, molecular mechanisms, in vitro methods and regulatory aspects. Current knowledge assembled at an international workshop at BfR, Germany. Cell Mol Life Sci 69(5):763–781
Pichler WJ (ed) (2007) Drug hypersensitivity. Karger, Basel
Price A, Ramachandran S, Smith GP, Stevenson ML, Pomeranz MK, Cohen DE (2015) Oral allergy syndrome (pollen-food allergy syndrome). Dermatitis 26(2):78–88
Quake C, Nadeau KC (2015) The role of epigenetic mediation and the furzure of food allergy research. Semin Cell Dev Biol 43:125–130
Radauer C, Breiteneder H (2007) Evolutionary biology of plant food allergens. J Allergy Clin Immunol 120(3):518–525
Ramsay HM, Goddard W, Gill S, Moss C (2003) Herbal creams used for atopic eczema in Birmingham, UK illegally contain potent corticosteroids. Arch Dis Child 88(12):1056–1057
Roschek B Jr, Fink RC, McMichael M, Alberte RS (2009) Nettle extract (*Urtica dioica*) affects key receptors and enzymes associated with allergic rhinitis. Phytother Res 23(7):920–926
Rosenstreich DL, Fried MP, de Vos GS, Jackmann AH (eds) (2016) Manual of allergy and clinical immunology for otolaryngologists. Plural Publishing, San Diego
Sabounchi S, Bollyky J, Nadeau K (2015) Review of environmental impact on the epigenetic regulation of atopic diseases. Curr Allergy Asthma Rep 15(6):533
SCCS (2012) Opinion on fragrance allergens in cosmetic products. Scientific Committee on Consumer Safety. Available at: http://ec.europa.eu/health/scientific_committees/consumer_safety/docs/sccs_o_102.pdf
Scheurer S, Toda M, Vieths S (2015) What makes an allergen? Clin Exp Allergy 45(7):1150–1161
Scholes EW, Basketter DA, Lovell WW, Sarll AE, Pendlington RU (1991) The identification of photoallergic potential in the local lymph node assay. Photodermatol Photoimmunol Photomed 8(6):249–254
Shin HS, Shon DH (2015) Food and natural materials target mechanisms to effectively regulate allergic responses. J Nutr Sci Vitaminol (Tokyo) 61(Suppl):S109–S111
Stein KR, Scheinfeld NS (2007) Drug-induced photoallergic and phototoxic reactions. Expert Opin Drug Saf 6(4):431–443
Sticher O, Heilmann J, Zündorf I (2015) Hänsel/Sticher – Pharmakognosier Phytopharmazie. Wissenschaftliche Verlagsgesellschaft mbH, Stuttgart
Thyssen JP, Maibach HI (2008) Drug-elicited systemic allergic (contact) dermatitis – update and possible pathomechanisms. Contact Dermatitis 59:195–202
Ulrich P, Homey B, Vohr HW (1998) A modified murine local lymph node assay for the differentiation of contact photoallergy from phototoxicity by analysis of cytokine expression in skin-draining lymph node cells. Toxicology 125(2–3):149–168
van Ree R, Hummelshøj L, Plantinga M, Poulsen LK, Swindle E (2014) Allergic sensitization: host-immune factors. Clin Transl Allergy 4:12
Verdier F, Chazal I, Descotes J (1994) Anaphylaxis models in the guinea-pig. Toxicology 93(1):55–61
Wang J, Li XM (2012) Chinese herbal therapy for the treatment of food allergy. Curr Allergy Asthma Rep 12:332–338
Wang J, Sampson HA (2011) Food allergy. J Clin Invest 121(3):827–835
Zeller Medical (2012) Tesalin® N Filmtabletten (package leaflet). Available at: https://zellermedical.ch/wp-content/uploads/sites/3/2012/12/tesalin_.pdf
Żukiewicz-Sobczak WA, Wróblewska P, Adamczuk P, Kopczyński P (2013) Causes, symptoms and prevention of food allergy. Postępy Dermatol Alergol 30(2):113–116

Chapter 12
Herbal Hepatotoxicity

Chit Shing Jackson Woo and Hani El-Nezami

Abstract The liver is probably the most important organ responsible for protecting humans from exposure to foreign compounds. With its strategic location between the intestinal tract and the circulation system, and its unique metabolic capacity, the liver acts as the first defense barrier metabolizing and detoxifying foreign compounds. In contrast, these features have rendered it highly susceptible to adverse and toxic effects. The current herbalism all over the globe has led to the increasing consumption of herbal products with an associated incidence of herbal hepatotoxicity. Herbal hepatotoxicity has been a well-recognized issue for years, but so far systematic studies are lacking. This chapter presents the current situation of liver damage induced by herbal products, the classification of herb-induced liver injuries (HILI), and their potential risk factors. The hepatotoxicity of selected well-known herbal products as reported in the literature is reviewed, and causality assessment for HILI is discussed.

Keywords Herbal hepatotoxicity • Herb-induced liver injury • Herbal medicine • Herbs • Herbal remedy

Abbreviations

ADP	Adenosine diphosphate
ADR	Adverse drug reaction
AG	Atractylis gummifera
ATP	Adenosine triphosphate
ATR	Atractyloside
CAM	Complementary and alternative medicine
CAT	Causality assessment tool
CIOMS	Council for the International Organization of Medical Sciences

C.S.J. Woo
Consumer and Retail, SGS Hong Kong Limited,
3/F, On Wui Centre 25, Lok Yip Rd, Fanling, Hong Kong

H. El-Nezami (✉)
School of Biological Sciences, Faculty of Science, The University of Hong Kong,
Kadoorie Biological Sciences Building, Pokfulam, Hong Kong
e-mail: elnezami@hkucc.hku.hk

O. Pelkonen et al. (eds.), *Toxicology of Herbal Products*,
DOI 10.1007/978-3-319-43806-1_12

CMV	Cytomegalovirus
CYP	Cytochrome P
DILI	Drug-induced liver injury
DILIN	Drug-Induced Liver Injury Network
EBV	Epstein-Barr virus
EPH	Ephedrine
FDA	Food and Drug Administration
HAV	Hepatitis A virus
HBV	Hepatitis B virus
HCV	Hepatitis C virus
HDS	Herbal and dietary supplements
HILI	Herb-induced liver injury
HSV	Herpes simplex virus
MV Scale	Maria and Victorino Scale
NDGA	Nordihydroguaiaretic acid
PA	Pyrrolizidine alkaloids
PSE	Pseudoephedrine
RUCAM	Roussel Uclaf Causality Assessment Method
TNF	Tumor necrosis factor
VZV	Vesicular stomatitis virus
WHO	World Health Organization

Introduction

Herbal and dietary supplements (HDS) are also known as "herbal medicine," "herbal remedies," or "botanical medicine." According to the World Health Organization (WHO), herbal medicines include herbs, herbal materials, herbal preparations, and finished herbal products that contain active ingredient parts of plants, or other plant materials, or combinations thereof (Robinson and Zhang 2011). Before the introduction of aspirin extracted from *Spiraea ulmaria*, which had been described in the Ebers papyrus for the treatment of fever and swelling, as well as recommended by Hippocrates of Kos (Jack 1997), indigenous peoples mainly relied on herbal remedies for medicinal purposes to retain vitality and cure diseases (Pan et al. 2014). The first written mention of the well-established medicinal uses of plants such as laurel, caraway, and thyme, dated back at least 5,000 years to the Sumerians (Falodun 2010). Herbalism can be traced back to 2100 B.C. in ancient China (Xia Dynasty) and India (Vedic Period) (Schuppan et al. 1999).

Due to the lack of availability and high cost of conventional synthetic drugs, as well as their association with unpleasant or even serious side effects, people nowadays prefer to take personal control over their health (Navarro and Seeff 2013). In addition, there is a general perception that herbal remedies must be safe and allow for better control and management of disease because they are "natural" and have been used for centuries (Seeff et al. 2014; Stickel et al. 2005). For these reasons the use of herbs worldwide has become more and more prevalent in recent years. Herbs – both in crude forms (such as teas, roots, seeds, or leaves) and commercial (such as capsules or tablets) products – are commonly available. Apart from improving well-being, bodybuilding purposes, and weight loss, herbal products have frequently been used as alternative therapies and complementary medicine. Instead of being confined to Asia, Africa, and the Middle East, herbal products are now widely used in Western societies (Teschke et al. 2013b). It is estimated that a significant percentage of the population in developed countries such as Canada (70 %), France (49 %), Australia (48 %), the USA (42 %), and Belgium (31 %) have used complementary and alternative medicine (CAM) at least once for health care (WHO 2002). With the globalization of herbalism, there is increasing consumption of herbal products with an associated incidence of herbal hepatotoxicity.

The liver may be considered to be the most important organ responsible for protecting humans from exposure to foreign compounds. With its strategic location between the intestinal tract and systemic circulation, and with its unique metabolic capacity, the liver acts as the first defense barrier metabolizing and detoxifying foreign compounds. However, on the other hand, these features have rendered it highly susceptible to adverse and toxic effects. Although the liver has regenerative capacity, major functions of the liver can still be altered by exposure to toxicants. Increasing numbers of herbal products have been reported to be associated with hepatotoxicity and a number of various mechanisms are shown to be involved. Herb-related liver injuries can range from mild and acute hepatitis to life-threatening liver failure requiring transplantation.

Epidemiology

Although the hepatotoxic potential of herbal products has been recognized for years, there is a definite lack of reliable population-based epidemiological studies specifically relating to the incidence of herbal hepatotoxicity. Unlike conventional synthetic drugs, information on herbal hepatotoxicity is obtained largely from anecdotal case reports, case series, retrospective databases and, more recently, from prospective registries of drug-induced liver injury (DILI) (Bunchorntavakul and Reddy 2013); so far, the actual incidence of herbal hepatotoxicity is unknown (Navarro 2009).

Herb-induced liver injury (HILI) is usually underestimated as patients are often reluctant to disclose the use of herbal products or do not consider them to be medications (Kennedy 2005). There is a lack of precise information about the frequency of hepatotoxicity resulting from the use of herbs. According to the U.S. Drug-Induced Liver Injury Network (DILIN), it is estimated that approximately 16 % of cases of DILI in the U.S. between 2004 and 2013 could be attributed to herbal remedies (Navarro et al. 2014). In Spain and Sweden, HILI accounted for 2–10 % and 5 % among DILI cases, respectively (Andrade et al. 2005; Ibanez et al. 2002; De Valle et al. 2006). Although herbal remedies have been commonly used as traditional medication in China and India as well as in Southeast Asia, Africa, and Central

America, epidemiological data with regard to herbal hepatotoxicity from these areas are even more scant and limited. Prospective studies from China, Korea, and Singapore have reported that HILI accounts for 18.6 %, 73 % and 71 % among all cases of DILI, respectively (Table 12.1). Surprisingly, HILI only accounted for 1.3 % of DILI in India, where Ayurvedic herbal products are commonly used (Devarbhavi et al. 2010). Such wide variations in incidences may reflect a local use of particular hepatotoxic herbs; in our opinion, they are more likely due to difficulties in causality assessments and widely varying pharmacovigilance reporting systems (Shaw et al. 2012; Zhang et al. 2012).

Table 12.1 Prevalence of herbal hepatotoxicity among drug-induced liver injury in various countries

Country	Study period	Number of DILI case	Prevalence of HILI	Reference
China	1994–2011	24,112	18.6 %	Zhou et al. (2013)
Hong Kong	–	1,701	0.2 %	Chan et al. (1992)
Iceland	2010–2012	96	16 %	Bjornsson et al. (2013)
India	1997–2008	313	1.3 %	Devarbhavi et al. (2010)
Japan	1979–1999	2,496	0.6 %	Mantani et al. (2002)
Korea	2005–2007	371	73 %	Suk et al. (2012)
Latin America[a]	2012–2013	73	12 %	(Bessone et al. 2013)
Spain	1994–2013	861	6 %	Medina-Cáliz et al. (2013), Navarro and Lucena (2014)
	1984–2004	446	2 %	Andrade et al. (2005)
	1993–1998	107	10 %	Ibanez et al. (2002)
Singapore	2004–2006	31	71 %	Wai et al. (2007)
Sweden	1995–2005	1164	5 %	De Valle et al. (2006)
USA	2004–2007	300	9 %	Chalasani et al. (2008)
	1998–2007	133	10.6 %	Reuben et al. (2010)
	2001–2002	20	50 %	Estes et al. (2003)
	1990–2002	270	2.6 %	Russo et al. (2004)

[a]Argentina, Uruguay, Chile, Brazil, Mexico, Peru, Venezuela, and Bolivia

Classification of Herbal Hepatotoxicities

Herbal hepatotoxicity, i.e., HILI, results from the human consumption of natural products (Teschke et al. 2013a). HILI shares similar mechanisms and symptoms with DILI. In most cases, the clinical, biochemical, and histological features of hepatotoxicity induced by herbal products or conventional synthetic drugs are similar or even identical (Teschke et al. 2008a). However, unlike conventional synthetic drugs for which the classification of a prevailing type of liver injury is relatively well established as it usually involves one single compound, the classification of herbal hepatotoxicity is challenging to define because a herbal product is often a combination of various constituents and involves various confounding variables (Teschke et al. 2011a).

HILI can be broadly categorized as intrinsic vs. idiosyncratic, and the latter further classified into allergic vs. non-allergic (Table 12.2) (Russmann et al. 2009; Teschke et al. 2013b). Intrinsic hepatotoxicity is dose-dependent and predictable above specific thresholds. The intrinsic toxicity of the herbs at a high dosage is the major cause of liver injury. In contrast, idiosyncratic hepatotoxicity is not characterized by dose-related response and presents in an unpredictable fashion. Allergic idiosyncratic hepatotoxicity involves adaptive immune reactions with the presence of typical symptoms such as fever, skin reactions, eosinophilia, and the formation of autoantibodies, and a short latency period in particular after rechallenge (Russmann et al. 2009). The risk of acute liver failure associated with idiosyncratic hepatotoxicity is usually less than 1 per 10,000 patients (Russmann et al. 2009).

Table 12.2 Pathogenetic classification of herbal hepatotoxicity

Pathogenetic classification	Features
Intrinsic	Predictable
	Dose-dependent
	Short and consistent latency period
	High incidence in humans
	Experimentally reproducible
Idiosyncratic	Unpredictable
	Dose-independent
	Long and variable latency period
	Low incidence in humans
	Experimentally irreproducible
Non-allergic or metabolic type	Variable exposure
	Absence of hypersensitivity symptoms
	Delayed response to rechallenge
Allergic or immunologic type	Short exposure time
	Hypersensitivity symptoms
	Prompt response to rechallenge

The pathophysiologic mechanism of drug-induced hepatotoxicity can be divided into hepatocellular and extracellular processes (Lee 2003; Jaeschke et al. 2002; Russmann et al. 2009):

A. Disruption of hepatocytes: Covalent binding of a substance to intracellular proteins results in a decrease in adenosine triphosphate (ATP) levels and actin disruption; hence, cell swelling and rupture.
B. Disruption of the transport proteins: Interruption of transport systems results in the impairment of canalicular transport of bile salts, leading to cholestasis.
C. Cytolytic T-cell activation: Covalent binding of substance to cytochrome P450 enzyme results in the formation of immunogens activating T-cells and cytokines, hence evoking immune response.
D. Apoptosis of hepatocytes: Activation of apoptotic pathways by tumor necrosis factor (TNF) and the Fas pathways results in programmed cell death.
E. Mitochondrial disruption: Inhibition of mitochondrial functions such as beta-oxidation and energy production results in the decline of ATP production and the generation of reactive oxygen species.
F. Bile duct injury: Toxic metabolites excreted in bile may damage the bile-duct epithelium.

Apart from mechanistic classification, HILI can also be classified based on descriptive clinical or histopathological criteria. Similar to DILI, the hepatic biochemistry pattern of hepatotoxicity associated with herbal products can be categorized into three types of liver injury; these include hepatocellular, cholestatic, or mixed pattern (Navarro and Lucena 2014). Herbal hepatotoxicity is often nonspecific and may result in acute or chronic liver disease. There is a broad spectrum of clinical features associated with herbal hepatotoxicity, including acute and chronic hepatitis with autoimmune features, hepatic fibrosis, cirrhosis, zonal or diffuse hepatic necrosis, microvesicular steatosis, giant cell hepatitis, cholestatic hepatitis, bile duct injury, sinusoidal obstruction syndrome, fulminant liver failure, and carcinogenesis (Estes et al. 2003; Navarro and Lucena 2014; Stedman 2002; Chitturi and Farrell 2000; Seeff et al. 2013).

Potential Risk Factors Associated with Herbal Hepatotoxicity

Risk factors associated with herbal hepatotoxicity have not been well identified, largely because there are few epidemiological studies available. Although most herbal hepatotoxic incidents have been published mainly as isolated case reports or small series, several trends of HILI are becoming apparent. *In vitro* and *in vivo* toxicodynamic studies have also revealed the risk factors potentially contributing to the hepatotoxicity induced by various herbal products.

Genetics

Genetics appears to play a significant role regarding the susceptibility to herbal hepatotoxicity. Cytochrome P450 (CYP) and major drug-metabolizing enzymes are encoded by particular genes. Different expressions of these enzymes would influence the drug metabolism capability and hence may contribute to adverse drug reactions resulting in liver damage. For instance, in two European case reports, kava-related hepatotoxicity was associated with a poor metabolizer phenotype based on CYP2D6 activity (Russmann et al. 2001).

Gender

Between 80 and 100 % of the cases of hepatotoxicity associated with Jin Bu Huan, chaparral, germander, and greater celandine have occurred in women. It may be due to women's propensity to consume such remedies; for example, germander has primarily been marketed as a weight-loss agent (Stedman 2002). Whereas an epidemiological study of liver disease clinics (USA, 1999) revealed that the use of CAM was more common in women (Seeff et al. 2001), another national survey of 2,055 people in the U.S. (1997) found no statistically significant gender difference in the overall usage of herbal medicine. As in most DILI cases, women appear to be more susceptible to HILI, although conflicting data were reported with regard to gender-related differences on the usage of herbal products (Stickel et al. 2005; Stedman 2002; Amacher 2014).

Age

Elderly people and children are generally suspected to be at a higher risk for HILI. Similar to DILI, the elderly are thought to be at higher risk because of decreased hepatic and renal clearance, drug-to-drug interactions, reduced hepatic blood flow, variation in serum drug binding, and lower hepatic volume (Zimmerman 1999). For children, either greater susceptibility in this age group or proportionately higher dosage exposure is suggested, based on a few studies of atractyloside (ATR) poisonings (Wainwright et al. 1977). In most cases of HILI, the mean age is between 45 and 58 years (Stedman 2002); apart from accidental exposure, HILI rarely occurs in children. The relatively few cases of herbal hepatotoxicity occurring among the elderly may suggest that the older age group would not be a critical risk factor for HILI. However, epidemiological data with age-specific usage are required to better understand possible links (Stedman 2002).

Herb-Herb/Drug Interaction

In addition to the direct hepatotoxic effect of a herbal product itself, herb-drug interaction also potentiates the risk of hepatotoxicity and increases the difficulty for causality assessment. For instance, the hepatotoxicity of diterpenoids from germander and pyrrolizidine alkaloids can be potentiated by pharmacologic induction of CYP3A4 (Fau et al. 1997; Loeper et al. 1994; Stickel et al. 2000). Also, it has been suggested that co-medication with St. John's wort may potentiate the hepatotoxicity of kava (Ernst 2007; Teschke 2006). Herb-herb interaction is of particular concern, as most herbal products are complex mixtures involving various ingredients.

Apart from this, it has been suggested that the concomitant ingestion of alcohol may contribute to hepatotoxicity of pennyroyal through CYP2E1induction (Khojasteh-Bakht et al. 1999). The concomitant consumption of herbal remedies and pharmaceutical drugs, as well as alcohol, should be considered to be a cause for concern.

Contamination and Adulteration

Herbs are usually harvested from the wild or specifically cultivated. There is a risk of contamination with potential toxicants such as mycotoxins (e.g., aflatoxins), heavy metals (lead, cadmium, mercury, or arsenic) and pesticides (chlorinated pesticides or organic phosphates) as well as microorganisms (*Bacillus cereus*) that may contribute to liver injury (Stickel et al. 2009; Wong et al. 1993; Gray et al. 2004; De Smet 2002; Kneifel et al. 2002). In addition, synthetic drugs (corticosteroids, benzodiazepines, or anti-inflammatory drugs) may be added to herbal products to provide or fortify the purported efficacy of the herbal remedies without notifying the consumer (Miller and Stripp 2007; De Smet 2002). This may result in the concomitant consumption of herbal remedies and synthetic drugs – hence, the potentiate hepatotoxic adverse effect. Although impurities and adulterants have been considered to be key problems (Zhang et al. 2012; Shaw et al. 2012), these issues are rarely addressed in publications that concern herbal hepatotoxicity (Teschke et al. 2013b).

Common Herbal Hepatotoxicants

There are several studies that show adverse hepatotoxic effects related to herbal products. Liver injuries induced by herbal products are similar to those attributed to conventional synthetic drugs. The clinical presentation of HILI can be variable even with the same herb. A review of herbal products associated with hepatotoxicity is provided in Table 12.3. Some of the more frequently reported hepatotoxic herbal products are discussed below.

Table 12.3 Table showing herbal remedies associated with liver injury

Herbal remedy	Common use	(Suspected) toxic constituent	(Suspected) toxic mechanism	Clinical pattern	Reference
Aloe (*Aloe vera* L.) Burm.f., *Aloe arborescens* Mill.)	Gastrointestinal ailments/topical emollient	Anthraquinones	Unknown	Acute hepatitis Portal/lobular inflammation with eosinophilic infiltrates and acidophilic bodies	Yang et al. (2010)
Aristolochia debilis Siebold & Zucc.	Multiple uses	Aristolochic acid	Unknown	Acute hepatitis	Levi et al. (1998)
Atractylis gummifera Salzm. ex L. (a synonym of *Carlina gummifera* (L.) Less.) (Distaff thistle, African remedy)	Antipyretic, antiemetic, abortifacient and diuretic	Atractyloside and carboxyatractyloside	Inhibition of oxidative phosphorylation. Triggers apoptosis by inducing mitochondrial permeability	Diffuse hepatic necrosis. Fulminant hepatic failure	Daniele et al. (2005), Stewart and Steenkamp (2000)
Black cohosh (*Cimicifuga racemose* (L.) Nutt., synonym of *Actaea racemose* L.)	Treatment of menopausal symptoms, joint pain and myalgia	Unknown	Liver cell apoptosis via mitochondrial damage	Acute hepatitis. Fulminant liver failure	Lude et al. (2007), Whiting et al. (2002), Levitsky et al. (2005), Lynch et al. (2006)
Boldo (*Peumus boldus* Molina)	Multiple uses	Ascaridole	Inhibition of oxidative phosphorylation.	Cholestatic hepatitis	Piscaglia et al. (2005), EMA (2009)

(continued)

Table 12.3 (continued)

Herbal remedy	Common use	(Suspected) toxic constituent	(Suspected) toxic mechanism	Clinical pattern	Reference
Callilepis laureola D.C. (Impila, Zulu remedy)	For stomach problems, tapeworm infestations, cough, impotence	Atractyloside and carboxyatractyloside	Inhibition of oxidative phosphorylation. Trigger apoptosis by inducing mitochondrial permeability	Diffuse hepatic necrosis	Sohni (2002)
Camphor (*Cinnamomum camphora* (L.) J. Presl)	Rubefacient	Cyclic terpenes	Unknown	Acute hepatitis. Hepatitis with necrosis	Uc et al. (2000)
Cascara sagrada (*Rhamnus purshiana* DC., cascara buckthorn, Sacred bark)	Laxative	Anthracene glycosides and anthraquinones	Unknown	Cholestatic hepatitis	Nadir et al. (2000)
Centella asiatica (L.) Urb.	Wound healing, leprosy, psychophysical regenerator, blood purifier	Triterpenoids	Promotion of apoptosis and alteration of cell membrane principles	Cellular necrosis and apoptosis (eosinophilic degeneration), and lymphoplasmocytic infiltrate	Jorge and Jorge (2005)
Chaparral (*Larrea tridentata* (Sessé & Moc. ex DC.) Coville)	Liver and health tonic	Nordihydroguaiaretic acid	Inhibition of cyclooxygenase pathways	Cholestasis, cholangitis, chronic hepatitis, cirrhosis	Sheikh et al. (1997)

Chaso pr Onshido	Weight loss	N-nitroso-fenfluramine	Inhibition of oxidative phosphorylation	Diffuse or massive necrosis with nonspecific inflammatory infiltrate. Ductular proliferation with bile stasis, and bridging fibrosis	Adachi et al. (2003)
Comfrey (*Symphytum officinale* L.)	Health tonic	Pyrrolizidine alkaloids	Metabolic activation of PA results in the formation of reactive pyrroles	Sinusoidal obstruction syndrome (veno-occlusive disease)	Stickel and Seitz (2000)
Dai-saiko-to (TJ-8)	Treatment of dyspepsia and gallstones	Unknown	Unknown	Autoimmune hepatitis	Kamiyama et al. (1997)
Dictamnus dasycarpus Turcz.	Treatment of jaundice, cough, rheumatism and skin diseases	Unknown	Unknown	Acute hepatitis	Lei et al. (2008), Abdualmjid and Sergi (2013)
Flavocoxid	Treatment of chronic osteoarthritis	Catechins, baicalin	Unknown	Acute hepatitis	Chalasani et al. (2012)
Germander (*Teucrium chamaedrys* L.)	Weight loss	Furane-containing neo-clerodane diterpenoids	Metabolites of the diterpenoids cause membrane disruption and apoptosis	Acute and chronic hepatitis, cirrhosis, fibrosis, massive necrosis	Gori et al. (2011)
Greater Celandine (*Chelidonium majus* L.)	Treatment of gastrointestinal disorders and dyspepsia	Alkaloids	Unknown	Chronic hepatitis, cirrhosis, cholestatic hepatitis, massive necrosis	Benninger et al. (1999)

(continued)

Table 12.3 (continued)

Herbal remedy	Common use	(Suspected) toxic constituent	(Suspected) toxic mechanism	Clinical pattern	Reference
Green tea (*Camellia sinensis* (L.) Kuntze)	Weight loss (80° ethanolic extract particularly rich in catechins)	Catechins and their gallic acid esters (Epigallocatechin-3-gallate)	Induction of oxidative stress	Inflammatory infiltrates cholestasis, occasional steatosis, and necrosis	Bonkovsky (2006)
Horse chestnut (*Aesculus hippocastanum* L.)	Venotonic	Unknown	Unknown	Liver injury with hepatocellular pattern	García-Cortés et al. (2008)
Isabgol	Laxative	Unknown	Unknown	Giant cell chronic hepatitis	Fraquelli et al. (2000)
Jin Bu Huan (*Lycopodium serratum* Thunb., a synonym of *Huperzia serrata* (Thunb.)	Sedative	Levo-tetrahydropalmatine	Unknown	Chronic hepatitis, portal and parenchymal lymphocytic inflammation, portal fibrosis, focal necrosis, steatosis, cholestatic hepatitis	Woolf et al. (1994), Picciotto et al. (1998), Horowitz et al. (1996), Divinsky (2002)
Kava (*Piper methysticum* G. Forst.)	Anxiolytic, sleeping aid	Kava lactones	Unknown	Acute and chronic hepatitis, cholestasis, liver failure	Stickel et al. (2003a)
Ma Huang (*Ephedra sp.*)	Weight reduction	Ephedrine	Unknown	Massive necrosis, polymorphonuclear infiltrate	Neff et al. (2004), Bajaj et al. (2003), Nadir et al. (1996), Borum (2001), Skoulidis et al. (2005)

Margosa oil (*Antelaea azadirachta* (L.) Adelb., a synonym of *Azadirachta indica* A.Juss.)	Health tonic	Unknown	Mitochondrial dysfunction and poisoning of electron transport pathway	Reye's syndrome	Sinniah and Baskaran (1981), Koga et al. (1987)
Mistletoe (*Viscum album* L.)	Complementary medicine in cancer therapy	Unknown	Unknown	Hepatitis	Harvey and Colin-Jones (1981), Doehmer and Eisenbraun (2012)
Noni (*Morinda citrifolia* L.)	Health tonic	Anthraquinones	Unknown	Idiosyncratic acute liver injury	Stadlbauer et al. (2008, 2005), Millonig et al. (2005), Yuce et al. (2006), Yu et al. (2011)
Pennyroyal oil (*Mentha pulegium* L.)	Abortifacient	Pulegone/Menthofuran	Glutathione depletion through electrophilic metabolites	Fulminant hepatic failure. Hepatic necrosis	Anderson et al. (1996), Bakerink et al. (1996)
Purple coneflower (*Echinacea angustifolia* DC., *E. pallida* (Nutt.) Nutt. and *E. purpurea* (L.) Moench)	Immunostimulant	Pyrrolizidine alkaloids	Metabolic activation of PA results in the formation of reactive pyrroles	Acute cholestatic autoimmune hepatitis. Liver necrosis	Kocaman et al. (2008), Abdualmjid and Sergi (2013), Jacobsson et al. (2009)

(continued)

Table 12.3 (continued)

Herbal remedy	Common use	(Suspected) toxic constituent	(Suspected) toxic mechanism	Clinical pattern	Reference
Sacaca (*Croton cajucara* Benth.)		Furano diterpenoids	Unknown	Acute, fulminant and chronic hepatitis	Soares Mdo (2004)
Sassafras (*Sassafras albidum* (Nutt.) Nees)	Herbal tea	Safrole	Unknown	Hepatitis	Segelman et al. (1976)
				Hepatic carcinogenesis (animals)	
Saw palmetto (*Serenoa repens* (W.Bartram Small)	Treatment of benign prostatic hyperplasia	Unknown	Estrogenic and anti-androgenic effect	Cholestatic hepatitis. Fibrosis	Hamid et al. (1997), Singh et al. (2007)
Senna (*Cassia acutifolia* Delile, and *C. angustifolia* M. Vahl., synonyms of *Senna alexandrina* Mill.)	Laxative	Anthraquinones	Idiosyncratic	Portal and lobular infiltration. Necrosis	Beuers et al. (1991)
Shou-wu-pian (*Polygonum multiflorum* Thunb., a synonym of *Reynoutria multiflora* (Thunb.) Moldenke)	Multiple uses	Anthraquinones	Unknown	Acute hepatitis	Park et al. (2001), But et al. (1996)

Sho-saiko-to (xiao-chai-hu-tang)	Liver and health tonic	Scutellaria	Unknown	Zonal/bridging necrosis. Fibrosis. Microvesicular steatosis	Itoh et al. (1995), Stedman (2002)
Skullcap (*Scutellaria lateriflora* L. and *Scutellaria baicalensis* Georgi)	Sedative, anti-inflammatory agents	Alkylating agents	Unknown	Hepatitis	Linnebur et al. (2010), Whiting et al. (2002), Yang et al. (2012)
Teucrium polium L.	Anti-inflammatory agents	Unknown	Unknown	Acute hepatitis	Mattei et al. (1995)
Usnea dasypoga (Ach.) Nyl. (mushroom)	Weight loss	Usnic acid	Uncoupling of oxidative phosphorylation	Acute hepatocellular necrosis and inflammation	Neff et al. (2004), Navarro and Lucena (2014), Favreau et al. (2002), Pramyothin et al. (2004)
Valerian (*Valeriana officinalis* L.)	Sedative	Unknown	Unknown	Mild fibrosis. Portal inflammation	Vassiliadis et al. (2009), Cohen and Del Toro (2008)

Atractylis gummifera *and* **Callilepsis laureola**

Atractylis gummifera Salzm. ex L. (a synonym for *Carlina gummifera* (L.) Less.) (AG) is a thistle found largely throughout the Mediterranean area (Daniele et al. 2005). It is used as an antipyretic, antiemetic, and diuretic. Because of its sweet-tasting juice, it is enjoyed by children as chewing gum (Larrey 1997). There have been more than 100 cases of *Atractylis gummifera*-induced liver and renal injury that frequently involved children (Bateman et al. 1998).

Atractylis gummifera-induced toxicity has been primarily ascribed to a toxic diterpenoid glucoside called atractyloside (ATR) (Daniele et al. 2005). ATR was first extracted from *Atractylis gummifera* in 1868 and characterized by Lefranc (Stewart and Steenkamp 2000; Obatomi and Bach 1998); it is a powerful inhibitor of oxidative phosphorylation in mitochondria that interacts with adenine nucleotide translocators (Roux et al. 1996). Its action is especially exerted in cells rich in mitochondria, such as hepatocytes (Daniele et al. 2005). ATR inhibits adenosine diphosphate (ADP) transport into the mitochondrial compartment, leading to the blockage of oxidative phosphorylation and Krebs cycle oxidative reactions (Obatomi and Bach 1998). Moreover, it has also been suggested that ATR would trigger apoptosis by inducing the mitochondrial permeability transition pores, invoking the release of soluble intermembrane proteins such as cytochrome c as well as caspase-activating proteases (Vancompernolle et al. 1998).

Another plant species that is commonly known to contain ATR is *Callilepis laureola* DC. ("*impila*" in Zulu). It is a herbaceous perennial plant indigenous to the KwaZulu-Natal region of South Africa and has been used as a traditional remedy by the Zulu Tribe (Bye and Dutton 1991). It has been associated with several cases of fulminant hepatitis and renal tubular necrosis in South Africa (Bye and Dutton 1991).

Chaparral

Chaparral (*Larrea tridentata* (Sessé & Moc. ex DC.) Coville), commonly known as creosote bush or greasewood, is a botanical dietary and "energy" supplement. It is prepared from the leaves of the evergreen desert shrub that can be found in the southwestern U.S. and Mexico (Cupp 2000). Native Americans have traditionally used it medicinally for the treatment of various ailments including respiratory tract infections, rheumatic pain, abdominal pain, chicken pox, and snakebite pain (Sheikh et al. 1997). Currently, chaparral has been marketed in the form of tea, tablets, capsules, and salves (for burns), weight loss, liver tonic, blood purifiers, cancer cures, and treatment of skin disorders (Sheikh et al. 1997). In addition, it has been proposed as an alternative treatment for AIDS (Kassler et al. 1991).

In the 1990s, a series of incidents regarding chaparral-associated hepatotoxicity were reported to the U.S. Food and Drug Administration (FDA) (Sheikh et al. 1997). Liver injuries ranging from mild hepatitis to cirrhosis and fulminant liver failure were reported (Sheikh et al. 1997). The predominant pattern of liver damage was cholestatic hepatitis with high serum transaminases and the elevation of bilirubin and alkaline phosphatase (Katz and Saibil 1990).

The hepatotoxicity of chaparral is generally attributed to lignane nordihydroguaiaretic acid (NDGA). NDGA is an antioxidant and was once used as a preservative in foods for humans and for pharmaceuticals. However, it was taken off the "generally recognized as safe" FDA list in 1968 because of animal toxicity data (Sheikh et al. 1997). NDGA can inhibit lipoxygenase and cyclooxygenase (Agarwal et al. 1991; Capdevila et al. 1988), and hence it affects many intrahepatic pathways.

Ephedra (Ma Huang)

Ephedra, or Ma Huang as it is known in traditional Chinese medicine, is one of the oldest medicinal herbs, having been used in Chinese medicine as a stimulant and anti-asthmatic agent for more than 5,100 years (Chen and Schmidt 1926; Abourashed et al. 2003; Karch 2000). It is derived from plants of the Ephedra species, which can be found in the temperate and subtropical regions of Asia, Europe, and North and Central America (Stedman 2002; Abourashed et al. 2003). In the early twentieth century, ephedra was advertised as a slimming aid and energy-level enhancer in the U.S. and Europe (Karch 2000).

Despite its apparent safe use as traditional Chinese medicine for centuries, ephedra has been associated with a number of liver injuries (Neff et al. 2004; Bajaj et al. 2003; Nadir et al. 1996; Borum 2001). Features of autoimmunity (positive antinuclear antibodies and anti-smooth muscle antibodies) were observed; diffuse severe hepatocellular necrosis and polymorphonuclear infiltrates with occasional eosinophils were revealed by biopsy. Ephedra-associated hepatotoxicities resulting in liver transplantation were also reported in the U.S. and in Europe (Skoulidis et al. 2005; Estes et al. 2003).

The mechanism behind ephedra-induced toxicity is still not known. Ephedra comprises various active ingredients, notably the sympathomimetic alkaloids ephedrine (EPH) and pseudoephedrine (PSE) (Abourashed et al. 2003). Cardiovascular adverse effects have been attributed to these alkaloids (Haller and Benowitz 2000). However, it is unclear whether or not hepatotoxicity is caused by EPH.

Germander

Germanders of the genus *Teucrium* are found in Europe and the Middle East, and the aerial parts of the plant are used (Korth 2014). The blossoms of germander have been used for centuries as traditional herbal remedies because of their diuretic, diaphoretic, antipyretic, antispasmodic, anti-inflammatory, antihypertensive, anorexic, analgesic, antibacterial, hypoglycemic, and hypolipidemic properties (Menichini et al. 2009; Hasani-Ranjbar et al. 2010). Germander has become popular as a slimming decoction and anorectic remedy over the past few decades (Gori et al. 2011). In 1986, germander (*Teucrium chamaedrys* L.)-containing capsules and tea preparation were approved in France as an adjuvant to weight control. However, several incidents of germander-associated acute, chronic, and even fulminant hepatitis resulted in withdrawing it from the market and banning it in 1992 (Larrey et al. 1992; Mostefa-Kara et al. 1992). Germander-induced hepatotoxicity has also been reported in Canada (Laliberté and Villeneuve 1996) and Spain (Perez Alvarez et al. 2001).

In general, germander-induced hepatotoxicity may occur after approximately 2–3 months of ingestion at the manufacturer's recommended doses (600–1,600 mg/day). Symptoms related to germander-induced hepatotoxicity are non-specific and typically include fatigue, nausea, and the development of jaundice associated with marked elevation of serum aminotransferase levels (Larrey et al. 1992; Gori et al. 2011; Laliberté and Villeneuve 1996; Mostefa-Kara et al. 1992; Castot and Larrey 1992), which are common features of hepatocellular adverse reactions.

Saponins, glycosides, flavonoids and several furane-containing neo-clerodane diterpenoids were identified in *T. chamaedrys* preparation by Piozzi et al. (1987). Furanic diterpenoids are well known to be cytotoxic and carcinogenic. In animal studies, the diterpenoids are metabolized by cytochrome P450 3A to reactive electrophilic metabolites that bind to proteins, deplete cellular glutathione and cytoskeleton-associated protein thiols, increase cytosolic [Ca^{2+}], and activate Ca^{2+}-dependent tissue transglutaminase and endonucleases, resulting in hepatocyte apoptosis (Fau et al. 1997; Loeper et al. 1994; Lekehal et al. 1996). On the other hand, the detection of anti-microsomal epoxide hydrolase autoantibodies in the sera of patients who drank germander teas for a long period of time has also suggested the potential involvement of immune mechanisms in germander-related hepatotoxicity (De Berardinis et al. 2000).

Apart from *T. chamaedrys*, several herbs in the same genus, such as *T. polium* L. (Mattei et al. 1995; Starakis et al. 2006; Savvidou et al. 2007), *T. capitatum* L. (Dourakis et al. 2002), and *T. viscidum* Blume (Poon et al. 2008) are also reported to be associated with hepatitis. These herbal products are often used as hypoglycemic agents to help treat diabetes. By chemical analysis, putative hepatotoxic neo-clerodane diterpenoids have also been isolated from four other species, including *T. alpestre* Sm., *T. cuneifolium* Sm., *T divaricatum subsp. villosum* (Celak.) Rech.f. (a synonym of *T. divaricatum* Sieber ex Heldr.) and *T. flavum subsp. hellenicum* Rech.f. (Piozzi et al. 1997).

Greater Celandine

Greater celandine (*Chelidonium majus* L.) is a plant of the poppy family (*Papaveraceae*) that grows wild in Asia and Europe (Moro et al. 2009). It has been used for centuries to treat gastrointestinal complaints, dyspepsia, and gallbladder disease. Several cases of liver injury associated with greater celandine have been reported, mostly from Germany (Benninger et al. 1999; Hardeman et al. 2008; Moro et al. 2009; Stickel et al. 2003b; Rifai et al. 2006; Greving et al. 1998; Crijns et al. 2002; Colombo and Bosisio 1996a). The largest case series of ten female patients from Germany revealed acute hepatitis with a cholestatic pattern and low titers of autoantibodies in most cases (Benninger et al. 1999). Lobular and portal inflammation with bridging fibrosis and eosinophilic infiltrates were observed in most of the liver histology (Benninger et al. 1999).

Greater celandine contains at least 20 different alkaloids, including sanguinarine, chelidonine, chelerythrine, berberine and coptisine (Colombo and Bosisio 1996b). However, none of them has been shown to be specifically hepatotoxic. The mechanism of greater celandine-induced hepatotoxicity is unclear. As there is no evidence of dose dependency, and with long and variable latency as well as the presence of eosinophilic infiltrate, it is likely that an idiosyncratic reaction is involved (Benninger et al. 1999; Teschke et al. 2012).

Jin Bu Huan

Jin Bu Huan (*Lycopodium serratum* Thunb., a synonym of *Huperzia serrata* (Thunb.) Trevis.), is a traditional Chinese herbal remedy that has been used as a sedative and analgesic for more than 1,000 years (Stedman 2002). It has morphine-like properties due to the alkaloid levo-tetrahydropalmatine (Liu et al. 1982). It was first available as Jin Bu Huan Anodune Tablets in the U.S. in the 1980s (Woolf et al. 1994). There have been more than ten cases of Jin Bu Huan associated hepatotoxicity including acute and chronic liver injuries (Woolf et al. 1994; Picciotto et al. 1998; Horowitz et al. 1996; Divinsky 2002).

The mechanism behind Jin Bu Huan-induced hepatotoxicity is not known, but the active ingredient, levo-tetrahydropalmatine, is believed to be responsible for its toxicity (Woolf et al. 1994). Levo-tetrahydropalmatine may contribute to a direct hepatotoxic effect (Woolf et al. 1994; Wang et al. 2010).

Kava

Kava, also known as kava-kava, is an herbal product derived from roots of the *Piper methysticum* G. Forst. plant. Being consumed for centuries as a recreational and ceremonial drink in Oceania, it has long been used as a traditional psychotropic remedy in Hawaii, Polynesia, and the Fiji Islands. It has also been marketed for the treatment of anxiety disorders, depression, and as a sleeping aid (Stickel et al. 2003a). Due to a few cases of hepatotoxicity related to kava in the 1990s, a label warning of possible hepatotoxicity has been mandated by the German government for acetone extracts of kava, and the American Herbal Products Association has required its members to label kava products as potentially hepatotoxic since 1997 (Yarnell and Abascal 2014). By 2005, the WHO had received notification of 55 cases of kava-associated liver injury, including three cases of liver failure and two of hepatic comas (Zhang et al. 2011). With about 100 cases associated with kava, international interest in kava hepatotoxicity has resulted in the restriction or ban of sales of kava products among European countries such as Austria, France, Germany, and the U.K. (Yarnell and Abascal 2014; Ernst 2007). However, kava products are still available in the U.S., Canada, Australia, New Zealand, and the South Pacific Islands (WHO 2007).

The cause of kava-induced hepatotoxicity is unclear. Clinical case reports have indicated liver injuries with predominantly hepatocellular pattern and variable latency (Stickel et al. 2003a; Teschke 2010); metabolic-type idiosyncratic pathogenesis appears to be the most likely explanation (Schulze et al. 2003; Teschke et al. 2008b; Teschke 2010). Previous studies on animal and tissue samples have shown that its hepatotoxicity may probably be due to various chemical entities such as kavalactones, pipermethystine, and flavokavain B (Rowe and Ramzan 2012). Although *in vitro* and *in vivo* studies have suggested that kava has the potential to cause drug interactions through inhibiting cytochromes P450 (Mathews et al. 2002, 2005), the range of reported hepatotoxic reactions is not compatible with usual adverse drug interactions (WHO 2007).

Pennyroyal

Pennyroyal oil, also known as "squawmint oil," is an herbal extract or oil derived from the leaves of either *Mentha pulegium* L. or *Hedeoma pulegioides* (L.) Pers. (Gordon and Khojasteh 2015). It was traditionally used as a natural abortifacient, insect repellent, or as a means to induce menstruation (Sullivan et al. 1979; Gordon et al. 1982). Due to its distinct spearmint odor, it was used in small quantities as a flavoring agent in food and alcoholic beverages as well as as a fragrance component in soaps and detergents (Gordon et al. 1982). However, the ingestion of large amounts of pennyroyal oil has resulted in intoxication (Vallance 1955; Early 1961; Gunby 1979; Gold and Cates 1980), and has long been identified as a cause of severe liver injury (Sullivan et al. 1979; Gunby 1979).

Pennyroyal contains a mixture of several monoterpenes. Among them, R-(+)-pulegone, the major constituent, is primarily considered to be responsible for the hepatotoxic effects observed in small rodents (Anderson et al. 1996; Gordon et al. 1982). Pulegone is converted by cytochrome P450 to hepatotoxic menthofuran, which is further oxidized to reactive electrophiles (Gordon et al. 1987; Thomassen et al. 1992; Nelson et al. 1992; Speijers 2001).

Although the exact mechanism of pennyroyal-induced hepatotoxicity is not clearly defined, a pathway similar to that responsible for acetaminophen toxicity has been suggested (Mitchell et al. 1973). It is postulated that pulegone extensively depletes glutathione, which is responsible for biotransformation and detoxification, resulting in the buildup of toxic metabolites such as menthofuran to cause direct tissue injury (Bakerink et al. 1996).

Pyrrolozidine Alkaloids

Pyrrolizidine alkaloids (PA) are present in numerous plants, most notably in the Boraginaceae (all genera), Compositae (Senecioneae and Eupatorieae tribes), and Leguminosae (genus *Crotalaria*) families (Smith and Culvenor 1981). It is estimated that more than 6,000 species (around 3 % of the world's flowering plants) contain these harmful alkaloids (Smith and Culvenor 1981). The hepatotoxic properties of PA have been well recognized for over 70 years, in particular the *Senecio* "disease," which was first described in South Africa (Willmot and Robertson 1920). The largest outbreak of human alkaloid poisoning, involving more than 2,000 people, originated in Afghanistan, where the causative plants were ingested as medicinal herbs (Smith and Culvenor 1981).

The key pattern of liver injury evoked by PA is veno-occlusive disease (VOD) (Stickel et al. 2000), recently renamed "sinusoidal obstruction syndrome" (SOS) because the condition begins with damage to sinusoidal endothelial cells (DeLeve et al. 2002). The mechanism behind PA-induced hepatotoxicity is not yet fully understood; it probably involves a direct toxic effect instead of an immunological response. Metabolic activation of 1,2-unsaturated alkaloids by cytochrome P450 enzymes results in the formation of reactive pyrrolic derivatives that bind with nucleophilic cellular macromolecules to cause liver damage (Ruan et al. 2014).

Causality Assessment

Causality assessment of HILI is the process of determining whether there is a reasonable likelihood that herbal products can be the cause of liver injury (Seeff et al. 2013). There are a great many varieties of herbal products available on the market; their multiple ingredients, as well as discrepancies between the ingredients label and actual content, make it a huge challenge to diagnose herbal-induced hepatotoxicity. The diagnosis of herbal hepatotoxicity requires excluding other causes, such as hepatitis A virus (HAV), hepatitis B virus (HBV), hepatitis C virus (HCV), cytomegalovirus (CMV), Epstein-Barr virus (EBV), herpes simplex virus (HSV), vesicular stomatitis virus (VZV), autoimmune hepatitis, alcoholic liver disease, ischemic liver injury/hemodynamic collapse, genetic liver diseases, biliary obstruction, and vascular injury (Rossi and Navarro 2014).

As in the case of liver injury induced by conventional synthetic drugs and xenobiotics, there is no universal standard for diagnosing HILI because of the wide variety of clinical features and time of onset. Several diagnostic scoring systems have been adopted for the causality assessment of HILI – for example, the Naranjo scoring system or Adverse Drug Reaction Probability Scale (Naranjo et al. 1981; Sarma et al. 2008), the Roussel Uclaf Causality Assessment Method (RUCAM) (Benichou 1990), Maria and Victorino (MV) Scale (Maria and Victorino 1997), and others. The causality assessment tools are predominantly based on clinical criteria such as age, alcohol use, the exclusion of underlying liver diseases, and temporal exposure to a drug (Rossi and Navarro 2014).

The Naranjo scoring system, an early causality assessment process, has been applied to assess the relationships between natural product exposure and disease (Sarma et al. 2008). However, as the Naranjo scale only relates adverse drug reactions (ADR) to conventional medication rather than specifically to idiosyncratic reactions like hepatotoxicity, it has limited specificity for hepatic reactions and has not been validated for hepatotoxicity (Teschke et al. 2011b; Teschke and Schulze 2012).

Among these scoring systems, the liver-specific RUCAM scale, in its original form, or, even better in its updated form, is probably the most widely used causality assessment tool for HILI (Licata et al. 2013; Danan and Benichou 1993; Teschke et al. 2008a). The RUCAM scale was validated by cases with positive re-exposure tests as the gold standard, and it has shown excellent sensitivity (86 %), specificity (89 %), positive predictive value (93 %), and negative predictive value (78 %) (Benichou et al. 1993). The RUCAM scale was developed in 1989 after an international meeting in Paris under the auspices of the Council for the International Organization of Medical Sciences (CIOMS). The RUCAM scale applies numerical weighing to key features across various domains as shown in Table 12.4 (Teschke et al. 2008a, 2013b). The total score is calculated by the summation of the individual scores for each domain so as to reflect the causality probability of HILI. However, since herbal products are less likely to be well characterized regarding information on the hepatotoxicity of active ingredients, the score may be compromised, since no points would be given for agents without previous information on hepatotoxicity (Bunchorntavakul and Reddy 2013).

Considering the limitations of the RUCAM scale, such as difficulty in interpreting definitions, a lack of accounting for HCV infection and low level of inter-observer reproducibility (Rockey et al. 2010; Rochon et al. 2008), an expert opinion process has been proposed and adopted as a standard approach by the U.S. DILIN (Fontana et al. 2009). This methodology relies on an expert opinion from three experienced hepatologists (Fontana et al. 2009). Unlike the above-mentioned scoring systems, the expert opinion process allows comprehensive consideration of all clinical details and a greater degree of latitude for the assessors in making causality judgment (Navarro and Lucena 2014). The causality assessment depends on independent grading (Table 12.5) by the three experts with regard to the likelihood of a causal relationship between the drug and liver injury, based on a narrative summary, summary of clinical findings, and sequential biochemical abnormalities (Fontana et al. 2009; Rockey et al. 2010). Causality adjudication is completed if there is complete agreement among the three expert assessors. In case of disagreement, the assessors will reconcile the differences to reach a final single score (Rockey et al. 2010). By this method, the likelihood of a causal relationship between the drug and liver injury is graded in one of five scores with definite (>95 % assurance), highly likely (75–95 % assurance), probable (50–74 % assurance), possible (25–49 % assurance) and unlikely (<25 % assurance) (Chalasani et al. 2008). This DILIN expert opinion process has produced higher agreement rates and likelihood scores than RUCAM in assessing causality (Rockey et al. 2010). Although this approach has been extended to herbal products, only a small subset of study samples (5 %) comprised HILI cases (Chalasani et al. 2008).

Recently, a novel Causality Assessment Tool (CAT) potentially applicable to herbal products was preliminarily developed by a group of experts from DILIN (Navarro et al. 2012). Greater emphasis has been placed on the herbal complexity, experts' perception of the quality of available literature, and personal experience

Table 12.4 RUCAM scale (Licata et al. 2013; Teschke et al. 2013b)

Type of liver injury	Hepatocellular		Cholestatic/mixed		Points
Time of onset of the event	First exposure	Second exposure	First exposure	Second exposure	–
Time from drug intake until reaction onset	5–90 days	1–15 days	5–90 days	1–90 days	+2
	<5 or >90 days	>15 days	<5 or >90 days	>90 days	+1
Time from drug withdrawal until reaction onset	≤15 days	≤15 days	≤30 days	≤30 days	+1
Risk factors	Alcohol		Alcohol or pregnancy		+1
	Age ≥55 years		Age ≥55 years		+1
Course of the reaction	>50 % improvement 8 days		–		+3
	>50 % improvement 30 days		>50 % improvement 180 days		+2
	–		<50 % improvement 180 days		+1
	Lack of information or no improvement		Lack of information or no improvement		0
	Worsening or <50 % improvement 30 days		–		−2
Concomitant therapy	None or no information				0
	None or information not available or time to onset incompatible				0
	Time to onset compatible but with unknown reaction				−1
	Time to onset compatible but known reaction				−2
	With evidence for its role in this case				−3

(continued)

Table 12.4 (continued)

Type of liver injury	Hepatocellular		Cholestatic/mixed		Points
Time of onset of the event	First exposure	Second exposure	First exposure	Second exposure	–
Exclusion of non drug-related causes	Group I (6 causes)				
	Anti-HAV-IgM				
	HBsAg, anti-HBc-IgM, HBV-DNA				
	Anti-HCV, HCV-RNA				
	Hepatobiliary sonography/color Doppler sonography of liver vessels/endosonography/CT/MRC				
	Alcoholism (AST/ALT ≥2)				
	Acute recent hypotension history (particularly if there is underlying heart disease)				
	Group II (5 causes)				
	Complications of underlying disease(s)				
	Infection suggested by PCR and titer change for				
	CMV (anti-CMV-IgM, anti-CMV-IgG)				
	EBV (anti-EBV-IgM, anti-EBV-IgG)				
	HSV (anti-HSV-IgM, anti-HSV-IgG)				
	VZV (anti-VZV-IgM, anti-VZV-IgG)				
	All causes – reasonably ruled out				+2
	6 causes of Group I ruled out				+1
	5 or 4 causes of Group I ruled out				0
	Less than 4 causes of Group I ruled out				−2
	Non-herb cause highly probable				−3
Previous information on hepatotoxicity	Reaction labeled in the product characteristics				+2
	Reaction published but unlabeled				+1
	Reaction unknown				0
Response to re-administration (rechallenge)	Doubling of ALT with the herb alone, provided ALT below 5 N before re-exposure				+3
	Doubling of ALT with the herb(s) and drug(s) already given at the time of first reaction				+1
	Increase of ALT but less than N in the same conditions as for the first administration				−2
	Other situations				0
Total points for patient					

ALP Alkaline phosphatase, *ALT* Alanine aminotransferase, *AST* Aspartate aminotransferase, *CMV* Cytomegalovirus, *CT* Computer tomography, *EBV* Epstein – Barr virus, *HAV* Hepatitis A virus, *HBc* Hepatitis B core, *HBsAg* Hepatitis B surface antigen, *HBV* Hepatitis B virus, *HCV* Hepatitis C virus, *HILI* Herb-induced liver injury, *HSV* Herpes simplex virus, *MRC* Magnetic resonance cholangiography, *N* Upper limit of the normal range, *VZV* Varicella zoster virus.

with herbal-induced injury (Navarro and Lucena 2014). Although a small test-retest reliability study was conducted with modest accuracy (Navarro et al. 2012), this tool has not yet been adopted, as further investigation and validation are needed.

Table 12.5 Causality assessment scoring in the DILIN expert opinion process (Fontana et al. 2009)

Score	Causality score	Likelihood (%)	Description
1	Definite	>95	Causality is "beyond a reasonable doubt"
2	Highly likely	75–95	Causality is supported by "clear and convincing evidence"
3	Probable	50–74	Causality is supported by "the preponderance of the evidence"
4	Possible	25–49	Less than the preponderance of the evidence supports causality, but it is nevertheless possible
5	Unlikely	<25	Causality is unlikely or excluded

Conclusions

The consumption of herbal products has become more and more popular nowadays as there is a general belief that herbs are "natural" and "safe." Apart from traditional medicines, herbal products are widely consumed as dietary supplements. Compared to conventional synthetic drugs, not only the efficacy of most herbal products lacks scientific evidence, but also their toxicity is not comprehensively studied. The liver acts as the first defense barrier metabolizing and detoxifying foreign compounds. Its strategic location and features have rendered it highly susceptible to adverse and toxic effects. With increasing consumption of herbal products, especially in Western countries where the trend is relatively new and quite untested, an increasing number of herbal hepatotoxicities are being reported. More and more herbal products, which are commonly consumed sometimes even daily, are now proven or suspected to elicit hepatotoxic effects. However, in the absence of systematic epidemiological and toxicological study, and due to a practically nonexistent pharmacovigilance system for products marketed as "food supplements" or "health tonics", there remains an enormous challenge to reveal the toxicological profile of herbal products. In addition, the inherent complexity of herbal products has further increased the challenge on causality assessment of HILI.

References

Abdualmjid RJ, Sergi C (2013) Hepatotoxic botanicals – an evidence-based systematic review. J Pharm Pharm Sci 16(3):376–404

Abourashed EA, El-Alfy AT, Khan IA, Walker L (2003) Ephedra in perspective – a current review. Phytother Res 17(7):703–712

Adachi M, Saito H, Kobayashi H, Horie Y, Kato S, Yoshioka M, Ishii H (2003) Hepatic injury in 12 patients taking the herbal weight loss AIDS Chaso or Onshido. Ann Intern Med 139(6):488–492

Agarwal R, Wang ZY, Bik DP, Mukhtar H (1991) Nordihydroguaiaretic acid, an inhibitor of lipoxygenase, also inhibits cytochrome P-450-mediated monooxygenase activity in rat epidermal and hepatic microsomes. Drug Metab Dispos 19(3):620–624

Amacher DE (2014) Female gender as a susceptibility factor for drug-induced liver injury. Hum Exp Toxicol 33(9):928–939

Anderson IB, Mullen WH, Meeker JE, Khojasteh Bakht SC, Oishi S, Nelson SD, Blanc PD (1996) Pennyroyal toxicity: measurement of toxic metabolite levels in two cases and review of the literature. Ann Intern Med 124(8):726–734

Andrade RJ, Lucena MI, Fernandez MC, Pelaez G, Pachkoria K, Garcia-Ruiz E, Garcia-Munoz B, Gonzalez-Grande R, Pizarro A, Duran JA, Jimenez M, Rodrigo L, Romero-Gomez M, Navarro JM, Planas R, Costa J, Borras A, Soler A, Salmeron J, Martin-Vivaldi R (2005) Drug-induced liver injury: an analysis of 461 incidences submitted to the Spanish registry over a 10-year period. Gastroenterology 129(2):512–521

Bajaj J, Knox JF, Komorowski R, Saeian K (2003) The irony of herbal hepatitis: Ma-Huang-induced hepatotoxicity associated with compound heterozygosity for hereditary hemochromatosis. Dig Dis Sci 48(10):1925–1928

Bakerink JA, Gospe SM Jr, Dimand RJ, Eldridge MW (1996) Multiple organ failure after ingestion of pennyroyal oil from herbal tea in two infants. Pediatrics 98(5):944–947

Bateman J, Chapman RD, Simpson D (1998) Possible toxicity of herbal remedies. Scott Med J 43(1):7–15

Benichou C (1990) Criteria of drug-induced liver disorders. Report of an international consensus meeting. J Hepatol 11(2):272–276

Benichou C, Danan G, Flahault A (1993) Causality assessment of adverse reactions to drugs – II. An original model for validation of drug causality assessment methods: case reports with positive rechallenge. J Clin Epidemiol 46(11):1331–1336

Benninger J, Schneider HT, Schuppan D, Kirchner T, Hahn EG (1999) Acute hepatitis induced by greater celandine (Chelidonium majus). Gastroenterology 117(5):1234–1237

Bessone F, Hernandez N, Sanchez A, di Pace M, Arrese M, Brahm JR, Ruiz A, Arancibia J, Kershenobich D, Loaeza del Castillo A, Girala M, Paraná R, Schinoni MI, Mendez-Sanchez N, Medina-Cáliz I, González-Jiménez A, Stephens C, Robles-Díaz M, Lucena MI, Andrade RJ (2013) The Spanish-Latin American DILI network: preliminary results from a collaborative strategic initiative. J Hepatol 58(Suppl 1):S212–S213

Beuers U, Spengler U, Pape GR (1991) Hepatitis after chronic abuse of senna. Lancet 337(8737):372–373

Bjornsson ES, Bergmann OM, Bjornsson HK, Kvaran RB, Olafsson S (2013) Incidence, presentation, and outcomes in patients with drug-induced liver injury in the general population of Iceland. Gastroenterology 144(7):1419–1425, 1425 e1411-1413; quiz e1419-1420

Bonkovsky HL (2006) Hepatotoxicity associated with supplements containing Chinese green tea (Camellia sinensis). Ann Intern Med 144(1):68–71

Borum ML (2001) Fulminant exacerbation of autoimmune hepatitis after the use of Ma Huang. Am J Gastroenterol 96(5):1654–1655

Bunchorntavakul C, Reddy KR (2013) Review article: herbal and dietary supplement hepatotoxicity. Aliment Pharmacol Ther 37(1):3–17

But PP, Tomlinson B, Lee KL (1996) Hepatitis related to the Chinese medicine Shou-wu-pian manufactured from Polygonum multiflorum. Vet Hum Toxicol 38(4):280–282

Bye SN, Dutton MF (1991) The inappropriate use of traditional medicines in South Africa. J Ethnopharmacol 34(2–3):253–259

Capdevila J, Gil L, Orellana M, Marnett LJ, Mason JI, Yadagiri P, Falck JR (1988) Inhibitors of cytochrome P-450-dependent arachidonic acid metabolism. Arch Biochem Biophys 261(2):257–263

Castot A, Larrey D (1992) Hepatitis observed during a treatment with a drug or tea containing Wild Germander. Evaluation of 26 cases reported to the Regional Centers of Pharmacovigilance. Gastroenterol Clin Biol 16(12):916–922

Chalasani N, Fontana RJ, Bonkovsky HL, Watkins PB, Davern T, Serrano J, Yang H, Rochon J (2008) Causes, clinical features, and outcomes from a prospective study of drug-induced liver injury in the United States. Gastroenterology 135(6):1924–1934

Chalasani N, Vuppalanchi R, Navarro V, Fontana R, Bonkovsky H, Barnhart H, Kleiner DE, Hoofnagle JH (2012) Acute liver injury due to flavocoxid (Limbrel), a medical food for osteoarthritis: a case series. Ann Intern Med 156(12):857–860, W297-300

Chan TY, Chan AY, Critchley JA (1992) Hospital admissions due to adverse reactions to Chinese herbal medicines. J Trop Med Hyg 95(4):296–298

Chen KK, Schmidt CF (1926) The action and clinical use of ephedrine: an alkaloid isolated from the Chinese drug Ma Huang. JAMA 87(11):836–842

Chitturi S, Farrell GC (2000) Herbal hepatotoxicity: an expanding but poorly defined problem. J Gastroenterol Hepatol 15(10):1093–1099

Cohen DL, Del Toro Y (2008) A case of valerian-associated hepatotoxicity. J Clin Gastroenterol 42(8):961–962

Colombo ML, Bosisio E (1996a) Pharmacological activities of Chelidonium majus L. (Papaveraceae). Pharmacol Res 33(2):127–134

Colombo ML, Bosisio E (1996b) Pharmacological activities of chelidonium majusl. Papaveraceae 33(2):127–134

Crijns AP, de Smet PA, van den Heuvel M, Schot BW, Haagsma EB (2002) Acute hepatitis after use of a herbal preparation with greater celandine (Chelidonium majus). Ned Tijdschr Geneeskd 146(3):124–128

Cupp MJ (2000) Toxicology and clinical pharmacology of herbal products, Forensic science and medicine. Humana Press, Totowa

Danan G, Benichou C (1993) Causality assessment of adverse reactions to drugs – I. A novel method based on the conclusions of international consensus meetings: application to drug-induced liver injuries. J Clin Epidemiol 46(11):1323–1330

Daniele C, Dahamna S, Firuzi O, Sekfali N, Saso L, Mazzanti G (2005) Atractylis gummifera L. poisoning: an ethnopharmacological review. J Ethnopharmacol 97(2):175–181

De Berardinis V, Moulis C, Maurice M, Beaune P, Pessayre D, Pompon D, Loeper J (2000) Human microsomal epoxide hydrolase is the target of germander-induced autoantibodies on the surface of human hepatocytes. Mol Pharmacol 58(3):542–551

De Smet PA (2002) Herbal remedies. N Engl J Med 347(25):2046–2056

De Valle MB, Av Klinteberg V, Alem N, Olsson R, Bjornsson E (2006) Drug-induced liver injury in a Swedish university hospital out-patient hepatology clinic. Aliment Pharmacol Ther 24(8):1187–1195

DeLeve LD, Shulman HM, McDonald GB (2002) Toxic injury to hepatic sinusoids: sinusoidal obstruction syndrome (veno-occlusive disease). Semin Liver Dis 22(1):27–42

Devarbhavi H, Dierkhising R, Kremers WK, Sandeep MS, Karanth D, Adarsh CK (2010) Single-center experience with drug-induced liver injury from India: causes, outcome, prognosis, and predictors of mortality. Am J Gastroenterol 105(11):2396–2404

Divinsky M (2002) Case report: jin bu huan – not so benign herbal medicine. Can Fam Physician 48:1640–1642

Doehmer J, Eisenbraun J (2012) Assessment of extracts from mistletoe (Viscum album) for herb-drug interaction by inhibition and induction of cytochrome P450 activities. Phytother Res 26(1):11–17

Dourakis SP, Papanikolaou IS, Tzemanakis EN, Hadziyannis SJ (2002) Acute hepatitis associated with herb (Teucrium capitatum L.) administration. Eur J Gastroenterol Hepatol 14(6):693–695

Early D (1961) Pennyroyal: a rare cause of epilepsy. Lancet 278(7202):580–581

EMA (2009) Assessment report on Peumus boldus Molina, folium. European Medicines Agency, London

Ernst E (2007) A re-evaluation of kava (Piper methysticum). Br J Clin Pharmacol 64(4):415–417

Estes J, Stolpman D, Olyaei A et al (2003) HIgh prevalence of potentially hepatotoxic herbal supplement use in patients with fulminant hepatic failure. Arch Surg 138(8):852–858

Falodun A (2010) Herbal medicine in Africa-distribution, standardization and prospects. Res J Phytochem 4(3):154–161

Fau D, Lekehal M, Farrell G, Moreau A, Moulis C, Feldmann G, Haouzi D, Pessayre D (1997) Diterpenoids from germander, an herbal medicine, induce apoptosis in isolated rat hepatocytes. Gastroenterology 113(4):1334–1346

Favreau JT, Ryu ML, Braunstein G, Orshansky G, Park SS, Coody GL, Love LA, Fong TL (2002) Severe hepatotoxicity associated with the dietary supplement LipoKinetix. Ann Intern Med 136(8):590–595

Fontana RJ, Watkins PB, Bonkovsky HL, Chalasani N, Davern T, Serrano J, Rochon J (2009) Drug-induced liver injury network (DILIN) prospective study: rationale, design and conduct. Drug Saf 32(1):55–68

Fraquelli M, Colli A, Cocciolo M, Conte D (2000) Adult syncytial giant cell chronic hepatitis due to herbal remedy. J Hepatol 33(3):505–508

García-Cortés M, Borraz Y, Lucena MI, Peláez G, Salmerón J, Diago M, MartínezSierra MC, Navarro JM, Planas R, Soria MJ, Bruguera M, Andrade RJ (2008) Liver injury induced by "natural remedies": an analysis of cases submitted to the Spanish Liver Toxicity Registry. Rev Esp Enferm Dig (Madrid) 100(11):688–695

Gold J, Cates W Jr (1980) Herbal abortifacients. JAMA 243(13):1365–1366

Gordon P, Khojasteh SC (2015) A decades-long investigation of acute metabolism-based hepatotoxicity by herbal constituents: a case study of pennyroyal oil. Drug Metab Rev 47(1):12–20

Gordon WP, Forte AJ, McMurtry RJ, Gal J, Nelson SD (1982) Hepatotoxicity and pulmonary toxicity of pennyroyal oil and its constituent terpenes in the mouse. Toxicol Appl Pharmacol 65(3):413–424

Gordon WP, Huitric AC, Seth CL, McClanahan RH, Nelson SD (1987) The metabolism of the abortifacient terpene, (R)-(+)-pulegone, to a proximate toxin, menthofuran. Drug Metab Dispos 15(5):589–594

Gori L, Galluzzi P, Mascherini V, Gallo E, Lapi F, Menniti-Ippolito F, Raschetti R, Mugelli A, Vannacci A, Firenzuoli F (2011) Two contemporary cases of hepatitis associated with *Teucrium chamaedrys* L. decoction use: case reports and review of literature. Basic Clin Pharmacol Toxicol 109(6):521–526

Gray SL, Lackey BR, Tate PL, Riley MB, Camper ND (2004) Mycotoxins in root extracts of American and Asian ginseng bind estrogen receptors alpha and beta. Exp Biol Med (Maywood) 229(6):560–568

Greving I, Meister V, Monnerjahn C, Muller KM, May B (1998) Chelidonium majus: a rare reason for severe hepatotoxic reaction. Pharmacoepidemiol Drug Saf 7(Suppl 1):S66–S69

Gunby P (1979) Plant known for centuries still causes problems today. JAMA 241(21):2246–2247

Haller CA, Benowitz NL (2000) Adverse cardiovascular and central nervous system events associated with dietary supplements containing ephedra alkaloids. N Engl J Med 343(25):1833–1838

Hamid S, Rojter S, Vierling J (1997) Protracted cholestatic hepatitis after the use of prostata. Ann Intern Med 127(2):169–170

Hardeman E, Van Overbeke L, Ilegems S, Ferrante M (2008) Acute hepatitis induced by greater celandine (*Chelidonium majus*). Acta Gastroenterol Belg 71(2):281–282

Harvey J, Colin-Jones DG (1981) Mistletoe hepatitis. Br Med J (Clin Res Ed) 282(6259):186–187

Hasani-Ranjbar S, Nayebi N, Larijani B, Abdollahi M (2010) A systematic review of the efficacy and safety of Teucrium species; from anti-oxidant to anti-diabetic effects. Int J Pharmacol 6(4):315–325

Horowitz RS, Feldhaus K, Dart RC, Stermitz FR, Beck JJ (1996) The clinical spectrum of Jin Bu Huan toxicity. Arch Intern Med 156(8):899–903

Ibanez L, Perez E, Vidal X, Laporte JR (2002) Prospective surveillance of acute serious liver disease unrelated to infectious, obstructive, or metabolic diseases: epidemiological and clinical features, and exposure to drugs. J Hepatol 37(5):592–600
Itoh S, Marutani K, Nishijima T, Matsuo S, Itabashi M (1995) Liver injuries induced by herbal medicine, syo-saiko-to (xiao-chai-hu-tang). Dig Dis Sci 40(8):1845–1848
Jack DB (1997) One hundred years of aspirin. Lancet 350(9075):437–439
Jacobsson I, Jonsson AK, Gerden B, Hagg S (2009) Spontaneously reported adverse reactions in association with complementary and alternative medicine substances in Sweden. Pharmacoepidemiol Drug Saf 18(11):1039–1047
Jaeschke H, Gores GJ, Cederbaum AI, Hinson JA, Pessayre D, Lemasters JJ (2002) Mechanisms of hepatotoxicity. Toxicol Sci 65(2):166–176
Jorge OA, Jorge AD (2005) Hepatotoxicity associated with the ingestion of *Centella asiatica*. Rev Esp Enferm Dig 97(2):115–124
Kamiyama T, Nouchi T, Kojima S, Murata N, Ikeda T, Sato C (1997) Autoimmune hepatitis triggered by administration of an herbal medicine. Am J Gastroenterol 92(4):703–704
Karch S (2000) Ma huang and the ephedra alkaloids. In: Cupp M (ed) Toxicology and clinical pharmacology of herbal products. Forensic Science and Medicine. Humana Press, Totowa, New Jersey. pp 11–30
Kassler WJ, Blanc P, Greenblatt R (1991) The use of medicinal herbs by human immunodeficiency virus-infected patients. Arch Intern Med 151(11):2281–2288
Katz M, Saibil F (1990) Herbal hepatitis: subacute hepatic necrosis secondary to chaparral leaf. J Clin Gastroenterol 12(2):203–206
Kennedy J (2005) Herb and supplement use in the US adult population. Clin Ther 27(11):1847–1858
Khojasteh-Bakht SC, Chen W, Koenigs LL, Peter RM, Nelson SD (1999) Metabolism of (R)-(+)-pulegone and (R)-(+)-menthofuran by human liver cytochrome P-450s: evidence for formation of a furan epoxide. Drug Metab Dispos 27(5):574–580
Kneifel W, Czech E, Kopp B (2002) Microbial contamination of medicinal plants – a review. Planta Med 68(1):5–15
Kocaman O, Hulagu S, Senturk O (2008) Echinacea-induced severe acute hepatitis with features of cholestatic autoimmune hepatitis. Eur J Intern Med 19(2):148
Koga Y, Yoshida I, Kimura A, Yoshino M, Yamashita F, Sinniah D (1987) Inhibition of mitochondrial functions by margosa oil: possible implications in the pathogenesis of Reye's syndrome. Pediatr Res 22(2):184–187
Korth C (2014) Drug-induced hepatotoxicity of select herbal therapies. J Pharm Pract 27(6):567–572
Laliberté L, Villeneuve JP (1996) Hepatitis after the use of germander, a herbal remedy. CMAJ 154(11):1689–1692
Larrey D (1997) Hepatotoxicity of herbal remedies. J Hepatol 26(Suppl 1):47–51
Larrey D, Vial T, Pauwels A, Castot A, Biour M, David M, Michel H (1992) Hepatitis after germander (*Teucrium chamaedrys*) administration: another instance of herbal medicine hepatotoxicity. Ann Intern Med 117(2):129–132
Lee WM (2003) Drug-induced hepatotoxicity. N Engl J Med 349(5):474–485
Lei J, Yu J, Yu H, Liao Z (2008) Composition, cytotoxicity and antimicrobial activity of essential oil from *Dictamnus dasycarpus*. Food Chem 107(3):1205–1209
Lekehal M, Pessayre D, Lereau JM, Moulis C, Fouraste I, Fau D (1996) Hepatotoxicity of the herbal medicine germander: metabolic activation of its furano diterpenoids by cytochrome P450 3A Depletes cytoskeleton-associated protein thiols and forms plasma membrane blebs in rat hepatocytes. Hepatology 24(1):212–218
Levi M, Guchelaar HJ, Woerdenbag HJ, Zhu YP (1998) Acute hepatitis in a patient using a Chinese herbal tea – a case report. Pharm World Sci 20(1):43–44

Levitsky J, Alli TA, Wisecarver J, Sorrell MF (2005) Fulminant liver failure associated with the use of black cohosh. Dig Dis Sci 50(3):538–539

Licata A, Macaluso FS, Craxi A (2013) Herbal hepatotoxicity: a hidden epidemic. Intern Emerg Med 8(1):13–22

Linnebur SA, Rapacchietta OC, Vejar M (2010) Hepatotoxicity associated with Chinese skullcap contained in move free advanced dietary supplement: two case reports and review of the literature. Pharmacotherapy 30(7):750, 258e-262e

Liu GQ, Algeri S, Garattini S (1982) D-L-tetrahydropalmatine as monoamine depletor. Arch Int Pharmacodyn Ther 258(1):39–50

Loeper J, Descatoire V, Letteron P, Moulis C, Degott C, Dansette P, Fau D, Pessayre D (1994) Hepatotoxicity of germander in mice. Gastroenterology 106(2):464–472

Lude S, Torok M, Dieterle S, Knapp AC, Kaeufeler R, Jaggi R, Spornitz U, Krahenbuhl S (2007) Hepatic effects *of Cimicifuga racemosa* extract *in vivo* and *in vitro*. Cell Mol Life Sci 64(21):2848–2857

Lynch CR, Folkers ME, Hutson WR (2006) Fulminant hepatic failure associated with the use of black cohosh: a case report. Liver Transpl 12(6):989–992

Mantani N, Kogure T, Sakai S, Goto H, Shibahara N, Kita T, Shimada Y, Terasawa K (2002) Incidence and clinical features of liver injury related to Kampo (Japanese herbal) medicine in 2,496 cases between 1979 and 1999: problems of the lymphocyte transformation test as a diagnostic method. Phytomedicine 9(4):280–287

Maria VA, Victorino RM (1997) Development and validation of a clinical scale for the diagnosis of drug-induced hepatitis. Hepatology 26(3):664–669

Mathews JM, Etheridge AS, Black SR (2002) Inhibition of human cytochrome P450 activities by kava extract and kavalactones. Drug Metab Dispos 30(11):1153–1157

Mathews JM, Etheridge AS, Valentine JL, Black SR, Coleman DP, Patel P, So J, Burka LT (2005) Pharmacokinetics and disposition of the kavalactone kawain: interaction with kava extract and kavalactones *in vivo* and *in vitro*. Drug Metab Dispos 33(10):1555–1563

Mattei A, Rucay P, Samuel D, Feray C, Reynes M, Bismuth H (1995) Liver transplantation for severe acute liver failure after herbal medicine (*Teucrium polium*) administration. J Hepatol 22(5):597

Medina-Cáliz I, González-Jiménez A, Bessone F, Hernández N, Sánchez A, Di Pace M, Arrese M, Brahm JR, Ruíz A, Arancibia J, Kershenobich D, Loaeza A, Girala M, Mendez-Sanchez N, Dávalos M, Lizarzabal M, Mengual E, Stephens C, Robles-Díaz M, Andrade RJ, Lucena MI (2013) Variations in drug-induced liver injury (DILI) between different prospective DILI registries. Clin Ther 35(8):e24

Menichini F, Conforti F, Rigano D, Formisano C, Piozzi F, Senatore F (2009) Phytochemical composition, anti-inflammatory and antitumour activities of four *Teucrium* essential oils from Greece. Food Chem 115(2):679–686

Miller GM, Stripp R (2007) A study of western pharmaceuticals contained within samples of Chinese herbal/patent medicines collected from New York City's Chinatown. Leg Med (Tokyo) 9(5):258–264

Millonig G, Stadlmann S, Vogel W (2005) Herbal hepatotoxicity: acute hepatitis caused by a Noni preparation (*Morinda citrifolia*). Eur J Gastroenterol Hepatol 17(4):445–447

Mitchell JR, Jollow DJ, Potter WZ, Davis DC, Gillette JR, Brodie BB (1973) Acetaminophen-induced hepatic necrosis. I. Role of drug metabolism. J Pharmacol Exp Ther 187(1):185–194

Moro PA, Cassetti F, Giugliano G, Falce MT, Mazzanti G, Menniti-Ippolito F, Raschetti R, Santuccio C (2009) Hepatitis from Greater celandine (*Chelidonium majus* L.): review of literature and report of a new case. J Ethnopharmacol 124(2):328–332

Mostefa-Kara N, Pauwels A, Pines E, Biour M, Levy VG (1992) Fatal hepatitis after herbal tea. Lancet 340(8820):674

Nadir A, Agrawal S, King PD, Marshall JB (1996) Acute hepatitis associated with the use of a Chinese herbal product, ma-huang. Am J Gastroenterol 91(7):1436–1438

Nadir A, Reddy D, Van Thiel DH (2000) Cascara sagrada-induced intrahepatic cholestasis causing portal hypertension: case report and review of herbal hepatotoxicity. Am J Gastroenterol 95(12):3634–3637

Naranjo CA, Busto U, Sellers EM, Sandor P, Ruiz I, Roberts EA, Janecek E, Domecq C, Greenblatt DJ (1981) A method for estimating the probability of adverse drug reactions. Clin Pharmacol Ther 30(2):239–245

Navarro VJ (2009) Herbal and dietary supplement hepatotoxicity. Semin Liver Dis 29(4):373–382

Navarro VJ, Lucena MI (2014) Hepatotoxicity induced by herbal and dietary supplements. Semin Liver Dis 34(2):172–193

Navarro VJ, Seeff LB (2013) Liver injury induced by herbal complementary and alternative medicine. Clin Liver Dis 17(4):715–735

Navarro VJ, Barnhart HX, Bonkovsky HL, Reddy KR, Seeff L, Serrano J, Talwalkar JA, Vega M, Vuppalanchi R (2012) Diagnosing hepatotoxicity attributable to herbal & dietary supplements (HDS): test-retest reliability of a novel causality assessment tool. J Hepatol 56(Suppl 2):S536

Navarro VJ, Barnhart H, Bonkovsky HL, Davern T, Fontana RJ, Grant L, Reddy KR, Seeff LB, Serrano J, Sherker AH, Stolz A, Talwalkar J, Vega M, Vuppalanchi R (2014) Liver injury from herbals and dietary supplements in the U.S. Drug-Induced Liver Injury Network. Hepatology 60(4):1399–1408

Neff GW, Reddy KR, Durazo FA, Meyer D, Marrero R, Kaplowitz N (2004) Severe hepatotoxicity associated with the use of weight loss diet supplements containing ma huang or usnic acid. J Hepatol 41(6):1062–1064

Nelson SD, McClanahan RH, Thomassen D, Gordon WP, Knebel N (1992) Investigations of mechanisms of reactive metabolite formation from (R)-(+)-pulegone. Xenobiotica 22(9–10):1157–1164

Obatomi DK, Bach PH (1998) Biochemistry and toxicology of the diterpenoid glycoside atractyloside. Food Chem Toxicol 36(4):335–346

Pan SY, Litscher G, Gao SH, Zhou SF, Yu ZL, Chen HQ, Zhang SF, Tang MK, Sun JN, Ko KM (2014) Historical perspective of traditional indigenous medical practices: the current renaissance and conservation of herbal resources. Evid Based Complement Alternat Med 2014:525340

Park GJ, Mann SP, Ngu MC (2001) Acute hepatitis induced by Shou-Wu-Pian, a herbal product derived from *Polygonum multiflorum*. J Gastroenterol Hepatol 16(1):115–117

Perez Alvarez J, Saez-Royuela F, Gento Pena E, Lopez Morante A, Velasco Oses A, Martin Lorente J (2001) Acute hepatitis due to ingestion of *Teucrium chamaedrys* infusions. Gastroenterol Hepatol 24(5):240–243

Picciotto A, Campo N, Brizzolara R, Giusto R, Guido G, Sinelli N, Lapertosa G, Celle G (1998) Chronic hepatitis induced by Jin Bu Huan. J Hepatol 28(1):165–167

Piozzi F, Rodriguez B, Savona G (1987) Advances in the chemistry of the furanoditerpenoids for Teucrium species. Heterocycles 25:807–841

Piozzi F, Bruno M, Ciriminna R, Fazio C, Vassallo N, Arnold NA, de la Torre MC, Rodriguez B (1997) Putative hepatotoxic neoclerodane diterpenoids from Teucrium species. Planta Med 63(5):483–484

Piscaglia F, Leoni S, Venturi A, Graziella F, Donati G, Bolondi L (2005) Caution in the use of boldo in herbal laxatives: a case of hepatotoxicity. Scand J Gastroenterol 40(2):236–239

Poon WT, Chau TL, Lai CK, Tse KY, Chan YC, Leung KS, Chan YW (2008) Hepatitis induced by *Teucrium viscidum*. Clin Toxicol (Phila) 46(9):819–822

Pramyothin P, Janthasoot W, Pongnimitprasert N, Phrukudom S, Ruangrungsi N (2004) Hepatotoxic effect of (+)usnic acid from Usnea siamensis Wainio in rats, isolated rat hepatocytes and isolated rat liver mitochondria. J Ethnopharmacol 90(2–3):381–387

Reuben A, Koch DG, Lee WM (2010) Drug-induced acute liver failure: results of a U.S. multicenter, prospective study. Hepatology 52(6):2065–2076

Rifai K, Flemming P, Manns MP, Trautwein C (2006) Severe drug hepatitis caused by Chelidonium. Internist (Berl) 47(7):749–751

Robinson MM, Zhang X (2011) Traditional medicines: global situation, issues and challenges. The World Medicines Situation 2011. World Health Organization, Geneva

Rochon J, Protiva P, Seeff LB, Fontana RJ, Liangpunsakul S, Watkins PB, Davern T, McHutchison JG (2008) Reliability of the Roussel Uclaf causality assessment method for assessing causality in drug-induced liver injury. Hepatology 48(4):1175–1183

Rockey DC, Seeff LB, Rochon J, Freston J, Chalasani N, Bonacini M, Fontana RJ, Hayashi PH, for the USD-ILIN, (2010) Causality assessment in drug-induced liver injury using a structured expert opinion process: comparison to the Roussel-Uclaf causality assessment method. Hepatology 51(6):2117–2126

Rossi S, Navarro VJ (2014) Herbs and liver injury: a clinical perspective. Clin Gastroenterol Hepatol 12(7):1069–1076

Roux P, Le Saux A, Fiore C, Schwimmer C, Dianoux AC, Trezeguet V, Vignais PV, Lauquin GJ, Brandolin G (1996) Fluorometric titration of the mitochondrial ADP/ATP carrier protein in muscle homogenate with atractyloside derivatives. Anal Biochem 234(1):31–37

Rowe A, Ramzan I (2012) Are mould hepatotoxins responsible for kava hepatotoxicity? Phytother Res 26(11):1768–1770

Ruan J, Liao C, Ye Y, Lin G (2014) Lack of metabolic activation and predominant formation of an excreted metabolite of nontoxic platynecine-type pyrrolizidine alkaloids. Chem Res Toxicol 27(1):7–16

Russmann S, Lauterburg BH, Helbling A (2001) Kava hepatotoxicity. Ann Intern Med 135(1):68–69

Russmann S, Kullak-Ublick GA, Grattagliano I (2009) Current concepts of mechanisms in drug-induced hepatotoxicity. Curr Med Chem 16(23):3041–3053

Russo MW, Galanko JA, Shrestha R, Fried MW, Watkins P (2004) Liver transplantation for acute liver failure from drug induced liver injury in the United States. Liver Transpl 10(8):1018–1023

Sarma DN, Barrett ML, Chavez ML, Gardiner P, Ko R, Mahady GB, Marles RJ, Pellicore LS, Giancaspro GI, Low Dog T (2008) Safety of green tea extracts: a systematic review by the US Pharmacopeia. Drug Saf 31(6):469–484

Savvidou S, Goulis J, Giavazis I, Patsiaoura K, Hytiroglou P, Arvanitakis C (2007) Herb-induced hepatitis by Teucrium polium L.: report of two cases and review of the literature. Eur J Gastroenterol Hepatol 19(6):507–511

Schulze J, Raasch W, Siegers CP (2003) Toxicity of kava pyrones, drug safety and precautions – a case study. Phytomedicine 10(Suppl 4):68–73

Schuppan D, Jia JD, Brinkhaus B, Hahn EG (1999) Herbal products for liver diseases: a therapeutic challenge for the new millennium. Hepatology 30(4):1099–1104

Seeff LB, Lindsay KL, Bacon BR, Kresina TF, Hoofnagle JH (2001) Complementary and alternative medicine in chronic liver disease. Hepatology 34(3):595–603

Seeff L, Stickel F, Navarro VJ (2013) Hepatotoxicity of herbals and dietary supplements. In: DeLeve NKD (ed) Drug-induced liver disease, 3rd edn. Academic Press, Boston, pp 631–657

Seeff LB, Bonkovsky HL, Navarro VJ, Wang G (2014) Herbals and the liver – a review of adverse effects and mechanisms. Gastroenterology 148:517–532

Segelman AB, Segelman FP, Karliner J, Sofia RD (1976) Sassafras and herb tea. Potential health hazards. JAMA 236(5):477

Shaw D, Graeme L, Pierre D, Elizabeth W, Kelvin C (2012) Pharmacovigilance of herbal medicine. J Ethnopharmacol 140(3):513–518

Sheikh NM, Philen RM, Love LA (1997) Chaparral-associated hepatotoxicity. Arch Intern Med 157(8):913–919

Singh YN, Devkota AK, Sneeden DC, Singh KK, Halaweish F (2007) Hepatotoxicity potential of saw palmetto (Serenoa repens) in rats. Phytomedicine 14(2–3):204–208

Sinniah D, Baskaran G (1981) Margosa oil poisoning as a cause of Reye's syndrome. Lancet 1(8218):487–489

Skoulidis F, Alexander GJ, Davies SE (2005) Ma huang associated acute liver failure requiring liver transplantation. Eur J Gastroenterol Hepatol 17(5):581–584

Smith LW, Culvenor CC (1981) Plant sources of hepatotoxic pyrrolizidine alkaloids. J Nat Prod 44(2):129–152

Soares Mdo C (2004) Would Sacaca, Croton cajucara Benth (Euphorbiaceae) be an hepatotoxic plant like Germander, Teucrium chamaedrys L. (Labiatae)? Rev Soc Bras Med Trop 37(Suppl 2):96–97

Sohni YR (2002) The toxicity of *Callilepis laureola*, a South African traditional herbal medicine. Clin Biochem 35(6):499

Speijers GJA (2001) WHO Food Additives Series 46: Pulegone and related substances. National Institute of Public Health and the Environment, Section on Public Health, Centre for Substances and Risk Assessment. http://www.inchem.org/documents/jecfa/jecmono/v46je10.htm. Accessed 23 Nov 2015

Stadlbauer V, Fickert P, Lackner C, Schmerlaib J, Krisper P, Trauner M, Stauber RE (2005) Hepatotoxicity of NONI juice: report of two cases. World J Gastroenterol 11(30):4758–4760

Stadlbauer V, Weiss S, Payer F, Stauber RE (2008) Herbal does not at all mean innocuous: the sixth case of hepatotoxicity associated with morinda citrifolia (noni). Am J Gastroenterol 103(9):2406–2407

Starakis I, Siagris D, Leonidou L, Mazokopakis E, Tsamandas A, Karatza C (2006) Hepatitis caused by the herbal remedy *Teucrium polium* L. Eur J Gastroenterol Hepatol 18(6):681–683

Stedman C (2002) Herbal hepatotoxicity. Semin Liver Dis 22(2):195–206

Stewart MJ, Steenkamp V (2000) The biochemistry and toxicity of atractyloside: a review. Ther Drug Monit 22(6):641–649

Stickel F, Seitz HK (2000) The efficacy and safety of comfrey. Public Health Nutr 3(4A):501–508

Stickel F, Egerer G, Seitz HK (2000) Hepatotoxicity of botanicals. Public Health Nutr 3(2):113–124

Stickel F, Baumuller HM, Seitz K, Vasilakis D, Seitz G, Seitz HK, Schuppan D (2003a) Hepatitis induced by Kava (Piper methysticum rhizoma). J Hepatol 39(1):62–67

Stickel F, Poschl G, Seitz HK, Waldherr R, Hahn EG, Schuppan D (2003b) Acute hepatitis induced by Greater Celandine (*Chelidonium majus*). Scand J Gastroenterol 38(5):565–568

Stickel F, Patsenker E, Schuppan D (2005) Herbal hepatotoxicity. J Hepatol 43(5):901–910

Stickel F, Droz S, Patsenker E, Bogli-Stuber K, Aebi B, Leib SL (2009) Severe hepatotoxicity following ingestion of Herbalife nutritional supplements contaminated with *Bacillus subtilis*. J Hepatol 50(1):111–117

Suk KT, Kim DJ, Kim CH, Park SH, Yoon JH, Kim YS, Baik GH, Kim JB, Kweon YO, Kim BI, Kim SH, Kim IH, Kim JH, Nam SW, Paik YH, Suh JI, Sohn JH, Ahn BM, Um SH, Lee HJ, Cho M, Jang MK, Choi SK, Hwang SG, Sung HT, Choi JY, Han KH (2012) A prospective nationwide study of drug-induced liver injury in Korea. Am J Gastroenterol 107(9):1380–1387

Sullivan JB Jr, Rumack BH, Thomas H Jr, Peterson RG, Bryson P (1979) Pennyroyal oil poisoning and hepatotoxicity. JAMA 242(26):2873–2874

Teschke R (2006) Acute hepatitis due to kava-kava and St. John's Wort: an immune-mediated mechanism? Dtsch Med Wochenschr 131(34–35):1880–1881; author reply 1882-1883

Teschke R (2010) Kava hepatotoxicity – a clinical review. Ann Hepatol 9(3):251–265

Teschke R, Schulze J (2012) Suspected herbal hepatotoxicity: requirements for appropriate causality assessment by the US Pharmacopeia. Drug Saf 35(12):1091–1097

Teschke R, Schwarzenboeck A, Hennermann KH (2008a) Causality assessment in hepatotoxicity by drugs and dietary supplements. Br J Clin Pharmacol 66(6):758–766

Teschke R, Schwarzenboeck A, Hennermann KH (2008b) Kava hepatotoxicity: a clinical survey and critical analysis of 26 suspected cases. Eur J Gastroenterol Hepatol 20(12):1182–1193
Teschke R, Qiu SX, Lebot V (2011a) Herbal hepatotoxicity by kava: update on pipermethystine, flavokavain B, and mould hepatotoxins as primarily assumed culprits. Dig Liver Dis 43(9):676–681
Teschke R, Schmidt-Taenzer W, Wolff A (2011b) Spontaneous reports of assumed herbal hepatotoxicity by black cohosh: is the liver-unspecific Naranjo scale precise enough to ascertain causality? Pharmacoepidemiol Drug Saf 20(6):567–582
Teschke R, Frenzel C, Glass X, Schulze J, Eickhoff A (2012) Greater Celandine hepatotoxicity: a clinical review. Ann Hepatol 11(6):838–848
Teschke R, Frenzel C, Glass X, Schulze J, Eickhoff A (2013a) Herbal hepatotoxicity: a critical review. Br J Clin Pharmacol 75(3):630–636
Teschke R, Schwarzenboeck A, Eickhoff A, Frenzel C, Wolff A, Schulze J (2013b) Clinical and causality assessment in herbal hepatotoxicity. Expert Opin Drug Saf 12(3):339–366
Thomassen D, Knebel N, Slattery JT, McClanahan RH, Nelson SD (1992) Reactive intermediates in the oxidation of menthofuran by cytochromes P-450. Chem Res Toxicol 5(1):123–130
Uc A, Bishop WP, Sanders KD (2000) Camphor hepatotoxicity. South Med J 93(6):596–598
Vallance WB (1955) Pennyroyal poisoning: a fatal case. Lancet 269(6895):850–851
Vancompernolle K, Van Herreweghe F, Pynaert G, Van de Craen M, De Vos K, Totty N, Sterling A, Fiers W, Vandenabeele P, Grooten J (1998) Atractyloside-induced release of cathepsin B, a protease with caspase-processing activity. FEBS Lett 438(3):150–158
Vassiliadis T, Anagnostis P, Patsiaoura K, Giouleme O, Katsinelos P, Mpoumponaris A, Eugenidis N (2009) Valeriana hepatotoxicity. Sleep Med 10(8):935
Wai CT, Tan BH, Chan CL, Sutedja DS, Lee YM, Khor C, Lim SG (2007) Drug-induced liver injury at an Asian center: a prospective study. Liver Int 27(4):465–474
Wainwright J, Schonland MM, Candy HA (1977) Toxicity of *Callilepis laureola*. S Afr Med J 52(8):313–315
Wang C, Zhou J, Wang S, Ye M, Fan G, Zou H, Jiang C (2010) Shotgun approach based comparative proteomic analysis of levo-tetrahydropalmatine-induced apoptosis in hepatocytes. Toxicol Lett 194(1–2):8–15
Whiting PW, Clouston A, Kerlin P (2002) Black cohosh and other herbal remedies associated with acute hepatitis. Med J Aust 177(8):440–443
WHO (2002) WHO traditional medicine strategy 2002–2005. World Health Organization, Geneva
WHO (2007) Assessment of the risk of hepatotoxicity with kava products. World Health Organization, Geneva
Willmot F, Robertson G (1920) Senecio disease or cirrhosis of the liver due to senecio poisoning. Lancet 196(5069):848–849
Wong MK, Tan P, Wee YC (1993) Heavy metals in some Chinese herbal plants. Biol Trace Elem Res 36(2):135–142
Woolf GM, Petrovic LM, Rojter SE, Wainwright S, Villamil FG, Katkov WN, Michieletti P, Wanless IR, Stermitz FR, Beck JJ, Vierling JM (1994) Acute hepatitis associated with the Chinese herbal product jin bu huan. Ann Intern Med 121(10):729–735
Yang HN, Kim DJ, Kim YM, Kim BH, Sohn KM, Choi MJ, Choi YH (2010) Aloe-induced toxic hepatitis. J Korean Med Sci 25(3):492–495
Yang L, Aronsohn A, Hart J, Jensen D (2012) Herbal hepatoxicity from Chinese skullcap: a case report. World J Hepatol 4(7):231–233
Yarnell E, Abascal K (2014) Hepatotoxicity of botanicals. Altern Complement Ther 20(3):136–144
Yu EL, Sivagnanam M, Ellis L, Huang JS (2011) Acute hepatotoxicity after ingestion of Morinda citrifolia (Noni Berry) juice in a 14-year-old boy. J Pediatr Gastroenterol Nutr 52(2):222–224

Yuce B, Gulberg V, Diebold J, Gerbes AL (2006) Hepatitis induced by Noni juice from Morinda citrifolia: a rare cause of hepatotoxicity or the tip of the iceberg? Digestion 73(2–3):167–170
Zhang LY, Rowe A, Ramzan I (2011) Does inflammation play a role in kava hepatotoxicity? Phytother Res 25(4):629–630
Zhang L, Yan J, Liu X, Ye Z, Yang X, Meyboom R, Chan K, Shaw D, Duez P (2012) Pharmacovigilance practice and risk control of Traditional Chinese Medicine drugs in China: current status and future perspective. J Ethnopharmacol 140(3):519–525
Zhou Y, Yang L, Liao Z, He X, Guo H (2013) Epidemiology of drug-induced liver injury in China: a systematic analysis of the Chinese literature including 21,789 patients. Eur J Gastroenterol Hepatol 25(7):825–829
Zimmerman H (1999) Hepatotoxicity: the adverse effects of drugs and other chemicals on the liver. Lippincott, Philadelphia

Chapter 13
Nephrotoxicity of Herbal Products

Thomas Baudoux and Joëlle L. Nortier

Abstract The kidney is one of the major routes of excretion of drugs and metabolites, and given its high blood flow and metabolic activity, it is highly susceptible to injury by toxic drugs and herbs. In addition, high concentrations of toxic metabolites are in close contact with tubular cells due to urine concentration, especially during a state of fluid deprivation. The incidence of nephrotoxicity related to alternative medicines is not known. However, regarding the large number of individuals consuming herbs as alternative remedies and regarding the lack of control of over-the-counter products, clinicians should consider alternative medicine consumption in the differential diagnosis of unexplained kidney injury.

This chapter will describe the major forms of renal involvement associated with the use of traditional medicinal products, including acute kidney injury, tubular function defects, electrolyte disturbances, systemic hypertension, chronic kidney disease, renal papillary necrosis, lithiasis, and urothelial cancer. The most famous example of kidney injury caused by herbal products is aristolochic acid nephropathy, which was first reported in 1993; it eventually involved hundreds of cases of chronic kidney disease and urothelial cancer in Europe and Asia, and raised global public health concerns. In addition, the so-called "Balkan endemic nephropathy" is obviously due to wheat contamination with the seeds of *Aristolochia clematitis* naturally growing in the fields, leading to chronic kidney failure and urothelial cancers. In addition to aristolochic acids included in *Aristolochia* species, several nephrotoxic herbs have also been written up in case reports and article reviews and are described in terms of their specific nephrotoxic pathways.

Finally, besides these mechanistic aspects, we will review and compare various methods proposed to assess the renal toxicity of herbs and natural products, i.e., *in silico*, *in vitro,* and *in vivo* approaches. Further developments in "omics" investigation would be helpful for screening potential subjects at risk of developing kidney injuries.

Keywords Nephrotoxicity • Renal toxic herbs • Aristolochic acid • Proximal tubular injury

---

T. Baudoux, MD • J.L. Nortier, MD, PhD (✉)
Nephrology Department, Erasme Hospital and Laboratory of Experimental Nephrology, Faculty of Medicine, Université Libre de Bruxelles,
808 route de Lennik, 1070 Brussels, Belgium
e-mail: Joelle.Nortier@erasme.ulb.ac.be

O. Pelkonen et al. (eds.), *Toxicology of Herbal Products*,
DOI 10.1007/978-3-319-43806-1_13

Physiopathological and Epidemiological Aspects of Herbal Products Nephrotoxicity

According to the definition of the World Health Organization (WHO) in 1991, nephrotoxicity refers to any renal disease or dysfunction directly or indirectly secondary to the exposure to drugs, chemicals, or industrial or environmental toxic agents, including herbal products or derivatives. Several factors make the kidneys particularly vulnerable to toxic insults (Perazella 2009). First of all, the kidneys are one of the major excretion organs of drugs, toxins, and their metabolites. Second, the kidneys are characterized by a large volume of blood supply (20–25 % of the cardiac output), which ensures an important delivery of nephrotoxic compounds. Third, the wide surface area of the tubular epithelium and endothelium provides sites for drug uptake, interactions, and potential damage mediated by vasoactive drugs. Fourth, active uptake by tubular cells, counter-current mechanisms in the Henle loop, and urine concentration in the distal tubules contribute to an increase in toxic drug levels locally within tubular epithelial cells, in the renal medulla and in the collecting tubules, respectively. Fifth, the high metabolic activity of renal tubular cells and their relative hypoxic environment make them especially vulnerable to metabolic inhibitors. And finally, pH variations of the urine may affect the solubility of a variety of excreted drugs, causing intra-tubular precipitation and physical damage.

Seventy-five percent of the world's population, mostly in developing countries, relies on botanical medicines for their basic health care needs (Barrett et al. 1999). In addition, herbal products are used more and more in Western countries; in Switzerland, the Federal Office of Statistics reports that 2.6 % of the population uses herbal medicines (Hess et al. 2009), and according to the WHO, over 100,000,000 Europeans are currently traditional and/or complementary and alternative medicine (CAM) users, one-fifth using them regularly. Briefly, CAM is often defined as a large and diverse set of systems of diagnosis, treatment and prevention based on techniques and information other than those used in conventional Western medicine and derived from traditional medical practice used in other non-Western cultures. For instance, the output of Chinese Materia Medica was estimated to account for US $83.1 billion in 2012, an increase of more than 20 % as compared to the previous year (WHO 2013). Consumers generally assume that herbs involved in CAM remedies are "natural" and therefore safe, with only few and minor adverse reactions. However, a cross-sectional national survey conducted in the U.S. between 1999 and 2008 concluded that 8 % of patients suffering from chronic kidney disease (CKD) were currently using dietary supplements containing at least one of the 37 herbs identified by the National Kidney Foundation as potentially harmful in the setting of CKD (Grubbs et al. 2013). In a small study conducted in Nigeria, herbal products were involved in 50 % of the cases of acute kidney injury (AKI), and 60 % of these cases required hemodialysis (Adelekun et al. 1999). In Belgium during the 1990s, about 150 patients developed rapidly progressing renal fibrosis due to the ingestion of root extracts of the Chinese herb *Aristolochia fangchi* (Y.C. Wu ex L.D.Chow and S.M.Hwang) (Vanherweghem et al. 1993).

Besides the direct nephrotoxicity of herbal products, consumption of herbal products by patients with compromised kidney function such as CKD, transplanted or dialyzed patients is also important. Indeed, these patients are especially at risk of developing systemic toxic injury from herbal preparations. Underlying mechanisms include possible drug interactions and unpredictable pharmacokinetics of the toxic compounds. One study suggested that the prevalence of complementary alternative medicine (CAM) used by kidney recipients was 11.8 %, with 23.8 % using Chinese medicine and 11 % using herbal medicine or St. John's wort (i.e., *Hypericum perforatum L.*) (Hess et al. 2009). The combination of CAM and Western medicine can endanger kidney disease prognosis due to underestimated interactions (see section "Nephrotoxicity Related to Interaction Between Herbal Products and Other Medications" for details). In another study conducted in Turkey, 25.2 % of the patients reported that they had used CAM therapy at least once after the onset of renal disease (Akyol et al. 2011). In a German study, 57 % of dialysis patients and 49 % of transplan patients were reported to be regular CAM users. Among them, 17 % of dialysis patients and 33 % of transplant patients were using phytotherapy; in addition, 16 % carried a theoretical risk of herb-drug interactions (Nowack et al. 2009).

Potential Factors Contributing to the Nephrotoxicity of Herbal Medicines

Several factors contributing to the nephrotoxicity of herbal medicines have been identified, such as the intrinsic toxicity of plant compounds, the contamination with nephrotoxic products including heavy metals, falsification or adulteration with drugs, the misidentification of herbs, interactions with other medications, or misuse (De Smet 2007; Lai et al. 2009; Zhao et al. 2006). Besides these factors, nine major herbs induced syndromes that were recognized as responsible for kidney injury (Colson and De Broe 2005; Isnard Bagnis et al. 2004): hypertension, acute tubular necrosis, acute interstitial nephritis, Fanconi syndrome, papillary necrosis, chronic interstitial nephritis, urinary retention, kidney stones, and urinary tract carcinoma. Acute tubular necrosis and acute interstitial nephritis are the main underlying mechanisms linked to nephrotoxicity. As the differential diagnosis between these two entities is often difficult on a clinical basis only, renal tissue samples should be obtained by percutaneous biopsy to achieve a histological confirmation (Fig. 13.1).

Reports on the nephrotoxicity of herbs are summarized in Tables 13.1, 13.2, 13.3, 13.4, and 13.5, and are classified according to the factors and the mechanisms implicated. However, caution is warranted regarding generalizability of these data because of differences in pharmacokinetics profiles, modalities of preparation, absorption, and metabolization of these herbal remedies. Data presented in Tables 13.1, 13.2, 13.3, 13.4, and 13.5 are largely drawn from case reports and small studies published in English, which limits general conclusions. Moreover, published data may largely underestimate the problem: according to the Food and Drug Administration (FDA)

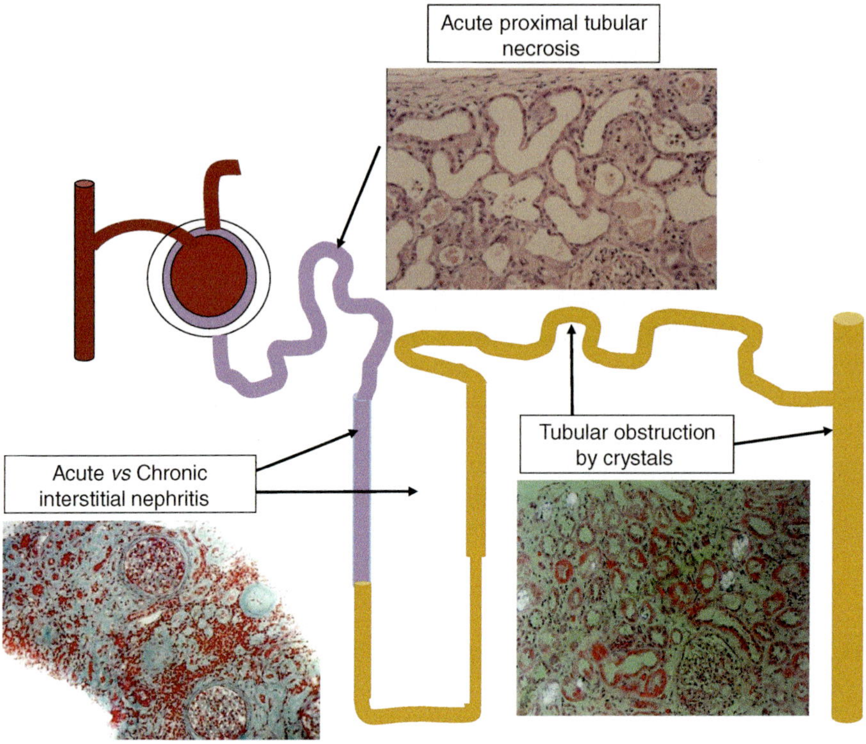

Fig. 13.1 Schematic representation of a nephron with the segments most often affected by herbal products, as shown by photomicrographs obtained from renal biopsy specimens (*courtesy of M. Depierreux, Pathology Dept, Erasme Hospital, Brussels, Belgium*): acute necrosis of the proximal tubule is classically responsible for acute kidney injury; in acute interstitial nephritis, the interstitium is infiltrated by inflammatory cells and in chronic interstitial nephritis (as represented here), it is the site of an extensive fibrosis (in *green* by the use of Goldner's staining); obstructive nephropathy is usually due to the precipitation of crystals into the lumen of distal tubules. Adapted with permission (From *Rev Med Brux* 32 (2011), 305–11)

statistics, only 10 % of serious adverse effects associated with the use of prescribed drugs, and less than 1 % of adverse events caused by dietary supplements, including herbs, are officially reported and registered (Marcus and Grollman 2002).

Finally, medical literature on herbal nephrotoxicity mostly focuses on AKI episodes rather than chronic intoxication consequences. Indeed, renal manifestations of AKI are often more obvious than the development of latent signs related to CKD. Moreover, there are very few studies that look at the long-term prognosis of AKI in developing countries after herbal intoxication (Ponce et al. 2016; Stanifer et al. 2014). This issue is particularly critical as studies have demonstrated that AKI is an important risk factor for CKD development or for CKD worsening, even in cases of short-term full renal function recovery (Chawla and Kimmel 2012). However at present, few patients at risk for subsequent decline in kidney function following AKI are seen by a nephrologist during the first year of discharge, even in Western countries (Ponce et al. 2016; Olowu et al. 2016; Siew et al. 2012).

Table 13.1 Intrinsic nephrotoxicity of herbal products: case studies

Names	Toxic compounds or metabolite	Route of administration	Indications	Renal manifestations	Mechanism of nephrotoxicity	Outcome	Country, Ref.
CKLS (stands for colon, kidney, liver, and spleen)	Unknown, CKLS contains aloe vera, chamomile, cascara sagrada, chaparral, mullien, uva ursi, fenugreek, cayenne, dandelion, and eucalyptus. According to authors, toxicity likely due to aloe vera and cascara sagrada (as they contain anthraquinone glycosides)	Oral	Colon, *k*idney, *l*iver, and *s*pleen purifier	AIN	Not specified	Dialysis for 2 days, then recovery	USA (Adesun loye 2003)
Cassia obtusifolia	Anthraquinone derivatives (chrysophenol, obtusin, physcion) in association with NSAIDS included in the slimming regimen)	Oral	Slimming regimen	ATN	Not specified	Recovery	Hong-Kong (Li et al. 2004)
Cat's claw, or Uno degatta (Unicarra tomentosa)	Alkaloids and flavonols	Oral	Cirrhosis, gastritis, gonorrhea, cancers of the female genital tract and rheumatism	AIN	Not specified	Partial recovery after discontinuation of the herbal preparation	Peru (Hilepo et al. 1997)

(continued)

Table 13.1 (continued)

Names	Toxic compounds or metabolite	Route of administration	Indications	Renal manifestations	Mechanism of nephrotoxicity	Outcome	Country, Ref.
Anatolian hawthorn (Crataegus orientalis)	Unknown, possibly due to flavonoids and oligomeric procyanthins	Oral	Cardiac tonic?	AIN	Not specified	Recovery after dialysis and steroids	Turkey (Horoz et al. 2008)
Bird flower (Crotalaria laburnifolia)	Pyrrolizidine alkaloids	Enemas, oral	Abortion, constipation, dysmenorrhea	ATN	Hepatic necrosis and intrahepatic veno-occlusion, with secondary acute tubular necrosis	Death or recovery	Zimbabwe, Sri Lanka (Colson et al. 2005; Luyckx 2012)
Cape aloes (leaves from Aloe ferox Miller)	Aloesin	Oral	Laxative	ATN or AIN	Interstitial nephritis, but the volume depletion caused by the laxative effect can play an additive role in the development of acute kidney injury	Recovery after dialysis required	South-Africa (Luyckx et al. 2002)
Chinese Yew (Taxus celebica)	Flavonoid sciadopitysin	Oral	Diabetes	ATN	Unknown. Probably related to hemoglobin or myoglobin tubular toxicity due to to rhabdomyolysis and hemolysis, however a direct nephrotoxic effect is also possible	Death or recovery after dialysis	Asia (Lin and Ho 1994)
Cone flower (echinacea)	Possible contamination by pyrrolizidine alkaloids or due to arabinogalactan	Oral	Immuno-stimulant	Distal renal tubular acidosis	Possibly related to contamination with pyrrolizidine alkaloids or due to exacerbation of autoimmune disease (Sjögren) by echinacea arabinogalactan and related tubular disorders	Recovery	Australia, USA (Mullins and Heddle 2002; Logan and Ahmed 2003)

Djenkol bean, jering (Pithecelobium lobatum)	Djenkolic acid	Oral. Some authors claim that toxicity may be prevented by cooking the beans properly but others argue that toxicity is independent of the method of preparation and of the number fruits consumed	Eaten as a delicacy but sometimes used for medical purposes	From asymptomatic hematuria to acute tubular obstruction and glomerular necrosis	Precipitation of djenkolic acid crystals in renal tubules resulting in obstructive nephropathy, sludge and stones	Recovery in most cases, 4 deaths reported	Malaysia, Indonesia, Thailand (Wong et al. 2007; Vachvanichsanong and Lebel 1997; Bunawan et al. 2014; Segasothy et al. 1995)
Horse chestnut seed extract (Aesculus hippocastanatum)	Aesculetin	Oral	Chronic venous insufficiency	Hemorrhage from renal angiomyolipoma	Hemorrhage due to anticoagulant effect of aesculetin	Survived	Ireland (Snow et al. 2012)

(continued)

Table 13.1 (continued)

Names	Toxic compounds or metabolite	Route of administration	Indications	Renal manifestations	Mechanism of nephrotoxicity	Outcome	Country, Ref.
Glycyrrhiza species contained in licorice or Chinese herbal teas (Boui-ougi-tou) and in gancao	Glycyrrhizic acid	Oral	Antibiotic, anti-inflammatory, gastro-intestinal disorders or to lose weight	From hypertension and hypokaliema to proximal tubulopathy (Fanconi) and ATN	Glycyrrhizic acid is hydrolyzed to glycyrrhetinic acid, which inhibits 11-hydroxysteroid dehydrogenase in the kidney. This enzyme catalyzes inactivation of cortisol to cortisone. Cortisol, unlike cortisone, has the same affinity for the mineralocorticoid receptors and induces an aldosterone "like" effect	Recovery after cessation of the remedy	Worldwide (Luyckx 2012; Gabardi et al. 2007; Isnard Bagnis et al. 2004; Luyckx and Naicker 2008)
					Renal tubular injury is secondary to rhabdomyolysis-induced renal dysfunction from prolonged hypokalemia		
					Fanconi's tubulopathy by inhibiting the Na,K ATPase activity of proximal tubular cells in 1 case report		
Ma-huang (Ephedra species)	Ephedrine, norephedrine and pseudoephedrine	Oral	Asthma, flu symptoms, edema, headaches and other aches	Hypertension, stones composed of ephedrine and ephedrine metabolites and ATN secondary to rhabdomyolysis	Hypertension precipitation of ephedrine and pseudoephedrine in urine	Full recovery in most cases, rarely transient dialysis and a few deaths	China, USA, UK (Powell et al. 1998; Haller and Benowitz 2000; Stahl et al. 2006; Rhidian 2011)

Oduvan (Cleistanthus collinus)	Arylnaphthalene lignan lactones (including cleistanthin A, cleistanthin B, collinusin, and diphyllin)	Oral	Cancer, abortion, poisoning or suicide	Type 1 renal tubular acidosis with severe hypokalemia followed by renal failure	Likely due to inhibition of vacuolar H+ATPase activity in the renal tubular brush border membrane	Death in 40 % of cases	India (Das et al. 2014; Chrispal 2012; Benjamin et al. 2006)
American pennyroyal (Hedeoma pulegioides)	Pulegone, menthofuran	Oral	Menstrual stimulant, to induce abortion, antispasmodic, anti-flatulent	ATN	Mechanism not found. Renal failure is associated with liver failure and necrosis, adrenal hemorrhage and lung consolidation	Death or survival with prolonged illness	USA (Luyckx 2012; Gabardi et al. 2007; Vanholder et al. 1999)
Rhubarb (Rhizoma Rhei)	Anthraquinone and oxalic acid	Oral	Laxative or slimming regimen	Chronic interstitial nephritis	Not specified	Recovery after cessation	Hong-Kong (Kwan et al. 2006)
Senna fruit (fruit from Cassia angustifolia)	Anthraquinone, possible adulteration by cadmium and mercury	Oral	Laxative	Proximal and distal renal tubular acidosis and diabetes insipidus	Experimental models suggest that anthraquinone may accumulate in the kidneys. Possible adulteration by cadmium that is toxic for proximal tubule	Recovery after intensive care unit	Belgium (Vanderperren et al. 2005)
"Sobi-lobi," "tornapple" or "Jimson weed" (Datura metel and stramonium)	Scopolamine and anticholinergic alkaloid esters	Oral or smoke	Hallucinogen, bronchitis or asthma	Acute urine retention	Anticholinergic substances causing urinary retention	Recovery	Niger, China and Zimbabwe (Colson and De Broe 2005; Luyckx 2012; Isnard Bagnis et al. 2004; Djibo and Bouzou 2000)

(continued)

Table 13.1 (continued)

Names	Toxic compounds or metabolite	Route of administration	Indications	Renal manifestations	Mechanism of nephrotoxicity	Outcome	Country, Ref.
Spurge (Euphorbia matabelensis and paralias)	Unknown, could be due to ingenanes	Oral, transdermal route	anti-inflammatory agent, purgative, local anesthetic, to facilitate the absorption of other active compounds of herbal remedies through the skin	ATN/AIN	Unknown, believed to be toxic and immunoallergic	Dialysis and steroids. Partial recovery after 2 months	Zimbabwe, Tunisia (Colson and De Broe 2005; Boubaker et al. 2013)
Thunder good vine, Lei Gong Teng (Tripterygium wilfordii)	Unknown, could be triptolide	Oral	Rheumatoid arthritis and other inflammatory conditions	AKI (not specified)	Unknown, experimental data suggest that lesions are limited to the proximal tubule and could be due to reactive oxygen species and destruction of organic anion transporters of the proximal tubular epithelial cells	Death	Taiwan (Chou et al. 1995; Huang et al. 2009; Dan et al. 2008)
Ting Kung Teng or "Bao Gong Teng" (Erycibe henryi Prain)	Tropane alkaloids	Oral	Rheumatoid arthritis	AKI (not specified)	Tropane alkaloids exhibit cholinergic activities. AKI could be related to transient hypotension or due to toxic effect	Recovery	Taiwan (Lin and Chen 2002; Hsu et al. 1998; Huang et al. 2006)
Tribulus terrestris	Unknown	Oral	Aphrodisiac and antiurolithiasic	ATN, AIN not excluded	Unknown	Recovery after cessation of the herb and transient dialysis	Iran (Talasaz et al. 2010)

Wild wisteria, violet tree (Securida longepedunculata)	Methylsalicylate, saponin and alkaloid that is identical to securinine	Toxic when taken intravaginally or intrarectally. Probably not when taken orally	Abortion, epilepsy, headache, rheumatism, emetic or purgative	ATN	Unknown, possibly due to intrarenal vasoconstriction caused by salicylates associated with volume depletion	May cause death	Congo, Zambia and Zimbabwe (Colson and De Broe 2005; Steenkamp and Stewart 2005)
Yellow oleander (Thevetia peruviana)	Cardiac glycosides (neriifolin, thevetin A, thevetin B, and oleandrin)	Oral	Anti-inflammatory, suicide	ATN	Not specified	May cause death	India and Sri-Lanka (Gabardi et al. 2007; Bandara et al. 2010)
Yohimbine (Pausinystalia Yohimbe)	Yohimbine	Oral	Impotence, aphrodisiac	Drug- induced systemic lupus erythematosus	Not specified	Improvement with steroids	USA and West Africa (Sandler and Aronson 1993)
Willow bark (Salix daphnoides)	Salicin	Oral	Analgesic, antirheumatic, and antipyretic properties	Renal papillary necrosis	Salicin is converted to salicylic acid and causes analgesic nephropathy	May lead to CKD	Europe (Gabardi et al. 2007; Schwarz 1993; Jha 2010)
Chaparral (Larrea tridentata)	Nordihydroguaiaretic acid (NDGA)	Oral	Tonic, internal skin cleanser, arthritis, nutritional supplement, weight loss, and anticancer agent	Cystic renal disease and cystic adenocarcinoma	NDGA is used experimentally to induce cystic renal disease in rats. NDGA metabolites increase the fragility of lysosomal membranes by lipid peroxidation, causing autolysis, followed by desquamation of necrotic proximal tubular epithelial cells	Nephrectomy required	USA and Mexico (Gabardi et al. 2007; Steenkamp and Stewart 2005; Smith et al. 1994)

(continued)

Table 13.1 (continued)

Names	Toxic compounds or metabolite	Route of administration	Indications	Renal manifestations	Mechanism of nephrotoxicity	Outcome	Country, Ref.
Glue thistle (Atractylis gummifera)	Potassium atractylate and gummiferin	Mainly oral	Chewing gum, antipyretic, purgative, emetic, diuretic, and abortion	ATN	Potassium atractylate and gummiferin are inhibitors of mitochondrial oxidative phosphorylation. Fulminant hepatitis and hypoglycemia are followed by kidney failure	Death in some cases	Mediterranean countries, especially North Africa (Gabardi et al. 2007; Daniele et al. 2005)
Cranberry juice (Vaccinium macrocarpon)	Oxalate	Oral	Prevention and treatment of urinary tract infection	Kidney stones	Massive ingestion of cranberry concentrate tablets is accompanied by an increase in oxalate urinary excretion and could thus promote calcium oxalate lithiasis in patients at risk	Recovery	USA, Europe (Terris et al. 2001)
King of bitters, "Chuan-Xin-Lian" or heart- piercing lotus (Andrographis paniculata)	Andrographolide	Intravenous	Flu, aphthous stomatitis, cough, diarrhea, skin sores and ulcers, snake bite	ATN	Unknown, could be due to andrographolide inhibitory effects on prostaglandin E2 production and secondary intrarenal vasoconstriction	Recovery in most cases, sometimes after transient dialysis	China (Zhang et al. 2014)
Alfalfa (Medicago Sativa) and Black cohosh (Cimicifuga Racemosa)	Could be linked to L-canavanine present in alfalfa	Oral	Menopausal symptoms	Acute graft rejection	L-canavanine has been shown to be a potent T-cell activator *in vitro*	Partial recovery with antirejection therapy	USA (Light and Light 2003)

ATN acute tubular necrosis, *AIN* acute interstitial nephritis, *CKD* chronic kidney disease, *NSAIDs* non-steroidal anti-inflammatory drugs, *AKI* acute kidney injury

Table 13.2 Nephrotoxicity related to misuse of herbal products

Names	Toxic compounds or metabolite	Route of administration	Indications	Renal manifestations	Mechanism of nephrotoxicity	Outcome	Country, Ref.
Yam (Dioscorea quartiniana or quinqueloba)	Dioscorine and dioscin	Oral. Yams need to be detoxified first by placing them in running water, soaking them in salted water and boiling them	As food or to cause poisoning or suicide (different preparation)	ATN or AIN	Drug induced AIN. *In vivo* experiments suggest that *Dioscorea* can trigger chronic kidney injury via direct toxicity and pro-fibrotic process in the kidney	May cause death	Africa and Asia (Luyckx 2012; Steenkamp and Stewart 2005; Kim et al. 2014; Jha and Parameswaran 2013; Kim et al. 2012; Kane and Heo 2015)
Noni juice, Och plant noni or nonu or cheese fruit (Morinda citrifolia)	Potassium	Oral. Higher risk in CKD patients.	Multiple purposes (e.g., Cancer, tonic, hypertension…)	Hyperkalemia	Potassium content of noni juice is 56.3 mEq/L	Transient hyperkalemia. No follow-up	India, Samoa, Tahiti, Southeast Asia, and Australia (Mueller et al. 2000)

(continued)

Table 13.2 (continued)

Impila (Callilepsis laureola)	Atractyloside	Oral or enema. Impila is recommended to be ingested with large amounts of water and regurgitated soon after ingestion to avoid toxic effect. Impila should not be used as an enema	Stomach problems, purgative vermifuge, impotence and infertility	ATN	Not fully understood. Induce apoptosis by opening the mitochondrial membrane permeability transition pore and by inhibiting oxidative phosphorylation	The majority of cases are fatal, particularly in children (up to 90 %). Mortality (within 24 h because of hypoglycemia and at 5 days because of acute kidney failure combined with hepatotoxicity) is 63 %	South Africa (Popat et al. 2001; Stewart and Steenkamp 2000; Seedat 1993; Seedat and Hitchcock 1971; Watson et al. 1979)
Starfruit (Averrhoacarambola)	Oxalic acid	Oral. Toxic if juice is undiluted or if fruits are consumed in large amounts. Higher risk in CKD patients	Food, for hangover or flu	ATN and oxalate nephropathy Nephrolithiasis, obstructive nephropathy	Oxalate crystals induce tubular obstruction after precipitation but also apoptosis of renal epithelial cells	Recovery in most cases	Brazil, Hong-Kong and Taiwan, south-east Asia (Luyckx 2012; Neto et al. 2009; Fang et al. 2008; Chen 2001)
Wormwood oil (Artemisia absinthium)	Terpene	Usual aromatherapy compound but orally ingested	Recreational, to stimulate digestion	ATN	Unknown, could be due to rhabdomyolysis and tubular necrosis as a result of seizure	Recovery with supportive therapy	USA and Europe (Luyckx 2012; Weisbord et al. 1997)

ATN acute tubular necrosis, *AIN* acute interstitial nephritis, *CKD* chronic kidney disease

Table 13.3 Nephrotoxicity related to the contamination of herbal products by drugs or heavy metals

Names	Toxic compounds or metabolite	Route of administration	Indications	Renal manifestations	Mechanism of nephrotoxicity	Outcome	Country, Ref.
Mustard oil	Adulterated with seeds of Argemone mexicana that contains sanguinarine, a toxic alkaloid	Oral or transdermal route	For cooking and massage	Acute tubular necrosis and vascular congestion	Sanguinarine, interferes with the oxidation of pyruvic acid, which accumulates and causes dilation of capillaries and small arterioles leading to extensive disturbances in the systemic circulation and renal hypoperfusion	Recovery in most cases, sometimes with transient dialysis, death is rare	India, Mauritius and South Africa (Prakash 1999; Sharma et al. 1999)
Takaout roumia	Adultered by paraphenylenediamine (PPD). "Takaout El Badia" (or "Takaout beldia"), is a nontoxic powder made of the seeds of *Tamaris orientalis* as a hair dye. Because the traditional mixture is less available, it is substituted by PPD under the name of "Takaout roumia"	Transdermal route	As a hair and body dye	Acute tubular necrosis	Poisoning by PPD induces severe rhabdomyolysis and acute renal failure because of hypovolemia and myoglobinuria	Death in some cases	Middle Eastern countries and Japan, Morocco, Sudan (Sir Hashim et al. 1992; Bourquia et al. 1988; Averbukh et al. 1989; Shemesh et al. 1995)

(continued)

Table 13.3 (continued)

Ayurvedic herbal product	Lead	Oral	Not specified	Acute kidney injury in a CKD patient	Related to lead toxicity	Return to baseline renal function	Canada (Prakash et al. 2009)
Mixture of Chinese herbs	Cadmium	Oral	Tonic	Fanconi's syndrome and a nephrogenic diabetes insipidus	Cadmium might exert its toxicity by inhibiting Na -K -ATPase and could also change the excretion of other nephrotoxic substances	Improvement without complete recovery. Bicarbonate and potassium supplementation still required	Taiwan (Wu et al. 1996)
Traditional herbal preparation	120 mg of phenylbutazone per tablet	Oral	Osteoarthritis	Papillary necrosis	Toxic effect of phenylbutazone	Not specified	Malaysia (Segasothy and Samad 1991)
Nyanga (traditional township healers) remedies	Sodium or potassium dichromate	Enema or oral	Constipation, flu, psychiatric disturbance and sexual dysfunction	AKI and hepatocellular dysfunction	AKI, gastrointestinal hemorrhage, and hepatocellular dysfunction	From renal failure to coma and death in some cases	South Africa (Wood et al. 1990)
Tung Shueh pills (Cow's Head brand)	Diazepam and mefenamic acid as additives	Oral	Arthralgia	AIN	Mefenamic acid is a nonsteroidal anti-inflammatory drug and has been reported to cause interstitial nephritis	Recovery after dialysis and steroids	Taiwan (Diamond and Pallone 1994)
Bladderwrack (Fucus vesiculosus)	Contaminated with arsenic and other heavy metals	Oral	Weight loss	AIN	Arsenic toxicity	Recovery after 1 year	Italy (Conz et al. 1998)

ATN acute tubular necrosis, *AIN* acute interstitial nephritis, *CKD* chronic kidney disease

Table 13.4 Nephrotoxicity related to a misidentification of herbal products

Names	Toxic compounds or metabolite	Route of administration	Indications	Renal manifestations	Mechanism of nephrotoxicity	Outcome	Country, Ref.
Mourning cypress (Cupressus funebris)	Flavonoids	Oral	Hot-water extract of sliced wood (C funebris) instead of Taxus species (yew)	ATN/AIN	Unknown, could be related to acute intravascular hemolysis and hemoglobin- related toxicity and tubular obstruction. Direct nephrotoxicity of flavonoid or/and an immune-mediated mechanism is also suspected. Flavonoids may induce oxidative injury through cytochrome P450	Recovery after transient dialysis	Taiwan (Lee and Chen 2006)
Guang fang chi	Aristolochic acids	Oral	Slimming regimen: Aristolochia species instead of Stephania tetrandra	Interstitial fibrosis and tubular atrophy or Fanconi syndrome	See appropriate section	Tubular atrophy, renal fibrosis and urothelial cancer	China, Europe, Taiwan, USA (Yang et al. 2011; Debelle et al. 2008; Chen et al. 2012)
Autumn crocus, wild saffron, meadow saffron (Colchicum autumnale)	Colchicine	Oral	Hypertension, heart and alcoholic liver disease; Colchicum autumnale instead wild garlic (Allium ursinum)	ATN	Colchicine is a well-known poison that binds to the microtubules and alters cellular processes such as cell shaping, division, mobility and phagocytosis	Recovery or death, depending on the dosage ingested	Croatia, Turkey (Brvar et al. 2004; Klintschar et al. 1999; Brncic et al. 2001)

ATN acute tubular necrosis, *AIN* acute interstitial nephritis

Table 13.5 Nephrotoxicity related to an interaction between herbal products and other medications

Names	Toxic compounds or metabolite	Route of administration	Indications	Renal manifestations	Mechanism of nephrotoxicity	Outcome	Country, Ref.
St. John's Wort (Hypericum perforatum)	Hyperforin as an inducer of CYP-450	Oral	Depression, anxiety	Acute graft rejection	By inducing the hepatic cytochrome P-450 microsomal oxidase enzyme system, SJW may decrease or increase cyclosporine blood level when it is initiated or stopped and may thus induce acute graft rejection or ciclosporine toxicity	Graft loss or ciclosporine toxicity	Germany, USA (Ernst 2002; Mai et al. 2003)
Herbal tea containing cimicifuga, isoliquiritigeni, oleanolic acid	Oleanolic acid according to authors	Oral	Food	Acute renal injury due to ciclosporine intoxication	Oleanolic acid, isoliquiritigeni and cimicifuga are potent inhibitors of CYP3A4	Recovery	Hong-Kong (Kwan et al. 2014)

Intrinsic Nephrotoxicity of Herbal Products

In the U.S., drug-induced AKI accounts for 18–27 % of all AKI cases in hospitalized patients (Taber and Pasko 2008). In developing countries, up to 80 % of the population uses folk remedies for primary health care, and similarly, the use of traditional remedies has been implicated in up to 35 % of AKI cases in Africa, for instance (Luyckx et al. 2005). It should be noted that many herbal remedies are either taken as a single herb or as mixtures of a variety of herbs, which may have multiple interactions with each other. Data regarding herbal nephrotoxic derived products from reviews and from case reports are compiled in Table 13.1 (Colson and De Broe 2005; Isnard Bagnis et al. 2004; Gabardi et al. 2007; Luyckx 2012). Results from experimental *in vitro* and *in vivo* studies have been excluded to focus on human clinically relevant cases only.

Nephrotoxicity Related to Misuse of Herbal Products

Due to variable weather and growing conditions as well as differences in harvesting procedures, or processing and modes of extraction, the biological activity of herbal products may vary substantially, possibly leading to overdose (see Table 13.2). As an example, concentrations of hyperforin, the biological active compound of St. John's wort, measured in various parts of the plant (leaves, stems, and flowers) vary considerably, depending on the harvesting time (Couceiro et al. 2006). In addition, the toxicity of the same compound may vary when administered orally or through a different route (e.g., enemas, vaginally, via cataplasm, etc.) or if the preparation was incorrectly made (Luyckx 2012; Luyckx and Naicker 2008). The lack of knowledge of proper preparation of local plants or hazardous use of substances thought to be medicinal remedies is encountered in urban areas, due to a breakdown of traditional communities (Luyckx and Naicker 2008). Additionally, potentially toxic substances, such as paint thinners, turpentine, chloroxylenol, ginger, pepper, soap, vinegar, copper sulphate, and potassium permanganate added to laxative to increase their effect has been described in the past (Dunn et al. 1991). Finally, several plants known to be toxic are used for suicide purposes.

Nephrotoxicity Related to the Contamination of Herbal Products by Drugs or Heavy Metals

Several case reports of kidney injury related to the consumption of herbal products that have been adulterated or contaminated with drugs or heavy metals have been published (see Table 13.3). Many herbal products contain undisclosed drugs available by prescription or over the counter. In 1998, the California Department of

Health reported that 32 % of Asian patent medicines contained undeclared pharmaceuticals or heavy metals such as ephedrine, chlorpheniramine, methyltestosterone, phenacetin, lead, mercury or arsenic (Marcus and Grollman 2002). In Japan, the use of certain imported Chinese dietary supplements was associated with hepatic failure and hyperthyroidism. After analysis, these products were proven to be adulterated with N-nitroso-fenfluramine, fenfluramine, and thyroid extracts. One hundred forty-nine patients required hospitalization, and three of them died (Marcus and Grollman 2002). In addition, medicinal herbs may absorb environmental contaminants such as heavy metals (e.g., arsenic, cadmium, lead, or mercury) leading to renal toxicity. In Nigeria, 20–100 % of herbal medicinal products contained elevated amounts of heavy metals (Obi et al. 2006; Nwoko and Mgbeahuruike 2011). In the U.S. and U.K., approximately 20–30 % of Ayurvedic herbal products were estimated to be contaminated with heavy metals (Saper et al. 2004; Aslam et al. 1979). In another evaluation, out of 54 samples of Asian medicines collected in the U.S. and Asia, 49 % contained toxic concentrations of heavy metals (Garvey et al. 2001).

In addition to the proven acute toxicity of heavy metals at high doses, evidence is accumulating regarding the role of low-dosage chronic exposure in the development of CKD. Low-level chronic exposure to lead has been proven to exacerbate underlying CKD and to accelerate the progression to advanced stages of CKD and dialysis requiring end-stage renal disease (ESRD) (Yu et al. 2004; Fadrowski et al. 2010). It has also been suggested that urinary and blood cadmium levels, even below accepted thresholds, were associated with a higher rate of kidney disease and albuminuria (Ferraro et al. 2010; Navas-Acien et al. 2009). Lead, cadmium and mercury nephrotoxicity mechanisms are not fully understood. Proximal tubular epithelial cells are the main targets of heavy metal. Physiopathology seems to involve oxidative stress and mitochondrial toxicity mechanisms (Sabath and Robles-Osorio 2012).

Nephrotoxicity Related to the Misidentification of Herbal Products

The lack of information on traditional herbs in urban areas, or confusing names of traditional herbal products, may lead to incorrect use of medicinal remedies. Chinese herb nephropathy or aristolochic acid nephropathy (AAN) outbreak is the most famous example of potential dramatic consequences of substitution of herbal products where *Stephania tetrandra* S. Moore was substituted for *Aristolochia fangchi*. Despite many publications, a warning, and a ban by the FDA in 2001, aristolochic acids (AA) containing herbal products are still sold on local markets by traditional healers; the substitution of one plant for another containing AA still occurs nowadays (Yamani et al. 2015; Ioset et al. 2003; Hsieh et al. 2008; Debelle et al. 2008), and several other cases where substitutions took place have been published (see Table 13.4 and Sect. 3.2). More recently, a study using DNA barcoding technique conducted in Canada to check the authenticity of 44 herbal products available in the U.S. or Canada found that approximately 60 % of them contained DNA from plant species that were not listed on the labels, confirming the substitution of 30 among 44 products (Newmaster et al. 2013). Another study conducted in Australia on traditional Chinese medicine (TCM) products that were seized by the customs authorities at the border identified plant species known to contain toxic compounds (e.g., *asarum* species that may contain aristolochic acids or *ephedra* species) as well as animal DNA from species that are currently protected by international laws (Coghlan et al. 2012).

Nephrotoxicity Related to Interaction Between Herbal Products and Other Medications

Botanical medicines can act through a variety of mechanisms and alter the pharmacokinetic profile of concomitantly administered drugs. Cytochrome P450 (enzyme CYP3A4) is one of the most famous targets of drug-herb interaction. Kidney recipients and patients suffering from CKD or ESRD are more prone to risky interaction; indeed, these patients often take many drugs that can interact with herbal compounds. The best known clinical situation concerns patients with renal transplants treated with cyclosporine having ingested regular doses of St. John's wort (i.e., *Hypericum perforatum*) for depression. The induction of CYP3A4 by hyperforin increased the metabolism of cyclosporine, which led to a dramatic drop of cyclosporine plasma concentration and to transplant rejection (Marcus and Grollman 2002; Mai et al. 2003). Finally, it should be taken into account that in CKD, renal drug metabolism and elimination capacity are affected. Liver metabolism and elimination are also profoundly altered by complex mechanisms involving uremic toxins accumulation (Ladda and Goralski 2016). Case reports are summarized in Table 13.5.

Aristolochia Species-Induced Nephropathy and Cancer

Introduction

AA are natural products derived from *Aristolochiaceae* and have been associated with a high risk of nephrotoxicity and urothelial carcinoma (in the upper urinary tract as well as the bladder). As the *Aristolochia* genus includes more than 500 species and are widespread in the warm regions of the Mediterranean, Africa and Asia, secondary nephropathies resulting from the toxicity of plants containing AA have been described worldwide (Debelle et al. 2008).

In France, birthwort (i.e., *Aristolochia clematitis* L. or *Aristolochia rotunda* L. and *Aristolochia pistolochia* L.) grows mainly in limestone soil and can be found on roadsides, in coppices, vineyards, and other agricultural areas (Laïs 2003; Bertrand

2009). In Asia, several species have also been described, such as *Aristolochia manshuriensis* Kom., *Aristolochia fangchi* and *Aristolochia indica* L. Finally, *Aristolochia clematitis* is a parasitic plant that grows in local wheat fields and meadows in warm and humid regions of the Danube tributaries.

In the past, *Aristolochia* species were widely used in Western medicine (Heinrich et al. 2009). In fact, their first use – to stimulate the expulsion of the placenta during childbirth – was responsible for coining the name "Aristos lokos" or "excellent delivery" (Nortier et al. 2015). Since the plants were also recommended for treating snakebites, *Aristolochia* plants were included in the preparation of theriac, and "Dutchman's pipe" (i.e., *Aristolochia macrophyllia* Lam,.) was prescribed for treating gout. Finally, *Aristolochia* species were also included in vulnerary ointments such as *Manus Dei*, which was used as part of the medical treatment for the anal fistula of Louis XIV. *Asarum europaeum* L., another species that contains AA, was also widely used in Europe for the treatment of gout and arthralgia (Adams et al. 2009).

More importantly as regards global public health issues, *Aristolochia* is considered to be an integral part of the herbology used in TCM, (Gokmen et al. 2013) Japanese *Kampo* (Hashimoto et al. 1999), and Ayurvedic medicine (Vanherweghem 1997). They are found within the same therapeutic family as the *Akebia*, *Asarum*, *Cocculus* and *Stephania* plants. Referred to by common names such as *Mu Tong*, *Mokutsu* and *Fang ji*, they are used in large numbers of herbal mixtures for therapeutic purposes (Debelle et al. 2008).

In 1969, Ivic suggested that *Aristolochia clematitis* could be responsible for the so-called "BEN" Balkan endemic nephropathy (Ivic 1969). This hypothesis has recently been confirmed by the detection of specific DNA adducts formed by aristolactams, the metabolites of AA present in the renal tissue as well as in the urothelial tumors of BEN patients (Grollman 2013; Jelakovic et al. 2012).

Today, the term AAN is used to include any form of toxic interstitial nephropathy that is caused either by the ingestion of plants containing AA, as part of traditional phytotherapies (formerly known as "Chinese herbs nephropathy"), or by the environmental contaminants in food (Balkan endemic nephropathy) (De Broe 2012).

Aristolochic Acids in Herbal Medicines

In 1993, the occurrence of a rapidly progressive form of renal interstitial fibrosis associated with a weight loss diet including the ingestion of pulverized plant root extracts used in traditional Chinese medicine, led to the description of a new toxic nephropathy initially called Chinese-herb nephropathy (CHN) (Fig. 13.2) (Vanherweghem et al. 1993). AA were identified in these plants' powdered roots and appeared to be the dramatic consequence of a substitution of *Stephania tetrandra* by *Aristolochia fangchi* rich in AA, because both herbs share the same common name in Mandarin or Pin Yin (*Han Fang Ji* and *Guang Fang Ji*), and they can be used interchangeably in traditional Chinese medicine (Vanherweghem et al. 1993; Vanhaelen et al. 1994; Wu et al. 2007). In all, more than 100 cases of AAN were reported in Belgium (Debelle et al. 2008). Following the Belgian outbreak, several

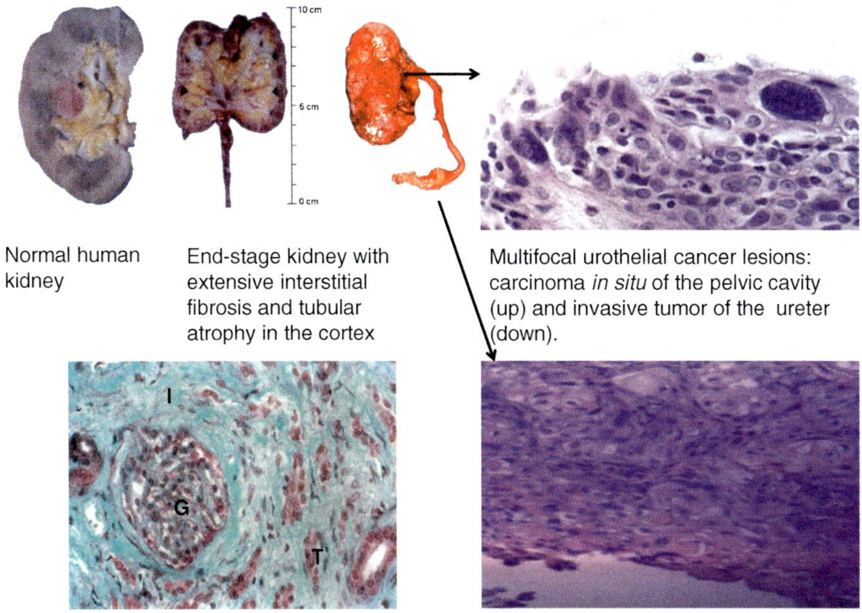

Fig. 13.2 Typical macroscopic aspects of the kidney from a patient suffering from end-stage aristolochic acid nephropathy (AAN): reduced size and atrophy of the cortex (external part responsible for plasma filtration); microscopically, the interstitium (I) is the site of an extensive fibrosis (colored in *green*) with severe atrophy of the tubules (T), remaining the glomeruli (G) almost intact but without any connection with functioning tubules. This patient developed multifocal urothelial carcinoma along the upper urinary tract with variable histological grading (from noninvasive to invasive cancer lesions). Reproduced with permission from Nortier J, Pozdzik A, Roumeguere T, Vanherweghem JL. Néphropathie aux acides aristolochiques (néphropathie aux herbes chinoises). EMC – *Néphrologie* 2013;10(2):1–14 [Article 18-040-J-10]. Copyright © 2013 Elsevier Masson SAS. Tous droits réservés

cases were also described in France, Spain, the U.K., Australia, the U.S., Korea, Japan, China, and Taiwan (Debelle et al. 2008; Gokmen et al. 2013). Despite the outbreak of AAN in Belgium in the 1990s, and despite numerous publications regarding AA toxicity and the warning from the FDA, the use of herbs containing AA is still allowed in China under certain conditions. In addition, it is still possible, anywhere in the world, to obtain plant extracts that may contain AA, particularly via parallel markets such as the Internet (Gokmen et al. 2013). In northeastern Morocco, the substitution of the *Bryonia* genus by the *Aristolochia* species is still current in stores and public markets selling traditional herbal remedies. Unexplained AKI and CKD cases in these regions are suspected to be linked to the regular use of these plants (Yamani et al. 2015). In Iran, the *aristolochia* species' role has also been suspected in CKD, as it is also used in traditional medicinc (Ardalan et al. 2015).

Today, it is estimated that exposure to AAN affects at least 8,000,000 people in Taiwan and more than 100,000,000 on mainland China (Grollman 2013). Given the fact that the nephrotoxic effect of AA is irreversible and that the carcinogenic effects may be very slow in manifesting themselves after the patient's initial exposure,

AAN and associated cancers are likely to become a major public health concern in the years to come (Debelle et al. 2008; Gokmen et al. 2013; Grollman 2013).

AAN has been successfully reproduced in various experimental models (Debelle et al. 2002; Baudoux et al. 2012; Cosyns et al. 2001). Experimental and clinical studies both indicate that the proximal tubule, particularly its S3 segment, is targeted by AA (Debelle et al. 2008). After acute tubular necrosis, the interstitium is infiltrated by activated macrophages, as well as B and T lymphocytes in association with profibrosing cytokine production by macrophages and tubular cells (such as transforming growth factor β) (Pozdzik et al. 2008a; Pozdzik et al. 2008b). These early and sustained events result in severe interstitial fibrosis and tubular atrophy associated with renal impairment.

Carcinogenicity of Aristolochic Acids

Under normal physiological conditions, AA are metabolized by the reduction of nitro compounds into active metabolites called "aristolactams," which are capable of forming covalent bonds with purine bases of DNA (Fig. 13.3). These AA-DNA adducts have demonstrated premutagenic properties; indeed, they lead to a mutation of A:T > T:A in the tumor suppressor TP 53 gene. This mutation is considered to be the most specific one in case of AA exposure, at least in urothelial cancers analyzed from patients in the Balkans and Taiwan (frequency among all TP53 mutational spectra: 64.5 % and 53 %, respectively), as compared to 5 % in non-AA related urothelial carcinoma cases (Grollman 2013; Arlt et al. 2002). In addition, as the AA-DNA adducts persist for several years after the patients' initial exposure, their discovery in renal or cancerous tissues constitute a biomarker of exposure to AA that possibly occurred much earlier (Schmeiser et al. 2014).

Natural History of the Disease

In the initial Belgian study, cohorts suffering from iatrogenic nephropathy due to AA, the majority of patients were described as exhibiting a rapid and progressive evolution towards end-stage kidney disease (ESKD) (Vanherweghem et al. 1993; Vanherweghem 1998). The progression rate in environmental Balkan endemic nephropathy due to AA is much slower, with the end-stage of renal failure occurring only after an evolution of 15–20 years (Pavlovi 2013). A few cases of iatrogenic nephropathy due to AA took the form of Fanconi syndrome or acute renal failure (Debelle et al. 2008).

Blood pressure is elevated in 50 % of the cases. Anemia is often more severe than one would expect from the level of kidney failure, probably because of the premature destruction of peritubular cells that secrete erythropoietin. Kidney size is often reduced asymmetrically (Nortier et al. 2015).

Whether iatrogenic or environmental, AAN is more frequently associated with urinary tract cancers. In cases where bilateral ureteronephrectomies were performed

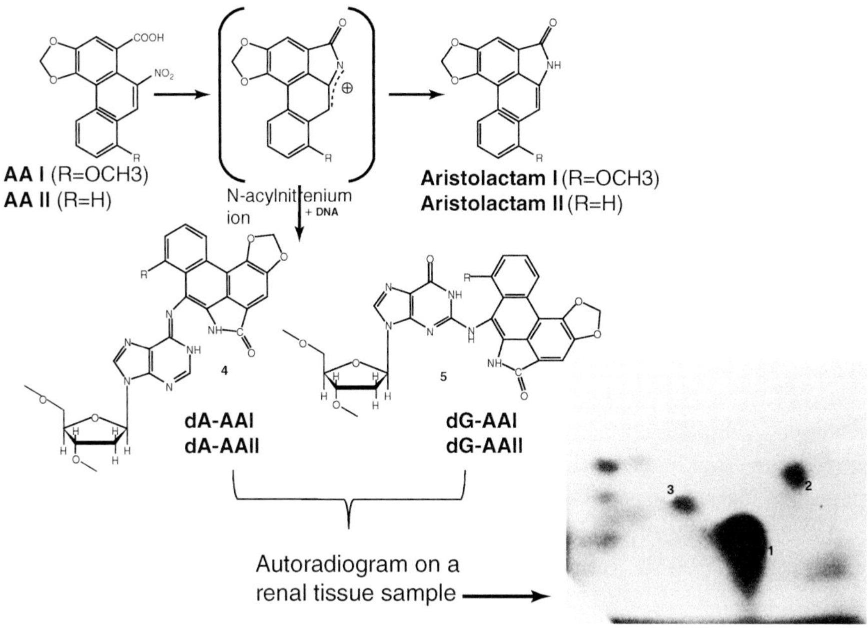

Fig. 13.3 Chemical structure of aristolochic acids I (AAI) and II (AAII), of corresponding aristolactams I and II (formed by nitroreduction) and DNA adducts. Representative autoradiogram of specific AA metabolites (aristolactams) related DNA adducts detected in the renal cortex from a patient suffering from end-stage aristolochic acid nephropathy (AAN): 1) 7-(deoxyguanosine-N^6-yl)-aristolactam I; 2) 7-(deoxyadenosine-N^6-yl)-aristolactam II; 3) 7-(deoxyadenosine-N^6-yl)-aristolactam I, the most abundant adduct detected in renal tissue from individuals exposed to AA. Reproduced with permission from Nortier J, Pozdzik A, Roumeguere T, Vanherweghem JL. Néphropathie aux acides aristolochiques (néphropathie aux herbes chinoises). EMC – *Néphrologie* 2013;10(2):1–14 [Article 18-040-J-10]. Copyright © 2013 Elsevier Masson SAS. Tous droits réservés

on female patients from the initial cohorts treated by dialysis or transplantation, 40 % of these women suffered from urothelial cancers that were often diagnosed as multifocal (Fig. 13.2) (Cosyns et al. 1999; Nortier et al. 2000). Cancers of the bladder appeared in female patients who had undergone transplants more than 15 years after exposure to the toxin had been stopped (Lemy et al. 2008). Frequent iatrogenic exposure to AA in Taiwan explains why this region has the world's highest level of urothelial cancers (Gokmen et al. 2013; Chen et al. 2012; Yang et al. 2002; Guh et al. 2007).

Diagnosis of AA Exposure

Histological examination of the kidney is a key element in the diagnosis of AAN. From a macroscopic point of view, in advanced cases of nephropathy, the kidneys are small and the renal cortex is considerably thinned. Microscopically, the most characteristic type of lesion consists of paucicellular interstitial fibrosis associated with tubular atrophy. The severity of these lesions decreases when moving from the external to the internal cortex (Cosyns et al. 1994; Depierreux et al. 1994). However, as histological lesions are not pathognomonic, there is a consensus regarding the definition of diagnostic criteria (Gokmen et al. 2013; Nortier et al. 2015). The positive diagnosis of AAN can be made in any person who suffers from renal failure in combination with any two of the following three criteria: a renal histology displaying interstitial fibrosis with a cortico-medullary gradient; a history of ingesting vegetal or herbal products whose phytochemical analysis has demonstrated the presence of AA; and the demonstrated presence of DNA adducts formed with aristolactams (or the specific mutation A:T > T:A of P53 gene) in a kidney tissue biopsy sample or a urothelial tumor (Fig. 13.3). Nevertheless, if only one of these three criteria is met, the diagnosis of AAN remains highly probable, and tests should continue in this direction. Whatever the case, the presence of either AA in plant extracts ingested by patients, or of DNA adducts formed with AA metabolites in a patient's renal tissue, are central to an absolutely certain diagnosis.

Treatment

The treatment of AAN is similar to that of any CKD and includes controlling the patient's blood pressure, symptomatic treatment of metabolic complications, and preparing for replacement therapy by means of dialysis or renal transplantation. Based on a pilot test, (Vanherweghem et al. 1996) treatments with steroids have also been proposed, but their efficacy is limited. Regarding the high incidence of urothelial cancer, screening for cancer of the urinary tract in patients with AAN is mandatory (Gokmen et al. 2013). In addition, patients with ESRD who have been treated with dialysis or kidney transplantation should undergo a prophylactic bilateral ureteronephrectomy and a follow-up for bladder cancer.

Balkan Endemic Nephropathy

Balkan endemic nephropathy is a chronic tubulointerstitial kidney disease found in farming villages close to the Danube River in Bosnia, Bulgaria, Croatia, Romania, and Serbia. In the most affected regions, it represents up to 70 % of kidney disease in dialyzed patients (Djukanovic et al. 2002), and at least 25,000 individuals are known to have the disease (Tatu et al. 1998). The similarity between the histological aspects of this particular nephropathy and the so-called "Balkan endemic nephropathy," (Vanherweghem et al. 1993; Cosyns et al. 1994; Depierreux et al. 1994) – combined with the association between these types of renal disease and urinary tract cancers, (Grollman 2013; Cosyns et al. 1999; Nortier et al. 2000) – contribute to the revival of an old hypothesis on the etiology of Balkan nephropathy.

In 1969, Ivic observed that what would be called "BEN" in the future was limited to families living in rural areas and some of the villages (sparing the towns) throughout the Danube Valley, suggesting environmental factors. He remarked that the wheat fields in these regions were heavily contaminated by *Aristolochia clematitis*, a common parasite, and suggested that BEN might be caused by the chronic ingestion of their seeds, finally contaminating the bread (Ivic 1969). More recently, definitive proof of AA exposure in patients with BEN has come from the detection of aristolactam-DNA adducts and the high incidence of urothelial tumors in patients suffering from Balkan nephropathy (Grollman 2013; Jelakovic et al. 2012; Pavlovi 2013). The accidental ingestion of these species of *Aristolochia* not only explains the existence of Balkan endemic nephropathy, but also clarifies incidents of severe livestock poisoning that occurs among horses in the Balkans (Grollman 2013) and goats in Africa (el Dirdiri et al. 1987).

Review of Applicable Methods for Testing and Evaluating Herbal Medicines Nephrotoxicity

Several established methods are available for assessing nephrotoxicity induced by drugs, chemicals, industrial, or environmental toxic agents. These methods are briefly reviewed and compared in this section. In addition, protocols and guidelines on safety and toxicity assessment of herbal products have been published by the International Life Sciences Institute (Schilter et al. 2003), the Committee on the Framework for Evaluating the Safety of the Dietary Supplements (Committee on the Framework for Evaluating the Safety of the Dietary Supplements and National Research 2004), the International Union of Pure and Applied Chemistry (Mosihuzzaman and Choudhary 2008), the World Health Organization (World Health Organization 2004, 2007), and the European Food Safety Authority (European Food Safety Authority 2009). An article reviewing these publications was published in 2010 (Jordan et al. 2010).

In Vivo Animal Models

Animal models represent the "gold standard" (De Broe and Porter George 2008) of toxicity assessment and are recommended by the European Medicines Agency (EMA) and the FDA for investigating the toxicity of new drugs. Indeed, the whole animal model is correlated to human toxicity as absorption, distribution, metabolism, and elimination (ADME) are reproduced more closely when administered by a route similar to its intended use. Acute, sub-acute and chronic toxicities are assessed in two animal species (one rodent and one non-rodent), with functional parameters (creatinine or blood urea nitrogen, classically) and histologically. However, there are several drawbacks to this method: It is expensive and time-consuming, so it is not possible to screen all herbal products this way. It is also sometimes difficult to conclude the toxicity or safety for humans from herbal products tested in animal models, as subtle differences within species can affect the type of effects that are observed. Finally, ethical considerations such as the "3R" – refinement (use of methods that alleviate or minimize potential pain, suffering, or distress, and enhance animal welfare), replacement (use of an alternative technique), and reduction (number of animals) should also be taken into consideration when using animals in an experiment.

Isolated Perfused Kidney

Since the isolated perfused kidney is an intact organ and maintains structural integrity, it is ideal for pathophysiologic studies (such as renal vascular, tubular, and glomerular functions) (De Broe and Porter George 2008) and drug metabolism studies (elucidation of mechanism of renal excretion, and assessment of drug interactions) (Taft 2004). However, isolated rat and mouse perfused kidneys have also been used to study toxicity for very acute toxicants (Gandolfi and Brendel 1990; Martins et al. 2002; Shanley et al. 1986) when hemodynamic changes due to the toxin were suspected, but to the best of our knowledge, never for subacute or chronic toxicity models, nor for toxicity screening purposes. In addition, this method requires extensive preparation, skillful handling, and is only viable for a few hours. For these reasons, this technical approach lacks validation and is not recommended for studying herbal toxicity.

Renal Cell Culture Models

Renal cell culture models offer a better way to understand biochemical and cellular mechanisms of cytotoxicity and are cheaper and less time-consuming than animal experimentation. For example, a recent study has investigated the renal toxicity of 47 herbs traditionally used for kidney and urinary disorders in two cell line models (Wojcikowski et al. 2009).

The epithelial cells from the proximal tubule are usual targets of toxic injury. However, depending on the nephrotoxic compound, other kidney cell types are available, allowing the study of the specific toxicity of different targets in the kidney tissue.

There are mainly two types of renal cell cultures: cell lines and primary cell cultures. Cell lines are easier to culture, have an unlimited life span, don't need an isolation procedure, and are more stable; however, they are less close than "*in vivo*" cells. Renal cell lines reproducing different renal cell types are available, such as cell lines with characteristics of the proximal tubule (normal rat kidney cells (NRK52E), Lewis Lung Carcinoma Porcine Kidney (LLC-PK1), opossum kidney (OK), or Human Kidney-2 (HK-2 cells)), cell lines with characteristics of the distal tubule or the collecting duct (Madin Darby Canine Kidney cells (MDCK), or Madin Darby Bovine Kidney cells (MDBK)).

Primary renal proximal tubular epithelial cells (RPTEC) are closer to *in vivo* cells, but the isolation procedure and the limited access to the whole organ limit their utilization. In addition, their limited life span, their changing phenotype over time, and the absence of standard culture conditions limit reproducibility of the experiment.

On the other hand, it is often difficult to extrapolate results from these models to in vivo situations for several reasons. Firstly, herbal products are tested as an extract in cell culture models instead of raw powders. Secondly, concentrations used in in vitro experiments may not reflect those seen after consumption by humans, as pharmacokinetic profiles are lacking. Finally, these experimental approaches do not allow studies related to ADME. *In vitro* toxicity studies on metabolites are indeed unfeasible in general, as metabolic pathways have not been previously described for most herbal products. To the best of our knowledge, taking into account prior data about AA metabolism in vivo, renal cell lines studies have been mainly applied to AA and their metabolites (aristolactams) obtained from *Aristolochia* species (Balachandran et al. 2005; Krumbiegel et al. 1987; Yu et al. 2011; Yuan et al. 2011).

Renal Slice Models

This model provides a heterogeneous cell population maintaining the organization of the renal cortex and the intercellular communication. Therefore, this technique is useful for evaluating renal drug metabolism and drug transport.

Slices from various species are commercially available or are prepared from the renal cortex. Renal cortical slices are generally prepared from kidneys of untreated rats and incubated with tested compounds or vehicles. However, in contrast to hepatotoxicity studies, only a few renal slice models are used to study nephrotoxicity related to iodinated contrast compounds, cisplatin, and sevoflurane (Catania et al. 2001; Harmon et al. 2009; Minigh and Valentovic 2003; Vickers and Fisher 2004; Vickers et al. 2004). As far as we know regarding herbal products nephrotoxicity, renal slice models have been used only for investigating AA (Dickman et al. 2011) and tripterygium (Dan et al. 2008) toxicity, but never for screening acute/chronic nephrotoxicity of herbal products, and should therefore not be used for chronic renal toxicity assessment.

Toxicogenomics: Transcriptomics, Metabolomics and Proteomics

Herbal toxicogenomics refers to the combination of toxicology with different "omics" tools that assess the potential toxicity of herbal products on selected organs such as the kidney. Metabolomics (or metabonomics) and proteomics have been widely used in renal toxicological studies that include drugs or chemicals (Boudonck et al. 2009) (gentamicin, cisplatin, cyclosporine, NiCl2 (Tyagi et al. 2013), ochratoxin A (Sieber et al. 2009) and indomethacin (Lv et al. 2011)), and also in traditional medicine compounds: for cinnabar and realgar (Lao et al. 2009; Wei et al. 2008, 2009) (minerals containing mercury or arsenic) or herbal products such as *Aristolochia manshuriensis*, (Sieber et al. 2009; Lao et al. 2009; Chen et al. 2006, 2008; Lin et al. 2010; Zhang et al. 2006), morning glory seeds (prepared from seeds of *Ipomoea* nil (L.) Roth, or *Ipomoea* purpurea (L.) Roth), (Ma et al. 2010) triterpenoid extracted from *Cimicifuga foetida* L. (He et al. 2011), and *Tripterygium wilfordii* Hook.f. (Xia et al. 2009). This approach would help distinguish different intoxications, would allow an earlier detection of kidney injury, and would identify the target in the kidney. However, two studies comparing traditional techniques for assessing xenobiotic-induced nephrotoxicity with proteomics and metabonomics techniques found mitigated results. In the first review – published in 2005 (Gibbs 2005) – the authors concluded that there was no consistent evidence that the novel methodologies were more sensitive than the traditional methods. The second, recently published by the Consortium for Metabolic Toxicology (COMET) group (Ebbels et al. 2007), was performed to build a model for predicting toxicity using a set of 80 toxicants. The sensitivity for kidney toxicity was only 41 % (with a specificity of 100 %). In our opinion, "omics" techniques are currently insufficiently validated to correctly assess the safety of herbal medicines. However, these techniques could be helpful in detecting new biomarkers of renal injury (Beger et al. 2010; Devarajan 2008; Rouse et al. 2011), and seem to be promising for the future. Moreover, "omics" techniques may be informative in selecting main potent nephrotoxic herbs.

Conclusions

Herbal nephrotoxicity is an underestimated problem and needs to be assessed worldwide. To date, scientific information comes almost exclusively from case reports and experimental studies. In many situations, only a possible causality link can be proposed. More clinical research (epidemiological data and systematic reporting of clinical cases) and basic research (*in vitro* and *in vivo* studies) are required to improve our knowledge of herbal nephrotoxicity. The use of integrated toxicological approaches remains a challenge for the toxicologist nowadays (Williamson et al. 2015). In our opinion, the only validated method for correctly assessing the potential nephrotoxicity of a chemical, drug, or traditional herbal medicine remains the elaboration of an *in vivo* animal model. The place of "omic" techniques still needs to be defined.

References

Adams M, Berset C, Kessler M et al (2009) Medicinal herbs for the treatment of rheumatic disorders – a survey of European herbals from the 16th and 17th century. J Ethnopharmacol 121:343–359

Adelekun TA, Ekwere TR, Akinsola A (1999) The pattern of acute toxic nephropathy in Ife, Nigeria. West Afr J Med 18:60–63

Akyol AD, Yildirim Y, Toker E et al (2011) The use of complementary and alternative medicine among chronic renal failure patients. J Clin Nurs 20:1035–1043

Ardalan MR, Khodaie L, Nasri H et al (2015) Herbs and hazards: risk of aristolochic acid nephropathy in Iran. Iran J Kidney Dis 9:14–17

Arlt VM, Stiborova M, Schmeiser HH (2002) Aristolochic acid as a probable human cancer hazard in herbal remedies: a review. Mutagenesis 17:265–277

Aslam M, Davis SS, Healy MA (1979) Heavy metals in some Asian medicines and cosmetics. Public Health 93:274–284

Balachandran P, Wei F, Lin RC et al (2005) Structure activity relationships of aristolochic acid analogues: toxicity in cultured renal epithelial cells. Kidney Int 67:1797–1805

Barrett B, Kiefer D, Rabago D (1999) Assessing the risks and benefits of herbal medicine: an overview of scientific evidence. Altern Ther Health Med 5:40–49

Baudoux TE, Pozdzik AA, Arlt VM et al (2012) Probenecid prevents acute tubular necrosis in a mouse model of aristolochic acid nephropathy. Kidney Int 82:1105–1113

Beger RD, Sun J, Schnackenberg LK (2010) Metabolomics approaches for discovering biomarkers of drug-induced hepatotoxicity and nephrotoxicity. Toxicol Appl Pharmacol 243:154–166

Bertrand B (2009) L'herbier toxique. Codes secrets pour plantes utiles, Plume de carotte edn. Plume de carotte, Toulouse

Boudonck KJ, Rose DJ, Karoly ED et al (2009) Metabolomics for early detection of drug-induced kidney injury: review of the current status. Bioanalysis 1:1645–1663

Catania JM, Parrish AR, Gandolfi AJ (2001) Toxicity of a sevoflurane degradation product incubated with rat liver and renal cortical slices. Drug Chem Toxicol 24:347–357

Chan W, Lee KC, Liu N et al (2008) Liquid chromatography/mass spectrometry for metabonomics investigation of the biochemical effects induced by aristolochic acid in rats: the use of information-dependent acquisition for biomarker identification. Rapid Commun Mass Spectrom 22:873–880

Chawla LS, Kimmel PL (2012) Acute kidney injury and chronic kidney disease: an integrated clinical syndrome. Kidney Int 82:516–524

Chen M, Su M, Zhao L et al (2006) Metabonomic study of aristolochic acid-induced nephrotoxicity in rats. J Proteome Res 5:995–1002

Chen CH, Dickman KG, Moriya M et al (2012) Aristolochic acid-associated urothelial cancer in Taiwan. Proceedings of the National Academy of Sciences of the United States of America

Coghlan ML, Haile J, Houston J et al (2012) Deep sequencing of plant and animal DNA contained within traditional Chinese medicines reveals legality issues and health safety concerns. PLoS Genet 8:e1002657

Colson CRD, De Broe ME (2005) Kidney injury from alternative medicines. Adv Chron Kid Dis 12:261–275

Committee on the Framework for Evaluating the Safety of the Dietary Supplements, National Research Council (2004) Dietary supplements: a framework for evaluating safety. Washington, DC

Cosyns JP, Jadoul M, Squifflet JP et al (1994) Chinese herbs nephropathy: a clue to Balkan endemic nephropathy? Kidney Int 45:1680–1688

Cosyns JP, Jadoul M, Squifflet JP et al (1999) Urothelial lesions in Chinese-herb nephropathy. Am J Kidney Dis 33:1011–1017

Cosyns JP, Dehoux JP, Guiot Y et al (2001) Chronic aristolochic acid toxicity in rabbits: a model of Chinese herbs nephropathy? Kidney Int 59:2164–2173

Couceiro M, Afreen F, Zobayed S et al (2006) Variation in concentrations of major bioactive compounds of St. John's wort: effects of harvesting time, temperature and germplasm. Plant Sci 170:128–134

Dan H, Peng RX, Ao Y et al (2008) Segment-specific proximal tubule injury in tripterygium glycosides intoxicated rats. J Biochem Mol Toxicol 22:422–428

De Broe ME (2012) Chinese herbs nephropathy and Balkan endemic nephropathy: toward a single entity, aristolochic acid nephropathy. Kidney Int 81:513–515

De Broe ME, Porter George A (2008) Clinical nephrotoxins: renal injury from drugs and chemicals. Kluwer Academic Publishers, Dordrecht, The Netherlands

De Smet PA (2007) Clinical risk management of herb-drug interactions. Br J Clin Pharmacol 63:258–267

Debelle FD, Nortier JL, De Prez EG et al (2002) Aristolochic acids induce chronic renal failure with interstitial fibrosis in salt-depleted rats. J Am Soc Nephrol 13:431–436

Debelle FD, Vanherweghem JL, Nortier JL (2008) Aristolochic acid nephropathy: a worldwide problem. Kidney Int 74:158–169

Depierreux M, Van Damme B, Vanden Houte K et al (1994) Pathologic aspects of a newly described nephropathy related to the prolonged use of Chinese herbs. Am J Kidney Dis 24:172–180

Devarajan P (2008) Proteomics for the investigation of acute kidney injury. Contrib Nephrol 160:1–16

Dickman KG, Sweet DH, Bonala R et al (2011) Physiological and molecular characterization of aristolochic acid transport by the kidney. J Pharmacol Exp Ther 338:588–597

Djukanovic L, Radovic M, Bakovic J et al (2002) Epidemiology of end-stage renal disease and current status of hemodialysis in Yugoslavia. Int J Artif Organs 25:852–859

Dunn JP, Krige JE, Wood R et al (1991) Colonic complications after toxic tribal enemas. Br J Surg 78:545–548

Ebbels TM, Keun HC, Beckonert OP et al (2007) Prediction and classification of drug toxicity using probabilistic modeling of temporal metabolic data: the consortium on metabonomic toxicology screening approach. J Proteome Res 6:4407–4422

el Dirdiri NI, Barakat SE, Adam SE (1987) The combined toxicity of Aristolochia bracteata and Cadaba rotundifolia to goats. Vet Hum Toxicol 29:133–137

European Food Safety Authority (2009) Guidance on safety assessment of botanicals and botanical preparations intended for use as ingredients in food supplements. EFSA J 7:1249–1268

Fadrowski JJ, Navas-Acien A, Tellez-Plaza M et al (2010) Blood lead level and kidney function in US adolescents: the third National Health and Nutrition Examination Survey. Arch Intern Med 170:75–82

Ferraro PM, Costanzi S, Naticchia A et al (2010) Low level exposure to cadmium increases the risk of chronic kidney disease: analysis of the NHANES 1999–2006. BMC Public Health 10:304

Gabardi S, Munz K, Ulbricht C (2007) A review of dietary supplement-induced renal dysfunction. Clin J Am Soc Nephrol 2:757–765

Gandolfi AJ, Brendel K (1990) In vitro systems for nephrotoxicity studies. Toxicol In Vitro 4:337–345

Garvey GJ, Hahn G, Lee RV et al (2001) Heavy metal hazards of Asian traditional remedies. Int J Environ Health Res 11:63–71

Gibbs A (2005) Comparison of the specificity and sensitivity of traditional methods for assessment of nephrotoxicity in the rat with metabonomic and proteomic methodologies. J Appl Toxicol 25:277–295

Gokmen MR, Cosyns JP, Arlt VM et al (2013) The epidemiology, diagnosis, and management of aristolochic acid nephropathy: a narrative review. Ann Intern Med 158:469–477

Grollman AP (2013) Aristolochic acid nephropathy: harbinger of a global iatrogenic disease. Environ Mol Mutagen 54:1–7

Grubbs V, Plantinga LC, Tuot DS et al (2013) Americans' use of dietary supplements that are potentially harmful in CKD. Am J Kidney Dis 61:739–747

Guh JY, Chen HC, Tsai JF et al (2007) Herbal therapy is associated with the risk of CKD in adults not using analgesics in Taiwan. Am J Kidney Dis 49:626–633

Harmon RC, Duffy SP, Terneus MV et al (2009) Characterization of a novel model for investigation of radiocontrast nephrotoxicity. Nephrol Dial Transplant 24:763–768

Hashimoto K, Higuchi M, Makino B et al (1999) Quantitative analysis of aristolochic acids, toxic compounds, contained in some medicinal plants. J Ethnopharmacol 64:185–189

He CC, Dai YQ, Hui RR et al (2011) NMR-based metabonomic approach on the toxicological effects of a Cimicifuga triterpenoid. J Appl Toxicol 32:88–97

Heinrich M, Chan J, Wanke S et al (2009) Local uses of Aristolochia species and content of nephrotoxic aristolochic acid 1 and 2 – a global assessment based on bibliographic sources. J Ethnopharmacol 125:108–144

Hess S, De Geest S, Halter K et al (2009) Prevalence and correlates of selected alternative and complementary medicine in adult renal transplant patients. Clin Transpl 23:56–62

Hsieh SC, Lin IH, Tseng WL et al (2008) Prescription profile of potentially aristolochic acid containing Chinese herbal products: an analysis of National Health Insurance data in Taiwan between 1997 and 2003. Chin Med 3:13

Ioset JR, Raoelison GE, Hostettmann K (2003) Detection of aristolochic acid in Chinese phytomedicines and dietary supplements used as slimming regimens. Food Chem Toxicol 41:29–36

Isnard Bagnis C, Deray G, Baumelou A et al (2004) Herbs and the kidney. Am J Kidney Dis 44:1–11

Ivic M (1969) Etiology of endemic nephropathy. Lijec Vjesn 91:1273–1281

Jelakovic B, Karanovic S, Vukovic-Lela I et al (2012) Aristolactam-DNA adducts are a biomarker of environmental exposure to aristolochic acid. Kidney Int 81:559–567

Jordan SA, Cunningham DG, Marles RJ (2010) Assessment of herbal medicinal products: challenges, and opportunities to increase the knowledge base for safety assessment. Toxicol Appl Pharmacol 243:198–216

Krumbiegel G, Hallensleben J, Mennicke WH et al (1987) Studies on the metabolism of aristolochic acids I and II. Xenobiotica 17:981–991

Ladda MA, Goralski KB (2016) The effects of CKD on cytochrome P450-mediated drug metabolism. Adv Chronic Kidney Dis 23:67–75

Lai MN, Lai JN, Chen PC et al (2009) Increased risks of chronic kidney disease associated with prescribed Chinese herbal products suspected to contain aristolochic acid. Nephrology (Carlton) 14:227–234

Laïs E (2003) Le grand livre des plantes magiques. Jardin de sorcières, Rustica edn. Rustica, Paris

Lao YM, Jiang JG, Yan L (2009) Application of metabonomic analytical techniques in the modernization and toxicology research of traditional Chinese medicine. Br J Pharmacol 157:1128–1141

Lemy A, Wissing KM, Rorive S et al (2008) Late onset of bladder urothelial carcinoma after kidney transplantation for end-stage aristolochic acid nephropathy: a case series with 15-year follow-up. Am J Kidney Dis 51:471–477

Lin S, Chan W, Li J et al (2010) Liquid chromatography/mass spectrometry for investigating the biochemical effects induced by aristolochic acid in rats: the plasma metabolome. Rapid Commun Mass Spectrom 24:1312–1318

Luyckx VA (2012) Nephrotoxicity of alternative medicine practice. Adv Chronic Kidney Dis 19:129–141

Luyckx VA, Naicker S (2008) Acute kidney injury associated with the use of traditional medicines. Nat Clin Pract Nephrol 4:664–671

Luyckx VA, Steenkamp V, Stewart MJ (2005) Acute renal failure associated with the use of traditional folk remedies in South Africa. Ren Fail 27:35–43

Lv H, Liu L, Palacios G et al (2011) Metabolomic analysis characterizes tissue specific indomethacin-induced metabolic perturbations of rats. Analyst 136:2260–2269

Ma C, Bi K, Su D et al (2010) Serum and kidney metabolic changes of rat nephrotoxicity induced by morning glory seed. Food Chem Toxicol 48:2988–2993

Mai I, Stormer E, Bauer S et al (2003) Impact of St. John's wort treatment on the pharmacokinetics of tacrolimus and mycophenolic acid in renal transplant patients. Nephrol Dial Transplant 18:819–822

Marcus DM, Grollman AP (2002) Botanical medicines – the need for new regulations. N Engl J Med 347:2073–2076
Martins AM, Toyama MH, Havt A et al (2002) Determination of Crotalus durissus cascavella venom components that induce renal toxicity in isolated rat kidneys. Toxicon 40:1165–1171
Minigh JL, Valentovic MA (2003) Characterization of myoglobin toxicity in renal cortical slices from Fischer 344 rats. Toxicology 184:113–123
Mosihuzzaman M, Choudhary I,M (2008) Protocols on safety, efficacy, standardization and documentation of herbal medicine. Pure Appl Chem 80:2195–2230
Navas-Acien A, Tellez-Plaza M, Guallar E et al (2009) Blood cadmium and lead and chronic kidney disease in US adults: a joint analysis. Am J Epidemiol 170:1156–1164
Newmaster SG, Grguric M, Shanmughanandhan D et al (2013) DNA barcoding detects contamination and substitution in North American herbal products. BMC Med 11:222
Nortier JL, Martinez MC, Schmeiser HH et al (2000) Urothelial carcinoma associated with the use of a Chinese herb (Aristolochia fangchi). N Engl J Med 342:1686–1692
Nortier J, Pozdzik A, Roumeguere T et al (2015) Néphropathie aux acides aristolochiques ("néphropathie aux herbes chinoises"). Nephrol Ther 11:574–588
Nowack R, Balle C, Birnkammer F et al (2009) Complementary and alternative medications consumed by renal patients in southern Germany. J Ren Nutr 19:211–219
Nwoko CO, Mgbeahuruike L (2011) Heavy metal contamination of ready-to-use herbal remedies in South Eastern Nigeria. Pak J Nutr 10:959–964
Obi E, Akunyili DN, Ekpo B et al (2006) Heavy metal hazards of Nigerian herbal remedies. Sci Total Environ 369:35–41
Olowu WA, Niang A, Osafo C et al (2016) Outcomes of acute kidney injury in children and adults in sub-Saharan Africa: a systematic review. Lancet Glob Health 4:e242–e250
Pavlovi NM (2013) Balkan endemic nephropathy – current status and future perspectives. Clin Kidney J 6:257–265
Perazella MA (2009) Renal vulnerability to drug toxicity. Clin J Am Soc Nephrol 4:1275–1283
Ponce D, Dias DB, Nascimento GR et al (2016) Long-term outcome of severe acute kidney injury survivors followed by nephrologists in a developing country. Nephrology (Carlton) 21:327–334
Pozdzik AA, Salmon IJ, Debelle FD et al (2008a) Aristolochic acid induces proximal tubule apoptosis and epithelial to mesenchymal transformation. Kidney Int 73:595–607
Pozdzik AA, Salmon IJ, Husson CP et al (2008b) Patterns of interstitial inflammation during the evolution of renal injury in experimental aristolochic acid nephropathy. Nephrol Dial Transplant 23:2480–2491
Rouse RL, Zhang J, Stewart SR et al (2011) Comparative profile of commercially available urinary biomarkers in preclinical drug-induced kidney injury and recovery in rats. Kidney Int 79:1186–1197
Sabath E, Robles-Osorio ML (2012) Renal health and the environment: heavy metal nephrotoxicity. Nefrologia 32:279–286
Saper RB, Kales SN, Paquin J et al (2004) Heavy metal content of Ayurvedic herbal medicine products. JAMA 292:2868–2873
Schilter B, Andersson C, Anton R et al (2003) Guidance for the safety assessment of botanicals and botanical preparations for use in food and food supplements. Food Chem Toxicol 41:1625–1649
Schmeiser HH, Nortier JL, Singh R et al (2014) Exceptionally long-term persistence of DNA adducts formed by carcinogenic aristolochic acid I in renal tissue from patients with aristolochic acid nephropathy. Int J Cancer 135:502–507
Shanley PF, Brezis M, Spokes K et al (1986) Differential responsiveness of proximal tubule segments to metabolic inhibitors in the isolated perfused rat kidney. Am J Kidney Dis 7:76–83
Sieber M, Wagner S, Rached E et al (2009) Metabonomic study of ochratoxin a toxicity in rats after repeated administration: phenotypic anchoring enhances the ability for biomarker discovery. Chem Res Toxicol 22:1221–1231
Siew ED, Peterson JF, Eden SK et al (2012) Outpatient nephrology referral rates after acute kidney injury. J Am Soc Nephrol 23:305–312

Stanifer JW, Jing B, Tolan S et al (2014) The epidemiology of chronic kidney disease in sub-Saharan Africa: a systematic review and meta-analysis. Lancet Glob Health 2:e174–e181
Taber SS, Pasko DA (2008) The epidemiology of drug-induced disorders: the kidney. Expert Opin Drug Saf 7:679–690
Taft DR (2004) The isolated perfused rat kidney model: a useful tool for drug discovery and development. Curr Drug Discov Technol 1:97–111
Tatu CA, Orem WH, Finkelman RB et al (1998) The etiology of Balkan endemic nephropathy: still more questions than answers. Environ Health Perspect 106:689–700
Tyagi R, Rana P, Gupta M et al (2013) Differential biochemical response of rat kidney towards low and high doses of NiCl2 as revealed by NMR spectroscopy. J Appl Toxicol 33:134–141
Vanhaelen M, Vanhaelen-Fastre R, But P et al (1994) Identification of aristolochic acid in Chinese herbs. Lancet 343:174
Vanherweghem JL (1997) Aristolochia sp and chronic interstitial nephropathies in Indians. Lancet 349:1399
Vanherweghem LJ (1998) Misuse of herbal remedies: the case of an outbreak of terminal renal failure in Belgium (Chinese herbs nephropathy). J Altern Complement Med 4:9–13
Vanherweghem JL, Depierreux M, Tielemans C et al (1993) Rapidly progressive interstitial renal fibrosis in young women: association with slimming regimen including Chinese herbs. Lancet 341:387–391
Vanherweghem JL, Abramowicz D, Tielemans C et al (1996) Effects of steroids on the progression of renal failure in chronic interstitial renal fibrosis: a pilot study in Chinese herbs nephropathy. Am J Kidney Dis 27:209–215
Vickers AE, Fisher RL (2004) Organ slices for the evaluation of human drug toxicity. Chem Biol Interact 150:87–96
Vickers AE, Rose K, Fisher R et al (2004) Kidney slices of human and rat to characterize cisplatin-induced injury on cellular pathways and morphology. Toxicol Pathol 32:577–590
Wei L, Liao P, Wu H et al (2008) Toxicological effects of cinnabar in rats by NMR-based metabolic profiling of urine and serum. Toxicol Appl Pharmacol 227:417–429
Wei L, Liao P, Wu H et al (2009) Metabolic profiling studies on the toxicological effects of realgar in rats by (1)H NMR spectroscopy. Toxicol Appl Pharmacol 234:314–325
WHO traditional medicine strategy 2014–2023 (2013) www.who.int/medicines/publications/traditional/trm_strategy14_23/en/(last access December 14, 2016)
Williamson EM, Chan K, Xu Q et al (2015) Evaluating the safety of herbal medicines: integrated toxicological approaches. Science 347:S47–S49
Wojcikowski K, Wohlmuth H, Johnson DW et al (2009) An in vitro investigation of herbs traditionally used for kidney and urinary system disorders: potential therapeutic and toxic effects. Nephrology (Carlton) 14:70–79
World Health Organization (2004) WHO guidelines on safety monitoring of herbal medicines in pharmacovigilance systems. See http://apps.who.int/medicinedocs/en/d/Js7148e/
World Health Organization (2007) WHO guidelines for assessing quality of herbal medicines with reference to contaminants and residues. See http://apps.who.int/medicinedocs/fr/m/abstract/Js14878e/
Wu KM, Farrelly JG, Upton R et al (2007) Complexities of the herbal nomenclature system in traditional Chinese medicine (TCM): lessons learned from the misuse of Aristolochia-related species and the importance of the pharmaceutical name during botanical drug product development. Phytomedicine 14:273–279
Xia S, Liu H, Zhu H et al (2009) NMR-based metabonomic study on rat's urinary metabolic response to dosage of triptolide. Chin J Chem 27:751–758
Yamani A, Bunel V, Antoine MH et al (2015) Substitution between Aristolochia and Bryonia genus in North-Eastern Morocco: toxicological implications. J Ethnopharmacol 166:250–260
Yang MH, Chen KK, Yen CC et al (2002) Unusually high incidence of upper urinary tract urothelial carcinoma in Taiwan. Urology 59:681–687
Yu CC, Lin JL, Lin-Tan DT (2004) Environmental exposure to lead and progression of chronic renal diseases: a four-year prospective longitudinal study. J Am Soc Nephrol 15:1016–1022

Yu FY, Wu TS, Chen TW et al (2011) Aristolochic acid I induced oxidative DNA damage associated with glutathione depletion and ERK1/2 activation in human cells. Toxicol In Vitro 25:810–816
Yuan SY, Yang CR, Cheng CL et al (2011) Comparative nephrotoxicity of aristolochic acid and tetrandrine in vitro and in vivo. Int J Toxicol 30:35–46
Zhang X, Wu H, Liao P et al (2006) NMR-based metabonomic study on the subacute toxicity of aristolochic acid in rats. Food Chem Toxicol 44:1006–1014
Zhao Z, Yuen JP, Wu J et al (2006) A systematic study on confused species of Chinese materia medica in the Hong Kong market. Ann Acad Med Singap 35:764–769

References cited in Tables 13.1 to 13.5

Adesunloye BA (2003) Acute renal failure due to the herbal remedy CKLS. Am J Med 115:506–507
Averbukh Z, Modai D, Leonov Y et al (1989) Rhabdomyolysis and acute renal failure induced by paraphenylenediamine. Hum Toxicol 8:345–348
Bandara V, Weinstein SA, White J et al (2010) A review of the natural history, toxinology, diagnosis and clinical management of *Nerium oleander* (common oleander) and *Thevetia peruviana* (yellow oleander) poisoning. Toxicon 56:273–281
Benjamin SP, Fernando ME, Jayanth JJ et al (2006) Cleistanthus collinus poisoning. J Assoc Physicians India 54:742–744
Boubaker K, Ounissi M, Brahmi N et al (2013) Acute renal failure by ingestion of Euphorbia paralias. Saudi J Kidney Dis Transpl 24:571–575
Bourquia A, Jabrane AJ, Ramdani B et al (1988) Systemic toxicity of paraphenylenediamine. 4 cases. Presse Med 17:1798–1800
Brncic N, Viskovic I, Peric R et al (2001) Accidental plant poisoning with Colchicum autumnale: report of two cases. Croat Med J 42:673–675
Brvar M, Ploj T, Kozelj G et al (2004) Case report: fatal poisoning with Colchicum autumnale. Crit Care 8:R56–R59
Bunawan NC, Rastegar A, White KP et al (2014) Djenkolism: case report and literature review. Int Med Case Rep J 7:79–84
Chen CL, Fang HC, Chou KJ et al (2001) Acute oxalate nephropathy after ingestion of star fruit. Am J Kidney Dis 37:418–422
Chou WC, Wu CC, Yang PC et al (1995) Hypovolemic shock and mortality after ingestion of Tripterygium wilfordii hook F.: a case report. Int J Cardiol 49:173–177
Chrispal A (2012) Cleistanthus collinus poisoning. J Emerg Trauma Shock 5:160–166
Conz PA, La Greca G, Benedetti P et al (1998) Fucus vesiculosus: a nephrotoxic alga? Nephrol Dial Transplant 13:526–527
Daniele C, Dahamna S, Firuzi O et al (2005) Atractylis gummifera L. poisoning: an ethnopharmacological review. J Ethnopharmacol 97:175–181
Das S, Hamide A, Mohanty MK et al (2014) Fatal Cleistanthus collinus toxicity: a case report and review of literature. J Forensic Sci 59:1441–1447
Diamond JR, Pallone TL (1994) Acute interstitial nephritis following use of tung shueh pills. Am J Kidney Dis 24:219–221
Djibo A, Bouzou SB (2000) Acute intoxication with "sobi-lobi" (Datura). Four cases in Niger. Bull Soc Pathol Exot 93:294–297
Ernst E (2002) St. John's wort supplements endanger the success of organ transplantation. Arch Surg 137:316–319
Fang HC, Lee PT, Lu PJ et al (2008) Mechanisms of star fruit-induced acute renal failure. Food Chem Toxicol 46:1744–1752
Haller CA, Benowitz NL (2000) Adverse cardiovascular and central nervous system events associated with dietary supplements containing ephedra alkaloids. N Engl J Med 343:1833–1838

Hilepo JN, Bellucci AG, Mossey RT (1997) Acute renal failure caused by "cat's claw" herbal remedy in a patient with systemic lupus erythematosus. Nephron 77:361
Horoz M, Gok E, Genctoy G et al (2008) Crataegus orientalis associated multiorgan hypersensitivity reaction and acute renal failure. Intern Med 47:2039–2042
Hsu HY, Lin CC, Chen JY et al (1998) Toxic effects of Erycibe obtusifolia, a Chinese medicinal herb, in mice. J Ethnopharmacol 62:101–105
Huang HH, Yen DH, Wu ML et al (2006) Acute erycibe henryi prain ("Ting Kung Teng") poisoning. Clin Toxicol (Phila) 44:71–75
Huang GZ, Li L, Liu L (2009) Pathological study on autopsy died of tripterygium intoxication – report of 4 cases. Zhongguo Zhong Xi Yi Jie He Za Zhi Zhongguo Zhongxiyi Jiehe Zazhi Chin J Integr Tradit West Med/Zhongguo Zhong Xi Yi Jie He Xue Hui Zhongguo Zhong Yi Yan Jiu Yuan Zhu Ban 29:165–168
Jha V (2010) Herbal medicines and chronic kidney disease. Nephrology (Carlton) 15(Suppl 2):10–17
Jha V, Parameswaran S (2013) Community-acquired acute kidney injury in tropical countries. Nat Rev Nephrol 9:278–290
Kang K-S, Heo ST (2015) A case of life-threatening acute kidney injury with toxic encephalopathy caused by Dioscorea quinqueloba. Yonsei Med J 56:304–306
Kim CS, Kim SM, Choi JS et al (2012) Dioscorea quinqueloba induces acute kidney injury. Clin Toxicol (Phila) 50:80
Kim HY, Kim SS, Bae SH et al (2014) Acute interstitial nephritis induced by Dioscorea quinqueloba. BMC Nephrol 15:143
Klintschar M, Beham-Schmidt C, Radner H et al (1999) Colchicine poisoning by accidental ingestion of meadow saffron (Colchicum autumnale): pathological and medicolegal aspects. Forensic Sci Int 106:191–200
Kwan TH, Tong MK, Leung KT et al (2006) Acute renal failure associated with prolonged intake of slimming pills containing anthraquinones. Hong Kong Med J 12:394–397
Kwan LP, Mok MM, Ma MK et al (2014) Acute drug toxicity related to drinking herbal tea in a kidney transplant recipient. Ren Fail 36:309–312
Lee JJ, Chen HC (2006) Flavonoid-induced acute nephropathy by Cupressus funebris Endl (mourning cypress). Am J Kidney Dis 48:e81–e85
Li FK, Lai CK, Poon WT et al (2004) Aggravation of non-steroidal anti-inflammatory drug-induced hepatitis and acute renal failure by slimming drug containing anthraquinones. Nephrol Dial Transplant 19:1916–1917
Light TD, Light JA (2003) Acute renal transplant rejection possibly related to herbal medications. Am J Transplant 3:1608–1609
Lin CC, Chen JC (2002) Medicinal herb Erycibe henri Prain ("Ting Kung Teng") resulting in acute cholinergic syndrome. J Toxicol Clin Toxicol 40:185–187
Lin JL, Ho YS (1994) Flavonoid-induced acute nephropathy. Am J Kidney Dis 23:433–440
Logan JL, Ahmed J (2003) Critical hypokalemic renal tubular acidosis due to Sjogren's syndrome: association with the purported immune stimulant echinacea. Clin Rheumatol 22:158–159
Luyckx VA, Ballantine R, Claeys M et al (2002) Herbal remedy-associated acute renal failure secondary to Cape aloes. Am J Kidney Dis 39:E13
Mueller BA, Scott MK, Sowinski KM et al (2000) Noni juice (Morinda citrifolia): hidden potential for hyperkalemia? Am J Kidney Dis 35:310–312
Mullins RJ, Heddle R (2002) Adverse reactions associated with echinacea: the Australian experience. Ann Allergy Asthma Immunol 88:42–51
Neto MM, Silva GE, Costa RS et al (2009) Star fruit: simultaneous neurotoxic and nephrotoxic effects in people with previously normal renal function. NDT Plus 2:485–488
Popat A, Shear NH, Malkiewicz I et al (2001) The toxicity of Callilepis laureola, a South African traditional herbal medicine. Clin Biochem 34:229–236
Powell T, Hsu FF, Turk J et al (1998) Ma-huang strikes again: ephedrine nephrolithiasis. Am J Kidney Dis 32:153–159
Prakash J (1999) Acute renal failure in epidemic dropsy. Ren Fail 21:707–711
Prakash S, Hernandez GT, Dujaili I et al (2009) Lead poisoning from an Ayurvedic herbal medicine in a patient with chronic kidney disease. Nat Rev Nephrol 5:297–300

Rhidian R (2011) Running a risk? Sport supplement toxicity with ephedrine in an amateur marathon runner, with subsequent rhabdomyolysis. BMJ Case Rep 2011, doi:10.1136/bcr.11.2011.5093
Sandler B, Aronson P (1993) Yohimbine-induced cutaneous drug eruption, progressive renal failure, and lupus-like syndrome. Urology 41:343–345
Schwarz A (1993) Beethoven's renal disease based on his autopsy: a case of papillary necrosis. Am J Kidney Dis 21:643–652
Seedat YK (1993) Acute renal failure in the black population of South Africa. Int J Artif Organs 16:801–802
Seedat YK, Hitchcock PJ (1971) Acute renal failure from Callilepsis laureola. S Afr Med J 45:832–833
Segasothy M, Samad S (1991) Illicit herbal preparation containing phenylbutazone causing analgesic nephropathy. Nephron 59:166–167
Segasothy M, Swaminathan M, Kong NC et al (1995) Djenkol bean poisoning (djenkolism): an unusual cause of acute renal failure. Am J Kidney Dis 25:63–66
Sharma BD, Malhotra S, Bhatia V et al (1999) Epidemic dropsy in India. Postgrad Med J 75:657–661
Shemesh IY, Mishal Y, Baruchin AM et al (1995) Rhabdomyolysis in paraphenylenediamine intoxication. Vet Hum Toxicol 37:244–245
Sir Hashim M, Hamza YO, Yahia B et al (1992) Poisoning from henna dye and paraphenylenediamine mixtures in children in Khartoum. Ann Trop Paediatr 12:3–6
Smith AY, Feddersen RM, Gardner KD Jr et al (1994) Cystic renal cell carcinoma and acquired renal cystic disease associated with consumption of chaparral tea: a case report. J Urol 152:2089–2091
Snow A, Halpenny D, McNeill G et al (2012) Life-threatening rupture of a renal angiomyolipoma in a patient taking over-the-counter horse chestnut seed extract. J Emerg Med 43:e401–e403
Stahl CE, Borlongan CV, Szerlip M et al (2006) No pain, no gain – exercise-induced rhabdomyolysis associated with the performance enhancer herbal supplement ephedra. Med Sci Monit 12:Cs81–Cs84
Steenkamp V, Stewart MJ (2005) Nephrotoxicity associated with exposure to plant toxins, with particular reference to Africa. Ther Drug Monit 27:270–277
Stewart MJ, Steenkamp V (2000) The biochemistry and toxicity of atractyloside: a review. Ther Drug Monit 22:641–649
Talasaz AH, Abbasi MR, Abkhiz S et al (2010) Tribulus terrestris-induced severe nephrotoxicity in a young healthy male. Nephrol Dial Transplant 25:3792–3793
Terris MK, Issa MM, Tacker JR (2001) Dietary supplementation with cranberry concentrate tablets may increase the risk of nephrolithiasis. Urology 57:26–29
Vachvanichsanong P, Lebel L (1997) Djenkol beans as a cause of hematuria in children. Nephron 76:39–42
Vanderperren B, Rizzo M, Angenot L et al (2005) Acute liver failure with renal impairment related to the abuse of senna anthraquinone glycosides. Ann Pharmacother 39:1353–1357
Vanholder R, De Smet R, Lesaffer G (1999) p-cresol: a toxin revealing many neglected but relevant aspects of uraemic toxicity. Nephrol Dial Transplant 14:2813–2815
Watson AR, Coovadia HM, Bhoola KD (1979) The clinical syndrome of Impila (Callilepis laureola) poisoning in children. S Afr Med J 55:290–292
Weisbord SD, Soule JB, Kimmel PL (1997) Poison on line – acute renal failure caused by oil of wormwood purchased through the Internet. N Engl J Med 337:825–827
Wong JS, Ong TA, Chua HH et al (2007) Acute anuric renal failure following jering bean ingestion. Asian J Surg 30:80–81
Wood R, Mills PB, Knobel GJ et al (1990) Acute dichromate poisoning after use of traditional purgatives. A report of 7 cases. S Afr Med J 77:640–642
Wu MS, Hong JJ, Lin JL et al (1996) Multiple tubular dysfunction induced by mixed Chinese herbal medicines containing cadmium. Nephrol Dial Transplant 11:867–870
Yang L, Su T, Li XM et al (2011) Aristolochic acid nephropathy: variation in presentation and prognosis. Nephrol Dial Transplant 27:292–298
Zhang WX, Zhang ZM, Zhang ZQ et al (2014) Andrographolide induced acute kidney injury: analysis of 26 cases reported in Chinese literature. Nephrology (Carlton) 19:21–26

Chapter 14
Herbal Neurotoxicity: An Introduction to Its Occurrence and Causes

Elizabeth M. Williamson

Abstract A number of neurotoxins have been found in herbal medicines and it is likely that others with more subtle or complex effects remain to be discovered. Severe poisoning by herbal products may be due to factors such as misidentification, adulteration, and poor processing of the plant material, and CNS "herbal" toxicity in particular may also be the result of the illicit use of recreational drugs. Any risk assessment must be considered in the context that herbal medicines are very variable both in quality and the way they are used, and also bearing in mind that a whole herb extract may produce test results quantitatively and qualitatively different than those of a constituent tested in isolation. The problems involved in assessing herbal neurotoxicity are similar to those in other areas of neurotoxicology, but are even more complex given the role of the CNS in controlling other systems of the body, and the fact that many neurotoxicity tests are not yet validated. Testing for developmental neurotoxicity is now recognized to be of crucial importance, but requires developing adapted methodologies. Increasing attention is being devoted to the development of *in vitro* systems for screening, but validation studies remain to be done to correlate *in vitro* results with neurotoxicological responses in whole animals.

Keywords Herbal neurotoxicity • Developmental neurotoxicity • Neurological abnormality

Introduction

Plant constituents can affect the central nervous system (CNS) and/or the autonomic nervous system (ANS) in a rather spectacular way – for example, strychnine, physostigmine, harmaline, and bicucculine. Plants containing these neurotoxins are not usually considered to be herbal medicines, and several natural compounds are even used to produce neurotoxicity in animal models, such as annonacin -1 from soursop,

E.M. Williamson
School of Pharmacy, University of Reading, Reading, UK
e-mail: e.m.williamson@reading.ac.uk

O. Pelkonen et al. (eds.), *Toxicology of Herbal Products*,
DOI 10.1007/978-3-319-43806 1_14

Annona muricata. Some traditional systems of medicine do use highly toxic species; notable examples are *Aconitum*, *Rhododendron*, and *Strychnos* in Chinese medicine, but these are known as such and treated accordingly (Liu et al. 2014). Poisonous and psychoactive plants are not usually found in commercialized Western herbal products (except by mistake or fraud), and are not permitted by law in the EU and the U.S. The ANS/CNS-active compounds they contain may be used therapeutically, as is the case with atropine from *Atropa belladonna*, and morphine and codeine from *Papaver somniferum*, but the botanical sources of these drugs are more commonly referred to as "medicinal plants" rather than "herbal medicines."

Despite the wealth of knowledge about herbal neurotoxins, they continue to cause clinical cases of CNS poisoning that require hospitalization and, in extreme cases, lead to the death of the patient; the *Aconitum* species are notable for this (Chan 2015). Severe poisoning is often due to the circumstances in which the herb is taken, which may include errors in dosing, misidentification, adulteration, preparation and processing – as is the case with aconite – of the plant material. CNS "herbal" toxicity encountered in practice is the result of complex factors, including the ingestion of plant products that are not herbal medicines in this context, being imported products from other systems of traditional medicine or the result of the illicit use of recreational drugs. Some of the circumstances surrounding the ingestion of herbal medicines that may result in CNS toxicity are listed in Table 14.1, with examples, in order to show that the causes of herbal toxicity can be complicated and may require more background information than clinical symptoms to be understood.

Surprisingly, there are few publications reviewing herbal neurotoxins as a group: database searches of "herbal and neurotoxin" almost exclusively report the use of herbal extracts for *reducing* neurotoxicity caused by other agents. There are exceptions for notorious toxins such as anisatin and aconitine, the effects of which are well documented, but in general, reports of neurotoxin herbal ingredients are scattered throughout the literature. They need to be considered in various contexts, such as the intended use of the herb, and especially its quality (see Table 14.1) and the mechanism of action of the constituent, as shown in Table 14.2. As CNS toxicity due to natural products is such a vast topic, covering many areas of neuropharmacology as well as botany, chemistry, and pharmacokinetics, this chapter will focus on those neurotoxins that have been responsible for human poisonings in practice, and which have been attributed to herbal medicines regardless of the circumstances in which the toxicity occurred (see Table 14.3). These issues of attribution of causality obviously apply to all types of herbal toxicity, and are discussed more fully in other chapters.

An Overview of CNS Toxicity and Its Manifestations

Toxicity to the CNS occurs when toxic lipophilic compounds cross the blood brain barrier or damage its permeability, permitting entry of hydrophilic molecules that would not normally enter the CNS (Wager et al. 2012). A common feature of many

Table 14.1 Circumstances surrounding ingestion of herbal medicines causing ANS/CNS toxicity

		Compounds involved		Type of CNS toxicity (details in Tables 14.2 and 14.3)
Circumstance	Plant species concerned	Chemical class/type	Major compounds	
CNS active responsible for therapeutic effect (overdose)	*Hypericum perforatum* (St. John's wort)	Naphthodianthrone and isoprenylated phloroglucinol	Hypericins and hyperforins	Serotonergic syndrome
CNS toxin in herb used for non-CNS therapeutic effect	*Annona muricata* (soursop) leaf and bark extracts (juice fairly safe)	Acetogenin	Annonacin-1	Atypical Parkinson's disease
CNS active overdose due to errors in processing	*Aconitum* species	Diterpene alkaloid	Aconitine	Muscle weakness, confusion, paralysis
CNS toxin not contributing to therapeutic effect	*Gingko biloba* (toxin mainly in fruit, not leaf extracts)	Pyridine alkaloid	Ginkgotoxin (4′-*O*-methylpyridoxine)	Seizures
Commercial adulteration with closely-related toxic species	*Illicium anisatum* instead of *Illicium verum* (star anise)	Sesquiterpene lactone	Anisatin	Convulsions
Recreational (illicit) use of medicinal plants	*Erythroxylum coca* (coca leaf)	Tropane alkaloid	Cocaine	Convulsions, cardiac arrest
Misidentification of herb due to superficial resemblance	*Atropa belladonna* leaf used instead of burdock, *Arctium lappa* (case report)	Tropane alkaloid	Atropine	Confusion, hallucinations
Use of distilled essential oil rather than herb	*Salvia*, *Rosmarinus* and many others where herb is safe used internally	Mono- and sesquiterpene	Thujone, borneol, camphor	Seizures
Toxic plants used only in traditional medicine systems	*Strychnos ignatii*	Indole alkaloid	Strychnine	Convulsions

Table 14.2 Drug classes of some herbal ANS/CNS toxins

Drug class	Example herb species	Example compounds	Symptoms → Clinical use (if any) ↓	CNS depression	Agitated delirium	Hallucinations	Convulsions/seizures
GABA antagonist	Water dropwort (*Oenanthe crocata* etc.)	Oenanthotoxin, cicutoxin	None	–	++	–	++
	Many essential oils: e.g., wormwood, thuja, sage (*Artemisia*, *Thuja*, *Salvia*)	Thujone, camphor, borneol	Digestive, decongestant, antiseptic, flavoring				
GABA agonist	Fly agaric (*Amanita muscaria*)	Muscimol	None; illicit use only	+	++	+	–
Muscarinic antagonist	Belladonna, henbane, angel's trumpet (*Atropa*, *Hyoscyamus*, *Datura*, *Scopolia* spp)	Atropine, hyoscyamine, hyoscine (= scopolamine)	Many uses in cardiology, gastroenterology, surgery	+	++	+	+
Sympatho-mimetic	Ma Huang (*Ephedra sinica*),	Ephedrine	Decongestant, stimulant		++	–	+
	Khat (*Catha edulis*)	Cathine, cathinone	Stimulant, euphoriant				
Opioid agonist	Opium poppy (*Papaver somniferum*)	Morphine	Strong analgesic; wide illicit use	++	–	+	–
Non-steroidal anti-inflammatory drug (NSAID)	Willow (*Salix* spp), meadowsweet (*Filipendula ulmaria*), poplar (*Populus* spp)	Salicylates, salicin, populin, etc.	Anti-inflammatory	–	–	–	+

Non-selective adenosine antagonists	Guarana (*Paullinia cupana*), *maté* (*Ilex paraguariensis*), coffee (*Coffea*), kola (*Cola*), tea (*Camellia sinensis*), cocoa (*Theobroma cacao*)	Methyl xanthines, caffeine (= guaranine), theophylline, theobromine	CNS, cardiac and respiratory stimulant	–	+	+	+
Serotonergic agent	St. John's wort (*Hypericum perforatum*)	Hypericin, hyperforin	Antidepressant	+	++	–	–
Na^+ channel blocker, TRI[a]	Coca leaf (*Erythroxylum* spp)	Cocaine	Local anesthetic; wide illicit use as stimulant	–	++	–	++

[a]Triple reuptake inhibitor (serotonin-noradrenaline-dopamine reuptake inhibitor)

Table 14.3 Herbs with documented ANS/CNS toxicity

Herb or herbal product	Intended use of herb	Toxin(s)	Mechanisms of toxicity
Apricot kernels (*Prunus armeniaca*)	Crushed kernels used as an alternative cancer treatment	Amygdalin, which releases hydrogen cyanide upon hydrolysis	Uncouples oxidative phosphorylation, causing anoxia, paralysis, death
Aconite (Monkshood) (*Aconitum* species, especially *A. carmichaelii*)	Used in TCM as a sedative and for pain relief. Raw herb processed to reduce toxicity	Aconitine, lycoctamine and many other diterpene alkaloids of varying toxicity	Binds to sodium channels, leading to reflux of potassium and efflux of calcium, also causing severe cardiac toxicity
Belladonna (deadly nightshade) (*Atropa belladonna*)	Sedative, antispasmodic, analgesic, also used to reduce secretions and gastrointestinal motility	Tropane alkaloids, atropine (racemic hyoscyamine), hyoscine and others	Anti-muscarinic effects leading to confusion, delirium, palpitations, flushing and mydriasis
Bitter orange (*Citrus aurantium*)	Weight loss, sports supplement to enhance performance	*p*-Synephrine, octopamine	α1- Adrenergic agonists, but toxicity disputed due to use with caffeine (q.v.)
Broom (Scotch broom) (*Cytisus scoparius*)	Formerly used as a sedative and for cardiac anti-arrhythmic effects	Sparteine and other quinolizidine alkaloids	Sodium channel blocker. Causes CNS sedation but cardiotoxicity more serious
Camphor (*Cinnamomum camphora*)	Decongestant, analgesic (external use). Former use for coughs, minor heart symptoms and fatigue	Camphor, a monoterpene. Natural essential oils also contain camphor	Convulsions. Inhibits nicotinic receptors, agonist of TRVP channels; also readily absorbed through the skin
Cannabis (Marijuana) (*Cannabis sativa*)	Wide illicit use for euphoria and relaxation; various medicinal uses	Δ^9-Tetrahydrocannabinol (a meroterpenoid)	Activation of cannabinoid-1 receptor causing confusion, disorientation etc.
Cocoa (*Theobroma cacao*)	Food, drink (chocolate)	Theobromine mainly, with other methyl xanthines (e.g., caffeine), and tyramine	Convulsions. Antagonist at adenosine receptors. Especially toxic to dogs
Coffee (*Coffea* spp), also tea, etc.	Stimulant, diuretic	Caffeine mainly, with other methyl xanthines (theophylline, theobromine)	Convulsions. Antagonist at all adenosine receptor subtypes

Table 14.3 (continued)

Herb or herbal product	Intended use of herb	Toxin(s)	Mechanisms of toxicity
Cowhage (*Mucuna pruriens*)	Anti-Parkinsonian agent in Ayurveda (due to presence of L-dopa)	*N*,*N*-dimethyltryptamine	Psychotomimetic; also causes hypertension
Ephedra (Ma Huang) (*Ephedra sinica* and others)	Anti-asthmatic, decongestant. More recent use as a weight-loss aid	Ephedrine, pseudoephedrine	Sympathomimetic, releases dopamine and noradrenaline. Cardiotoxic
Ergot (*Claviceps purpurea*)	Oxytocic, anti-migraine. Rarely used in herbal medicine but may be a mycotoxin contaminant of cereals	Alkaloids including ergotamine, ergometrine and many others	Ergotism: hallucinations, seizures, abortion, gangrene. Alkaloids are agonists for various neurotransmitters
Ginkgo (*Ginkgo biloba*)	Enhancement of memory and cognition and blood circulation	Ginkgotoxin (4′-*O*-methylpyridoxine, found mainly in fruits)	Pyridoxal kinase inhibition, causing vitamin B_6 deficiency and seizures
Guarana *Guarana cupana*	Used as an energy supplement	Caffeine (= guaranine)	See coffee
Henbane (*Hyoscyamus niger*)	Used illicitly as well as medicinally. See Belladonna	Tropane alkaloids: hyoscyamine, hyoscine	See Belladonna
Hyssop (*Hyssopus officinalis*)	Used for coughs and colds as an expectorant. Edible herb	Camphor, pinocarvone, thujone	Convulsions. See camphor
Khat (Qat) *Catha edulis*	Mild stimulant, leaves chewed as a social and recreational drug	Cathine, cathinone (phenethylamines)	Releases noradrenaline. Causes insomnia, anorexia, psychosis with excessive use
Lobelia (Indian tobacco) (*Lobelia inflata*)	Respiratory stimulant expectorant. Formerly used for smoking cessation	Lobeline, lobelanidine and other piperidine alkaloids	Nicotine receptor agonist. Overdose can cause nausea, sedation, convulsions
Nutmeg (*Myristica fragrans*)	Digestive and carminative, culinary spice. Illicit use in high (toxic) doses	Myristicin, safrole, elemicin	Hallucinogenic in overdose. May also cause nausea, convulsions

(continued)

Table 14.3 (continued)

Herb or herbal product	Intended use of herb	Toxin(s)	Mechanisms of toxicity
Rauvolfia (Rauwolfia) (*Rauvolfia serpentina*)	Used in India as a sedative in mental illness; elsewhere for hypertension	Indole alkaloids including reserpine, ajmaline, serpentine	Depletes noradrenaline and serotonin, may cause depression, hypotension
Soursop (Graviola) *Annona muricata*	Leaf and bark used traditionally as sedative; extracts used as "cancer treatment." They contain more toxins; fruit/juice considered safe	Acetogenins: annonacins Benzylisoquinoline alkaloids: reticuline, N-methylcocularine. Also in custard apple and American paw-paw	Annonacin-1 destroys dopaminergic neurons, causing atypical Parkinson's syndrome. Alkaloids cause neuronal cell death and also inhibit dopamine uptake
Star anise, Japanese *Illicium anisatum*	Digestive, especially in infants, used in error for true star anise *I. verum*	Anisatin, a sesquiterpene	Convulsions. Non-competitive antagonist of gamma-aminobutyric acid (GABA) A receptors
Syrian rue (Esfand, Harmel) *Peganum harmala*	Traditional use in cardiovascular and depressive disorders	Beta-carboline alkaloids: harmalol, harmaline, and harmine	Hallucinations, bradycardia, hypotension, agitation, tremors, and vomiting
Willow *Salix* species	Analgesic, anti-inflammatory and anti-pyretic effects	Salicylate glycosides, especially salicin	Salicylate overdose causes convulsions, but no cases reported for willow
Wormwood (Absinthe) *Artemisia absinthium*		Sesqui- and monoterpenes, absinthin, thujone, especially in essential oil. Normal use of herb safe	Convulsions, hallucinations. "Absinthism," the toxic syndrome of absinthe, but may be due to alcohol or copper (adulterants)
Yohimbe (*Pausynistalia johimba*)	Aphrodisiac	Yohimbine and other indole alkaloids	α-Adrenergic agonist, causing agitation, anxiety, hypertension, and tachycardia

References: General references Harborne and Baxter (1996) and Williamson (2003); specific references as listed in text under the relevant herb

types of CNS damage is anoxia, resulting from reduced blood flow to a specific area of the brain (ischemic anoxia), or uncoupling oxidative phosphorylation (chemical anoxia). Neurones are extremely sensitive to anoxia, and neuronal death is irreversible, but if neuronal death has occurred only on a limited basis, undamaged neurones may acquire some of the functions of those lost. If the neuronal injury is not lethal, it may repair, via nerve growth factors.

CNS toxicity may manifest as agitation, confusion, tremor, seizures, psychosis, or depressed levels of consciousness, and these may coexist (Ruhe and Levine 2014). Convulsions are the most common result of excessive CNS stimulation – for example, with caffeine or cocaine overdose. Impairment of sight, hearing, and perception of pain or parasthesia can occur if sensory neurones or the area of the brain involved are damaged or dead. This may result in neuropathy, decreasing neuronal conduction. Toxic insult can also cause increased neuronal excitability, for example with acetylcholine esterase inhibition. In addition to deficits in sensory perception, motor function deficit may occur, presenting as muscle weakness and paralysis, whereas over-excitation or rhythmic firing of neurons can cause tremor or more pronounced motor dysfunction. Damage to peripheral neurons may progress in a retrograde manner back to the CNS, causing neuronal death within the brain.

Damage to the CNS or peripheral nervous system may be subtle and observed only after a period of time, or only with specific behavioral tests for coordination, memory, and learning. This more cryptic type of toxicity is a feature of many types of herbal poisoning, where low doses of herbal toxins may be ingested over a long period of time. There is also the problem that patients may not associate their symptoms with the herbal product in the illogical belief that it is "natural" and therefore safe.

Classes of Herbal Drugs Causing CNS Toxicity

CNS active compounds found in herbs fall into most classes of CNS active drugs, as shown in Table 14.2, in addition to non-therapeutic categories, so it is necessary to differentiate between an overdose of a CNS-active compound and a CNS toxin with no therapeutic CNS application. Cyanide poisoning, for example, occurs when excessive doses of cyanogenic glycosides are ingested, such as with apricot kernels, which are not used for any CNS effects but as an unproven and dangerous alternative cancer treatment. Soursop (also known as graviola) is also used as an unproven cancer treatment, and although consumption of its fruit and juice is widespread and considered fairly safe, there are toxic acetogenins, the annonacins (which are also the anti-cancer compounds) abundant in the leaf and bark that cause atypical Parkinson's disease (PD) in animal models by destroying dopaminergic neurons (Champy et al. 2009). To complicate matters, there are also neurotoxic aporphine alkaloids such as reticularine and N-methylcocularine present in this herb (Kotake et al. 2004). In regions such as Guadeloupe, where the leaf and bark are consumed as herbal medicines (and not only for cancer), there is a significantly higher incidence of atypical PD than elsewhere. Other potent toxins, for example cicutoxin and oenanthotoxin, found in hemlock water dropwort (*Oenanthe crocata*, sometimes referred to as the most poisonous plant in Britain), have no medicinal use but cause poisoning due to misidentification of the plant. The water dropworts are members of the *Apiacae*, the parsley family, and are easily mistaken for other innocuous species. The same problem may arise with hemlock, *Coniium maculatum*, which is also a member of this family and contains the lethal alkaloid coniine; it has a history of use as a rather risky sedative.

There are many cases in which herbs are used for their CNS therapeutic effects, and their constituents cannot be called toxins in this context. Valerian and skullcap (*Scutellaria* species) contain mild GABA agonists and are used as mild sedatives and anxiolytics, and toxic effects have never been reported from these plants in this respect. However, muscimol, also a GABA agonist, has repeatedly caused CNS toxicity and its source, the fly agaric mushroom, is only used illicitly. Hyperforin, found in St. John's wort, is a GABA-reuptake inhibitor, but its mild sedative effect is different from the serotonergic effect that has been reported for *Hypericum* (Russo et al. 2014).

The ergot alkaloids have extensive therapeutic use in conventional medicine, but many of them (and their derivative, lysergic acid diethylamide (LSD)) are hallucinogenic and even induce epileptiformic seizures ("St. Anthony's fire"), and thus can be considered as neurotoxins. Ergot was formerly used in herbal medicine to aid childbirth, but is now considered far too toxic, with individual alkaloids and derivatives being preferred; cases of ergot poisoning are more likely to occur due to contamination.

Caffeine is an adenosine antagonist and a widely used CNS stimulant, used for this purpose by a great many people very frequently, in tea and in coffee. It is also found in herbal products such as guarana (where it was initially called "guaranine" before being found to be identical to caffeine), chocolate, and maté tea. The other methylxanthines, such as theobromine and theophylline, also show toxic effects in the CNS – mainly insomnia and tremor – but can also cause palpitations and convulsions in large doses.

Table 14.2 is an overview showing the wide range of CNS effects produced by plant toxins; in many cases they may involve a more complex mechanism of action than this brief description, and there are also usually over-lapping effects in other pharmacological systems. The pragmatic therapeutic categories used in Table 14.2 are those used in conventional drug classification and are intended to show the wide range of possible CNS toxicity that may arise from the ingestion of certain herbs.

These examples illustrate the fact that toxicity in the ANS and CNS is inextricably linked to other symptoms and cannot be viewed in isolation. The cardiovascular system is obviously immediately affected by drugs that alter neurotransmission. Ephedrine is a constituent of the Chinese medicine *Ma huang*, which is used to treat asthma and bronchitis. Ephedrine is a sympathomimetic agent with $\alpha 1$-adrenergic agonist activity, and like its isomer, pseudoephedrine, is used as isolated compounds in decongestant products. Recently, the use of ephedra supplements for weight loss and enhancing athletic performance has raised fears about their effects on the immune system (Senchina et al. 2014) in addition to their cardiovascular and ANS/CNS effects. Another sympathomimetic amine is *p*-synephrine, a constituent of bitter orange, which is also taken as a supplement to aid weight loss. There have been concerns that *p*-synephrine is toxic, based on its chemical structural similarity to the α_1-adrenergic receptor agonist phenylephrine (also known as *m*-synephrine), but these have been disputed (Stohs et al. 2011). In several studies, *p*-synephrine was taken in conjunction with caffeine, which appears to have enhanced its toxicity considerably (Hansen et al. 2012).

Camphor inhibits nicotinic receptors, and is also an agonist at several TRVP channels. Its CNS stimulatory effects may cause many severe side effects, such as nausea, vomiting, headache, dizziness, muscular excitability causing tremor and twitching, and convulsions and delirium (depending on the dosage). In a severe overdose, status epilepticus persisting for several hours occurs, ultimately causing coma and death by asphyxia or exhaustion (Chen et al. 2013). Camphor is used as a local analgesic due to its effects on de-sensitization of TRPV1 and blocking of TRPA1. One of the main issues involving its toxicity is the fact that it is very easily absorbed through the skin.

Table 14.3 provides a summary of herbs that have been documented to contain neurotoxins, together with the constituent(s) responsible, the most common uses of the herb – whether for CNS active or other reasons – and some of the mechanisms and toxic clinical outcomes of using the herb.

Figure 14.1 shows the structures of some of the more important neurotoxins, particularly those which cause particular problems (e.g., aconitine), those with limited therapeutic applications (e.g., camphor) and those with no therapeutic use at all (e.g., anisatin).

Testing Herbal Products for CNS Toxicity

Testing for neurotoxicity remains a very complex issue. There are many ways in which herbs may be responsible for causing neurotoxicity, but there are also cases of contradictory but subtle effects that are difficult to evaluate. For example, *Acanthopanax senticosus*, Siberian ginseng, used in Chinese and Russian herbal medicine for its neuroprotective effects, has recently been found to produce at least seven metabolites (including pipecolic acid) with known neurotoxic effects (Zhang et al. 2014). Clinical manifestations of CNS toxicity from this herb have not been described, and the study cited is an analysis of cerebral metabolomics, but it illustrates how important it is to develop new methods for measuring toxicokinetics, developmental toxicity, and delayed toxicity. Neurotoxicology has been defined as "the study of the adverse effects of chemical, biological, and certain physical agents on the nervous system and/or behavior during development and in maturity" (e.g., Harry et al. 1998; AltTox 2015). Neurotoxicity must be evaluated for regulatory applications; *in utero* exposure, which may cause developmental neurotoxicity (DNT), should also be investigated. Neurotoxicity, as with other poisoning, can result from different routes of exposure including oral, dermal, or pulmonary, and may be seen after a single (acute) dose or repeated (chronic) dosing. Given the considerable challenges of neurotoxicity testing, new strategies are being devised, as outlined below.

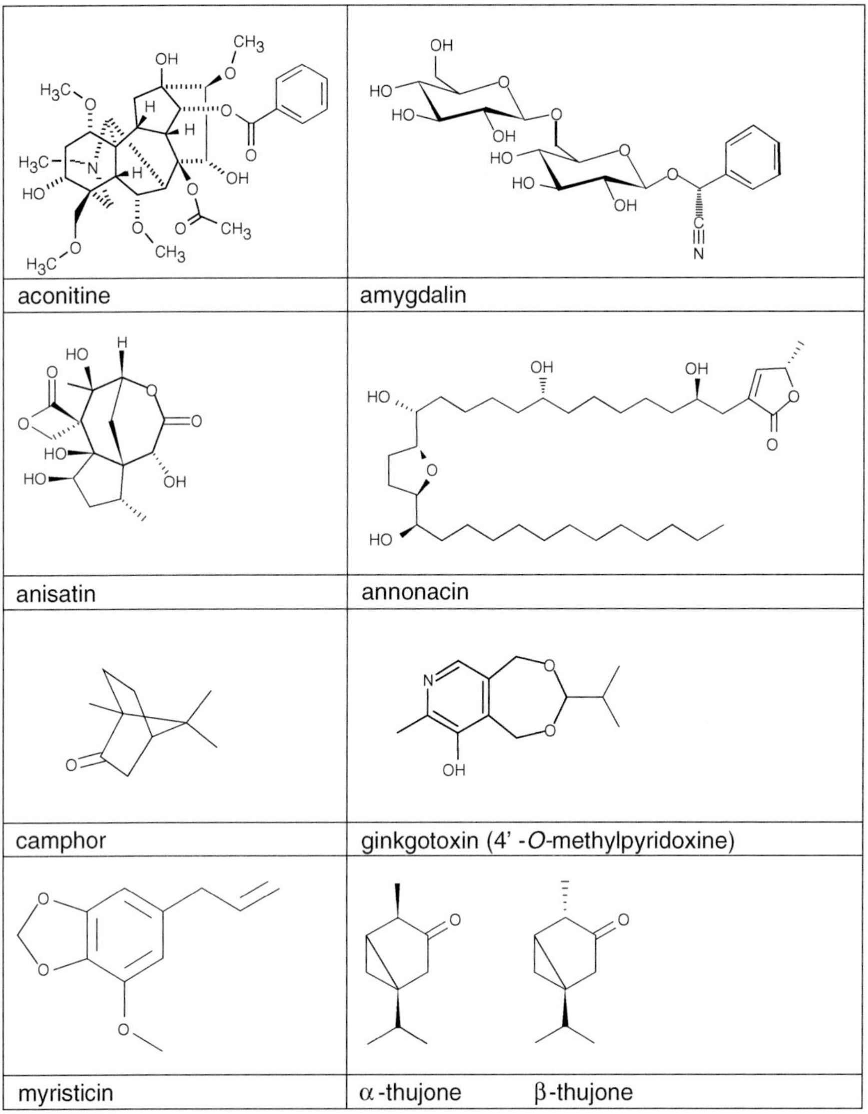

Fig. 14.1 Structures of some important herbal neurotoxins

Strategies for Neurotoxicity Testing

Most authorities agree that better methods of testing for neurotoxicity are required – in all areas, not just for herbal medicines (AltTox 2015; Wager et al. 2012; Llorens et al. 2012; FDA 2015). To date, testing has been based mainly on general pathological evaluation of neuronal tissue and a rather casual observation of test animals, in their cages, for overt signs of toxicity. This only detects severe forms of neurotoxicity, so the recommendations now are that assessment is carried out through a structured process of testing starting with a screen that is sufficiently comprehensive to detect pathological changes and functional disorders of the peripheral, central, and autonomic nervous systems (FDA 2015). Substances that show evidence of adverse effects during screening are then subjected to special neurotoxicity testing, to characterize these effects and determine whether there are any other, possibly more subtle, effects on the nervous system, in both adult and developing organisms. The assessment should include a core battery of tests designed to detect adverse changes to the cognitive, sensory, motor, and autonomic aspects of the mature and developing nervous system, and the results will eventually determine more specialized tests. If, for example, a substance induces convulsions during screening, its seizure potential and pro-convulsing properties should be determined. These second-line tests are all *in vivo*, and at present regulatory authorities do not accept purely non-animal methods or alternative testing strategies for neurotoxicity testing, for reasons detailed by Bal-Price et al. (2010a).

Animal Testing

In vivo animal test methods usually involve daily oral dosing for acute, sub-chronic, or chronic assessments over various periods of time. Primary observations include histopathology assessment, and while these provide insight into changes in neurons, behavioral and physiological methods are needed to assess their consequences, so a basic neurotoxicity screen should use both in conjunction. Specific histopathological examination should include tissue samples of major areas of the brain, spinal cord, and peripheral nervous system. *In vivo* evaluation should be conducted using a clearly defined battery of tests and observations selected to detect signs of neurological disorders, behavioral abnormalities and any other sign of nervous system toxicity. Incidence and severity of seizure, tremor, and paralysis, levels of motor activity and alertness, reactivity to stimuli, changes in motor coordination and gait, excessive lacrimation or salivation, piloerection, diarrhea, polyuria, and any other signs of abnormal behavior or nervous system toxicity should be assessed. Developmental studies and age-appropriate methods must be included, using measures of postnatal development and functional milestones in the experimental offspring. More sensitive and objective indices of neurotoxicity, such as tests of learning and memory, could be included as part of the screen or introduced later if appropriate.

A range of mouse models (transgenic and knock-out) is now available, and these can provide information on behavioral changes resulting from altered pathology using functional tests. These tests have an advantage over pathology methods in that a single animal can be reassessed over time to determine the onset, progression, duration, and reversibility of the neurotoxicity, as reviewed by Moser (2011). However, animal testing remains prohibitively expensive and extremely controversial among the general public, and steps are being taken towards non-animal methods (e.g., see AltTox 2015). Llorens et al. (2012) suggest that "A change in paradigm is needed . . . moving from its present reliance solely on data from animal testing to a prediction model mostly based on *in vitro* toxicity testing and *in silico* modelling . . ." and also describe strategies for doing this.

Dose Selection

The issues involved with dose selection, and with minimizing animal use, have been comprehensively discussed by the National Centre for the Replacement, Refinement and Reduction of Animals in Research (NC3Rs) and the Laboratory Animal Science Association (LASA) (LASA 2009). As with all toxicological assessments, the first *in vivo* step in a neurotoxicity test is usually a dose range finding (DRF) study in rodents. The starting dose in DRF is selected from the literature, efficacy models, pharmacokinetic studies and/or those used for closely related compounds (Crofton et al. 2011). In the case of herbal materials, the variability of plant material can be so great as to make accurate dosage measurements impossible unless the plant extract is chemically investigated and standardized. It is also useful to compare the effects of the extract, a complex mixture, with those of known individual pure constituents. Synergistic and antagonistic effects may result from interactions between components in the extract, giving an overall effect greater or less than expected. Some of these have been documented for single-herb extracts and multi-ingredient herbal products (e.g., Williamson 2001). However, if a single constituent is responsible for the activity, the results should correlate with those produced by a matching dose of the extract in the same system.

For herbal extracts, selecting the doses to be used initially involves a wide range of concentrations. Half log intervals may be employed (e.g., 1, 3, 10, 30, 100, 300, 1,000 mg/kg etc.), but the interval will depend on factors such as the expected linearity and slope of the dose response (LASA 2009). The high dose for further studies should produce signs of toxicity that are compatible with the study duration and are tolerated by the animal, i.e., the maximum tolerated dose (MTD), in addition to those based on period of exposure and maximum feasible dose. The MTD for 7 days may be greater than that for 28 days dosing, which may in turn be greater than for 90 days' dosing. For studies longer than 90 days, the MTD usually remains unchanged (Crofton et al. 2011). In neurotoxicological testing, longer studies are usually needed, and especially in the case of herbal medicines that may have insidious or delayed effects. The impact of environmental conditions on experimental animals is well recognized, and in the case of neurological testing, where behavioral issues are included in the outcomes being measured, it is crucial to provide animals with an environment that satisfies their behavioral as well as physiological needs (Lasa 2009).

In summary, neurotoxicity testing for herbal materials is challenging from the point of view of both experimental *in vivo* design, and quality assurance and chemical identity of the material being tested.

Selection of Control Materials

The assessment of the neurotoxicity of herbal extracts, as with any other drug, usually involves comparison with control agents of known mechanism of action in specific systems. These controls are the selected on that basis, and in fact many of them are natural products (e.g., annonacin, bicuculline, caffeine, nicotine, muscimol, aconitine). Others are chemical neurotoxicological agents (e.g., MPTP, 1-methyl-4-phenyl-1,2,3,6-tetrahydropyridine). MPTP and annonacin are both used to induce experimental Parkinson's syndrome in animal models.

"Non-animal" Methods

There are many challenges in deciding the mechanisms of neurotoxicity that need to be investigated, and in the development and validation of *in vitro* assays that encompass relevant endpoints for a test battery (AltTox 2015). To try and replicate neurotoxicity in the complex CNS – inextricably linked to every other major organ and system in the body – using non-animal methods is a massive undertaking, especially since cellular models usually cannot distinguish "pharmacological actions" from "toxicity responses," which is needed for risk assessment. Indeed, many useful publications describe all kinds of CNS-active compounds as "toxins" without discriminating between desired and unwanted medicinal CNS effects (e.g., Harborne and Baxter 1996), but of course all medicines are toxins in the wrong place or in the wrong dose. In the context of neurodevelopmental problems in children due to environmental neurotoxic agents, the obstacles to using the existing tests – economic, ethical and scientific – have been discussed by Smirnova et al. (2014).

The *in vitro* tests normally used in neurotoxicity testing are not "non-animal" tests – they are just not *in vivo* tests. Primary cells (neurons and glia from different regions of the brain), cell lines (e.g., neuroblastoma, astrocytoma, and glioma), mixed cell cultures, neural stem cells and brain slices are provided by animals. However, continuous cell lines, originally derived from human or animal tissues, can often be maintained for research and testing purposes for many years and may be more ethically acceptable than *in vivo tests* for that reason (e.g., AltTox 2015). However, since many of these immortalized cell lines are of cancerous origin, their relevance to testing mechanisms occurring in normal neurons can be questioned. More recently, tests using non-mammalian animal models have been established. The nematode *Caenorhabditis elegans* (Meyer and Williams 2014), a model lab organism with most of its neurotransmitters corresponding to human homologs, has been shown to have great potential for screening for neurotoxicity and a zebrafish model of Parkinson's disease has been validated by Babu et al. (2016).

Developmental Neurotoxicity

The developing brain is much more vulnerable to injury than the adult brain. This is partly due to the complex developments that happen over a relatively short period of time, which also means that it is susceptible at different points in time. Additionally, the adult brain is protected by the blood brain barrier (BBB), whereas the placenta only delays and partially protects against harmful exposure *in utero*, and the BBB is not entirely formed until about 6 months after birth.

Developmental neurotoxicity methods evaluate *in utero* and early postnatal effects of daily dosing of pregnant animals, from conception to weaning. Offspring are also evaluated for neurological and behavioral abnormalities, and brain weights and neuropathology are assessed at different times through adulthood. The effects of chemicals (including phytochemicals) on the development of the CNS are investigated using an extensive battery of tests (Makris et al. 2009). Primary neuronal cultures of cerebellar granule cells (CGCs) and 3-D aggregate models are now used and are being integrated with new approaches such as gene expression and metabonomics (Bal-Price et al. 2010b).

Conclusions

The problems involved in assessing herbal neurotoxicity are similar to those in other areas of toxicology (e.g., metabolic activation, toxicokinetics, etc.), but even more complex, given the role of the CNS in controlling other systems of the body, and the fact that many neurotoxicity tests are not yet validated. Increasing attention is being devoted to the development of *in vitro* systems for screening, but validation studies remain to be carried out to correlate *in vitro* results with neurotoxicological responses in whole animals. Such systems, once appropriately validated, may have particularly useful application in screening for potential neurotoxicity and in helping to elucidate the mode of action or mechanistic information.

As can be seen from Table 14.3, there are a number of well-documented neurotoxins present in herbal medicines. It is likely that there are many others with more subtle and/or mixed effects that remain to be found. Any risk assessment must be considered in the context that herbal medicines are very variable in quality, and also in the way they are used, and bearing in mind that a whole herb extract may produce test results quantitatively and qualitatively different compared to a constituent tested in isolation (see Williamson (2001) for examples). A comparison of the effect of a single compound at similar doses to those found in the whole herb (or product being evaluated) should therefore be carried out to ensure the relevance of the test, and if the isolated compound produces a different result to the extract, the reasons must be investigated.

Finally, the subject must be kept in perspective. Many common herbs, spices, and foods contain substances that in high doses can cause neurotoxicity, as shown in Table 14.3. The use of distilled essential oils, which contain high amounts of terpenes, is much more likely to lead to clinical symptoms than the herbs from which they were extracted, and cases of poisoning by ingesting peppermint oil have been reported, for example (Nath et al. 2012). However, no one would seriously consider banning it.

References

AltTox.org: http://alttox.org/mapp/toxicity-endpoints-tests/neurotoxicity/. Accessed May 2015

Babu NS, Murthy CL, Kakara S, Sharma R, Swamy CV, Idris MM (2016) MPTP induced Parkinson's disease in zebrafish. Proteomics. doi:10.1002/pmic.201500291 [Epub ahead of print]

Bal-Price AK, Hogberg HT, Buzanska L et al (2010a) Relevance of in vitro neurotoxicity testing for regulatory requirements: challenges to be considered. Neurotoxicol Teratol 32(1):36–41

Bal-Price AK, Hogberg HT, Buzanska L et al (2010b) In vitro developmental neurotoxicity (DNT) testing: relevant models and endpoints. Neurotoxicology 31(5):545–554

Champy P, Guérineau V, Laprévote O (2009) MALDI-TOF MS profiling of Annonaceous acetogenins in *Annona muricata* products for human consumption. Molecules 14:5235–5246

Chan TYK (2015) Incidence and causes of aconitum alkaloid poisoning in Hong Kong from 1989 to 2010. Phytother Res 29(8):1107–1111

Chen W, Vermaak I, Viljoen A (2013) Camphor – a fumigant during the Black Death and a coveted fragrant wood in ancient Egypt and Babylon – a review. Molecules 18(5):5434–5454

Crofton KM, Mundy WR, Lein PJ, Bal-Price A, Coecke S, Seiler AE, Knaut H, Buzanska L, Goldberg A (2011) Developmental neurotoxicity testing: recommendations for developing alternative methods for the screening and prioritization of chemicals. ALTEX 28(1):9–15

FDA(2015)http://www.fda.gov/Food/GuidanceRegulation/GuidanceDocumentsRegulatoryInformation/IngredientsAdditivesGRASPackaging/ucm078323.htm. Accessed May 2015

Hansen DK, George NI, White GE et al (2012) Physiological effects following administration of *Citrus aurantium* for 28 days in rats. Toxicol Appl Pharmacol 261(3):236–247

Harborne JB, Baxter H (1996) Dictionary of plant toxins. Wiley, Chichester

Harry GH, Billingsley M, Bruinink A et al (1998) *In vitro* techniques for the assessment of neurotoxicity. Environ Health Perspect 106(Suppl 1):131–158

Kotake Y, Okuda K, Kamizono M et al (2004) Detection and determination of reticuline and N-methylcocularine in the Annonaceae family using liquid chromatograpy-tandem mass spectrometry. J Chromatogr B 806:75–78

Lasa (2009) Guidance on dose level selection for regulatory general toxicology studies for pharmaceuticals. http://www.lasa.co.uk/PDF/LASA-NC3RsDoseLevelSelection.pdf

Liu X, Wong Q, Song Q et al (2014) The classification and application of toxic Chinese material medica. Phytother Res 28:334–347

Llorens J, Li AA, Ceccatelli S, Suñol C (2012) Strategies and tools for preventing neurotoxicity: to test, to predict and how to do it. Neurotoxicology 33(4):796–804

Makris SL, Raffaele K, Allen S et al (2009) A retrospective performance assessment of the developmental neurotoxicity study in support of OECD test guideline 426. Environ Health Perspect 117(1):17–25

Meyer D, Williams PL (2014) Toxicity testing of neurotoxic pesticides in *Caenorhabditis elegans*. J Toxicol Environ Health B Crit Rev 17(5):284–306

Moser VC (2011) Functional assays for neurotoxicity testing. Toxicol Pathol 39(1):36–45

Nath SS, Pandey C, Roy D (2012) A near fatal case of high dose peppermint oil ingestion – lessons learnt. Indian J Anaesth 56(6):582–584

Ruhe AM, Levine M (2014) Central nervous system toxicity. Emerg Med Clin N Am 32:205–221

Russo E, Scicchitano F, Whalley BJ et al (2014) *Hypericum perforatum*: pharmacokinetics, mechanism of action, tolerability, and clinical drug-drug interactions. Phytother Res 28(5):643–655

Senchina DS, Hallam JE, Kohut ML et al (2014) Alkaloids and athlete immune function: caffeine, theophylline, gingerol, ephedrine, and their congeners. Exerc Immunol Rev 20:68–93

Smirnova L, Hogberg HT, Leist M et al (2014) Developmental neurotoxicity – challenges in the 21st century and in vitro opportunities (2014). ALTEX 31(2):129–156

Stohs SJ, Preuss HG, Shara M (2011) The safety of *Citrus aurantium* (bitter orange) and its primary protoalkaloid p-synephrine. Phytother Res 25(10):1421–1428

Wager TT, Liras JL, Mente S, Trapa P (2012) Strategies to minimize CNS toxicity: in vitro high-throughput assays and computational modelling. Expert Opin Drug Metab Toxicol 8(5):531–542

Williamson EM (2001) Synergy and other interactions in phytomedicines. Phytomedicine 8(5):401–409

Williamson EM (2003) Potter's herbal cyclopedia. CW Daniels, Saffron Walden

Zhang SN, Li XZ, Wang Y et al (2014) Neuroprotection or neurotoxicity? New insights into the effects of *Acanthopanax senticosus* harms on nervous system through cerebral metabolomics analysis. J Ethnopharmacol 56:290–300

Chapter 15
Cardiovascular Toxicities of Herbal Products: An Overview of Selected Compounds

Pieter van der Bijl Jr. and Pieter van der Bijl Sr.

Abstract The use of herbal products for a wide variety of health and medicinal purposes by all ethnic groups worldwide is prevalent and rising, despite a conspicuous lack of rigorous scientific evidence regarding their safety and efficacy. While these products are frequently considered safe by patients and healthcare practitioners alike, they may cause adverse effects that frequently involve the cardiovascular system. A spectrum of chemical compounds is usually present in these products, some of which cause direct cardiovascular toxicity, and others to which clinically relevant herb-allopathic drug interactions may be attributed when taken concomitantly with conventional therapies. The objective of this chapter is to provide an overview of selected herbs that manifest cardiovascular toxicity and those for which herb-allopathic drug interactions affecting the cardiovascular system have been described. Furthermore, the general principles of diagnosis and management of cardiovascular toxicity of these remedies are also discussed.

Keywords Herbal products • Cardiovascular toxicity • Herb-allopathic drug interactions

Introduction

It has been known for centuries that some plants possess medicinal properties. The ancient Egyptians and Romans recognized the fact that extracts of *Urginea maritima* had diuretic, cardiotonic, expectorant, and emetic properties (Naudé 1997). Furthermore, the medicinal value of *Digitalis purpurea* (foxglove, which contains

P. van der Bijl Jr.
Division of Cardiology, Department of Internal Medicine, Faculty of Medicine and Health Sciences, Stellenbosch University, Tygerberg, Cape Town, South Africa
e-mail: pieter.vanderbijl@gmail.com

P. van der Bijl Sr. (✉)
Formerly, Department of Pharmacology, Faculty of Medicine and Health Sciences, Stellenbosch University, Tygerberg, Cape Town, South Africa
e-mail: pietervanderbijlcpt@gmail.com

O. Pelkonen et al. (eds.), *Toxicology of Herbal Products*,
DOI 10.1007/978-3-319-43806-1_15

cardiac glycosides) was reported by William Withering in 1785 after observing that patients with dropsy (edema secondary to cardiac failure) could be treated with an extract from this plant (Bessen 1986).

Since time immemorial, herbal medicines have been used as a mainstay of complementary and alternative medicine (CAM), which has experienced a resurgence in Western societies over the past two decades (Hunt et al. 2010; Merrit-Charles 2011; Mugabo et al. 2012). These plant-derived products are used for their beneficial (real or purported) effects in improving health and treating a wide range of clinical conditions including asthma, malignancies, dermatological ailments, epilepsy, acquired immunodeficiency syndrome (AIDS), cardiovascular diseases, diabetes mellitus, coryza, and influenza – as well as pain. Generally, their mechanisms of action are uncertain or unknown, and there is a dearth of clinical efficacy and safety data for these products (Ernst 2007; Sahoo et al. 2010; Cravotto et al. 2010).

It is estimated that worldwide, approximately 25 % of adults in developed countries and >80 % of the population of developing countries use herbal medicines, which are derived from >11,000 species of plants (Chen et al. 2012). Data from the National Health Interview Survey show that the use of non-vitamin, non-mineral dietary supplements was greater than any other complementary health approach used by adults in the U.S. in 2012 (Peregoy et al. 2014). In Europe, the use of CAM is even more prevalent, where studies from 20 countries (representing 69 % of the continent's population) estimate that 56 % of the general population and 52 % of children had used CAM at least once in the year prior to which the survey had been conducted (Zuzak et al. 2013). These figures resemble data from the U.K. obtained from a systematic review in which the average 1-year prevalence of use of herbal medicines was 64.2 % (Posadzki et al. 2012 and 2013). In South Africa, the use of plant-derived products is widespread in the practice of traditional medicine, and it is estimated that some 80 % of the South African population consults traditional healers regularly (Mugabo et al. 2012). These traditional medicines ("muti") are usually administered orally or given as enemas. Laboratory analyses of muti have shown that these preparations often consist of aqueous plant materials, e.g., roots, bark stem, or leaves, sometimes mixed with metallic salts, mushrooms and insects (McVann et al. 1992). Plant components are sometimes pulverized or sliced into small pieces, making botanical identification difficult or impossible.

Notwithstanding many assertions that herbal remedies are safe and lack adverse effects, this is untrue, particularly when they are used in the management of serious conditions (Singh 2009; Hunt et al. 2010). It is well documented that some herbal medicines contain toxic chemical compounds that have direct toxic effects, among others those involving the cardiovascular system (Van der Bijl and Van der Bijl 2012). Furthermore, during the co-administration with allopathic drugs, certain herbal medicines have the potential for herb-drug interactions that can be of significant clinical importance (Chen et al. 2012; Posadzki et al. 2012). Both the direct cardiovascular toxicity and herb-allopathic drug-interactive effects range from being merely inconvenient to life-threatening.

Direct Cardiovascular Toxicity

A selection of some important medicinally used plants that contain cardiovascular toxins are:

Digitalis lanata and *purpurea* (foxgloves), *Convallaria majalis* (Lily of the valley), *Nerium oleander* (common or pink oleander), *Thevetia peruviana* (yellow oleander), *Acokanthera oppositifolia* and *schimperi* (bushman poison bush), *Urginea maritima* and *indica* (squill), *Drimia sanguinea* (sekanama), *Bowiea volubilis* (climbing potato), *Asclepias curassavica* (milkwood), *Strophantus gratus*, *Apocynum cannabinum* (dogbane) and *Cheiranthus cheiri* (wallflower). The above-mentioned plants all contain cardiac glycosides, and these chemical compounds can be lethal to both livestock and humans (Botha and Penrith 2008, 2009; Snyman et al. 2011).

The cardiac glycosides, which are highly toxic chemical compounds, are found in a number of plants. These phytochemicals, which occur in the highest concentrations in the plant seeds, consist of an aglycone (structurally related to steroid hormones) linked to one or more sugar molecules. The aglycones of cardiac glycosides can be divided into two chemical groups – the cardenolides and bufadienolides. It is assumed that the general principles of digoxin toxicity also hold true for other glycosides, even though the latter have been less well studied.

The primary pharmacological effect of cardiac glycosides is to inhibit the Na^+/K^+-ATPase exchanger of the myocardiocyte that increases intracellular Na^+ concentration, thus reducing the amount of Ca^{2+} pumped out of the cell by the Na^+/Ca^{2+} exchanger (Hauptman and Kelly 1999; Kumar et al. 2013). Glycosides bind to the extracellular α-subunit (the enzyme being a heterotrimer consisting of α-, β- and γ-subunits) of the Na^+/K^+-ATPase exchanger (Hauptman and Kelly 1999). Consequently, the intracellular Ca^{2+} concentration rises, thereby occasioning positive inotropy (Fig. 15.1); this also appears to be the mechanism of tachydysrhythmogenesis. Excess Ca^{2+} remains intracellularly after the cell has repolarized, and this elicits a transient, inward Na^+ current, known as I_{ti} (via nonspecific cation channels), which in turn leads to a delayed afterdepolarization (during phase 4 of the cardiac action potential) (Hauptman and Kelly 1999) (Fig. 15.2).

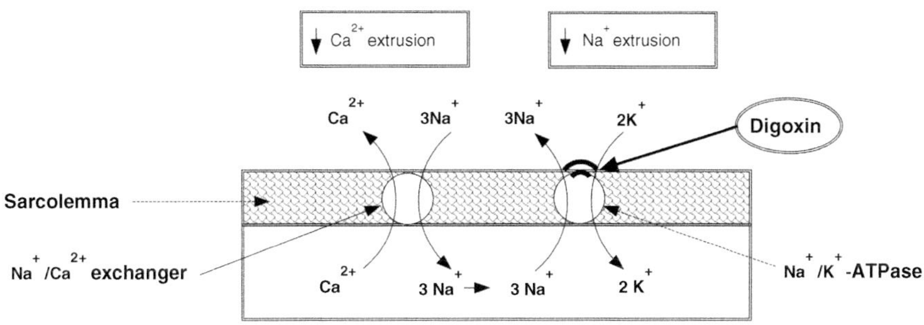

Fig. 15.1 The mechanism of action of cardiac glycosides, e.g., digoxin

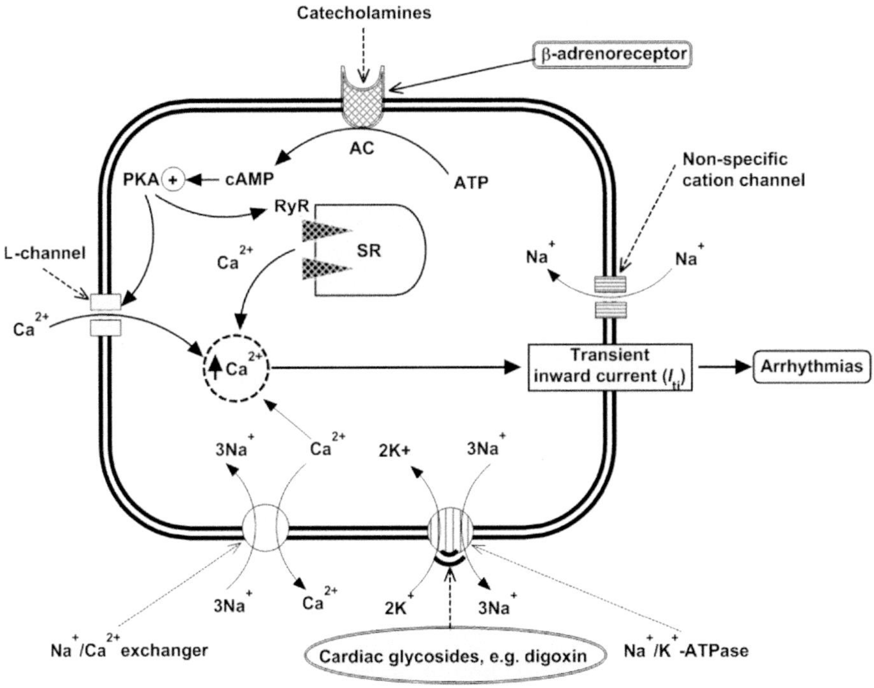

Fig. 15.2 The role of catecholamines, cardiac glycosides and Ca^{2+}-overload in the genesis of delayed afterdepolarizations. Cyclic adenosine monophosphate (cAMP); adenyl cyclase (AC); adenosine triphosphate (ATP); protein kinase A (PKA); ryanodine receptor (RyR); sarcoplasmic reticulum (SR); ligand-gated cation channel (L-channel)

A secondary pharmacological effect is a depressant effect on the atrioventricular node via central vagal stimulation (causing decreased dromotropy of the atrioventricular node, as well as an increase in its refractory period), which contributes to pathological bradycardias, together with a direct, depressant effect on the atrioventricular node. Catecholamines potentiate glycoside toxicity via stimulation of β-adrenoreceptors, which lead to the production of cyclic adenosine monophosphate (cAMP) by means of adenylate cyclase. cAMP in turn activates protein kinase A, which phosphorylates the α-subunit of the L-type Ca^{2+} channel (embedded in the cell membrane), and the ryanodine receptor in the sarcoplasmic reticulum, leading to elevated intracellular Ca^{2+} levels, which then follow the final common pathway of delayed afterdepolarizations (Lubbe et al. 1992) (Fig. 15.2). Glycoside toxicity is potentiated (via delayed afterdepolarization) by electrolyte disturbances (especially hypokalemia, but also hypomagnesemia and hypercalcemia), ischemia, reperfusion injury, and increased ventricular wall stress.

Cardiac glycosides have very narrow therapeutic indices, and acute toxicity is most commonly associated with ingestion of plant material, although chronic toxicity (manifesting with anorexia, for example) may also be seen. In cases of acute intoxication, nausea, emesis (effected via the chemoreceptor trigger zone, rather than due to a direct effect on the gastrointestinal system, but also attributed to parasympathomimetic effects) and abdominal pain typically occur, as well as central nervous system effects including lethargy, weakness, and visual disturbances – primarily chromatopsia, especially xanthopsia – but also scotomas and halos. A host of arrhythmias, including conduction disturbances and/or bradycardias (Wenckebach A & B-sinoatrial block and atrioventricular block), tachycardias (ectopic atrial tachycardia, atrial fibrillation, junctional tachycardia, bifascicular ventricular tachycardia, ventricular fibrillation) and other electrical disturbances (accelerated idioventricular rhythm and bigeminy) can manifest. Some are challenging to diagnose and are often overlooked, e.g., ectopic atrial tachycardia, where the focus is usually in the superior right atrium, producing P-waves that are similar in morphology to those of sinus rhythm, and also depolarizing the atria in the same direction. Others are fairly unique and easier to recognize, e.g., bifascicular ventricular tachycardia (a life-threatening rhythm with alternating left- and right axes due to the focus originating in the anterior and posterior fascicles of the left bundle branch, respectively) and unexpected regularization of ventricular impulses due to a junctional tachycardia that appears on a background of atrial fibrillation (Hauptman and Kelly 1999).

Aconitum carmichaeli *("chuanwu")*, Aconitum kuznezoffii *("caowu")*

Aconitine and related alkaloids of aconite (mesaconite and hyperconitine) are constituents of *Aconitum* species of plants, and they are used as analgesics/anti-inflammatories, particularly in China and Japan. The roots and root tubers of these plants are typically consumed as vegetables and used in the preparation of herbal soups and meals (Singhuber et al. 2009; Kang et al. 2012). The latter are, in contrast to aqueous decoctions (mashing and boiling), a more agreeable way of ingesting the herbal product. The wild plant, especially the raw roots and root tubers, are very toxic due to the presence of high concentrations of *Aconitum* alkaloids, which are hydrolyzed into less toxic and non-toxic derivatives by soaking and boiling (Chan 2009). However, even after processing, concentrations of these alkaloids (which are potent cardio- and neurotoxins) can remain high enough to cause poisoning.

Both the cardio- and neurotoxicity of aconitine and its related alkaloids are due to their effects on voltage-sensitive Na^+-channels of myocardial, neural, and muscle cells. Aconitine and mesaconitine bind strongly to the open state of voltage-sensitive Na^+-channels at site 2, causing a persistent activation of these channels and rendering them refractory to excitation (Chan 2009; Friese et al. 1997). Delayed after-depolarizations (i.e., triggered activity) induce arrhythmias via the downstream inhibition of the Na^+/Ca^{2+}-exchanger by excessive intracellular Na^+, causing intracellular Ca^{2+} overload (similar to glycosides). Tachycardias (ventricular tachycardia – including bifascicular tachycardia, *torsades de pointes*, and ventricular fibrillation), bradycardias/conduction disturbances (sinus bradycardia, asystole) and other rhythm disturbances (ventricular ectopics, junctional rhythms) have been linked to aconitine toxicity, and are often refractory to treatment (Lu and De Clerck 1993; Tai et al. 1992; Chan 2009). The ventromedial nucleus of the hypothalamus is affected by aconitine in rats, causing modulation of the autonomic nervous system with bradycardia and hypotension (Yamanaka et al. 2002; Hirasawa et al. 1998).

Following ingestion of toxic doses of aconitine and/or related alkaloids of aconite, the toxidrome manifests in the neurological (facial and limb paresthesia/weakness), gastrointestinal (nausea, emesis and abdominal cramps), and cardiovascular systems.

Hyoscyamus niger *(henbane)*

All parts of this plant contain tropane alkaloids in varying quantities (e.g., atropine, hyoscyamine and scopolamine), which competitively inhibit the muscarinic effects of acetylcholine and block impulse transmission in the parasympathetic nervous system, resulting in the classic anticholinergic syndrome (Spoerke et al. 1987). However, other potentially toxic natural compounds, e.g., coumarins, flavonoids, sterols, tannins, and terpenes have also been found in *Hyoscyamus niger* extracts (Khan and Gilani 2008). The toxidrome comprises central nervous system manifestations (e.g., seizures) and peripheral vasodilatation (dry, warm skin).

The primary cardiac effect of Henbane is sinus tachycardia due to the vagolytic effect of the alkaloids. This is not dangerous per se, but might be poorly tolerated by those with underlying heart disease. Even though the toxidrome includes peripheral vasodilatation, paradoxical hypertension (without a specific mechanistic explanation) has been described in overdose (Li et al. 2011; Urkin et al. 1991).

Lycopodium serratum *("jin bu huan")*

Jin bu huan is a popular Chinese herbal medication that has been used for centuries as a mild sedative, a decongestant, and as a treatment for conditions ranging from asthma and bronchitis to nictalopia, delirium, epilepsy, vertigo, pyrexia and inflammation, arthritic and orthopedic pain, and gastrointestinal complaints. While the basis for the sedative, analgesic, and anti-inflammatory properties of this plant material is unclear, it contains levo-tetrahydropalmitine and pyrrolozidine alkaloids. The former has sedative effects, possibly due to it being a dopamine receptor and a Ca^{2+}-channel antagonist (Larrey 1997). Unintentional overdoses have been shown to cause central nervous system and respiratory depression with rapid onset of transient, severe sinus bradycardia (Centers for Disease Control and Prevention 1993; Horowitz et al. 1996).

Mitragyna speciosa *("kratom"/"ketum")*

The traditional use of this tropical herb plant dates back many centuries and has its origins in Southeast Asian countries e.g., Thailand and Malaysia; it is known as "ketum" in Malaysia and "kratom" in Thailand. Natives of these countries traditionally consume the leaves by masticating, smoking, or drinking them (as a tea) for their stimulant and euphoric effects (Babu et al. 2008). In recent times, kratom has become popular for recreational purposes and as a substitute in cases of opioid dependency, as well as a treatment for systemic hypertension. The plant contains more than 40 compounds in its leaves, including many alkaloids such as mitragynine (once thought to be the primary active constituent), mitraphylline, and 7-hydroxymitragynine (which is currently the most likely candidate for the primary active chemical in the plant), and mitragynine pseudoindoxyl (Adkins et al. 2011; Chittrakarn et al. 2010; Prozialeck et al. 2012). Other active compounds in *Mitragyna speciosa* include raubasine (best known as a constituent of *Rauwolfia serpentina*), rhyncophylline, and corynantheidine among others (Takayama et al. 2002). Acute toxic effects of kratom appear to be related to its stimulant (e.g., anxiety and aggression) and opioid (e.g., nausea and sedation) activities.

Withdrawal symptoms that have been found are similar to those of other opioids; they include irritability, dysphoria, nausea, insomnia, oscitation, rhinorrhea, myalgia, diarrhea, arthralgia, and hypertension (Prozialeck et al. 2012). The fear of kratom products being adulterated or interacting with other drugs has been raised.

The primary cardiovascular manifestation appears to be systemic hypertension (the mechanism of which is unknown), which may be especially detrimental when kratom is taken for lowering blood pressure.

Tussilago farafara *(coltsfoot)*

This herb, also known as "coltsfoot," belongs to the family *Asterracea* and is commonly found in Europe, Asia, and the Americas. It has been ingested as a tea or syrup and topically applied for respiratory and cutaneous complaints, viral infections, influenza, coryza, and rheumatic conditions (Vogl et al. 2013). Toxic pyrrolizidine alkaloids are present in the plant. Apart from hepatotoxicity, these alkaloids are associated with cor pulmonale (likely due to its inhibition of nitric oxide release and subsequent pulmonary vasoconstriction), while left ventricular dysfunction and medial thickening of the coronary arteries have been described in rats after ingestion (Joint FAO/WHO Food Standards Programme Codex Committee on Contaminants in Food 2011). Furthermore, the administration of monocrotaline (a toxic pyrrolizidine alkaloid of plant origin) to rats caused myocarditis, independent of the degree of pulmonary hypertension (Akhavein et al. 2007). Coltsfoot has been found to cause a hypertensive response in humans (Li and Wang. 1988).

Erythroxylum coca *(Coca)*

This plant is one of two species of cultivated Coca that are native to the Andean region in western South America. The leaves of the *Erythroxylum coca* bush have been chewed by native South American tribes for thousands of years for their analgesic, anorexogenic, and stimulatory effects (Schwartz et al. 2010). The active ingredient, cocaine (an alkaloid), was purified over a century ago and used in tonics and elixirs. It is a highly addictive stimulant, making this alkaloid one of the most popular drugs of abuse. A freebase form ("crack") became a sought-after drug some 30 years ago. Although cocaine use occurs primarily in the Americas, Europe, and Oceania, the evidence regarding the extent of its use in Africa and Asia is unestablished (United Nations Office on Drugs and Crime 2014). However, some pockets of emerging use in these regions may be developing. Worldwide, there seems to be a slight decline in its use due to a decrease in the overall global availability of the alkaloid (United Nations Office on Drugs and Crime 2014).

Cocaine (benzoylmethylecgonine, $C_{17}H_{21}NO_4$) is a potent sympathomimetic and local anesthetic. The alkaloid is dissolved in hydrochloric acid to form a water-soluble hydrochloride salt that can exist in a crystalline, powder, or granular form. When a solution of the hydrochloride salt is alkalinized and extracted with ether, and the latter has evaporated, a non-salt (freebase) form results. The freebase form melts at 98 °C, making a crackling sound and has therefore been given the "street" name of "crack" (Maraj et al. 2010).

All mucous membranes of the body absorb cocaine well, the compound typically being inhaled intranasally ("snorted"), but the alkaloid can also be administered by intramuscular and intravenous routes. The onset of action is rapid (seconds to minutes), depending on the route of administration, and peak effects as well as the duration of action may range from several minutes to 1.5 h (Maraj et al. 2010). In humans, elimination $t_{1/2}$ ranges from 30 to 60 min, the metabolism of cocaine being mainly by plasma and hepatic cholinesterases. The water-soluble metabolites, benzoylecgonine and ethylmethylecgonine, as well as 5-10 % of unchanged cocaine are excreted in the urine. While unchanged cocaine is usually not found in urine after 6 h, the metabolites are, with benzoylecgonine being detected in urinary samples of chronic abusers for as long as 22 days after their last dose (Maraj et al. 2010). This may be important clinically, for example in diagnosing a myocardial infarct following recent abuse of this alkaloid.

By inhibiting catecholamine re-uptake, cocaine powerfully stimulates the sympathetic nervous system – it also sensitizes adrenergic nerve endings to norepinephrine (Riezzo et al. 2012) (Fig. 15.3). Cocaine releases endothelin-1 from endothelium and inhibits nitric oxide production – a combination that leads to vasoconstriction and a rise in systemic blood pressure (Wilbert-Lampen et al. 1998; Mo et al. 1998). It causes coronary artery vasoconstriction (more pronounced in atherosclerotic than normal vessels), smooth-muscle cell plaque rupture (in contrast to

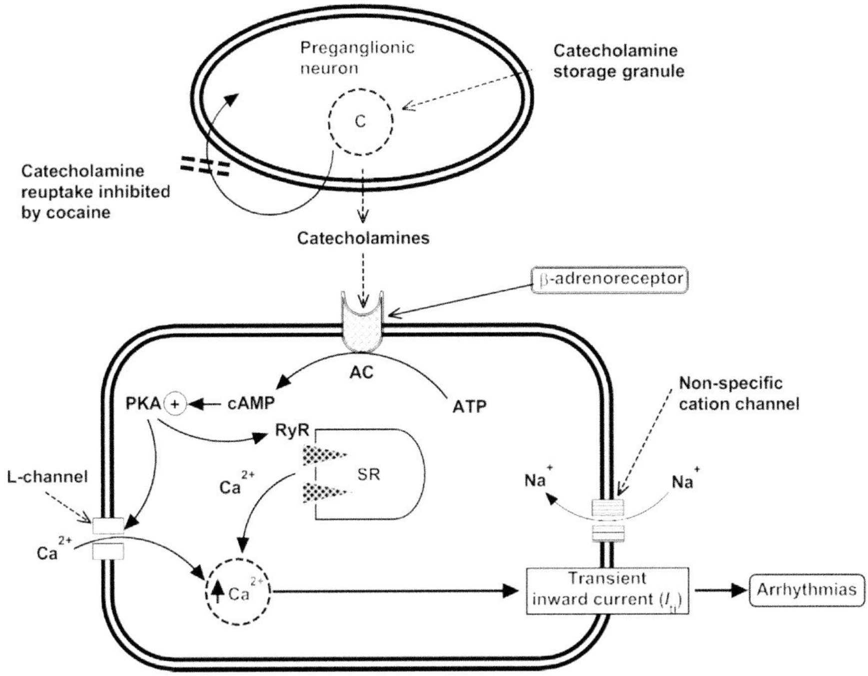

Fig. 15.3 The role of cocaine and Ca^{2+}-overload in the genesis of delayed afterdepolarizations. Cyclic adenosine monophosphate (*cAMP*); adenyl cyclase (*AC*); adenosine triphosphate (*ATP*); protein kinase A (*PKA*); ryanodine receptor (*RyR*); sarcoplasmic reticulum (*SR*); ligand-gated cation channel (*L-channel*); catecholamine (*C*)

atherosclerotic plaque in those with traditional risk factors for atherogenesis), thrombocyte activation, aggregation, degranulation, and thrombus formation, which lead to coronary syndromes and life-threatening arrhythmias (Heesch et al. 2000; Flores et al. 1990; Schwartz et al. 2010). Long-term abuse causes endothelial dysfunction of the coronary arteries, a known sensitizer for catecholamine-induced vasoconstrictor effects (Havranek et al. 1996; Vita et al. 1992). Cardiac ischemia is induced by the supply-demand mismatch, which reflects coronary vasoconstriction and increased O_2-requirements due to increased heart rate (tachycardia) and blood pressure (Lange et al. 1989).

By blocking K^+-channels and Na^+-channels, as well as by intracellular Ca^{2+}-overload (on the basis of sympathetic stimulation of β-adrenoreceptors, leading to formation of protein kinase A, which translates into intracellular Ca^{2+}-overload and delayed afterdepolarizations – similar to glycoside toxicity), cocaine has proarrhythmic effects in the absence of ischemia (Riezo et al. 2012). It is documented to cause a spectrum of arrhythmias, e.g., monomorphic ventricular tachycardia, *torsades de pointes*, ventricular fibrillation, atrioventricular block and asystole (Bauman et al. 1994; Hsue et al. 2007; Schwartz et al. 2010) (Fig. 15.3).

Cocaine causes myocardial dysfunction via (1) ischemia and infarction, (2) myocarditis (possibly engendered by infectious agents or adulterants that are co-administered, and (3) direct toxic effects. Direct cardiac damage is caused by mitochondrial toxicity, which in turn is due to two mechanisms, i.e., oxidative stress and Ca^{2+}-overload. Sympathetic stimulation of β-adrenoreceptors leads to formation of protein kinase A, causing intracellular Ca^{2+}-overload (similar to cardiac glycoside toxicity). Oxidative stress is occasioned by transformation of catecholamines into aminochromes, which partake in redox reactions in mitochondria and lead to the formation of free radicals. Mitochondrial permeability increases due to the above-mentioned mechanisms, and apoptotic and necrotic myocardiocyte death follows (Liaudet et al. 2014). Chronic use of this alkaloid may lead to left ventricular hypertrophy (likely due to systemic hypertension), dilated cardiomyopathy, and Takutsubo cardiomyopathy and a final common pathway of systolic as well as diastolic cardiac dysfunction (Schwartz et al. 2010; Arora et al. 2006; Daniel et al. 2007; Chambers et al. 1987). Reversible myocardial depression can also manifest after acute intoxication (Schwartz et al. 2010).

Aortic dissection is temporally related to cocaine abuse, and it is assumed to be a reflection of increased systemic arterial pressure. It should be considered in the differential diagnosis of a patient presenting with recent cocaine use and chest pain, in addition to coronary syndromes. Infective endocarditis occurs more frequently than what may be attributed to the intravenous route of administration per se. This may be due to direct endothelial damage caused by high arterial pressures and tachycardia and/or the direct immunosuppressive effects of cocaine, on which intravenous injection is superimposed.

Citrus aurantium *(bitter orange)*

The peel, flower, leaf, fruit, and fruit juice from this citrus tree are used to prepare medicine, and the oil is prepared from the peel. The plant and its products are used as a herbal medicine for a wide variety of conditions ranging from gastrointestinal complaints and obesity (as an anorexigen), to lowering blood sugar (as an antidiabetic agent).

The extracts from *Citrus aurantium* contain tyramine metabolites N-methyltyramine, octopamine, and synephrine (Gange et al. 2006). These compounds are chemically similar to synephrine and stimulate α_1-adrenergic receptors, causing peripheral vasoconstriction, systemic hypertension and tachycardia. After the banning of ephedra in the U.S. and Canada, bitter orange was substituted into "ephedra-free" herbal weight-loss products (FDA 2004). There have been reports of bitter orange causing acute coronary syndromes and cerebrovascular accidents (National Center for Complementary and Alternative Medicine 2012). After an incident of a healthy young man who suffered an ST-elevation myocardial infarct that was linked to the ingestion of bitter orange, it was discovered that certain manufacturers of dietary supplements substituted ephedra with its chemical congeners from bitter orange (Thomas et al. 2009). Like most other dietary supplement ingredients, *Citrus aurantium* has not yet undergone proper safety testing, leading to the report by the National Center for Complementary and Alternative Medicine's statement that "there is currently little evidence that bitter orange is safer to use than ephedra" (National Center for Complementary and Alternative Medicine 2012).

Glycyrrhiza glabra *(licorice)*

The genus *Glycyrrhiza* comprises roughly 30 species, of which *Glycyrrhiza glabra* is popularly recognized as licorice due to its sweet taste. Licorice is extracted from the root of this legume, which is similar to peas and beans, and is found in southern Europe, India, and some parts of Asia. Licorice was used as a medicinal herb in ancient Egypt and Greece to relieve symptoms in individuals with adrenal insufficiency, chronic hepatitis, cystitis, gastric ulcers, urolithiasis, and diabetes.

Glycyrrhizin, a triterpenoid consisting of the K^+, Ca^{2+} and Mg^{2+}-salts of glycerrhizic acid, is one of the main active constituents of licorice and, being up to 50 times sweeter than sucrose, is responsible for its sweet taste. The glycyrrhizin content in the roots of the plant is between 2 and 25 %, depending on the species of legume. Chemically, the glycyrrhizin molecule comprises a hydrophobic 5-ring structure (glycyrrhetinic acid) linked to two hydrophilic glucuronic acid molecules. Flavonoids (liquiritin, isoliquiritin, isoflavones, glabridin and hispaglabridins) are responsible for the yellow color of licorice. While the hispaglabridins A and B are antioxidants, glabridin and glabrene possess estrogen-like activity (Omar et al. 2012).

The active ingredient in licorice is glycyrrhizic acid, and together with its hydrolytic product, glycyrrhetinic acid, which is a 200–1,000 times more potent inhibitor of 11-ß-hydroxysteroid dehydrogenase 2 than glycyrrhizic acid itself, have well-known mineralocorticoid activity (Ruiz-Granados et al. 2012). The inhibition of 11-ß-hydroxysteroid dehydrogenase 2 prevents the physiological conversion of cortisol (which has activity at the mineralocorticoid receptor) to cortisone (which does not), and therefore results in excessive systemic cortisol levels (Fig. 15.4).

This can lead to a syndrome known as "apparent mineralocorticoid excess." Mineralocorticoids bind to the mineralocorticoid receptor of the principal cells of the distal nephron, where they translocate to the nucleus and initiate mRNA transcription and translation of so-called aldosterone-induced proteins. These proteins include luminal Na^+ channels known as epithelial, sodium channels (ENaC) (which are synthesized, redistributed from the cytosol to the luminal membrane and activated), and Na^+/K^+-ATPase (which is also synthesized, redistributed from the cytosol but to the basolateral membrane and then activated). Na^+ is more readily absorbed luminally by the ENaC channel, and extruded by the Na^+/K^+-ATPase, i.e., transepithelial Na^+-transport is enhanced. K^+ is transported in the opposite direction via the renal outer medullary potassium channel (ROMK), and H_2O as well as Cl^- follow Na^+, i.e., Na^+ and H_2O are absorbed, and K^+ is depleted (Bhalla and Hallows 2008; Wang 2006; Hebert et al. 2005) (Fig. 15.5). Licorice in quantities of as little as 50 g/day may cause mineralocorticoid-induced hypertension (resulting from Na^+ and H_2O retention) and arrhythmias (e.g., ventricular fibrillation, due to severe hypokalemia). A combination of high dietary NaCl intake in salted licorice can therefore result in significant systemic hypertension.

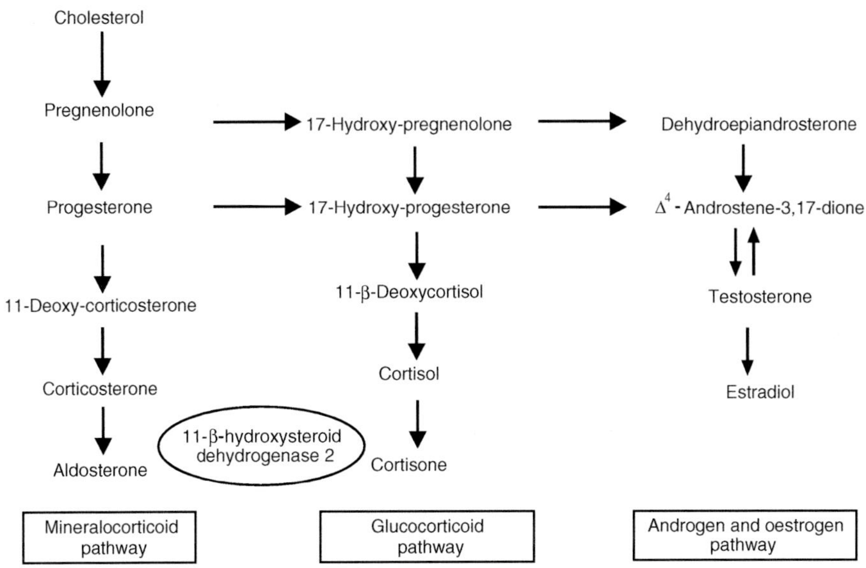

Fig. 15.4 Schematic representation of adrenocortical hormone biosynthesis pathways

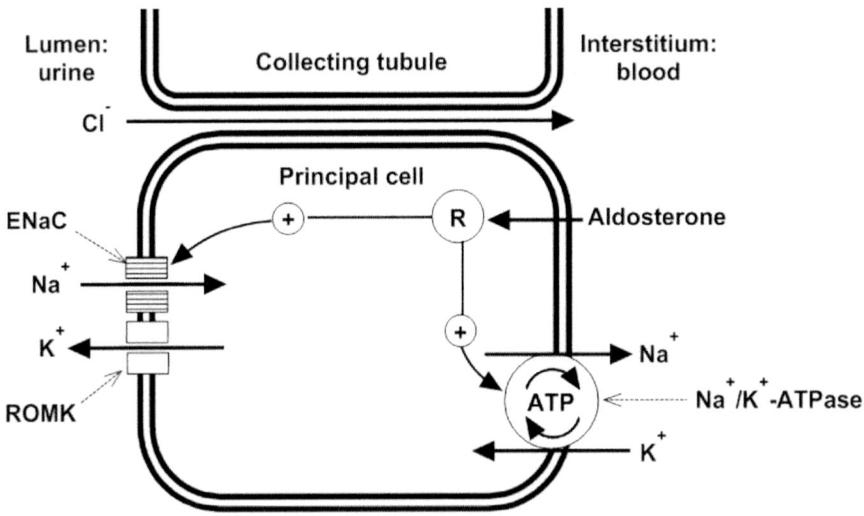

Fig. 15.5 Aldosterone-effects in the principal cells of the distal nephron. Mineralocorticoid receptor (*R*); epithelial sodium channel (*ENaC*); renal outer medullary potassium channel (*ROMK*); adenosine triphosphate (*ATP*)

Herb-Allopathic Drug Interactions and Cardiovascular Toxicity

A selection of plants for which clinically significant interactions with allopathic drugs have been documented, are discussed below.

Hypericum perforatum *(St. John's wort)*

This is a plant, the flowers and leaves of which are used primarily for mild to moderate depression and some of its symptoms, e.g., fatigue, anorexia, and insomnia. It has been used for a plethora of other conditions including coryza, herpes infections, and acquired immunodeficiency syndrome (AIDS) (Tachjian et al. 2010). *Hypericum perforatum* is indigenous to Europe, but is also found in southern and eastern Europe and parts of Asia.

St. John's wort is an inducer of the hepatic, drug-metabolising cytochrome P450 enzyme system, particularly CYP1A2, CYP3A4, but also CYP2C19 (Moses and McGuire 2010). CYP3A4 is involved in the metabolism of many allopathic drugs, including cardiovascular therapies, and St. John's wort should be avoided by patients using the prodrug clopidogrel, which is metabolized by both CYP3A4 and CYP2C19 systems. The concomitant use of clopidogrel (a $P2Y_{12}$ inhibitor) and the herbal product has been proven to decrease thrombocyte aggregation, which may lead to a risk of excessive bleeding (Lau et al. 2011).

When warfarin and St. John's wort are co-ingested, prothrombin times can be reduced and suboptimal anticoagulation may lead to a higher thromboembolic risk. This can be devastating, for example in those with metallic cardiac prostheses or atrial fibrillation on oral anticoagulation with a coumarin, and such patients should avoid this herbal product (Cohen and Ernst 2010). The concomitant intake of statins (HMG-CoA reductase inhibitors) and St. John's wort is not recommended, since the latter lowers statin blood concentrations, resulting in elevated cholesterol levels that may translate into an increased risk for cardiovascular events. Although the relationship between St. John's wort and hypertension is not well understood, it can inhibit re-uptake of serotonin, which may lead to the potentially life-threatening serotonin syndrome (which includes severe hypertension) (Cohen and Ernst 2010).

St. John's wort can induce the extensively distributed and expressed P-glycoprotein (P-gp), also known as multi-drug resistance protein. This is an ATP-dependent membrane efflux pump with a broad substrate specificity that has a protective action by pumping nonphysiological compounds, including drugs, out of cells. This may result in lowered blood levels and decreased efficacy of e.g. digoxin, a drug excreted from cells by P-gp. There are also potential interactions with dabigatran and dronedarone (both drugs being substrates for P-gp) (University of Washington 2014; Mendell et al. 2013).

Ginko biloba *(ginko)*

Extracts from the leaves from this tree, native to China, are said to have memory-enhancing, cognition-improving, antioxidant, neuroprotective, and cardio- and cerebrovascular benefits (Chen et al. 2012). The major components of this popular remedy include flavonoids, terpenoids, and organic acids. To date, results from trials regarding ginko's therapeutic efficacy regarding its beneficial effects on cognition have been ambiguous (Tachjian et al. 2010).

Ginko extracts may interact with warfarin and aspirin and increase the risk of hemorrhage (Tachjian et al. 2010). In several case reports, the use of ginko products and warfarin resulted in the development of intracerebral hemorrhage. There has been a case report on the development of spontaneous hyphema when the herbal extract was taken together with aspirin. Fatal intracerebral bleeding has also been recorded with the combined use of ginko and ibuprofen (Chen et al. 2012).

Ginko extracts may induce CYP3A4, thus enhancing the oxidative metabolism of substrates of this enzyme (Lau et al. 2011). This interaction may relate to the reduced efficacy of nicardipine, a Ca^{2+}-channel antagonist (Tachjian et al. 2010).

Allium sativum *(garlic)*

Garlic is an herb from the onion genus, which is native to central Asia and has been used for thousands of years for culinary and medicinal purposes. It has been widely used for treating infections because of its reputed antimicrobial and immunostimulatory effects (Tachjian et al. 2010). This herbal supplement is also commonly taken by persons with AIDS in the belief that it can bolster their immune response. It may have hypocholesterolemic, antihypertensive, and other anti-atherosclerotic effects; however, one clinical study showed no significant effects on low-density lipoprotein cholesterol or other plasma lipid levels (Tachjian et al. 2010). *Ajoene*, an unsaturated disulfide compound isolated from garlic does, however, inhibit collagen-induced platelet aggregation. Other organosulphur constituents in garlic have also exhibited inhibitory effects on human platelet aggregation both *in vitro* and *in vivo* (Chen et al. 2012).

Because the risk of bleeding is increased in anticoagulated patients or those on antiplatelet therapy, the concurrent use of garlic should be avoided. In cases where patients have been taking garlic-containing supplements, cessation is recommended before elective surgical procedures, particularly when anticoagulants (e.g., warfarin) or antiplatelet agents (e.g., aspirin) are consumed.

Panax ginseng *(ginseng)*

This plant is found in North America and in eastern Asia. The roots – and sometimes the leaves – have been used in folk medicine for many centuries, and the various extraction methods employed allow for a wide variation in the product composition. Immunostimulant, aphrodisiac, and antidiabetic properties as well as longevity have been attributed to these extracts (Tachjian et al. 2010). Other uses of ginseng are for its purported antihypertensive, hypolipidaemic, cognition-enhancing, anti-ulcerogenic, and anti-cancer effects (Chen et al. 2012). Its major constituents include ginsenosides, sterols, flavonoids, peptides, vitamins, polyacetylenes, minerals, β-elemine and choline.

Ginseng lowers blood pressure, apparently via effects on nitric oxide synthesis. Paradoxical hypertension, as well as psychomotor stimulation may occur, and caution is advised against its use in systemic hypertension (Lee et al. 2012). A nephrotoxic compound (germanium) in the extract is harmful to the ascending loop of Henle, antagonizing the action of loop diuretics. Ginseng extract, when co-administered with warfarin, has been reported to reduce prothrombin time (Tachjian et al. 2010; Izzo and Ernst 2009). Siberian ginseng interferes with the assay for digoxin levels, leading to false elevation in therapeutic drug monitoring levels (Tachjian et al. 2010).

Diagnosis and Management of Toxicity

It is challenging to establish the diagnosis of herbal poisoning or herb-allopathic drug interactions, because patients are often ill informed and view herbal products as safe, natural entities that are irrelevant to their conventional medical care, consequently omitting their mention in the anamnesis (Spoerke et al. 1987; Tachjian et al. 2010; Cohen and Ernst 2010). Diagnosing herbal toxicity primarily relies on a history of ingestion of cardiotoxic plant material and/or a suspicion generated by manifestations of direct toxicity, e.g., cardiac arrhythmias or inefficacy of allopathic drugs despite a reasonable certainty of compliance. Obtaining details of the constituents of many traditional medicines may be difficult, since they are often tightly guarded secrets not shared with patients or third parties (Van der Bijl and Van der Bijl 2012). Laboratory analyses for cardiac glycosides are available, and an immunoassay developed for the detection of digoxin also cross-reacts with other cardiac glycosides, such as oleandrin. However, more specific tissue and biological fluid assays for oleandrin have been developed (Poppenga 2010). The plasma level is not always a reliable indicator of toxicity, since it is only an indirect reflection of the myocardial tissue level (the true determinant of cardiac toxicity). An extreme example hereof is cardiac amyloidosis, where glycosides are concentrated in the myocardium (bound directly to myocardial fibres), leading to toxicity at therapeutic plasma levels (Lawler et al. 2014). Other compounds and their metabolites can sometimes be identified in plasma and urine samples by chromatography and mass spectrometry (e.g., aconitine), but this is limited by availability of these techniques, and is complicated by the fact that the toxidrome might be caused by a mixture of compounds present in a herbal remedy (Goldfrank et al. 2006).

The management of plant-intoxicated patients includes measures common to all clinical toxicology, i.e., immediate discontinuation of further exposure to the herbal products, general resuscitative and life-support measures such as the administration of activated charcoal, gastric lavage (caveat: within 1 h of ingestion) as well as electrocardiographic and other (e.g. hemodynamic) monitoring methods for arrhythmias and other signs of cardiorespiratory compromise.

Specific antidotes for toxidromes caused by plants are non-existent, but digoxin-specific antibody fragments appear to cross-react with at least some other cardiac glycosides, and therefore have a potential application in the treatment of poisoning in humans with the latter phytochemicals (Bandara et al. 2010). These antibody fragments have a much higher affinity for glycosides than the Na^+/K^+-ATPase exchanger, and to chemical concentration gradient is created that allows the antibody fragments to bind much of the extracellular glycoside. The antibody fragment-glycoside complex is then excreted renally. The antibody fragments are supplied in powder form that has to be reconstituted with sterile water before intravenous injection. They may be administered *ex juvantibus* when a patient with suspected glycoside toxicity is *in extremis*, especially when it is uncertain whether a plasma assay will detect the compound in question. The effect of this antidote is usually seen within 30 min, but it might take longer (up to 4 h) to have the desired effect. Therapeutic drug monitoring is inaccurate in the first 3 weeks after administration of this antidote, since most assays measure both the free and antibody fragment-bound glycoside. Even though these antibodies are polyclonal, they have been safely administered to the same individuals on more than one occasion (Hauptman and Kelly 1999). Charcoal hemoperfusion has also been used to remove toxins, e.g., aconitine, but data are mostly limited to case reports (Lin et al. 2004; Fatovich 1992).

Arrhythmias may be treated with the complete armamentarium available to cardiac rhythm disturbances generally, i.e., temporary, transvenous pacing, overdrive pacing, antiarrhythmic drugs (e.g., phenytoin in glycoside toxicity – it may reverse heart block, possibly via a central mechanism), electrical cardioversion and defibrillation – as deemed appropriate to the specific rhythm disturbance in question. Hypotension can be treated with intravenous fluid (such as crystalloids) and/or vasopressors (such as phenylephrine), while hypertension can be managed with antihypertensive drugs (e.g., intravenous nitroprusside). There have been case reports (e.g., in aconitine toxicity) where persistent hypotension was treated with mechanical techniques, e.g., percutaneous cardiopulmonary bypass, extracorporeal membrane oxygenation, and ventricular assist devices (Fatovich 1992; Lin et al. 2004). Acute coronary syndromes should be treated by means of standard protocols, although when these are cocaine-induced, β-adrenoreceptor-antagonists should be avoided (which will allow unopposed, α-adrenoreceptor-mediated vasoconstriction) and phentolamine (an α-adrenoreceptor-antagonist) can be administered therapeutically (Lange et al. 1989).

Treatment of the cardiovascular manifestations of plant toxicity is mostly not evidence-based. The old adage that prevention (including education) is better than cure, is particularly applicable to this scenario.

Conclusions

There is a conspicuous lack of scientific data pertaining to the efficacy and safety of herbal products. Currently, randomized, controlled clinical trials are not required by regulatory agencies, nor demanded by consumers and healthcare practitioners alike before marketing. Many thousands of these products, often labelled as dietary supplements, are sold worldwide, and consumers are under the impression that they are natural and safe. Only when herbal remedies have caused serious harm do regulatory agencies act, as demonstrated by the banning of ephedra by the Food and Drug Administration in the U.S. in 2004. Rather than waiting for adverse effects to occur, it is the opinion of the authors that manufacturers of herbal products should be compelled by law to prove that their products are efficacious and safe. Biological variability (geographical location, climate and soil conditions, etc.), manufacturing and storage techniques, contamination (pesticides, metals, intentional and unintentional adulteration with allopathic drugs), and general lack of consistency in the quality of these herbal products are of great concern. Furthermore, aggressive marketing techniques, in which bold as well as unsubstantiated claims for products are frequently made, are being used by manufacturers.

All healthcare practitioners, especially cardiologists, should be aware of the potential dangers, in the form of either the direct toxicity or drug-herb interactions that these herbal products pose. To this end, probing questions must be asked (patients anticipate doctors' disapproval when they report using alternative remedies) during the anamnesis regarding the use of these products, and a high index of suspicion should be maintained whenever patients present with unexplained symptoms/toxidromes or atypical responses to allopathic medicines.

References

Adkins JE, Boyer EW, McCurdy CR (2011) *Mitragyna speciosa*, a psychoactive tree from Southeast Asia with opioid activity. Curr Top Med Chem 11(9):1165–1175

Akhavein F, St-Michel EJ, Seifert E, Rohlicek CV (2007) Decreased left ventricular function, myocarditis, and coronary arteriolar medial thickening following monocrotaline administration in adult rats. J Appl Physiol 103(1):287–295

Arora S, Alfayoumi F, Srinivasan V (2006) Transient left ventricular apical ballooning after cocaine use: is catecholamine cardiotoxicity the pathologic link? Mayo Clin Proc 81(6): 829–832

Babu KM, McCurdy CR, Boyer EW (2008) Opioid receptors and legal highs: *Saliva divinorum* and Kratom. Clin Toxicol (Phila) 46(2):146–152. doi:10.1080/15563650701241795

Bandara V, Weinstein SA, White J, Eddleston M (2010) A review of the natural history, toxicology, diagnosis and clinical management of *Nerium oleander* (common oleander) and *Thevetia peruviana* (yellow oleander) poisoning. Toxicon 56(3):273–281. doi:10.1016/j.toxicon.2010.03.026

Bauman JL, Grawe JJ, Winecoff AP, Hariman RJ (1994) Cocaine-related sudden cardiac death: a hypothesis correlating basic science and clinical observations. J Clin Pharmacol 34(9): 902–911

Bessen HA (1986) Therapeutic and toxic effects of digitalis: William Withering, 1785. J Emerg Med 4(3):243–248

Bhalla V, Hallows KR (2008) Mechanisms of EnaC regulation and clinical implications. J Am Soc Nephrol 19(10):1845–1854. doi:10.1681/ASN.2008020225

Botha CJ, Penrith ML (2008) Poisonous plants of veterinary and human importance in southern Africa. J Ethnopharmacol 119(3):549–558. doi:10.1016/j.jep.2008.07.022

Botha CJ, Penrith ML (2009) Potential plant poisonings in dogs and cats in southern Africa. J S Afr Vet Assoc 80(2):63–74

Centers for Disease Control and Prevention (CDC) (1993) Jin bu huan toxicity in adults-Los Angeles. MMWR Morb Mortal Wkly Rep 42(47):920–922

Chambers HF, Morris DL, Täuber MG, Modin G (1987) Cocaine use and the risk for endocarditis in intravenous drug users. Ann Int Med 106(6):833–836

Chan TY (2009) Aconite poisoning. Clin Toxicol (Phila) 47(4):279–285. doi:10.1080/15563650902904407

Chen X, Sneed KB, Pan S, Pan SY, Cao C, Kanwar JR, Chew H, Zhou SF (2012) Herb-drug interactions and mechanistic and clinical considerations. Curr Drug Metab 13(5):640–651

Chittrakarn S, Keawpradub N, Sawangjaroen K, Kansenalak S, Janchawee B (2010) The neuromuscular blockade produced by pure alkaloid, mitragynine and methanol extract of kratom leaves (*Mitragyna speciosa* Korth.). J Ethnopharmacol 129(3):344–349. doi:10.1016/j.jep.2010.03.035

Cohen PA, Ernst E (2010) Safety of herbal supplements: a guide for cardiologists. Cardiovasc Ther 28:246–253. doi:10.1111/j.1755-5922.2010.00193.x

Cravotto G, Boffa L, Genzini L, Garella D (2010) Phytotherapeutics: an evaluation of the potential of 1,000 plants. J Clin Pharm Ther 35(1):11–48. doi:10.1111/j.1365-2710.2009.01096.x

Daniel JC, Huynh TT, Zhou W, Kougias P, El Sayed HF, Huh J, Coselli JS, Lin PH, LeMaire SA (2007) Acute aortic dissection associated with use of cocaine. J Vasc Surg 46(3):427–433

Ernst E (2007) Herbal medicines: balancing benefits and risks. Novartis Found Symp 282:154–167, discussion 167-172, 212-218

Fatovich DM (1992) Aconite: a lethal Chinese herb. Ann Emerg Med 21(3):309–311

FDA (2004) FDA acts to remove ephedra-containing dietary supplements from market. http://www.fda.gov/NewsEvents/Newsroom/PressAnnouncements/2004/ucm108379.htm. Accessed Nov 2014

Flores ED, Lange RA, Cigarroa RG, Hillis LD (1990) Effect of cocaine on coronary artery dimensions in atherosclerotic coronary artery disease: enhanced vasoconstriction at sites of significant stenoses. J Am Coll Cardiol 16(1):74–79

Friese J, Gleitz J, Gutser UT, Heubach JF, Matthiesen T, Wilffert B, Selve N (1997) *Aconitum* sp. alkaloids: the modulation of voltage-dependent Na^+ channels and antinociceptive properties. Eur J Pharmacol 337(2–3):165–174

Gange CA, Madias C, Felix-Getzik EM, Weintraub AR, Estes NA (2006) Variant angina associated with bitter orange in a dietary supplement. Mayo Clin Proc 81(4):545–548

Goldfrank LR, Flomenbaum NE, Lewin NA, Howland MA, Hoffman RS, Nelson LS (2006) Goldfrank's Toxicologic Emergencies, 8th edn. McGraw-Hill, New York

Hauptman PJ, Kelly RA (1999) Digitalis. Circulation 99(9):1265–1270

Havranek EP, Nademanee K, Grayburn PA, Eichhorn EJ (1996) Endothelium-dependent vasorelaxation is impaired in cocaine-arteriopathy. J Am Coll Cardiol 28(5):1168–1174

Hebert SC, Desir GV, Giebisch G, Wang WH (2005) Molecular diversity and regulation of renal potassium channels. Physiol Rev 85(1):319–371

Heesch CM, Wilhelm CR, Ristich J, Adnane J, Bontempo FA, Wagner WR (2000) Cocaine activates platelets and increases the formation of circulating platelet containing microaggregates in humans. Heart 83(6):688–695

Hirasawa M, Nishihara M, Takahashi M (1998) Activity of ventromedial hypothalamic neurons suppressing heart rate is associated with paradoxical sleep in the rat. Brain Res 797(1): 103–108

Horowitz RS, Feldhaus K, Dart RC, Stermitz FR, Beck JJ (1996) The clinical spectrum of Jin Bu Huan toxicity. Arch Intern Med 156(8):899–903. doi:10.1001/archinte.1996.00440080101012

Hsue PY, McManus D, Selby V, Ren X, Pillutla P, Younes N, Goldschlager N, Waters DD (2007) Cardiac arrest in patients who smoke crack cocaine. Am J Cardiol 99(6):822–824

Hunt KJ, Coelho HF, Wider B, Perry R, Hung SK, Terry R, Ernst E (2010) Complementary and alternative medicine use in England: results from a national survey. Int J Clin Pract 64(11):1496–1502. doi:10.1111/j.1742-1241.2010.02484.x

Izzo AA, Ernst E (2009) Interactions between herbal medicines and prescribed drugs: an updated systematic review. Drugs 69:1777–1798. doi:10.2165/11317010-000000000-00000

Joint FAO/WHO food standards programme codex committee on contaminants in food, 5th session. The Hague 21–25 March 2011. Discussion paper on pyrrolizidine alkaloids ftp://ftp.fao.org/codex/meetings/cccf/cccf5/cf05_14e.pdf. Accessed Nov 2014

Kang YX, Luczaj LJ, Ye S (2012) The highly toxic *Aconitum carmichaelii* Debeaux as a root vegetable in the Qinling Mountains (Shaanxi, China). Genet Resour Crop Evol 59(7): 1569–1575

Khan AU, Gilani AH (2008) Cardiovascular inhibitory effects of *Hyoscyamus niger*. Methods Find Exp Clin Pharmacol 30(4):295–300. doi:10.1358/mf.2008.30.4.1186075

Kumar A, De T, Mishra A, Mishra AR (2013) Oleandrin: a cardiac glycosides with potent cytotoxicity. Pharmacogn Rev 7(14):131–139. doi:10.4103/0973-7847.120512

Lange RA, Cigarroa RG, Yancy CW, Willard JE, Popma JJ, Sills MN, McBride W, Kim AS, Hillis LD (1989) Cocaine-induced coronary-artery vasoconstriction. N Engl J Med 321(23): 1557–1562

Larrey D (1997) Hepatotoxicity of herbal medicines. J Hepatol 26(Suppl 1):48–49

Lau WC, Welch TD, Shields T, Rubenfire M, Tantry US, Gurbel PA (2011) The effect of St. John's wort on the pharmacodynamic responses of clopidogrel in hyporesponsive volunteers and patients: increased platelet inhibition by enhancement of CYP3A4 metabolic activity. J Cardiovasc Pharmacol 57(1):86–93. doi:10.1097/FJC.0b013e3181ffe8d0

Lawler PR, Bergmark BA, Laubach JP, Lakdawala NK (2014) Having a heavy heart: approaches to infiltrative cardiomyopathy. Circulation 129(16):1703–1711. doi:10.1161/CIRCULATIONAHA.113.006932

Lee NH, Yoo SR, Kim HG, Cho JH, Son CG (2012) Safety and tolerability of *Panax ginseng* root extract: a randomized, placebo-controlled, clinical trial in healthy Korean volunteers. J Altern Complement Med 18(11):1061–1069. doi:10.1089/acm.2011.0591

Li YP, Wang YM (1988) Evaluation of tussilagone: a cardiovascular-respiratory stimulant isolated from Chinese herbal medicine. Gen Pharmacol 19(2):261–263

Li J, Shi J, Yu XW, Sun J, Men Q, Kang T (2011) Chemical and pharmacological researches on Hyoscyamus niger. Chin Herb Med 3(2):117–126. doi:10.3969/j.issn.1674-6384.2011.02.006

Liaudet L, Calderari B, Pacher P (2014) Pathophysiological mechanisms of catecholamine and cocaine-mediated cardiotoxicity. Heart Fail Rev 19:815–824

Lin CC, Chan TY, Deng JF (2004) Clinical features and management of herb-induced aconitine poisoning. Ann Emerg Med 43(5):574–579

Lu HR, De Clerck F (1993) R56 865, a Na^{+}/Ca^{2+}-overload inhibitor, protects against aconitine-induced cardiac arrhythmias in vivo. J Cardiovasc Pharmacol 22(1):120–125

Lubbe WF, Podzuweit T, Opie LH (1992) Potential arrhythmogenic role of cyclic adenosine monophosphate (AMP) and cytosolic calcium overload: implications for prophylactic effects of beta-blockers in myocardial infarction and proarrhythmic effects of phosphodiesterase inhibitors. J Am Coll Cardiol 19(7):1622–1633

Maraj S, Figueredo VM, Lynn Morris D (2010) Cocaine and the heart. Clin Cardiol 33(5):264–269. doi:10.1002/clc.20746

McVann A, Havlik I, Joubert PH, Monteagudo FS (1992) Cardiac glycoside poisoning involved in deaths from traditional medicines. S Afr Med J 81(3):139–141

Mendell J, Zahir H, Matsushima N, Noveck R, Lee F, Chen S, Zhang G, Shi M, Merritt-Charles L (2013) Drug-drug interaction studies of cardiovascular drugs involving P-Glycoprotein, an efflux transporter, on the pharmacokinetics of edoxaban, an oral factor Xa inhibitor. Am J Cardiovasc Drugs 13:331–342. doi:10.1007/s40256-013-0029-0

Merritt-Charles L (2011) The prevalence of herbal medicine use among surgical patients in Trinidad. Focus Altern Complement Ther 16:266–270

Mo W, Singh AK, Arruda JA, Dunea G (1998) Role of nitric oxide in cocaine-induced acute hypertension. Am J Hypertens 11(6 Pt 1):708–714

Moses GM, McGuire TM (2010) Drug interactions with complementary medicines. Aust Prescr 33(6):177–180

Mugabo P, Khan F, Burger A (2012) Effects of *Leonotis leonuris* aqueous extract on the isolated perfused rat heart. Int J Med Aromat Plants 2:281–292

National Center for Complementary and Alternative Medicine (NCCAM) (2012) http://nccam.nih.gov/health/bitterorange. Accessed Nov 2014

Naudé TW (1997) The occurrence and significance of South African cardiac glycosides. J S Afr Biol Soc 18:7–20

Omar HR, Komarova I, El-Ghonemi M, Fathy A, Rashad R, Abdelmalak HD, Yerramadha MR, Ali Y, Helal E, Caporesi EM (2012) Licorice abuse: time to send a warning message. Ther Adv Endocrinol Metab 3(4):125–138. doi:10.1177/2042018812454322

Peregoy JA, Clarke TC, Jones LI, Stussman BJ, Nahin RL (2014) NCHS data brief, no. 146, April 2014. http://www.cdc.gov/nchs/data/databriefs/db146.pdf. Accessed Nov 2014

Poppenga RH (2010) Poisonous plants: molecular, clinical and environmental toxicology. In: Luch A (ed) Clinical toxicology, vol 2. Birkhäuser Verlag AG, Basel

Posadzki P, Watson LK, Ernst E (2012) Herb-drug interactions: an overview of systematic reviews. Br J Clin Pharmacol 75(3):603–618. doi:10.1111/j.1365-2125.2012.04350.x

Posadzki P, Watson LK, Alotaibi A, Ernst E (2013) Prevalence of use of complementary and alternative medicine (CAM) by patients/consumers in the UK: systematic review of surveys. Clin Med 13(2):126–131. doi:10.7861/clinmedicine.13-2-126

Prozialeck WC, Jivan JK, Andurkar SV (2012) Pharmacology of kratom: an emerging botanical agent with stimulant, analgesic and opioid-like effects. J Am Osteopath Assoc 112(12):792–799

Riezzo I, Fiore C, De Carlo D, Pascale N, Neri M, Turillazzi E, Fineschi V (2012) Side effects of cocaine abuse: multiorgan toxicity and pathological consequences. Curr Med Chem 19(33):5624–5646

Ruiz-Granados ES, Shouls G, Sainsbury C, Antonios T (2012) A salty cause of severe hypertension. BMJ Case Rep. doi:10.1136/bcr.12.2011.5336, pii: bcr1220115336

Sahoo N, Manchikanti P, Dey S (2010) Herbal drugs: standards and regulation. Fitoterapia 81(6):462–471. doi:10.1016/j.fitote.2010.02.001

Schwartz BG, Rezkalla S, Kloner RA (2010) Cardiovascular effects of cocaine. Circulation 122(24):2558–2569. doi:10.1161/CIRCULATIONAHA.110.940569

Singh S (2009) The hidden risks of alternative medicine. Focus Alt Complement Ther 14:175–176

Singhuber J, Zhu M, Prinz S, Kopp B (2009) Aconitum in traditional Chinese medicine: a valuable drug or an unpredictable risk? J Ethnopharmacol 126:18–30
Snyman LD, Schulz RA, Van den Berg H (2011) Variability of yellow tulp (*Moraea pallida* Bak.) toxicity. J S Afr Vet Assoc 82:131–132
Spoerke DG, Hall AH, Dodson CD, Stermitz FR, Swanson CH, Rumack BH (1987) Mystery root ingestion. J Emerg Med 5:385–388
Tachjian A, Maria V, Jahangir A (2010) Use of herbal products and potential interactions in patients with cardiovascular diseases. J Am Coll Cardiol 55(6):515–525. doi:10.1016/j.jacc.2009.07.074
Tai Y, Lau C, Young K, But PP (1992) Cardiotoxicity after herb-induced aconite poisoning. Lancet 340(8830):1254–1256. doi:10.1016/0140-6736(92)92951-B
Takayama H, Ishikawa H, Kurihara M, Kitajima M, Aimi N, Ponglux D, Koyama F, Matsumoto K, Moriyama T, Yamamoto LT, Watanabe K, Murayama T, Horie S (2002) Studies on the synthesis and opioid agonistic activities of mitragynine-related indole alkaloids: discovery of opioid agonists structurally different from other opioid ligands. J Med Chem 45(9):1949–1956
Thomas JE, Munir JA, McIntyre PZ, Ferguson MA (2009) STEMI in a 24-year-old man after use of a synephrine-containing dietary supplement: a case report and review of the literature. Tex Heart Inst J 36(6):586–590
United Nations Office on Drugs and Crime (UNODC) (n.d.) World drug report 2014. http://www.unodc.org/wdr2014/. Accessed Nov 2014
University of Washington (n.d.) Dabigatran drug interaction potential. Published on anticoagulation services. http://depts.washington.edu/anticoag/home. Accessed Nov 2014
Urkin J, Shalev H, Sofer S, Witztum A (1991) Henbane (Hyoscyamus reticulatus) poisoning in children in the Negev. Harefuah 120(12):714–716
Van der Bijl P Jr, Van der Bijl P Sr (2012) Cardiotoxicity of plants in South Africa. Cardiovasc J Afr 23(9):476–477
Vita JA, Treasure CB, Yeung AC, Vekshtein VI, Fantasia GM, Fish RD, Ganz P, Selwyn AP (1992) Patients with evidence of coronary endothelial dysfunction as assessed by acetylcholine infusion demonstrate marked increase in sensitivity to constrictor effects of catecholamines. Circulation 85(4):1390–1397
Vogl S, Picker P, Mihaly-Bison J, Fakhrudin N, Atanasov AG, Heiss EH, Wawrosch C, Reznicek G, Dirsch VM, Saukel J, Kopp B (2013) Ethnopharmacological in vitro studies on Austria's folk medicine – an unexplored lore in vitro anti-inflammatory activities of 71 Austrian traditional herbal drugs. J Ethnopharmacol 149(3):750–771. doi:10.1016/j.jep.2013.06.007
Wang W (2006) Regulation of ROMK (Kir1.1) channels: new mechanisms and aspects. Am J Physiol Renal Physiol 290(1):F14–F19. doi:10.1152/ajprenal.00093.2005
Wilbert-Lampen U, Seliger C, Zilker T, Arendt RM (1998) Cocaine increases the endothelial release of immunoreactive endothelin and its concentrations in human plasma and urine: reversal by coincubation with sigma-receptor antagonists. Circulation 98(5):385–390
Yamanaka H, Doi A, Ishibashi H, Akaike N (2002) Aconitine facilitates spontaneous transmitter release at rat ventromedial hypothalamic neurons. Br J Pharmacol 135(3):816–822
Zuzak TJ, Boňková J, Careddu D, Garami M, Hadjipanayis A, Jazbec J, Merrick J, Miller J, Ozturk C, Persson IA, Petrova G, Saz Peiró P, Schraub S, Simões-Wüst AP, Steinsbekk A, Stockert K, Stoimenova A, Styczynski J, Tzenova-Savova A, Ventegodt S, Vlieger AM, Längler A (2013) Use of complementary and alternative medicine by children in Europe: published data and expert perspectives. Complement Ther Med Suppl 1:S34–S47. doi:10.1016/j.ctim.2012.01.001

Chapter 16
Clinical Perspectives in Diagnostic-omics and Personalized Medicine Approach to Monitor Effectiveness and Toxicity of Phytocomplexes

Alessandro Buriani, Stefano Fortinguerra, and Maria Carrara

Abstract The advances of systems medicine and network pharmacotoxicology, as well as 'omics diagnostic techniques, or "diagnostic-omics," nowadays provide formidable new strategies and tools for a guided and assisted use of phytocomplexes. At the same time, personalized medicine might promote a larger and better use of medicinal plants, especially for prevention and wellness. Despite the increasingly information available from systems medicine, research is not yet fully exploited in the clinic, although the introduction of diagnostic-omics in routine medical care is becoming more relevant and meaningful every day, especially in pharmacotoxicology, and clinical pharmaco/toxico-genomics are increasingly performed for a more effective and safer use of drugs and medicinal plants.

In the last 20 years, at least two major scientific advances provided a major contribution to the present situation: the simultaneous detection of entire molecular families in a given biological system, and the ability to collect, classify, network, and visualize an unexpectedly large amount of analytical data through bio-informatics. The genomics area has been at the vanguard of this evolution. Currently, whole genome sequencing allows the identification of clinically actionable genetic information and is included in all major prospective studies that focus on personalized and precision medicine. Other "'omics" techniques, such as proteomics and metabonomics (except for a few meaningful examples such as lipidomics), although developed to high-quality standards and considered to be key operating procedures in systems medicine research, are still in the pipeline to be applied routinely in a clinical setting.

A. Buriani (✉) • S. Fortinguerra
Center Maria Paola Belloni Regazzo for Pharmacogenomics and Personalized Medicine, Gruppo Data Medica, Padova, Italy

Department of Pharmacological and Pharmaceutical Sciences, University of Padova, Padova, Italy
e-mail: alessandro.buriani@gmail.com

M. Carrara
Department of Pharmacological and Pharmaceutical Sciences, University of Padova, Padova, Italy

O. Pelkonen et al. (eds.), *Toxicology of Herbal Products*,
DOI 10.1007/978-3-319-43806-1_16

"'Omics" techniques are particularly appropriate for analyzing the biological effects of herbal drugs, which have multiple concomitant effects on several molecular targets. 'Omics techniques can assess simultaneous molecular effects, and bioinformatics allows researchers to look at such effects with a global view of the biological system. This approach is even more relevant when using multi-herbal mixtures, as in traditional medicines where plants are often used as blended herbal preparations – sometimes consisting of mixtures of mixtures – with each herbal component exerting its specific role, either as an effector, an enhancer, or a mitigator.

Systems biology-oriented P4 personalized (and precision) medicine can be envisioned as an ultra-advanced holistic approach to the patients who, based on their individual characteristics, can be monitored for prominent risk factors and treated using targeted therapies. The ability to identify individual health risk factors from the molecular to the environmental level is progressively leading to a shift from medicine to proactive medicine, and preventive and pre-emptive medicine. In this context, traditional herbal medicines, highly personalized in their approach and with the information that they provide on preventative strategies, can be exploited for integrative strategies, aimed at combining the best of deterministic and holistic medical traditions. The increase in the use of herbal medicines, while introducing new therapeutic strategies and potentiating those available, at the same time raises safety concerns that need to be addressed and managed to assure a positive balance between benefits and risks when using herbal products. In a fully personalized context, genetic profiling and a pharmaco-toxicological characterization of the patient should be performed before prescribing or administering any herbal product – especially if other drugs or herbs are being taken – given the potential inter-molecular pharmacokinetic and toxicologic interference.

A step-by-step description of the patient management in a personalized medicine context has finally been suggested, to explain how diagnostic-omics of pharmaco-toxicological interest can be applied when using herbal prescriptions. Key moments are highlighted, from taking a family history, to risk evaluation, to medication reconciliation. An example of how a future systems medicine approach could be introduced is also given, as well as how various diagnostic-omics profiling can be performed longitudinally in the same subject, so as to characterize the pharmaco-toxicological networks of the patient and then use them for personalization of therapy, toxicity prediction, and monitoring when using phytocomplexes.

Keywords 'Omics • Companion diagnostics • Personalized medicine • Pharmacogenomics • Genomics • Whole genome sequencing • Natural drugs • Personalized medicine • Proteomics • Metabonomics • Metabolomics • Microbiome • Systems biology • Network pharmacology • Network toxicology • Herbal medicines • Phytocomplex • Precision medicine • P4 medicine • Holistic • Proactive • Prevention • Traditional Chinese medicine • Health risk assessment • Family health history • Genetic profiling • Medication reconciliation

Abbreviations

BD2K	Big data to knowledge initiative
BMH	Best medication history
CDSS	Clinical decision support systems
CGD	Clinical genomic database
DILI	Drug-induced liver injury
DSHEA	Dietary Supplement Health and Education Act
EBM =	Evidence-based medicine
ESF	European Science Foundation
FDA	Food and Drug Administration
HER	Electronic health records
HPWP	Hundred person wellness project
HRA	Health risk assessment
iHMP	Integrative Human Microbiome Project =
iPOP	Integrative personal omics profile
KEGG	Kyoto Encyclopedia of Genes and Genomes
NHS	UK National Health System
NIH	US National Institutes of Health
NMR	Nuclear magnetic resonance
N-of-1 RCTs	Single case randomized controlled trials
P4 medicine	Predictive, preventive, personalized, participatory medicine
PMC	Personalized medicine coalition
SNPs	Single nucleotide polymorphisms
TCM	Traditional Chinese medicine
TCMSP	Traditional Chinese medicine systems pharmacology database
WES	Whole exome sequencing
WGS	Whole genome sequencing
WHO	World Health Organization

Introduction

The clinical use of plants has always been an important part of every culture. Despite the availability of drugs for an ever-growing number of illnesses and symptoms, the popular use of herbs and derivatives has never really been totally replaced by such pharmaceutical progress, and not just in Asian countries where traditional medicine has been systematically developed for thousands of years. Experience-based traditional medicines are slowly but steadily enjoying an increase in their consensus in the international medical and scientific community, and the 2015 Nobel Prize in Physiology or Medicine to Prof. Youyou Tu, for her pioneering work on TCM formulas leading to the identification of the antimalarial artemisine (Tu 2011; Li and Wu 2003), besides recognizing her significant contribution to global health, is also a clear sign of a high-level debate on inclusiveness towards traditional medical knowledge (Su and Miller 2015; Normile 2015). Today, thanks mostly to the demystifying power of systems biology deciphering complex pharmaco-toxicological patterns, the use of herbal drugs is being re-evaluated to introduce them into the clinic. This trend is driven by a new cultural integrative approach between Western reductionist and Oriental holistic medicine, but it is also supported by a novel concept of evidence-based medicine that is opening minds towards effective treatments despite their poorly understood mechanisms of action. This is happening, for instance, with many therapies based on phytocomplexes. Novelties brought by systems biology and the technical and bio-informatic evolution that accompany personalized medicine have thus introduced fundamental new concepts that need to be mastered before approaching the state of the art and the perspectives of pharmaco-toxicological aspects in the clinical application of medicinal plants. However, the picture would not be complete without taking into consideration the use of herbal preparations outside the clinic, sometimes lacking an Eastern or Western valid medical rationale. These preparations certainly account for a much larger proportion and variety of phytocomplexes than those being officially introduced into the clinic and, while often perceived as safe, are regulated as dietary supplements or integrators. Thus, the impact of their use (and misuse or abuse) on pharmaco-toxicological issues such as adverse events and drug-herb interactions, represents a challenge for the day-to-day practice of the physician using phytocomplexes in the management of patients, often outside a controlled healthcare environment. Today, on the one hand, a new personalized medicine approach promotes a larger and better use of medicinal plants, especially in long-term use for prevention and health balance maintenance; on the other hand, systems medicine and network pharmaco-toxicology, as well as 'omics diagnostic techniques, or, as we call them in this chapter, "diagnostic-omics," are providing formidable new strategies and tools to assist the physician in the correct use of phytocomplexes.

For decades, pharmaco-toxicology has been considered an integral part of the curriculum of various medical specialties, with regulatory differences in individual countries and international entities. However, as scientific knowledge has rapidly advanced into molecular biology first, and into the systems biology-driven use of 'omics soon after, pharmaco-toxicology has changed at a pace too rapid for generations of practicing physicians, and even for medical educators to efficiently keep up to date – not just with the latest news, but also with the emerging concepts that are needed to even grasp the meaning of such news. Fields such as bioinformatics or molecular genetics, whose basics are today indispensible and introductory to systems medicine and 'omics techniques, have long been considered theoretical fields of a specialist's area, and as such have been overlooked by many, creating a fundamental need for translation that must be addressed in order to fully exploit the therapeutic and diagnostic developments in the clinic.

This fast-growing knowledge addresses, for the first time, the complexity for what it really is – not just a list of individual biochemical reactions, but a unique and undivided body of mutually interrelated biological events, occurring simultaneously and causally intermingled. Thanks to genomic, proteomic and metabonomic techniques, the complexity of such multilayer, three-dimensional networks is beginning to be revealed, and the large number of relations can be taken into consideration thanks to advanced and dedicated software, which can do what the mind cannot, i.e., simultaneously take into account many different concepts and place each piece of the 'omics knowledge in its proper place, in a dynamic multi-dimensional network of molecules, functions, and conditions, continuously growing and updating as further molecular and clinical pieces of information flood in from experimental data and clinical practice.

Although translation of such knowledge to the clinic is continuous and consistent, it is probably safe to say that the healthcare system, and especially its final arm, the physician, are not quite ready to exploit the amount of actionable data that could be used to make better clinical choices. Efforts are being made both at the educational and regulatory levels to accelerate the evolution from the classical concepts of pharmaco-toxicology to the new model of networks and systems-oriented pharmaco-toxicology. Such a complexity-oriented approach can be particularly important in the field of pharmacognosy, where the use of phytocomplexes has always eluded the simple molecular concept of "one drug, one target" specificity and, together with the limited standardization of products and practices, has been one important reason that the use of herbal medicinal products has been considered by many a "minor option" compared to single, purified drugs. The unfeasibility of the simplification of the reductionist approach is intrinsic in the nature of herbal drugs and is linked to the multiplicity of bioactive components present in the phytocomplex and the corresponding molecular targets. On the toxicological side, one should pay attention to a plethora of potential undesired effects, both specific, due to high-affinity interactions with selected receptor molecules, and unspecific low-affinity effects, usually occurring at higher dosages, which are typical of acute or chronic intoxications.

Today the limits of the reductionist approach are felt not just in the pharmaco-toxicological fields, but in all biological and medical sciences. Simplification has even led to the extreme compartmentalization of organs and tissues, so that entire classes of endogenous mediators have been initially classified according to the context where they had been discovered, creating a cultural reductionist bias, just to realize later that they were not exclusive of one particular system, and that the systems crosstalk often shares the same mediators. Today we know that the same receptor molecules are expressed in various systems where they exert specific actions, making the world of biological signals rather complex, where actions, redundancies, feedback, and buffering effects, all together, affect the targets and their networks. Drugs are not different from endogenous mediators, considering the fact that they act on the same multiple targets, so if the response to a single drug is complex, in the case of herbal drugs, the identification of the interactions of each single component with its molecular targets may rapidly get out of control. The biological effect of a phytocomplex is the collective effect of all its components, some of which will cooperate towards the therapeutic aim, and some even against it; some others will act on other, distantly connected targets, generating a number of biological events, most of which will probably never overcome the redundancy threshold of the biological system that will therefore be buffered naturally and never become evident in clinical terms. Multiple receptors, enzymes, redox systems, adhesion molecules, channels, transporters, structural components, transcription factors, nucleic acids, lipids, metabolites and all the imaginable molecules networking within, between, and beyond individual biological systems, can be affected and can affect each component of a phytocomplex, leading to pharmacological and toxicological effects. Purified drugs have the obvious advantage of limiting the potential interactions and, at least apparently, make the clinical use simpler; on the other hand, the healing and beneficial properties of herbal medicines have often not been fully appreciated and reproduced with their purified active principles. This suggests that at least some of the healing and health-promoting properties lie in the network of simultaneous actions elicited by the combination of multiple components of the phytocomplex itself, and cannot be reproduced by simplification. The final balance of all the single interconnected activities is hard to predict; this has always been a limitation in the use of herbal drugs, which are also characterized by an extreme degree of intrinsic variability in chemical contents, depending on the species, environmental conditions, contaminations, mode of collection and preparation – to mention just a few variables.

Pros and cons in the use of herbal drugs need to be balanced when a clinical choice is to be made, and this is when the criteria and guidelines should be clear for the responsible physician; this is also where the limits for using herbal medicines are felt more strongly. The situation is made even more complicated by the popular belief that herbs are milder-acting and safer to use than commercial drugs, which is also reflected in a less strict regulatory legislation compared to commercial drugs. In Western countries, many herbs can be administered as dietary supplements or adjuvants of pharmacological therapies; in some cases, physicians even informally recommend them to patients, who often take them without a prescription anyway, under no real professional control. On the other hand, the lack of precise knowledge of the molecular mechanisms of action keeps the herbal drugs outside the clinical setting, where a defensive medicine approach often prevails and the physician prefers to remain on the safe side and not use something not thoroughly known. This has drawn an even wider chasm between the popular uncontrolled use of herbal drugs by the public and the clinical, controlled settings.

Evidence-based medicine (EBM), including the recent more personalized approaches, has given an unexpected positive boost to the use of herbal medicines. Generated from the need to verify the clinical effectiveness of therapies with well-characterized molecular mechanisms and side effects, EBM has actually developed into a "proof-of-fact" on the effectiveness of all therapies, regardless of their more or less established mechanism of action. This result and patient-centered EBM approach has unexpectedly opened new possibilities for herbal drugs, despite their often poor molecular characterization and the complexity of their molecular targets, potentially weighing in favor of their clinical use.

This approach has given new strength to well-structured traditional medicines such as traditional Chinese medicine (TCM) and Ayurveda, with medical practice backgrounds of thousands of years. As a matter of fact, today we see a trend, rather new in Western countries, towards integration between Western and Oriental medicines, placing the patient at the center of a paradigm that is more complex than a disease-driven assistance model, characterized by a holistic view of healthcare intervention that aims at curing the patient rather than the disease or the affected organ. This healthcare model goes beyond the classic reactive approach to disease and promotes the development of a proactive, personalized approach where prevention and continuous monitoring of health conditions represent the main framework. Herbal drugs and dietary interventions can have a central role in this emerging wellness-oriented, proactive approach to health, and their effective and safe use should be carefully guided and closely monitored.

At the same time, understanding the mechanisms of action of complex mixtures has become possible thanks to the current systems biology approach to herbal pharmaco-toxicology. This sheds light on a subject that has been left on the side of the mainstream for many years and now has a new opportunity to be reintroduced, with precise guidelines and standard procedures, leading to a more effective and safer use of herbal drugs in and outside the clinic. This is a shared responsibility – both at the research level and at the patient care level – so it is clear how important it is for the physician to be introduced to the new systems medicine and approach bioinformatics, as well as network pharmaco-toxicology, and to become somewhat familiar with 'omics applied to diagnostics, or diagnostic-omics. Although this knowledge has been somewhat integrated in the official educational curricula of most countries, as well as in standard continuing education programs for physicians, the new dawn of personalized and precision medicine has given a significant impetus and motivation to all this, so it is safe to say that today it is the major driving force of the medical community towards twenty-first century medicine. Thanks to this new holistic, yet at the same time molecular approach to the management of patients and to wellness and disease prevention, advances in pharmaco- and toxicogenomics, proteomics, and metabonomics are being introduced in the clinical practice, within a systems medicine framework. Basically, despite the peculiarities of pharmaco-toxicology of phytocomplexes, today the fast evolution in the field is part of the more general evolution of medicine, and in order to approach its clinical application, it is necessary to review some of the general innovations coming from high throughput – 'omics techniques and the bioinformatic tools of systems biology as well as the new personalized and precision medicine approach.

While many of the subjects mentioned above are the focus of other chapters in this book, the main goal of this one is to show some of the state-of-the-art applications of systems biology, diagnostic-omics and personalized medicine in the clinical pharmaco-toxicological practice. In the first few paragraphs, however, the general subjects, their backgrounds and their recent evolution, are briefly but systematically introduced and reviewed, and references are provided to help the reader through the already large and increasing amount of available literature.

P4-Personalized (and Precision) Medicine and the Clinical Actionability of Big Data

From Systems Biology to Personalized Medicine

The new vision of systems biology (see also Chap. 4 of this book), well equipped with 'omics techniques and bioinformatics tools, has not taken too long to start translating into the clinical field, and the step from systems biology to systems medicine has been a rather natural one. Holistic and integrative systems biology has given rise to a systems approach to health and disease that focuses on the complexity of biological systems and aims to define the components of the system as a whole, determining how they interact, and with which dynamics, in a homeostatic, healthy context, as well as under disease-induced perturbations. These can be linked to genetic or epigenetic factors, and the complex interconnected molecular networks and pathways underlining the phenotypic events are the new level of understanding that is being pursued. This systems approach to disease aims at providing comprehensive molecular information at the individual level so that each patient can be managed according to his or her specificities, and personalized strategies can be used (Weston and Hood 2004; Hood and Flores 2012; Tammen et al. 2013; Wang et al. 2015; Bauer et al. 2015). The uniqueness of systems medicine is the global vision of the patient and its use of a top-down approach, focusing on complexity and functions rather than single molecular entities. It analyzes globally the dynamic data of the patient and identifies actionable information so that the best informed choices are made for therapy or prevention (Boissel et al. 2015). One important consequence of this new approach is a better management of complex, chronic diseases, where a single factor is unlikely to be the sole factor responsible for the dysfunctions, and multiple factors are usually involved. In this case, a perspective in which the interactions and dynamics are integrated in the bio-analytical model, has better chances of properly addressing the causes of a clinical problem. A holistic, systems perspective – unlike reductionisms – focuses on these interrelationships and therefore has better chances of success. Reductionism is helpful instead, when one or several components overwhelmingly influence the systems behavior, and a quick and effective solution is needed (Ahn et al. 2006).

The systems-driven vision from a reductionist to a holistic one, in medical terms, has also brought a paradigm shift from a reactive to a proactive approach. This is generated from the convergence of the holistic approach and the technical and digital innovations, with their ability to generate and analyze large amounts of data, potentially transforming them into clinically relevant and actionable information. This is especially relevant when investigating risk factors. If identified early, when the molecular dysfunctions – although present – have not yet developed into out-and-out disease conditions, a proactive, preventative strategy may be used to avoid, slow down, or at least reduce the intensity of the pathologic conditions. The result is a shift from a disease-oriented, reactive medical approach to one that includes wellness-oriented proactivity. Thanks to systems biology and systems medicine, multilevel biological networks are being established that can help decipher how elements in biological systems interact to produce healthy and diseased states, thus providing an unprecedented tool that can be used not only to demystify disease, but also to quantify what it means to be healthy (Flores et al. 2013).

A systems approach to medicine and health requires that enormous amounts of data be deciphered and integrated into a "network of networks" model that includes network interactions and integrations at many levels. The digital revolution is thus behind the rather rapid evolution of systems medicine. Cutting edge informatics tools and electronic devices make big data sets manageable through computational integration and analyses, providing information that is useful for improving the health of the individual patient, or making it "actionable" (Hood and Flores 2012; Alyass et al. 2015). Dedicated databases and software are being produced that provide physicians and other healthcare providers with the information they need to deliver personalized care (Wei and Denny 2015).

Like systems biology, systems medicine uses high throughput technologies to produce global data sets tracking multiple dimensions of dynamic networks, and it can provide patients and physicians with personalized information at the molecular, cellular, and organ levels (Flores et al. 2013; Alyass et al. 2015; Bauer et al. 2015). 'Omics techniques are also extensively used to carry on systems medicine studies both for characterizing disease-related molecular networks and pharmaco-toxicological actions. Among 'omics, metabonomics, by measuring the metabolic end points that link directly to whole system activity, takes into account the specific contributions and interactions with environmental factors such as gut microbial metabolites, chemicals, or dietary compounds. For this reason, metabonomics is considered by many to be a key technique for evaluating the possible outcomes, both pharmacological and toxicological, of a drug or dietary intervention (Nicholson 2006)The application of 'omics is rapidly expanding to all medical disciplines, and examples are available for all the techniques, but genomic testing, especially pharmacogenomics, is still the leader in clinical use, with applications to many different specialities ranging from genome-wide approaches, to the analysis of pharmacogenomic relevant SNPs (Sookoian and Pirola 2015; Fox et al. 2015). In order to keep up with the amount of information flowing into systems medicine databases, Web-based tools that integrate genomic variation data with phenotypic information are increasingly available. Stimulus-response signalling circuit activities can be inferred from 'omics data, thus providing biomarkers that can be further used for predictive purposes (Fryburg et al. 2014). To deal more efficiently with the large amounts of various types of data, new systems medicine databases and software are being developed. A new generation of genomic data analysis methods is emerging that allows studying variants found with next-generation sequencing experiments (Gonzalez-Garay 2014), in the context of signalling pathways, like PATHiVar, a computer-aided tool aiming at providing clues on interactions among the proteins that compose signalling pathways and how these account for cell functionalities and disease-related perturbations (Hernansaiz-Ballesteros et al. 2015). Several other projects aim at improving the functionality and usability of databases, such as the "Big Data to Knowledge Initiative" (BD2K), by the US.National Institutes of Health

(NIH). The aim of BD2K is to provide indications on how to extract value from the available data by improving the ability to locate, access, share, and use biomedical big data, developing and sharing data analysis methods and software, enhancing training in biomedical big data and data science, and establishing centers of excellence in data science (Margolis et al. 2014).

Praised for its vision to improve healthcare, the systems approach to medicine has only recently begun to be rigorously explored and its advantages verified on a large scale. Besides those projects involving large numbers of subjects (mentioned in Chap. 4 of this book), such as the British 100,000 Genomes Project, the Human Variome Project, and the Personal Genome Project (Siva 2015; Smith et al. 2015; Ball et al. 2012), other major projects can be mentioned for their specificities, like the large-scale, whole genome sequencing of the Icelandic population (Gudbjartsson et al. 2015), the international 1,000 Genomes Project, aimed at characterizing the human genome sequences with a focus on rare variations (Gibson et al. 2013), and the whole genome sequencing of super-centenarians, in the quest of the genetic relations to extreme longevity (110 years or older) (Gierman et al. 2014).Genomics is not the only molecular framework where large-scale projects are being developed. Human proteome variation in the various tissues and organs is the object of a proteomic-centered project mostly developed in Sweden. The resulting map of the human tissue proteome comprises 32 different tissues and organs and is freely available in an interactive Web-based database, as part of the Human Protein Atlas www.proteinatlas.org (Uhlén et al. 2015). A recent example of metabonomics is the large-scale study of samples from 1,200 healthy individuals that provided a comprehensive measurement of their serum metabolomes. The results of this study can be used as a reference dataset, available at http://www.husermet.org/ and at MetaboLights (http://www.ebi.ac.uk/metabolights/) (Dunn et al. 2015). Large-scale studies have also been carried out for the microbiome; an example is the Integrative Human Microbiome Project (iHMP, http://hmp2.org), an NIH project that analyzes microbiome and host activities in disease-specific cohorts, thus creating integrated data sets of microbiome and host functional properties (Integrative HMP Res Net Cons 2014). In addition, epigenetics has a central role in personalized medicine (Rasool et al. 2015), and the epigenome is the subject of the NIH Roadmap Epigenomics Program, developed with the goal of investigating epigenetic mechanisms, such as DNA methylation and a variety of post-translational histone modifications, playing an important role in establishing gene expression patterns. The consortium of institutions collaborating has already made substantial progress, and with more than 120 epigenomes characterized to date, has created a public resource of disease-relevant data (http://roadmapepigenomics.org) (Chadwick 2012; Skipper et al. 2015).

The large amount of data generated is sometimes hard to exploit by physicians and even by research scientists, and the development of user-friendly software is a key issue. Integrating multiple 'omics data from various databases, especially when dealing with a diverse array of 'omics – which requires integrating information from various types of molecules – is also a major problem for their usability, and dedicated strategies and software are being developed (Mason et al. 2014; Henry et al. 2014; Gibbs et al. 2014; Shukla et al. 2015).

While systems medicine is the disciplinary framework of this new holistic, individualized medical approach, its best expression in the medical practice has been "personalized medicine," characterized by a new central role of the individual characteristics in guiding medical action, rather than protocols often standardized on groups of patients sharing little more than a disease and its sub-classifications.

"A Rose Is a Rose Is a Rose" (Gertrude Stein), but Defining Personalized Medicine Is Not Just a Matter of Semantics

> For more than two millennia, medicine has not wavered from its aspiration of being personalized. In ancient times, Hippocrates combined an assessment of the four humors – blood, phlegm, yellow bile, and black bile – to determine the best course of treatment for each patient. Today, the sequence of the four chemical building blocks that comprise DNA, coupled with telltale proteins in the blood, enable more accurate medical predictions. These include whether an individual is developing an illness now or will develop it many years in the future, will respond positively to treatment, or will suffer a serious reaction to a drug. But what is different about medicine today – and the reason the word 'personalized' has been added for emphasis – is that technology has brought us much closer to exquisite precision in disease diagnosis and treatment. (from Personalized Medicine Coalition (2014). *The Case for Personalized Medicine.* 4th Edition. Washington, D.C.: Personalized Medicine Coalition)
>
> http://www.personalizedmedicinecoalition.org/Userfiles/PMC-Corporate/file/pmc_case_for_personalized_medicine.pdf

Personalized medicine is changing our approach to healthcare and we are now moving towards an approach in which patients, according to their individual characteristics, can be monitored for their prominent risk factors and can be treated using targeted therapies. At the same time, the new ability to identify individual health risk factors, from the molecular to the environmental level, is progressively leading to a shift from reactive medicine to proactive, preventive, and pre-emptive medicine. This new emphasis on proactive prevention is expected to contribute meaningfully to a new concept of healthcare, where the target is not only the cure of the patient, but also the wellness of each individual aiming to significantly decrease the incidence and prevalence of diseases. In terms of clinical utility and relevance, it is also essential to assure that the diagnostic information obtainable with diagnostic-omics is indeed actionable information leading to effective therapeutic choices or to well- focused proactive prevention. This new healthcare model, commonly referred to as "personalized medicine," has wide-ranging implications for all stakeholders, from individuals to community, and correct communication within and outside the medical environment has a central role.

Expressing in one single original word the full meaning of a new trend or view is always hard, but with systems biology/omics-driven personalized medicine, this has been intensely debated, and high-level meetings on the subject have engaged scientists, physicians, and regulators for some time. Personalized medicine can be broadly described as a customization of healthcare that accommodates individual differences as far as possible at all stages in the process, from prevention, through diagnosis and treatment, to post-treatment follow-up. Various terms, each with valid reasons such as "genomic medicine" (or molecular medicine), "stratified medicine" and "precision medicine," have been utilized to indicate it (ESF Forward Look 2012). Even so, a name largely shared by the medical-scientific community and by the public needs to be used as a unique identifier to avoid miscommunications and misunderstandings. The debate on the issue is not simply a formal one; behind each definition lies a different cultural background, specific objectives, and sometimes even special interest groups – academic or industrial – trying to shape the future of healthcare. At the same time, each definition has been criticized by the same groups and by those who try to resist altogether the changes brought in the health care world by the new vision. Many physicians react with initial scorn when personalized medicine is presented to them – as if their practices are not already based on personalized care. They are not wrong, but there is clearly a misunderstanding that can mine the very bases for the diffusion of personalized medicine and the introduction of its innovations in healthcare. Thus, professional education needs to be addressed, but the name issue is also on the table.

Genomic, or molecular medicine, is one of the earlier names, linked to the initial source of much of the biological data that arose from genome sequencing. It is rather clear that with the later advancements especially brought by bioinformatics, the term "genomic" becomes too narrow and almost reductionistic, thus contradicting the holistic, systems vision that characterizes the latest developments. Even stratified medicine is focused on a specific, although important, aspect of the new approach, focusing the attention on the identification of subgroups of patients with a particular disease who respond to a particular drug; although it is an important step towards personalized medicine, the term recalls a plural concept and seems to lead away from the main aspect: individualization.

Precision medicine is an alternative term that reflects the targeting of the specific elements responsible for pathology in a given individual at a particular point in time. Nevertheless, the concept of precision, focusing on the active role of the healthcare system, is still suggestive of a model where patients are seen as passive recipients. Thus, the best term might be a collective one, where all the aspects of personalized medicine are clearly spelled as predictive, preventive, personalized and participatory, a view that can be labelled as proactive "P4 medicine," suggested by Lee Hood (2008), Auffray et al. (2010), Hood and Flores (2012). This incorporates several different aspects and P4 medicine is thus predictive, as it uses molecular diagnostic tools to precisely predict individual health risks and individual treatment responses and outcomes; preventive, a concept that emphasizes wellness and prevention to stop a disease before it occurs and/or progresses; personalized, in that it is informed by each person's unique clinical, genomic, molecular and environmental data, thus taking an integrated approach to individualizing patient care across the continuum from health to disease; participatory, as it aims at empowering each patient to participate in his/her own care and make informed choices, and also takes into account the importance of individual psycho-cognitive components in health and disease. Some add a fifth P – "pre-emptive" – to indicate an action-oriented, individualized health planning. The P4 medicine term thus indicates a global clinical and social view, where the healthcare system works closely with each patient to promote health and wellness, patient education and satisfaction, and customized disease prevention, detection, and treatment (Hood and Flores 2012; Sookoian and Pirola 2015; Yang 2014; Cutica et al. 2014). Additional "Ps" have been suggested, standing for "population science," "psycho-cognitive," and "public," referring to population epidemiology and social aspects, psychological components and e-health-oriented data management, respectively, but the ensuing debate has not developed into a consensus strong enough for P5, P6 or P7 medicine concepts (Khoury et al. 2012; Gorini and Pravettoni 2011; Bragazzi 2013; Li and Meyre 2014).

The term "precision medicine" was recently revamped by a program publicly launched by U.S. President Barack Obama in early 2015 (http://www.nih.gov/precisionmedicine/) (Fox 2015; Shukla et al. 2015). The initiative aims at accelerating progress towards a new era of personalized medicine, which is indicated as precision medicine, meaning prevention and treatment strategies that take individual variability into account. The program is research-oriented and encourages creative approaches to precision medicine, for detecting, measuring, and analyzing a wide range of biomedical information, including molecular, genomic, cellular, clinical, behavioral, physiological, and environmental parameters (Jameson and Longo 2015). A longitudinal "cohort" of at least 1,000,000 volunteers will be organized and the research results will ultimately be used to build the evidence base needed to guide clinical practice (Collins and Varmus 2015; Ashley 2015). "P4 Personalized and Precision Medicine" could then be a new comprehensive name, or the two could be fused into a "P5 Medicine." The point is that the new medical approach is so appealing but at the same time much debated, thereby attracting proposals and criticism in unprecedented numbers of labels.

The above initiative is only the last one of many, but probably the first with such public resonance. Personalized medicine has been delivering benefits at increasing speed over the last few years. The Personalized Medicine Coalition (PMC), a not-for-profit organization based in the U.S. that represents a wide range of stakeholders, in its 2014 edition of *The Case for Personalized Medicine* reported that more than 20 % of the new drugs approved by the FDA that year could be classified as personalized medicines, defined as "therapeutic products for which the label includes reference to specific biological markers, identified by diagnostic tools, that help guide decisions and/or procedures for the product's use in individual patients" (Personalized Medicine Coalition 2014).

In 2006, the British "Biobank" project was initiated; it is a very large, detailed prospective study with over 500,000 participants aged 40-69 that aims at finding correlations between genomic and non-genomic risk factors and complex diseases of middle and old age. This provided an opportunity to study a wide array of known and novel risk factors for a wide range of illnesses and, given that the project is still going on and phenotypic data are continuously being collected, comprehensive follow-up and characterization of many different health-related outcomes are continuously being produced (Sudlow et al. 2015). Other examples, mostly focused on genomic data collection, but still important for the implementation of personalized medicine, have been mentioned previously in this chapter (see also Chap. 4 of this book).

Recently (early in 2014), the Institute for Systems Biology in Seattle, led by Leroy Hood, initiated the "Hundred Person Wellness Project" (HPWP), the first organized real-world test to demonstrate the effectiveness of the P4 medicine paradigm. Unlike the precision medicine initiative, whose focus is disease management (mainly cancer), the project focuses on wellness and its maintenance, aiming at creating wellness metrics that can be used to monitor individuals to identify early disease transitions for the most common diseases so that individuals can ideally be guided from a disease trajectory back to a wellness one. Initially, HPWP was a 10-month pilot study of 100 apparently healthy individuals whose data from whole-genome sequencing, gut microbiome, clinical laboratory tests, and quantified self measures were collected and individually integrated to provide actionable results for health coaching, with the goal of optimizing wellness and minimizing disease. Preliminary results are encouraging, and the study is being scaled up to 100,000 participants over the next few years (Hood et al. 2015).

All these projects have developed dedicated Web sites with databases continuously fed with 'omics and clinical data, some of which are accessible and even public and represent formidable tools for research and clinical decisions alike (Chen et al. 2014), while others are not accessible for researchers, clinicians, and the public outside the research projects. Other resources are available on the Web; an example is the Clinical Genomic Database (CGD) (available at http://research.nhgri.nih.gov/CGD/), a searchable, free-access database of genetic diagnosis of clinical utility associated with specific conditions (Solomon et al. 2013). The openSNP (hosted at http://www.opensnp.org) is a free, online platform of Single Nucleotide Polymorphisms (SNPs) linked to specific phenotypic variations, where anybody can share genotypic and phenotypic information, even direct-to-consumer genetic testing customers, and they can obtain more information on their genetic data (Greshake et al. 2014).

Integrating 'omics data with other biomarkers and clinical data is a challenging issue when a systems approach is used, given the complexity of the information and the often less-than-friendly informatic visualization methods. Transforming this multitude of data into actionable information for clinical decisions is one of the central issues in personalized/precision medicine. Back in 2010, Ashley et al. proposed the first integrated analysis of a complete human genome in a clinical context (Ashley et al. 2010). Based on family medical history, the whole genome sequence was correlated with the identified risks, and a network visualization approach for gene variants and diseases was used. The model was then further developed for the integration with other 'omics data to produce an "integrative Personal Omics Profile" (iPOP), using a network-integrated 'omics analysis of the transcriptome, the proteome ,and the metabolome, and using various visualization methods for the 'omics networks; it also includes a Circos plot summarizing iPOP, with the chromosome ideogram, genomic data, structural variants, transcriptomic data and proteomic data

(Chen et al. 2012). In the study, the different 'omics data were collected over a period of time during which there was the opportunity to view the profiles under normal and disease conditions. In a first step, risk estimation was performed, based on genomic variants in combination with complete medical information and family history; after that, the appropriate dynamic profiling of multiple 'omics was done at different times, and finally data were integrated and the biological impact assessed. The clinical applicability of the information was such that the study concluded that longitudinal iPOP can provide actionable health information (Mias and Snyder 2013).

P4 personalized and precision medicine is increasingly perceived as a more efficacious approach to health and disease, and other projects are focused on the issue of clinical actionability of 'omics-integrated data. Among them, the above-mentioned "Hundred Person Wellness Project" (Hood et al. 2015), of the Institute for Systems Biology today probably represents one of the leading projects, not only for its originality (given its special focus on wellness metrics and proactive prevention), but also for its development along all of the four Ps of personalized medicine, addressing in one project all the aspects that together are important for its application, from the subject's role and empowerment, to the challenge of handling the "big data" produced with 'omics, but also its integration with electronic patient records, imaging, and other clinical information and data, envisioning its management with cloud computing, a vision today shared by many (Panahiazar et al. 2014; Calabrese and Cannataro 2015).

The Emerging Medical Use of Medicinal Plants, from Integrative Medicine to Personalized Medicine and Beyond

A Focus Shift from Reactive to Proactive Pharmacology

While more and better biomarkers are being developed for the prediction of individual health risk factors, tools to help health care personnel in decision-making need to be bolstered. Diagnostic profiles to keep under control and closely monitor the possible shift of molecular patterns towards a disease state need to be further characterized by exploiting 'omics disciplines and a systems biology approach, as in the projects mentioned above. Acting at the onset of the disease, before it can clinically affect the subject, or even before the pathogenic events take place, represents a major challenge towards a proactive wellness-driven medicine. A long and successful history of reactive medicine tradition has led to a disease-driven medicine with a well endowed, single target-oriented pharmacology, but at the same time proactive disease prevention has been underestimated. Not knowing what to do in case of identification of risk factors, besides monitoring pathognomonic parameters, is often a major problem encountered by clinicians and healthy subjects alike when approaching personalized medicine, raising the question of whether one should really know his or her disease susceptibility, without having an appropriate pre-emptive tool.

Shifting the focus towards a predictive approach changes one of the prevalent paradigms of medicine from being reactive to being pre-emptive and proactive, and has the potential to significantly decrease the incidence and prevalence of diseases. Sustaining the individual's health balance before a disease has developed should play a new central role in personalized medicine, but at the same time it highlights a lack of tools and effective strategies in Western medicine, which is mostly built around the reactive approach of curing the disease. While predictive and actionable models for health and disease are a central aspect of personalized medicine, proactive prevention is still in its infancy, and chemoprevention or pharmaco-prevention strategies have been limited to specific areas, with oncology probably the most advanced; the rapidly increasing focus on primary and secondary cancer chemoprevention suggests it as a forefront field (Wu et al. 2011; Landis-Piwowar and Lyer 2014).

Maintaining the health balance of the individual, and tailoring a preventative measure that can be particularly suitable, might require a different vision of health and disease, and in this perspective, systems biology-oriented personalized medicine can be seen as a modern holistic approach to the patient; this is a rather new vision in Western medicine, strongly anchored to its deterministic approach to disease and therapeutics. Interestingly, in this perspective, traditional medicines like Chinese herbal medicine, traditional Iranian medicine, and Ayurveda, highly personalized in their approach and with their profound knowledge of preventative strategies, are finding themselves at the forefront of a new impetus towards integrative strategies aimed at bridging together deterministic and holistic medical traditions (Schroën et al. 2015; Jafari et al. 2014). A growing number of scientists working in the field of integrative medicine are indeed applying holistic systems biology approaches (Buriani et al. 2012) that provide new perspectives for understanding complex, multiple pathways effects, such as in traditional Chinese medicine, in an innovative pharmacological view that goes beyond target specificity and single molecule pharmacology, and embraces the entire equilibrium of a biological system undergoing simultaneous perturbations on primary and secondary multiple molecular targets. The special value of these studies lies in the understanding of a traditional medicine rich in formulas that have been used for thousands of years for disease prevention, especially considering the paucity of such tools in Western medicine, and their potential impact on a new personalized medicine approach, for which prevention will have a central role (Wang et al. 2010; Hun Lee et al. 2013). The focus of TCM has been to identify transitions from health to disease and act upon them in an "evidence-based experience" basis, mostly resorting to herbal medicine, and thereby developing one of the largest experience-based knowledge in prevention. Today the use of 'omics technology and a systems biology "network-centric" approach applied to investigate effects and mechanisms of Chinese herbal medicine, as well as other holistic medicines, such as Ayurveda as mentioned above, opens the way to unprecedented insights into the complex simultaneous multi-target interactions of multicomponent drugs that can be used in disease prevention, bringing a potential unexpected ally to the implementation of personalized medicine (Witt et al. 2015; Buriani and Fortinguerra 2015; Gupta 2015).

Holistic, but Not Mystic: A Global Vision and a Sight of Evidence Are Both Needed for Integrating the Best of Eastern Medical Tradition into the Western Medical Framework

There is no doubt that herbalism, structured traditional medicines and even the so-called non-conventional and alternative medicines have been on the rise in the last 30 years. Whatever the reason, the trend is meaningful, and even the World Health Organization is trying to exploit traditional medicines for their potential benefits for wellness, people-centered health care, and universal health, while promoting their safe and effective use and their integration into the health system, to help address some of the aspects of medical globalization (http://www.who.int/medicines/publications/traditional/trm_strategy14_23/en/) (Qi and Kelley 2014). Appropriate integration with mainstream medical practice is the objective that should be aimed at, in order to overcome the skepticism and the criticisms that have hampered any attempt to identify good practices in traditional medical traditions for so long and to introduce them into the Western medical mainstream. A Western medical reductionist vision, as opposed to a holistic one – characteristic of the most important traditional medicines – is one of the reasons that this task has been so hard. The scarcity of clear evidence of effectiveness and safety is the second most important reason behind the exclusion of traditional medical medicines from the clinic, a lack of evidence according to the evaluation criteria used, heavily oriented by the same reductionist vision that contradicts the very basis of any holistic approach. Reductionism, as already discussed in the previous sections of this chapter, has its limits, and we know that today it would not be possible to advance in the field of systems medicine and 'omics without a holistic approach. Also, although powerful in biomedical sciences or analysis of disease mechanisms, reductionism can lead to an excessive concentration on the diseased organ or even molecules instead of keeping the patient as the central focus. Western medicine has evolved from the concept of organ pathology to a progressively further reduction of its focus down to the cellular and then to the molecular level. Although a body's function depends on the function of each cell that constitutes it, it is not reducible to that of any constituent cell and depends on the interactions among cells, operating independently or as functional units of various complexities. This concept was nicely developed in the beginning of this century by Tomio Tada; it finds a synthesis in the idea of the human super-system and the science that can study it, for which he proposed the name "epimedical science." He presented it as "a new biomedicine that should include the best of modern analytical medicine, but stands above it to view the human individual as an irreducible super-system" (Tada 2004). Even though the term did not reverberate, the importance of a holistic or systems view to look at a complex organization, or super-system is the same shared today by systems medicine in the lab, and by personalized medicine in the clinic.

Systems biology today gives new strength to the concept of holism in medicine, and provides the opportunity to explore scientifically the biomolecular functional networks underlying the effects of traditional medicines, especially those of herbal remedies

(Lin et al. 2012). The pleiotropic effects of herbal drugs are now being investigated with the conscience of the multi-factorial nature of disease, causal or contextual, for which phytotherapy, whose therapeutic efficacy is based on the combined simultaneous action of a mixture of constituents, offers new treatment opportunities and at the same time may have unexpected effects that cause toxic reactions. This new pharmaco-toxicological perspective brings novel opportunities and awareness to herbal drugs, beyond a simple revival of interest in natural products for the identification and purification of new drugs. The renaissance of traditional herbal medicines is thus beginning to be well felt outside the so-called "alternative" and "non-conventional" circles of health care operators and users, and its integration with the Western health care mainstream is developing at an increasing speed (Efferth and Koch 2011; Harvey et al. 2015; van Galen 2014). At the same time, in Eastern countries like China and Japan, where integration between medical traditions is a well-established practice, a new surge of research applications of systems biology and 'omics, to characterize the molecular circuits behind millennial holistic medical practices, is on the way, involving a rapidly growing number of research and clinical centers. TCM is at the head of this new wave, and 'omics techniques are now frequently used as the driving force for translating the traditional Chinese medical formulas into medical practice. This often goes beyond the simple demonstration of efficacy of specific medicines and aims at the integration of TCM as a whole into the medical mainstream, with the belief that to treat complex diseases, phytocomplexes, acting as combination medicines and affecting multiple pathways, can be superior and have better efficacy than a single drug acting alone. 'Omics techniques, providing a comprehensive assessment of endogenous metabolites of a biological system in a holistic context, can provide the companion diagnostic tool to monitor and characterize the use of phytocomplexes (Wang et al. 2012; Jiang et al. 2012; Wang and Xu 2015).

Some of these studies actually provide an unprecedented increasing amount of data associating the use of herbal medicines with molecular and clinical effects, thus building a body of evidence that can be used for their clinical assessment. A holistic vision, in fact, although necessary to understand and appreciate systems medicine, by itself could not allow the re-evaluation of traditional medicines for the medical mainstream without the support of clinical evidence. A transition from experience-based medicine to evidence-based medicine (EBM) is indeed necessary and needs to be at the center of the whole process of integration of traditional medicines, with an evidence-based model of systematic evaluation of the research evidence accompanying each step forward (Wang and Xiong 2012; Xu and Chen 2010; Chiappelli et al. 2005). To strengthen the drive towards EBM, more EBM tools are now available and include, besides clinical trials and meta-analysis, evidence collected with a patient-centered, more personalized approach, using mixed methods research methodologies, and combining systems biology-based 'omics. Clinical trials can thus include co-morbidities and can use co-medications, so that more realistic situations and conditions closer to routine clinical care can improve the impact of the research outcomes (Witt et al. 2015). Single-case, randomized controlled trials (N-of-1 RCTs) are also being developed, given their feasibility for a personalized medicine context, so that single-patient N-of-1-pathways can be obtained, and single or "Integrative Personal Omics Profiles" (iPOP) can be monitored versus their own internal control;

this way, the limits of standardization of large population-driven clinical studies, where individual characteristics are typically overlooked, can be overcome (Huang et al. 2014; Gardeux et al. 2014; Chen et al. 2012). In summary, there has been a surge in clinical research activities of phytocomplexes, and a significant amount of clinical evidence has been collected over the last few years, sustaining the use of botanical medicines, especially from traditional use in many different clinical fields.

The integration of Western and Eastern medicine has developed rather cautiously, so that herbal remedies from traditional medicines, with few exceptions, were initially considered to be, at best, adjuvant to the more orthodox treatments. Today, thanks to advances in understanding complex pharmaceutical agents brought by systems medicine, phytocomplexes are also used alone, as their therapeutic value is being recognized, especially in proactive, preventative interventions for which "reactive" drugs, normally developed to counteract a specific damage or dysfunction, are usually unfit. Besides specific preventive traditional medicines, the mix of tradition and 'omics research has helped develop new strategies for personalized nutrition in the prevention of disease and wellness maintenance (Banerjeea et al. 2015; Shankar et al. 2013).

The increase in the use of herbal medicines, while introducing new therapeutic strategies and potentiating those available, at the same time raises some safety concerns that need to be addressed and managed to assure a positive balance between benefits and risks when using herbal products. Toxicological concerns can arise both from the intrinsic potential toxic effects of the phytocomplexes, and from the interactions with drugs and other herbal medicines. What makes the issue more problematic is the fact that, unlike pharmaceutical prescription drugs, herbal remedies are often perceived as innocuous and are thus used without the supervision of professional caregivers. Phytochemicals may not be of good quality to begin with, might be stored inappropriately, and are often self-prescribed and self-administered in an outpatient context, where toxic reactions, especially those linked to chronic consumption and interactions with drugs can be difficult to identify. As in the case of their multi-component, multi-target pharmacological effect, toxicological effects of plants are being studied – combining 'omics and a systems biology approach – so that networks of toxicological pathways can be established (Williamson et al. 2015; Fasinu et al. 2012; Liu et al. 2015).

Phytocomplexes in the Age of Diagnostic-omics and Personalized Medicine

Dissemination of Phytocomplexes Among the Public and in Ambulatory Care: Pros and Cons

The use of phytocomplexes and phytonutrients, or nutraceuticals, continues to expand rapidly. People frequently opt for these products to treat various health problems and for wellness, often without medical supervision. In developing countries, where pharmaceutical costs are a major limiting factor for democratization of healthcare, the WHO strongly encourages the use of traditional herbal medicines in primary health care (Qi and Kelley 2014). Depending on the various national health care settings, with the exception of some selected plants described in official pharmacopoeias, these products are often classified as dietary supplements and can even be available in supermarkets and via the Internet, contributing to the increase in popularity of herbal medications for curing ailments (Hu et al. 2012). Furthermore, with the increase of a personalized medicine-oriented approach, maintaining the health balance of the individual and tailoring preventative measures is becoming an increasingly important priority for health professionals as well. As people live longer and accumulate a greater burden of chronic conditions, they become more frail and susceptible to disease, thus needing efficacious pre-emptive strategies (Schroën et al. 2015). This requires a different approach towards medications, historically mainly intended as treatments to alleviate symptoms, target pathogenic dysfunctions, or fight pathologic agents. A preventive therapeutic approach should focus on agents that can modify or reduce health risks and counter pathophysiological unbalances that might slowly drift into frank disease states. Herbal medicines have traditionally found applications well beyond the narrow pharmaceutical paradigm, into the intervention areas of disease prevention and wellness. This poses herbal remedies, especially those included in the paradigm of traditional medicines, usually highly personalized and holistic, in a forefront position to benefit prevention (Wang et al. 2010; Hun et al. 2013).

As the trend towards the use of preventive medications proceeds, so has the number of classes of medications taken by individual patients typically seen for ambulatory care, thus increasing complexity in managing these patients with conventional diagnostic tools. Unfortunately, the effectiveness and toxicity of medicinal herbs is not always well known by health professionals; indeed, in contrast with conventional drug research and development, the effectiveness and toxicity of traditional herbal medicines is not always well documented. However, although herbal medicinal products are widely considered to be of lower risk compared with synthetic drugs, they are not completely free from the possibility of toxicity or other adverse effects (Cock 2015; Lv et al. 2012). While inherent effectiveness and toxicity of certain herbs are usually known, adverse effects from the use of herbal medicinal products may also result from lack of standardization, contamination of products with toxic metals, adulteration with pharmacologically active synthetic compounds, misidentification or substitution of herbal ingredients, and improperly processed or preserved products. Interactions may also occur between drugs, foods, and other herbal medicinal products (Hu et al. 2012; Cock 2015). Although limited, these risks are significant in the absence of strict regulations. In the U.S., dietary supplements, including many herbal products, are regulated under the Dietary Supplement Health and Education Act (DSHEA) as food products. This act does not require the same effectiveness and safety evaluation of pharmaceuticals prior to marketing. The FDA has little control over the marketing of herbal products, but may prohibit sales of herbal products containing pharmaceutical agents, or that have been proven to have serious or unreasonable risks. However, prohibition of an herbal product generally does not occur until after marketing and extensive distribution to the public, and the burden of proof lies with the FDA and consumer reporting (Abdel-Rahman et al. 2011). Case reports and studies have documented that herbal products may contain ingredients, sometimes toxic, not listed on the label. In addition, the quantities of ingredients listed on the label can vary greatly, hindering the definition of toxic ingredients and unsafe products for public consumption. In a situation that is presently so uncertain, the possibility of developing companion diagnostics for the use of complementary and alternative medications in the cure or prevention of disease would have an important impact on the dissemination of good practices in the clinical use of herbal medicines (Cock 2015). In the era of functional genomics and bioinformatics, innovations in genomics, proteomics and metabonomics play an important role in the clinical assessment of the effectiveness and toxicity of plant-based medicinal products, and in monitoring the efficacy and toxicity of phytocomplexes, suggesting diagnostic-omics as a potentially effective player in the game.

The Electronic Record: A Patient's Personal Health Diary with a Window on Worldwide Knowledge

Diagnostic information needs a continuous integration with clinical and pharmaco-toxicological information, requiring a rationalized environment for large amounts of data of a different nature and from different sources that need to be collected, stored, and reviewed in a friendly informatics interface for healthcare providers (Castaneda et al. 2015). The correct management of individual information is a crucial step for achieving precision medicine, allowing personalized diagnosis, prognosis, and treatment. To this end, electronic health records (EHR) are being widely implemented, so that information accumulated from clinical data can be easily handled and updated with a virtually continuous flow of clinical information that can later be translated into actionable data for patient management (Castaneda et al. 2015; Ritchie et al. 2015). Electronic records can also connect to diagnostic-omics databases and software dedicated to network pharmaco-toxicology, so that clinically relevant information can be shared bi-directionally and receive up-to-date information for the interpretation of clinical data for patient management and for the resolution of clinical cases. At the same time, a contribution is given to the medical and scientific community, reaching out to others for the treatment of their own patients in other parts of the world, thus speeding up the information translational step of biomedical innovation (Ghosh et al. 2011; Sudlow et al. 2015; Ritchie et al. 2015).

The application of EHRs in the clinic provides a wealth of information to researchers and physicians, and early adopters have begun to integrate clinical decision support systems (CDSS) that benefit from the network of information provided by EHRs. Implemented CDSSs include reminder boxes for patient follow-up, warning systems for deadlines for data submission, and diagnostic suggestions. CDSSs have been implemented in pharmacology, pharmacy, pharmacogenomics, and pathology (Wei and Denny 2015). Well-characterized CDSSs that assess renal function, pregnancy status, duplicate order entry, drug allergy checking, and alerting if the choice of a drug and its dosage are unusual in the context of a specific diagnosis, are currently used to prevent errors in pharmaceutical dosing, drug-drug, and herb-drug interactions, drug-pregnancy interactions, and other medication-related parameters. These systems serve critical needs, accounting for constantly changing pharmaceutical guidelines and knowledge and can help the connection of the frequently substantial numbers of subspecialty physicians interacting with an individual patient (Castaneda et al. 2015; Ritchie et al. 2015). CDSS are particularly well suited to predicting drug dosing complications associated with drugs and herbal medicines that interact with the cytochrome P450 system, metabolizing drugs that have narrow therapeutic windows (e.g.,, warfarin). Future applications of CDSS will optimize drug and dosage selections using genomic data to predict drug metabolism, including herbal active principles based on individual patient genetic makeup (Wei and Denny 2015). A CDSS tailored to this purpose will require a large initial data set, but once established will permit safer, evidence-based optimized treatment and dosing regimens for patients (Mias and Snyder 2013; Castaneda et al. 2015; Ritchie et al. 2015).

Health Risk Assessment

A standard Health Risk Assessment (HRA) to evaluate an individual's likelihood of developing chronic diseases, or disease events, is a fundamental component of personalized medicine (Wu et al. 2015). Evidence-based HRAs coupled with predictive models facilitate assessment and prioritization of a patient's risk of disease and can predict response to drugs. HRAs form the basis for prediction and risk stratification that is fundamental to personalized health care. The characterization of the patient exploits the ever-growing -omic techniques for the molecular understanding of disease states, and optimize preventive health care strategies and drug therapies, while people are still well or at the earliest stages of disease. The overarching goal of personalized medicine is to optimize medical care and outcomes for each individual, guiding choices of treatments or prevention strategies, medication types and dosages that may differ from person to person, thus achieving a customization of patient care (Mias and Snyder 2013).

Individuals who use health services can be divided into two broad categories: patients seeking medical care for acute or chronic conditions, and patients seeking clinicians' advice about maintaining their health, inquiring about good nutrition, exercise programs, and other wellness strategies to prevent disease. In both cases, for a correct HRA, it is initially important to take into consideration the genetic components of disease. However, the extent to which genes contribute to multifactorial diseases is variable and is linked to specific genetic variations. The use of early genomic diagnostic tests can thus give personalized indications for treatment, or interventions to prevent the onset of disease or minimize its severity (Weston and Hood 2004; Mias and Snyder 2013).

The choice of the genetic risks or characteristics that need to be analyzed is usually guided by a complete physical examination of the patient with the compilation of possible clinical signs and of family history. All multi-factorial diseases have a genetic causal component, associated with other biological, environmental, and psychosocial causal components; the ultimate goal is to integrate all information to treat, cure, or prevent the disease.

A standard comprehensive clinical examination composed of three major elements is always required:

1. Detailed compilation of family medical history and of medical and environmental history;
2. Physical examination;
3. Clinical and laboratory testing, including genomic testing and pharmacogenomic evaluation.

By integrating these three important elements, it is possible to obtain an initial risk estimation of disease, to set a pre-emptive strategy, and to predict response to drugs and phytocomplexes.

Family Health History

A family health history is a simple, yet invaluable tool for the compilation of personal health risk information. Reflecting the complex combination of shared genetic, environmental, and lifestyle factors, family histories can suggest genetics/genomic risk information. A robust family history assessment helps identify persons at higher risk for disease and represents a valid record of health information about a person and his or her close relatives that can also be passed on to subsequent generations. A complete record includes information from three generations of relatives, including children, brothers and sisters, parents, aunts and uncles, nieces and nephews, grandparents, and cousins. Family members have many factors in common besides genes; in fact, they can share environment and lifestyle. Together, these factors can provide clues to medical conditions that may run in a family. While a family medical history provides information on the risk of specific health concerns, having relatives with a medical condition does not mean that an individual will definitely develop that condition; on the other hand, a person with no family history of a disorder may still be at risk of developing that disorder. Nevertheless, taking a family history of adults who may be at risk of a common chronic disease can suggest specific genetic tests and motivate people to undergo recommended screenings (Wang et al. 2015).

The easiest way to obtain information on a family's medical history is to talk to patients, making sure that they have spoken to relatives and close acquaintances. It is important to know whether there have been any medical problems, and when they occurred. Additionally, it can be helpful to obtain medical records and other documents that can help complete a family medical history. The integration of electronic

CDS-supported family history systems is a vitally important step in the advancement of personalized health care (Castaneda et al. 2015; Ritchie et al. 2015).

Besides being a fundamental step in building the patient's profile, the collection of a family medical history requires participation from the subject a commitment to his or her own health; therefore, it is an integral part of the "participatory" side of a P4 personalized medicine paradigm, leading to increased consciousness about one's health risks and improving the personal compliance with respect to the medical or lifestyle choices that the empowered individual will later be called upon to share, including paying special attention to the correct use of drugs and herbal medicines (Weston and Hood 2004). A list of Web sites dedicated to family medical history is provided in Table 16.1.

Genetic Profiling and Pharmaco-toxicological Characterization of the Patient

There are several factors that raise the possibility of a genetic diathesis in disease or in response to drugs or phytocomplexes. One is the recurrence of a condition among family members. The occurrence of the same condition or response in more than one family member (particularly first-degree relatives) are suggestive of a genetic diathesis. Collecting a family history that is suggestive of genetic predispositions is thus a fundamental step to guiding the choice of the genomic or pharmacogenomic variants associated with a certain disease or drug response.

Today the individual Whole Exome Sequencing (WES), as well as the Whole Genome Sequencing (WGS), are becoming more affordable and accessible to the public and are soon expected to replace the single SNPs search. WGS can be obtained even before birth, thanks to the presence of circulating fetal DNA in the mother's blood (Allyse et al. 2015). An even more affordable alternative to WGS today is still represented by DNA chip analysis with large panels of probes for detecting thousands of genetic variations, like that offered by direct to consumer testing companies which, only if used ethically and under the supervision of a medical professional, can be exploited for identifying DNA variations of actionable medical value (Roberts and Ostergren 2013). Once the sequence or the genetic information is obtained and stored in the appropriate platform, it can be interrogated by the physician for any SNP predictive of susceptibility to disease or response to substances, based on the information gathered during the health history and physical examination. It is then possible to take appropriate preventative measures and, if and when needed, use the therapeutic agent(s) with precision. Genetic variants are selected based on Evidence Based Medicine and clinical validation, and can be found in dedicated databases often freely accessible on the Internet (e.g.,, pharmaGKB; https://www.pharmgkb.org). An example of this approach, combining WGS with the complete medical and family history and classical clinical risk factor profiling, has been implemented in the Personal Genome Project (Ball et al. 2014; Mias and Snyder 2013).

Genetic variations can be informative in various ways and can accordingly be used for various purposes. In addition to their widespread use for identifying individual

Table 16.1 Open access web sites dedicated to family medical history

Content	Link
NIH Senior Health, a service of the NIH, provides information and tools for documenting family health history	http://nihseniorhealth.gov/creatingafamilyhealthhistory/whycreateafamilyhealthhistory/01.html
Educational resources related to family health history available from GeneEd	http://geneed.nlm.nih.gov/topic_subtopic.php?tid=5&sid=13
The CDC of Public Health Genomics provides information about family medical history, including links to publications, reports, and tools for recording family health information	http://www.cdc.gov/genomics/famhistory/famhist.htm
The US Surgeon General provides a tool called My Family Health Portrait allowing to enter, print, and update family health history	https://familyhistory.hhs.gov/FHH/html/index.html
Information about collecting and recoding a family medical history from the National Society of Genetic Counselors	http://nsgc.org/p/cm/ld/fid=52
The American Medical Association provides family history tools, including questionnaires and forms for collecting medical information	http://www.ama-assn.org/ama/pub/physician-resources/medical-science/genetics-molecular-medicine/family-history.page
The National Genetics and Genomics Education Centre of the UK NHS explains how to collect information about family health history	http://www.geneticseducation.nhs.uk/for-practitioners-62/identifying-patients/taking-and-recording-a-family-history
The Genetic Alliance has a list of links to family history resources	http://www.geneticalliance.org/programs/genesinlife/fhh
The Italian Data Medica Padova Group provides free information and tools for documenting family health	History. http://www.medicina-personalizzata.it/?pid=2

risks of developing a disease prior to the onset of symptoms, they can also be used to confirm or improve a diagnosis in a symptomatic individual (Meisel et al. 2015), but the most interesting application in the pharmaco-toxicological context remains the pharmacogenetic testing. This can help physicians to personalize drug therapies based on people's genetic make-up, taking into account all the pharmacokinetic, pharmacodynamic, and toxicological aspects known to be associated with genetic variations. Among the numerous genomic variants of special interest for natural products, the so-called "VIP" ones constitute a selected subgroup of the best clinically validated variants that are highlighted in the PharmGKB database (https://www.pharmgkb.org) (Whirl-Carrillo et al. 2012). The analysis of these molecular variants can provide information on the functions of the specific pathways that the phytocomplex is supposed to interact with in order to exert its effects or to cause toxicity (Table 16.2). Testing people prior to initiating therapy with drugs or phytocomplexes allows for determining their likely response to various classes of substances. For many drugs and phytocomplexes, specific pharmacogenetic testing can be considered a "companion diagnostic" that can be used to predetermine proper therapies and dosages to maximize the effect, while reducing the likelihood of adverse events (Agarwal et al. 2015; Gupta 2015; Fortinguerra et al. 2015; Cheng et al. 2015).

Pharmacogenomics can also help guide complicated pharmaco-toxicological choices when multiple medical conditions suggest the use of several drugs or phytocomplexes. In such cases harmful drug-drug and drug-herb interactions can be avoided by identifying the genetic characteristics (Hu et al. 2012; Ashley et al. 2010; Scott 2011).

Genetic factors influence the effects of many common medications or the rate at which a patient can metabolize a drug, affecting its activity and life span. It is known that members of some ethnic groups lack the glucose-6-phosphate dehydrogenase (G6PD) enzyme, which is required for the metabolism of some phytocomplexes (e.g., *Salix caprea*, which is commonly used in Ayurvedic medicine), as well as some medicines (e.g., the antimalarial drugs chloroquine and primaquine). Individuals lacking G6PD should not take these medicines, as the drugs will remain in their bodies and may cause hemolysis, resulting in serious health problems (Cock 2015). The incidence of oral cancer in betel (*Piper betel*) chewers is also genetically influenced. Some individuals lack the ability to produce the enzyme CYP P450 2A6. This enzyme metabolizes compounds in the betel leaf to produce carcinogenic metabolites; therefore, individuals with low expression of this enzyme produce low levels of carcinogens from the betel, while individuals with high expression levels of the enzyme produce high levels of carcinogens and have a greater risk of developing oral cancer. Other people may bear a genetic variation of another gene, the CYP1A1 gene, expressing an enzyme that produces higher levels of the carcinogen, correlating with a higher incidence of oral cancer (Cock 2015; Chatterjee et al. 2015; Mahapatra et al. 2015).

The existence of large population differences is consistent with the inheritance as a determinant of drug response; it is estimated that genetics can account for 20-95 % of variability in drug disposition and effects. Although many non-genetic factors influence the effects of medications, including age, organ function, concomitant therapy, drug interactions, and the nature of the disease, it is now well recognized that inter-individual differences in drug response are largely due to sequence variants in genes' encoding drug-metabolizing enzymes, drug transporters, or drug targets. The interplay of these factors determines the profile of the plasma concentration over time for a drug and, as a result, its elicited pharmacologic effect at the site of interaction with targets (such as receptors and enzymes). Too little exposure leads to an ineffective drug regimen, and too much creates the potential for adverse effects. Recognition of such general relationships is long-standing, and information about some drugs is extensive. Moreover, the inherited determinants, unlike other factors influencing drug response, remain stable throughout the life of a person and can in turn be transmitted to subsequent generations. Genetic polymorphisms have long been studied and are known to affect the functioning of pharmacokinetic and pharmacodynamic pathways, thus accounting for individual differences in drug responses (Evans and McLeod 2003; Wang et al. 2011; Swen et al. 2011; Yip et al. 2015). Herbal medicines are a combination of biologically active compounds with various inherent pharmacological activities and specific metabolic pathways; thus, functional agonism/antagonism and pharmacokinetic interactions mediated by drug-metabolizing enzymes, or transporters, are involved in many herb-drug interactions, and may be genetically traceable. In some cases, the genetic individual assets are such that specific herb-drug and herb-herb associations can be beneficial, helping the pharmacological outcome. Knowing the genetic characteristics can thus also help guide the positive association of various herbal products or drugs, exploiting their overall synergic effect (Hu et al. 2012).

Some selected and clinically validated SNPs are being used by physicians to predict clinical response or potential toxicity. Polygenic determinants of drug effects have become increasingly important in pharmacogenomics and in toxicogenomics and are used in clinical diagnostics to optimize therapy and prevent adverse reactions to medications and herbal compounds. SNPs affecting metabolic enzymes are of particular interest in the use of herbal drugs since their pathways can be shared by many different classes of drugs and thus increase the variability of individual responses and toxicity to phytocomplexes, especially in association with other drugs (Hu et al. 2012; Waters and Fostel 2004). In particular, the concomitancy of these competing substances might affect the pharmacokinetic mechanisms, determining alterations in their concentrations or half lives, thus leading to altered effects. Pharmacokinetic interferences may cause effects on absorption, on distribution pattern, and on changes or competition in the metabolic and excretory pathways. Specific SNPs can alter the activity of the shared enzymes or transporters, making them more or less functional, thus amplifying or reducing the molecular competition between the substrates, leading to increased or decreased efficacy or toxicity. The pharmacokinetic mechanisms of herbal medications often involve drug transporters and intestinal or hepatic drug-metabolizing enzymes, particularly the cytochrome P450 (CYP) enzymes, a super-family of microsomal drug-metabolizing enzymes. The characteristics of the various cytochrome P-450 enzymes are well established, and the involvement of these enzymes in the metabolism of the most commonly used drugs is well known. This knowledge provides a basis for understanding and predicting individual differences in drug response, which can be caused by genetic variability and drug interactions. The CYP super-family is generally involved in oxidative, peroxidative, and reductive biotransformation of xenobiotics and endogenous compounds. They metabolize many chemicals present in the diet and environment as well as medications. Cytochrome P-450 enzymes reduce or alter the pharmacologic activity of many drugs and facilitate their elimination. They are conventionally divided into families and subfamilies based on nucleotide sequence homology and are designated by a family number, a subfamily letter, a number for an individual enzyme within the subfamily, and an asterisk followed by a number and a letter for each genetic (allelic) variant (more information is available at www.imm.ki.se/CYPalleles/). There is a high degree of substrate specificity among the various families. CYP belonging to the families 1, 2, and 3 are principally involved in xenobiotic metabolism, while others play a major role in the formation and elimination of endogenous compounds such as hormones, bile acids, and fatty acids. There are many families of drug-metabolizing enzymes in humans, and essentially all have genetic variants, many of which translate into functional changes in the proteins encoded. The most important CYP subfamilies responsible for drug metabolism in humans are 1A2, 2A6, 2B6, 2C8, 2C9, 2C19, 2D6, 2E1, 3A4, and 3A5. Most constituents of phytocomplexes undergo Phase I and/or II

Table 16.2 Clinically relevant genomic variants in pharmacognosy, selected from VIP pharmaco-toxicogenomic variants

Phenotype of pharmacognostic interest	Pharmacognosy-relevant pathways clinically shown to be affected	Genotype and variants
Mostly pharmacodynamics		
Adrenoceptor beta 1	Sympathetic nerve pathway (neuroeffector junction) Antiarrhythmic pathway, pharmacodynamics	**ADRB1** – rs1801252; rs1801253
Adrenoceptor beta 2, surface	Antiarrhythmic pathway, pharmacodynamics Beta-agonist/beta-blocker pathway, pharmacodynamics Sympathetic nerve pathway (neuroeffector junction)	**ADRB2** – rs1042713; rs1042714; rs1800888
Alcohol dehydrogenase 1A (class I), alpha polypeptide		**ADH1A** – rs975833
Alcohol dehydrogenase 1B (class I), beta polypeptide		**ADH1B** – rs1229984; rs2066702
Alcohol dehydrogenase 1C (class I), gamma polypeptide		**ADH1C** – rs698
Aldehyde dehydrogenase 1 family, member A1		**ALDH1A1** – rs6151031; rs72554629
Angiotensin I converting enzyme (peptidyl-dipeptidase A) 1	ACE inhibitor pathway, pharmacodynamics Agents acting on the renin-angiotensin system pathway, pharmacodynamics	**ACE** – rs1799752
Aryl hydrocarbon receptor	AHR signal transduction pathway	**AHR** – rs2066853
Arachidonate 5-lipoxygenase	Leukotriene modifiers pathway, pharmacodynamics Eicosanoid metabolism Leukotriene synthesis	**ALOX5** – chr10:45869552; rs2115819
Breast cancer 1, early onset	ATM mediated phosphorylation of repair proteins and ATM signaling pathway Aurora A signaling BARD1 signaling events Co-regulation of androgen receptor activity FOXA1 transcription factor network Recruitment of repair and signaling proteins to double-strand breaks	**BRCA1** – chr17:41276046; chr17:41209080

Table 16.2 (continued)

Phenotype of pharmacognostic interest	Pharmacognosy-relevant pathways clinically shown to be affected	Genotype and variants
Catechol-O-methyltransferase	Estrogen metabolism pathway	**COMT** – rs4680
Coagulation factor V (proaccelerin, labile factor)	Extrinsic prothrombin activation pathway – (BioCarta via pathway interaction database) Intrinsic prothrombin activation pathway – (BioCarta via pathway interaction database)	**F5** – rs6025
Cystic fibrosis transmembrane conductance regulator (ATP-binding cassette sub-family C, member 7)	Cystic fibrosis transmembrane conductance regulator (cftr) and beta 2 adrenergic receptor (b2ar) pathway (BioCarta via pathway interaction database) Transmembrane transport of small molecules (reactome via pathway interaction database)	**CFTR** – rs113993960; rs121908755; rs121908757; rs121909005; rs121909013; rs121909041; rs193922525; rs267606723; rs74503330; rs75527207; rs80282562
Dopamine receptor D2	Nicotine pathway (dopaminergic neuron), pharmacodynamics	**DRD2** – rs1799732; rs1800497; rs1801028; rs6277
Epidermal growth factor receptor	EGFR inhibitor pathway, pharmacodynamics	**EGFR** – rs121434568; rs121434569; rs2227983; rs28929495; rs712829
Glucose-6-phosphate dehydrogenase	Methylene blue pathway, pharmacodynamics Oxidative stress regulatory pathway (erythrocyte) Pentose phosphate pathway (erythrocyte)	**G6PD** – rs1050828; rs1050829; rs5030868; rs72554665
Glutathione S-transferase pi 1		**GSTP1** – rs1138272; rs1695
Glutathione S-transferase theta 1		**GSTT1** – chr22:24343276
3-hydroxy-3-methylglutaryl-CoA reductase	Statin pathway, pharmacodynamics	**HMGCR** – rs17238540; rs17244841; rs3846662
Major histocompatibility complex, class I, B	Immunoregulatory interactions between a lymphoid and a non-lymphoid cell (reactome via pathway interaction database)	**HLA-B** – HLA-B *15:02:01; HLA-B *57:01:01; HLA-B *58:01

(continued)

Table 16.2 (continued)

Phenotype of pharmacognostic interest	Pharmacognosy-relevant pathways clinically shown to be affected	Genotype and variants
Methylenetetrahydrofolate reductase (NAD(P)H)	Antimetabolite pathway – folate cycle, pharmacodynamics	**MTHFR** – rs1801131; rs1801133
Potassium voltage-gated channel, subfamily H (age-related), member 2	Antiarrhythmic pathway, pharmacodynamics	**KCNH2** – rs12720441; rs1805123; rs36210421; rs3807375; rs3815459
Potassium inwardly-rectifying channel, subfamily J, member 11	Anti-diabetic drug potassium channel inhibitors pathway, pharmacodynamics Antiarrhythmic pathway, pharmacodynamics	**KCNJ11** – rs5219
NAD(P)H dehydrogenase, quinone 1		**NQO1** – rs1800566
Nuclear recept. subfamily 1, group I, member 2	Taxane pathway, pharmacokinetics	**NR1I2** – rs12721608; rs3814055
Prostaglandin I2 (prostacyclin) synthase	Eicosanoid metabolism	**PTGIS** – chr20:48184659; rs5629
Prostaglandin-endoperoxide synthase 2 (prostaglandin G/H synthase and cyclooxygenase)	Glucocorticoid pathway – transcription regulation, pharmacodynamics Calcineurin-regulated NFAT-dependent transcription in lymphocytes – (http://pid.nci.nih.gov/) Calcium signaling in the CD4+ TCR pathway – (http://pid.nci.nih.gov/) COX reactions – (http://pid.nci.nih.gov/) Prostanoid hormones – (http://pid.nci.nih.gov/) S1P1 pathway – (http://pid.nci.nih.gov/) Signaling mediated by p38-alpha and p38-beta – (http://pid.nci.nih.gov/)	**PTGS2** – rs20417; rs5275; rs689466
Purinergic receptor P2Y, G-protein coupled, 1	Platelet aggregation inhibitor pathway, pharmacodynamics	**P2RY1** – rs1065776; rs701265
Purinergic receptor P2Y, G-protein coupled, 12	Platelet aggregation inhibitor pathway, pharmacodynamics	**P2RY12** – rs2046934
Sodium channel, voltage-gated, type V, alpha subunit	Antiarrhythmic pathway, pharmacodynamics	**SCN5A** – rs1805124; rs6791924; rs7626962
Solute carrier family 19 (folate transporter), member 1	Metabolism of folate and pterines – (http://pid.nci.nih.gov/)	**SLC19A1** – rs1051266; rs1051296; rs1051298; rs1131596; rs12659

Table 16.2 (continued)

Phenotype of pharmacognostic interest	Pharmacognosy-relevant pathways clinically shown to be affected	Genotype and variants
Sulfotransferase family, cytosolic, 1A, phenol-preferring, member 1	Aromatase inhibitor pathway (multiple tissues), pharmacodynamics Estrogen metabolism pathway	**SULT1A1** – rs1801030; rs3760091; rs750155; rs9282861
Thiopurine S-methyltransferase		**TPMT** – rs1142345; rs1800460; rs1800462; rs1800584
Thymidylate synthetase	Antimetabolite pathway – folate cycle, pharmacodynamics	**TYMS** – rs34489327; rs34743033
Vitamin D (1,25- dihydroxyvitamin D3) receptor	Control of gene expression by vitamin D receptor – (http://pid.nci.nih.gov/) Regulation of nuclear SMAD2/3 signaling – (http://pid.nci.nih.gov/) Retinoic acid receptors-mediated signalling – (http://pid.nci.nih.gov/) RXR and RAR heterodimerization with other nuclear receptor-(http://pid.nci.nih.gov/)	**VDR** – rs11568820; rs1540339; rs1544410; rs2228570; rs2239179; rs2239185; rs3782905; rs4516035; rs731236; rs7975232
Vitamin K epoxide reductase complex, subunit 1		**VKORC1** – rs7294; rs9923231; rs9934438
Mostly pharmacokinetics		
ATP-binding cassette, sub-family B (MDR/TAP), member 1	Lovastatin pathway, pharmacokinetics Codeine and morphine pathway, pharmacokinetics Statin pathway – generalized, pharmacokinetics Taxane pathway, pharmacokinetics Vinca alkaloid pathway, pharmacokinetics	**ABCB1** – rs1045642; rs1128503; rs2032582
Cytochrome P450, family 1, subfamily A, polypeptide 2	Caffeine pathway, pharmacokinetics Estrogen metabolism pathway Platelet aggregation inhibitor pathway, pharmacodynamics Theophylline pathway, pharmacokinetics	**CYP1A2** – rs12720461; rs2069514; rs762551

(continued)

Table 16.2 (continued)

Phenotype of pharmacognostic interest	Pharmacognosy-relevant pathways clinically shown to be affected	Genotype and variants
Cytochrome P450, family 2, subfamily A, polypeptide 6	Artemisinin and derivatives pathway, pharmacokinetics Caffeine pathway, pharmacokinetics Nicotine pathway, pharmacokinetics	**CYP2A6** – rs1801272; rs28399433; rs28399444; rs28399454; rs28399468; rs5031016; rs8192726
Cytochrome P450, family 2, subfamily B, polypeptide 6	Artemisinin and derivatives pathway, pharmacokinetics Nicotine pathway, pharmacokinetics Platelet aggregation inhibitor pathway, pharmacodynamics	**CYP2B6** – rs2279343; rs28399499; rs3211371; rs3745274
Cytochrome P450, family 2, subfamily C, polypeptide 19	Lovastatin pathway, pharmacokinetics Glucocorticoid pathway – transcription regulation, pharmacodynamics Platelet aggregation inhibitor pathway, pharmacodynamics Proton pump inhibitor pathway, pharmacokinetics Statin pathway – generalized, pharmacokinetics	**CYP2C19** – rs12248560; rs4244285; rs4986893
Cytochrome P450, family 2, subfamily C, polypeptide 8	Lovastatin pathway, pharmacokinetics Caffeine pathway, pharmacokinetics Statin pathway – generalized, pharmacokinetics Taxane pathway, pharmacokinetics	**CYP2C8** – rs10509681; rs1058930; rs11572080; rs11572103; rs17110453; rs7909236
Cytochrome P450, family 2, subfamily C, polypeptide 9	Lovastatin pathway, pharmacokinetics Caffeine pathway, pharmacokinetics Platelet aggregation inhibitor pathway, pharmacodynamics Statin pathway – generalized, pharmacokinetics	**CYP2C9** – rs1057910; rs1799853
Cytochrome P450, family 2, subfamily D, polypeptide 6	Lovastatin pathway, pharmacokinetics Codeine and morphine pathway, pharmacokinetics Statin pathway – generalized, pharmacokinetics	**CYP2D6** – rs1065852; rs16947; rs28371706; rs28371725; rs35742686; rs3892097; rs5030655; rs5030656; rs59421388; rs61736512

Table 16.2 (continued)

Phenotype of pharmacognostic interest	Pharmacognosy-relevant pathways clinically shown to be affected	Genotype and variants
Cytochrome P450, family 2, subfamily E, polypeptide 1	Caffeine pathway, pharmacokinetics Theophylline pathway, pharmacokinetics	**CYP2E1** – CYP2E1 *1B; CYP2E1 *5B; CYP2E1 *5A; CYP2E1 *6; CYP2E1 *1A
Cytochrome P450, family 2, subfamily J, polypeptide 2	Eicosanoid metabolism Fatty acids – (reactome via pathway interaction database)	**CYP2J2** – rs890293
Cytochrome P450, family 3, subfamily A, polypeptide 4	Artemisinin and derivatives pathway, pharmacokinetics Lovastatin pathway, pharmacokinetics Caffeine pathway, pharmacokinetics Codeine and morphine pathway, pharmacokinetics Glucocorticoid pathway – transcription regulation, pharmacodynamics Platelet aggregation inhibitor pathway, pharmacodynamics Proton pump inhibitor pathway, pharmacokinetics Statin pathway – generalized, pharmacokinetics Taxane pathway, pharmacokinetics Theophylline pathway, pharmacokinetics Vinka alkaloid pathway, pharmacokinetics	**CYP3A4** – rs12721627; rs12721634; rs2740574; rs4986907; rs4986908; rs4986909; rs4986910; rs4986913; rs4987161
Cytochrome P450, family 3, subfamily A, polypeptide 5	Artemisinin and derivatives pathway, pharmacokinetics Lovastatin pathway, pharmacokinetics Glucocorticoid pathway – transcription regulation, pharmacodynamics Platelet aggregation inhibitor pathway, pharmacodynamics Statin pathway – generalized, pharmacokinetics Taxane pathway, pharmacokinetics Vinka alkaloid pathway, pharmacokinetics	**CYP3A5** – rs10264272; rs76293380; rs776746
Cytochrome P450, family 4, subfamily F, polypeptide 2		**CYP4F2** – rs2108622

(continued)

Table 16.2 (continued)

Phenotype of pharmacognostic interest	Pharmacognosy-relevant pathways clinically shown to be affected	Genotype and variants
Dihydropyrimidine dehydrogenase		**DPYD** – rs1801158; rs1801159; rs1801160; rs1801265; rs3918290; rs72549303
N-acetyltransferase 2 (arylamine N-acetyltransferase)	Caffeine pathway, pharmacokinetics Acetylation – (reactome via pathway interaction database)	**NAT2** – rs1041983; rs1208; rs1495741; rs1799929; rs1799930; rs1799931; rs1801279; rs1801280; rs4271002; rs4646244
Solute carrier family 22 (organic cation transporter), member 1		**SLC22A1** – rs12208357; rs34059508; rs34130495; rs72552763
Solute carrier organic anion transporter family, member 1B1	Lovastatin pathway, pharmacokinetics Codeine and morphine pathway, pharmacokinetics Statin pathway – generalized, pharmacokinetics	**SLCO1B1** – rs2306283; rs4149015; rs4149056
UDP glucuronosyltransferase 1 family, polypeptide A1	Lovastatin pathway, pharmacokinetics Codeine and morphine pathway, pharmacokinetics Statin pathway – generalized, pharmacokinetics	**UGT1A1** – rs4148323; rs8175347

Among the numerous genomic variants of special interest for natural products and derivatives, the so-called "VIP" ones constitute a growing, selected subgroup of the best clinically validated variants. Pathways and genomic variants are from the PharmGKB database (https://www.pharmgkb.org) (Whirl-Carrillo et al. 2012)

metabolism, yielding inactive or active metabolites. The CYP enzymes are usually considered the most important Phase I drug metabolizing enzymes and are responsible for the oxidative metabolism of over 90 % of drugs and herbal medications. Drugs may be metabolized by a variety of sequential or competitive chemical processes involving oxidation, reduction, and hydrolysis (Phase I reactions) or glucuronidation, sulfation, acetylation, and methylation (Phase II reactions). In general, the water solubility of the resulting metabolites is greater, thus enhancing their removal (Evans and McLeod 2003; Wang et al. 2011; Swen et al. 2011; Yip et al. 2015). A partial collection of herb-drug interactions and the SNPs known to affect them is listed in Table 16.3. Following are some examples of herb-drug interferences.

Table 16.3 Genetic variants affecting pharmaco-toxicology of herbal medicines – key polymorphisms that can be used for pharmacogenomics in the clinical setting to help plan a personalized therapy with phytocomplexes and to recognize and prevent adverse drug reactions from drug-herb interactions

Herbal medicine	Herb-drug interactions	Toxicological mechanism	Clinically validated genes and variants affecting toxicity and drug-herb, herb-herb interactions
Acacia senegal, Acacia greggii	Digoxin	Inhib. cardiomuscular Na,K-adenosine phosphatase (ATPase)	**ABCB1** – rs1045642; rs1128503; rs2032582 **KCNH2** – rs12720441; rs1805123; rs36210421; rs3807375; rs3815459 **KCNJ11** – rs5219 **SCN5A** – rs1805124; rs6791924; rs7626962
Aconitum carmichaelii spp	Cardiovascular drugs	Hypotension Arrhythmias Diarrhea Vomiting	**ABCB1** – rs1045642; rs1128503; rs2032582 **CYP3A4** – rs12721627; rs12721634; rs2740574; rs4986907; rs4986908; rs4986909; rs4986910; rs4986913; rs4987161 **CYP3A5** – rs10264272; rs76293380; rs776746 **CYP2D6** – rs1065852; rs16947; rs28371706; rs28371725; rs35742686; rs3892097; rs5030655; rs5030656; rs59421388; rs61736512 **CYP2C19** – rs12248560; rs4244285; rs4986893 **CYP2E1** – CYP2E1 *1B; CYP2E1 *5B; CYP2E1 *5A; CYP2E1 *6; CYP2E1 *1A **KCNH2** – rs12720441; rs1805123; rs36210421; rs3807375; rs3815459 **KCNJ11** – rs5219 **SCN5A** – rs1805124; rs6791924; rs7626962

(continued)

Table 16.3 (continued)

Herbal medicine	Herb-drug interactions	Toxicological mechanism	Clinically validated genes and variants affecting toxicity and drug-herb, herb-herb interactions
Actaea racemosa (black cohosh)	Aspirin Cisplatin Clopidogrel Dipyridamole Docetaxel Doxurubicin Heparin Ticlopidine Warfarin Anesthetics Antihypertensives	Coumarin compounds may increase bleeding Hypotension	**ABCB1** – rs1045642; rs1128503; rs2032582 **CYP1A2** – rs12720461; rs2069514; rs762551 **CYP2B6** – rs2279343; rs28399499; rs3211371; rs3745274 **CYP2C19** – rs12248560; rs4244285; rs4986893 **CYP2C8** – rs10509681; rs1058930; rs11572080; rs11572103; rs17110453; rs7909236 **CYP2C9** – rs1057910; rs1799853 **CYP3A4** – rs12721627; rs12721634; rs2740574; rs4986907; rs4986908; rs4986909; rs4986910; rs4986913; rs4987161 **CYP3A5** – rs10264272; rs76293380; rs776746 **CYP4F2** – rs2108622 **VKORC1** – rs7294; rs9923231; rs9934438 **P2RY1** – rs1065776; rs701265 **P2RY12** – rs2046934 **GSTP1** – rs1138272; rs1695 **GSTT1** – chr22:24343276 **NQO1** – rs1800566 **NR1I2** – rs12721608; rs3814055
Adonis vernalis (Adonis)	Digoxin Quinidine	Inhib. cardiomuscular Na,K-adenosine phosphatase (ATPase)	**ABCB1** – rs1045642; rs1128503; rs2032582 **KCNH2** – rs12720441; rs1805123; rs36210421; rs3807375; rs3815459 **KCNJ11** – rs5219 **SCN5A** – rs1805124; rs6791924; rs7626962

Aesculus hippocastanum (horse chestnut)	Aspirin Warfarin	Increase bleeding	**ABCB1** – rs1045642; rs1128503; rs2032582 **CYP1A2** – rs12720461; rs2069514; rs762551 **CYP2C19** – rs12248560; rs4244285; rs4986893 **CYP2C8** – rs10509681; rs1058930; rs11572080; rs11572103; rs17110453; rs7909236 **CYP2C9** – rs1057910; rs1799853 **CYP3A4** – rs12721627; rs12721634; rs2740574; rs4986907; rs4986908; rs4986909; rs4986910; rs4986913; rs4987161 **CYP4F2** – rs2108622 **VKORC1** – rs7294; rs9923231; rs9934438 **P2RY1** – rs1065776; rs701265 **P2RY12** – rs2046934
Allium sativum (garlic)	Warfarin Saquinavir Ritonavir Chlorpropamide	Increase of the time for clot formation during bleeding Hypoglycemia Inductor CYP3A4 enzyme	**ABCB1** – rs1045642; rs1128503; rs2032582 **CYP1A2** – rs12720461; rs2069514; rs762551 **CYP2C19** – rs12248560; rs4244285; rs4986893 **CYP2C8** – rs10509681; rs1058930; rs11572080; rs11572103; rs17110453; rs7909236 **CYP2C9** – rs1057910; rs1799853 **CYP3A4** – rs12721627; rs12721634; rs2740574; rs4986907; rs4986908; rs4986909; rs4986910; rs4986913; rs4987161 **CYP2E1** – CYP2E1 *1B; CYP2E1 *5B; CYP2E1 *5A; CYP2E1 *6; CYP2E1 *1A **CYP4F2** – rs2108622 **VKORC1** – rs7294; rs9923231; rs9934438 **SLC22A1** – rs12208357; rs34059508; rs34130495; rs72552763 **SLCO1B1** – rs2306283; rs4149015; rs4149056 **KCNJ11** – rs5219

(continued)

Table 16.3 (continued)

Herbal medicine	Herb-drug interactions	Toxicological mechanism	Clinically validated genes and variants affecting toxicity and drug-herb, herb-herb interactions
Aloe: various aloe species including aloe vera, aloe ferox aloe barbadensis	Cardiac glycosides (e.g., digoxin) Diuretics Tolbutamide Glyburide	Decrease blood potassium levels Diarrhea Decrease drug absorption, Increase concentration drugs in blood	**ABCB1** – rs1045642; rs1128503; rs2032582 **CYP2C8** – rs10509681; rs1058930; rs11572080; rs11572103; rs17110453; rs7909236 **CYP2C9** – rs1057910; rs1799853 **CYP3A4** – rs12721627; rs12721634; rs2740574; rs4986907; rs4986908; rs4986909; rs4986910; rs4986913; rs4987161 **KCNJ11** – rs5219 **SLCO1B1** – rs2306283; rs4149015; rs4149056
Angelica dahurica (Baizhl)	Tolbutamide Insulin Benzodiazepines	Inhibition of CYP2E1	**CYP2E1** – CYP2E1 *1B; CYP2E1 *5B; CYP2E1 *5A; CYP2E1 *6; CYP2E1 *1A **CYP1A2** – rs12720461; rs2069514; rs762551 **CYP2C19** – rs12248560; rs4244285; rs4986893 **CYP2C8** – rs10509681; rs1058930; rs11572080; rs11572103; rs17110453; rs7909236 **CYP2C9** – rs1057910; rs1799853 **CYP2D6** – rs1065852; rs16947; rs28371706; rs28371725; rs35742686; rs3892097; rs5030655; rs5030656; rs59421388; rs61736512 **CYP3A4** – rs12721627; rs12721634; rs2740574; rs4986907; rs4986908; rs4986909; rs4986910; rs4986913; rs4987161 **CYP3A5** – rs10264272; rs76293380; rs776746 **SLCO1B1** – rs2306283; rs4149015; rs4149056
Andrographis panicolata (andrografis)	Theophylline	Diarrhea Colds Fever	**ABCB1** – rs1045642; rs1128503; rs2032582 **CYP2C9** – rs1057910; rs1799853 **CYP3A4** – rs12721627; rs12721634; rs2740574; rs4986907; rs4986908; rs4986909; rs4986910; rs4986913; rs4987161 **CYP1A2** – rs12720461; rs2069514; rs762551

Angelica sinensis (Dang gui)	Warfarin	Coumarin compounds may increase bleeding Inhibition CYP3A4	**ABCB1** – rs1045642; rs1128503; rs2032582 **CYP1A2** – rs12720461; rs2069514; rs762551 **CYP2C19** – rs12248560; rs4244285; rs4986893 **CYP2C8** – rs10509681; rs1058930; rs11572080; rs11572103; rs17110453; rs7909236 **CYP2C9** – rs1057910; rs1799853 **CYP3A4** – rs12721627; rs12721634; rs2740574; rs4986907; rs4986908; rs4986909; rs4986910; rs4986913; rs4987161 **CYP3A5** – rs10264272; rs76293380; rs776746 **CYP4F2** – rs2108622 **VKORC1** – rs7294; rs9923231; rs9934438
Arnica montana	Warfarin	Coumarin compounds may increase bleeding	**ABCB1** – rs1045642; rs1128503; rs2032582 **CYP1A2** – rs12720461; rs2069514; rs762551 **CYP2C19** – rs12248560; rs4244285; rs4986893 **CYP2C8** – rs10509681; rs1058930; rs11572080; rs11572103; rs17110453; rs7909236 **CYP2C9** – rs1057910; rs1799853 **CYP3A4** – rs12721627; rs12721634; rs2740574; rs4986907; rs4986908; rs4986909; rs4986910; rs4986913; rs4987161 **CYP4F2** – rs2108622 **VKORC1** – rs7294; rs9923231; rs9934438
Artemisia annua spp	Chloroquine Drugs that prolong the QT interval	Hypoglycemia Arrhythmias Neurotoxicity	**ABCB1** – rs1045642; rs1128503; rs2032582 **CYP1A2** – rs12720461; rs2069514; rs762551 **CYP3A4** – rs12721627; rs12721634; rs2740574; rs4986907; rs4986908; rs4986909; rs4986910; rs4986913; rs4987161 **CYP3A5** – rs10264272; rs76293380; rs776746 **CYP2B6** – rs2279343; rs28399499; rs3211371; rs3745274 **KCNH2** – rs12720441; rs1805123; rs36210421; rs3807375; rs3815459 **KCNJ11** – rs5219 **SCN5A** – rs1805124; rs6791924; rs7626962

(continued)

Table 16.3 (continued)

Herbal medicine	Herb-drug interactions	Toxicological mechanism	Clinically validated genes and variants affecting toxicity and drug-herb, herb-herb interactions
Astragalus propinquus	Immunosuppressants	Immunostimulation with possibility of tissue graft rejection	**ABCB1** – rs1045642; rs1128503; rs2032582 **CYP2C19** – rs12248560; rs4244285; rs4986893 **CYP3A4** – rs12721627; rs12721634; rs2740574; rs4986907; rs4986908; rs4986909; rs4986910; rs4986913; rs4987161 **CYP3A5** – rs10264272; rs76293380; rs776746 **MTHFR** – rs1801131; rs1801133 **PTGS2** – rs20417; rs5275; rs689466 **SLC19A1** – rs1051266; rs1051296; rs1051298; rs1131596; rs12659 **SLCO1B1** – rs2306283; rs4149015; rs4149056 **TYMS** – rs34489327; rs34743033
Atropa belladonna	Atropine and anticholinergics	Anticholinergic effects	**ABCB1** – rs1045642; rs1128503; rs2032582 **ACHE** – rs2571598
Averrhoa carambola (Carambola)	Benzodiazepines Carbamazepine	Headache Nausea Cough Insomnia, hypertension Diabetes	**ABCB1** – rs1045642; rs1128503; rs2032582 **CYP1A2** – rs12720461; rs2069514; rs762551 **CYP2A6** – rs1801272; rs28399433; rs28399444; rs28399454; rs28399468; rs5031016; rs8192726 **CYP2D6** – rs1065852; rs16947; rs28371706; rs28371725; rs35742686; rs3892097; rs5030655; rs5030656; rs59421388; rs61736512 **CYP2C8** – rs10509681; rs1058930; rs11572080; rs11572103; rs17110453; rs7909236 **CYP2C9** – rs1057910; rs1799853 **CYP2E1** – CYP2E1 *1B; CYP2E1 *5B; CYP2E1 *5A; CYP2E1 *6; CYP2E1 *1A **CYP3A4** – rs12721627; rs12721634; rs2740574; rs4986907; rs4986908; rs4986909; rs4986910; rs4986913; rs4987161

Azadirachta indica (Neem)	Azathioprine Imuran Glimepiride Glucotrol Micronase Orinase Prednisolone Tolinase Zenapax	Hypoglycemia	**ABCB1** – rs1045642; rs1128503; rs2032582 **CYP1A2** – rs12720461; rs2069514; rs762551 **CYP2C19** – rs12248560; rs4244285; rs4986893 **CYP2C8** – rs10509681; rs1058930; rs11572080; rs11572103; rs17110453; rs7909236 **CYP2C9** – rs1057910; rs1799853 **CYP2D6** – rs1065852; rs16947; rs28371706; rs28371725; rs35742686; rs3892097; rs5030655; rs5030656; rs59421388; rs61736512 **CYP3A4** – rs12721627; rs12721634; rs2740574; rs4986907; rs4986908; rs4986909; rs4986910; rs4986913; rs4987161 **SLC22A1** – rs12208357; rs34059508; rs34130495; rs72552763 **SLCO1B1** – rs2306283; rs4149015; rs4149056 **KCNJ11** – rs5219
Berberis vulgaris (barberry)	Paclitaxel Cyclosporine	Dyspepsia, diarrhea, gastritis, abdominal distension, flatulence	**ABCB1** – rs1045642; rs1128503; rs2032582
Borago officinalis	Warfarin Aspirin	Increase bleeding	**ABCB1** – rs1045642; rs1128503; rs2032582 **CYP1A2** – rs12720461; rs2069514; rs762551 **CYP2C19** – rs12248560; rs4244285; rs4986893 **CYP2C8** – rs10509681; rs1058930; rs11572080; rs11572103; rs17110453; rs7909236 **CYP2C9** – rs1057910; rs1799853 **CYP3A4** – rs12721627; rs12721634; rs2740574; rs4986907; rs4986908; rs4986909; rs4986910; rs4986913; rs4987161 **CYP4F2** – rs2108622 **VKORC1** – rs7294; rs9923231; rs9934438 **P2RY1** – rs1065776; rs701265 **P2RY12** – rs2046934

(continued)

Table 16.3 (continued)

Herbal medicine	Herb-drug interactions	Toxicological mechanism	Clinically validated genes and variants affecting toxicity and drug-herb, herb-herb interactions
Boswellia serrata (Indian Olibanum)	Warfarin	Increase bleeding	**ABCB1** – rs1045642; rs1128503; rs2032582 **CYP1A2** – rs12720461; rs2069514; rs762551 **CYP2C19** – rs12248560; rs4244285; rs4986893 **CYP2C8** – rs10509681; rs1058930; rs11572080; rs11572103; rs17110453; rs7909236 **CYP2C9** – rs1057910; rs1799853 **CYP3A4** – rs12721627; rs12721634; rs2740574; rs4986907; rs4986908; rs4986909; rs4986910; rs4986913; rs4987161 **CYP4F2** – rs2108622 **VKORC1** – rs7294; rs9923231; rs9934438
Caffeine containing herbs (*Coffea spp, Camelia sinensis, Ilex paraguariensis, Paullinia cupana, Theobroma cacao*)	Clozapine Lithium Theophylline	Increase xenobiotics absorption and reduction xenobiotics metabolism	**ABCB1** – rs1045642; rs1128503; rs2032582 **CYP1A2** – rs12720461; rs2069514; rs762551 **CYP2A6** – rs1801272; rs28399433; rs28399444; rs28399454; rs28399468; rs5031016; rs8192726 **CYP2C8** – rs10509681; rs1058930; rs11572080; rs11572103; rs17110453; rs7909236 **CYP2C9** – rs1057910; rs1799853 **CYP2E1** – CYP2E1 *1B; CYP2E1 *5B; CYP2E1 *5A; CYP2E1 *6; CYP2E1 *1A **CYP3A4** – rs12721627; rs12721634; rs2740574; rs4986907; rs4986908; rs4986909; rs4986910; rs4986913; rs4987161 **NAT2** – rs1041983; rs1208; rs1495741; rs1799929; rs1799930; rs1799931; rs1801279; rs1801280; rs4271002; rs4646244
Camellia sinensis (green tea)	Clozapine Theophylline Warfarin	Palpitations Inductor CYP1A2 enzyme	**CYP1A2** – rs12720461; rs2069514; rs762551
Capsicum spp.	ACE inhibitors Theophylline	Increase xenobiotics Absorption	**ABCB1** – rs1045642; rs1128503; rs2032582 **ACE** – rs1799752

Carica papaya	Warfarin	Increase bleeding	**ABCB1** – rs1045642; rs1128503; rs2032582 **CYP1A2** – rs12720461; rs2069514; rs762551 **CYP2C19** – rs12248560; rs4244285; rs4986893 **CYP2C8** – rs10509681; rs1058930; rs11572080; rs11572103; rs17110453; rs7909236 **CYP2C9** – rs1057910; rs1799853 **CYP3A4** – rs12721627; rs12721634; rs2740574; rs4986907; rs4986908; rs4986909; rs4986910; rs4986913; rs4987161 **CYP4F2** – rs2108622 **VKORC1** – rs7294; rs9923231; rs9934438
Catha edulis (Khat)	Penicillin Ampicillin Amoxicillin	Reduction absorption Formation of antibiotic-tannin complex	**ABCB1** – rs1045642; rs1128503; rs2032582
Caulophyllum thalictroides	Oxytocin Antidiabetic drugs	Hypertension Hyperglycemia	**ABCB1** – rs1045642; rs1128503; rs2032582 **CYP2C19** – rs12248560; rs4244285; rs4986893 **CYP2D6** – rs1065852; rs16947; rs28371706; rs28371725; rs35742686; rs3892097; rs5030655; rs5030656; rs59421388; rs61736512 **CYP2E1** – CYP2E1 *1B; CYP2E1 *5B; CYP2E1 *5A; CYP2E1 *6; CYP2E1 *1[A] **SLC22A1** – rs12208357; rs34059508; rs34130495; rs72552763 **SLCO1B1** – rs2306283; rs4149015; rs4149056 **CYP2C8** – rs10509681; rs1058930; rs11572080; rs11572103; rs17110453; rs7909236 **CYP2C9** – rs1057910; rs1799853 **KCNJ11** – rs5219
Citrus aurantium (bitter orange)	Midazolam Felodipine Antiviral MAO inhibitors	Coumarin compounds increase secretion of stomach chlorhydropeptic	**ABCB1** – rs1045642; rs1128503; rs2032582 **CYP3A4** – rs12721627; rs12721634; rs2740574; rs4986907; rs4986908; rs4986909; rs4986910; rs4986913; rs4987161 **CYP1A2** – rs12720461; rs2069514; rs762551

(continued)

Table 16.3 (continued)

Herbal medicine	Herb-drug interactions	Toxicological mechanism	Clinically validated genes and variants affecting toxicity and drug-herb, herb-herb interactions
Citrus paradise (grapefruit)	Amiodarone, amlodipine, atorvastatin, buspirone Benzodiazepines, cisapride, Carbamazepine, clomipramine, cyclosporine, digoxin Erythromycin, ethinylestradiol Fluvoxamine, indinavir Losartan, lovastatin Nicardipine, nisoldipine Praziquantel, quidine Simvastatin, sildenafil Verapamil	Inhibition of CYP enzymes activity	**ABCB1** – rs1045642; rs1128503; rs2032582 **CYP1A2** – rs12720461; rs2069514; rs762551 **CYP2A6** – rs1801272; rs28399433; rs28399444; rs28399454; rs28399468; rs5031016; rs8192726 **CYP2B6** – rs2279343; rs28399499; rs3211371; rs3745274 **CYP2C19** – rs12248560; rs4244285; rs4986893 **CYP2C8** – rs10509681; rs1058930; rs11572080; rs11572103; rs17110453; rs7909236 **CYP2C9** – rs1057910; rs1799853 **CYP2D6** – rs1065852; rs16947; rs28371706; rs28371725; rs35742686; rs3892097; rs5030655; rs5030656; rs59421388; rs61736512 **CYP2E1** – CYP2E1 *1B; CYP2E1 *5B; CYP2E1 *5A; CYP2E1 *6; CYP2E1 *1A **CYP3A4** – rs12721627; rs12721634; rs2740574; rs4986907; rs4986908; rs4986909; rs4986910; rs4986913; rs4987161 **CYP3A5** – rs10264272; rs76293380; rs776746 **HMGCR** – rs17238540; rs17244841; rs3846662 **SLCO1B1** – rs2306283; rs4149015; rs4149056

Commiphora wightii (gugul)	Diltiazem Estrogens Warfarin Aspirin Propranolol Tamoxifen Thyroid hormone	Increase bleeding Hormone-sensitive conditions Hypothyroidism or hyperthyroidism	**ABCB1** – rs1045642; rs1128503; rs2032582 **CYP3A4** – rs12721627; rs12721634; rs2740574; rs4986907; rs4986908; rs4986909; rs4986910; rs4986913; rs4987161 **CYP3A5** – rs10264272; rs76293380; rs776746 **CYP1A2** – rs12720461; rs2069514; rs762551 **CYP2C19** – rs12248560; rs4244285; rs4986893 **CYP2C8** – rs10509681; rs1058930; rs11572080; rs11572103; rs17110453; rs7909236 **CYP2C9** – rs1057910; rs1799853 **CYP4F2** – rs2108622 **VKORC1** – rs7294; rs9923231; rs9934438 **P2RY1** – rs1065776; rs701265 **P2RY12** – rs2046934 **COMT** – rs4680 **UGT1A1** – rs4148323; rs8175347 **SULT1A1** – rs1801030; rs3760091; rs750155; rs9282861
Coptis chinensis	Cyclosporine CYP3A4 substrates	Increase the amount of bilirubin in newborn	**ABCB1** – rs1045642; rs1128503; rs2032582 **CYP3A4** – rs12721627; rs12721634; rs2740574; rs4986907; rs4986908; rs4986909; rs4986910; rs4986913; rs4987161 **CYP3A5** – rs10264272; rs76293380; rs776746 **CYP1A2** – rs12720461; rs2069514; rs762551 **CYP2D6** – rs1065852; rs16947; rs28371706; rs28371725; rs35742686; rs3892097; rs5030655; rs5030656; rs59421388; rs61736512

(continued)

Table 16.3 (continued)

Herbal medicine	Herb-drug interactions	Toxicological mechanism	Clinically validated genes and variants affecting toxicity and drug-herb, herb-herb interactions
Corydalis spp	Hypnotics Phenothiazine-type antipsychotics	Sedative effect	**ABCB1** – rs1045642; rs1128503; rs2032582 **CYP3A4** – rs12721627; rs12721634; rs2740574; rs4986907; rs4986908; rs4986909; rs4986910; rs4986913; rs4987161 **CYP2C9** – rs1057910; rs1799853 **CYP2D6** – rs1065852; rs16947; rs28371706; rs28371725; rs35742686; rs3892097; rs5030655; rs5030656; rs59421388; rs61736512 **CYP2C19** – rs12248560; rs4244285; rs4986893 **CYP1A2** – rs12720461; rs2069514; rs762551 **SLCO1B1** – rs2306283; rs4149015; rs4149056 **CYP3A5** – rs10264272; rs76293380; rs776746 **NAT2** – rs1041983; rs1208; rs1495741; rs1799929; rs1799930; rs1799931; rs1801279; rs1801280; rs4271002; rs4646244

Crataegus spp. (Hawthorn)	Cardiac glycosides Antihypertensives Nitrates	Inhibition of the membranous Na,K-adenosine triphosphatase (ATPase) of cardiomuscular tissue Hypotension	**ABCB1** – rs1045642; rs1128503; rs2032582 **KCNH2** – rs12720441; rs1805123; rs36210421; rs3807375; rs3815459 **KCNJ11** – rs5219 **SCN5A** – rs1805124; rs6791924; rs7626962 **ACE** – rs1799752 **ADRB1** – rs1801252; rs1801253 **ADRB2** – rs1042713; rs1042714; rs1800888 **CYP1A2** – rs12720461; rs2069514; rs762551 **CYP2A6** – rs1801272; rs28399433; rs28399444; rs28399454; rs28399468; rs5031016; rs8192726 **CYP2C19** – rs12248560; rs4244285; rs4986893 **CYP2C8** – rs10509681; rs1058930; rs11572080; rs11572103; rs17110453; rs7909236 **CYP2C9** – rs1057910; rs1799853 **CYP2D6** – rs1065852; rs16947; rs28371706; rs28371725; rs35742686; rs3892097; rs5030655; rs5030656; rs59421388; rs61736512 **CYP3A4** – rs12721627; rs12721634; rs2740574; rs4986907; rs4986908; rs4986909; rs4986910; rs4986913; rs4987161 **CYP3A5** – rs10264272; rs76293380; rs776746 **CYP4F2** – rs2108622 **PTGIS** – chr20:48184659; rs5629
Cyamopsis spp. (Guar Gum)	Bumetanide Digoxin Glibenclamide, metformin Penicillin antibiotics	Reduction absorption	**ABCB1** – rs1045642; rs1128503; rs2032582 **SLC19A1** – rs1051266; rs1051296; rs1051298; rs1131596; rs12659 **SLC22A1** – rs12208357; rs34059508; rs34130495; rs72552763 **SLCO1B1** – rs2306283; rs4149015; rs4149056

(continued)

Table 16.3 (continued)

Herbal medicine	Herb-drug interactions	Toxicological mechanism	Clinically validated genes and variants affecting toxicity and drug-herb, herb-herb interactions
Cytisus scoparius (Broom)	Digoxin Beta blockers Tricyclic antidepressants	Arrhythmias	**ABCB1** – rs1045642; rs1128503; rs2032582 **KCNH2** – rs12720441; rs1805123; rs36210421; rs3807375; rs3815459 **KCNJ11** – rs5219 **SCN5A** – rs1805124; rs6791924; rs7626962 **ADRB1** – rs1801252; rs1801253 **ADRB2** – rs1042713; rs1042714; rs1800888 **CYP1A2** – rs12720461; rs2069514; rs762551 **CYP2C19** – rs12248560; rs4244285; rs4986893 **CYP2C9** – rs1057910; rs1799853 **CYP2D6** – rs1065852; rs16947; rs28371706; rs28371725; rs35742686; rs3892097; rs5030655; rs5030656; rs59421388; rs61736512
Curcubita spp.	Warfarin	Increase bleeding	**ABCB1** – rs1045642; rs1128503; rs2032582 **CYP1A2** – rs12720461; rs2069514; rs762551 **CYP2C19** – rs12248560; rs4244285; rs4986893 **CYP2C8** – rs10509681; rs1058930; rs11572080; rs11572103; rs17110453; rs7909236 **CYP2C9** – rs1057910; rs1799853 **CYP3A4** – rs12721627; rs12721634; rs2740574; rs4986907; rs4986908; rs4986909; rs4986910; rs4986913; rs4987161 **CYP4F2** – rs2108622 **VKORC1** – rs7294; rs9923231; rs9934438

Curcuma longa and C. zodearia (tumeric)	Digoxin Cyclosporine	Inhibition P glycoprotein	**ABCB1** – rs1045642; rs1128503; rs2032582 **CYP1A2** – rs12720461; rs2069514; rs762551 **CYP2B6** – rs2279343; rs28399499; rs3211371; rs3745274 **CYP2C9** – rs1057910; rs1799853 **CYP3A4** – rs12721627; rs12721634; rs2740574; rs4986907; rs4986908; rs4986909; rs4986910; rs4986913; rs4987161 **CYP2D6** – rs1065852; rs16947; rs28371706; rs28371725; rs35742686; rs3892097; rs5030655; rs5030656; rs59421388; rs61736512 **CYP2C19** – rs12248560; rs4244285; rs4986893 **KCNH2** – rs12720441; rs1805123; rs36210421; rs3807375; rs3815459 **KCNJ11** – rs5219 **SCN5A** – rs1805124; rs6791924; rs7626962
Digitalis purpúrea and *digitalis lanata*	Digoxin and cardiac glycosides	Inhib. cardiomuscular Na,K-adenosine phosphatase (ATPase)	**ABCB1** – rs1045642; rs1128503; rs2032582 **KCNH2** – rs12720441; rs1805123; rs36210421; rs3807375; rs3815459 **KCNJ11** – rs5219 **SCN5A** – rs1805124; rs6791924; rs7626962
Echinacea purpurea	Methotrexate Azathioprine Corticosteroids Cyclosporin Tacrolimus	Immunostimulation with possibility of tissue graft rejection	**ABCB1** – rs1045642; rs1128503; rs2032582 **CYP2C19** – rs12248560; rs4244285; rs4986893 **CYP3A4** – rs12721627; rs12721634; rs2740574; rs4986907; rs4986908; rs4986909; rs4986910; rs4986913; rs4987161 **CYP3A5** – rs10264272; rs76293380; rs776746 **MTHFR** – rs1801131; rs1801133 **PTGS2** – rs20417; rs5275; rs689466 **SLC19A1** – rs1051266; rs1051296; rs1051298; rs1131596; rs12659 **SLCO1B1** – rs2306283; rs4149015; rs4149056 **TYMS** – rs34489327; rs34743033

(continued)

Table 16.3 (continued)

Herbal medicine	Herb-drug interactions	Toxicological mechanism	Clinically validated genes and variants affecting toxicity and drug-herb, herb-herb interactions
Ephedra sinica	Beta blockers Decongestants MAO inhibitors Caffeine Guarana	Hypertension Palpitations Tachycardia	**ABCB1** – rs1045642; rs1128503; rs2032582 **ACE** – rs1799752 **ADRB1** – rs1801252; rs1801253 **ADRB2** – rs1042713; rs1042714; rs1800888 **CYP1A2** – rs12720461; rs2069514; rs762551 **CYP2A6** – rs1801272; rs28399433; rs28399444; rs28399454; rs28399468; rs5031016; rs8192726 **CYP2C19** – rs12248560; rs4244285; rs4986893 **CYP2C8** – rs10509681; rs1058930; rs11572080; rs11572103; rs17110453; rs7909236 **CYP2C9** – rs1057910; rs1799853 **CYP2D6** – rs1065852; rs16947; rs28371706; rs28371725; rs35742686; rs3892097; rs5030655; rs5030656; rs59421388; rs61736512 **CYP3A4** – rs12721627; rs12721634; rs2740574; rs4986907; rs4986908; rs4986909; rs4986910; rs4986913; rs4987161 **CYP3A5** – rs10264272; rs76293380; rs776746 **CYP4F2** – rs2108622 **PTGIS** – chr20:48184659; rs5629
Foeniculum vulgare	Contraceptive drugs Ciprofloxacin Tamoxifen	Bleeding disorders Estrogen effects	**ABCB1** – rs1045642; rs1128503; rs2032582 **CYP3A4** – rs12721627; rs12721634; rs2740574; rs4986907; rs4986908; rs4986909; rs4986910; rs4986913; rs4987161 **CYP2C9** – rs1057910; rs1799853 **CYP1A2** – rs12720461; rs2069514; rs762551 **COMT** – rs4680 **UGT1A1** – rs4148323; rs8175347 **SULT1A1** – rs1801030; rs3760091; rs750155; rs9282861
Eurycoma longifolia (tongkat ali)	Propranolol and beta-blockers	P glycoprotein inductor	**ABCB1** – rs1045642; rs1128503; rs2032582 **ADRB1** – rs1801252; rs1801253 **ADRB2** – rs1042713; rs1042714; rs1800888

Fucus spp. (kelp)	Thyroxine	Hyperthyroidism	**ABCB1** – rs1045642; rs1128503; rs2032582 **CYP3A4** – rs12721627; rs12721634; rs2740574; rs4986907; rs4986908; rs4986909; rs4986910; rs4986913; rs4987161
Gingko biloba	Aspirin, alprazolam Digoxin, diltiazem Haloperidol Ibuprofen Nicardipine Nifedipine Omeprazole Thiazide diuretics, ticlopidine, tolbutamide, trazodone Valproate Warfarin	Inhibitor of platelet activating factor (PAF) Increase bleeding Increase GABA activity Inhibition of P-gp transport proteins	**ABCB1** – rs1045642; rs1128503; rs2032582 **CYP1A2** – rs12720461; rs2069514; rs762551 **CYP2C19** – rs12248560; rs4244285; rs4986893 **CYP2C8** – rs10509681; rs1058930; rs11572080; rs11572103; rs17110453; rs7909236 **CYP2C9** – rs1057910; rs1799853 **CYP2D6** – rs1065852; rs16947; rs28371706; rs28371725; rs35742686; rs3892097; rs5030655; rs5030656; rs59421388; rs61736512 **CYP3A4** – rs12721627; rs12721634; rs2740574; rs4986907; rs4986908; rs4986909; rs4986910; rs4986913; rs4987161 **CYP4F2** – rs2108622 **P2RY1** – rs1065776; rs701265 **P2RY12** – rs2046934
Glycine max (soy)	Warfarin Tamoxifen	Inductor CYP3A4 enzyme Estrogens effects	**CYP3A4** – rs12721627; rs12721634; rs2740574; rs4986907; rs4986908; rs4986909; rs4986910; rs4986913; rs4987161 **CYP1A2** – rs12720461; rs2069514; rs762551 **COMT** – rs4680 **UGT1A1** – rs4148323; rs8175347 **SULT1A1** – rs1801030; rs3760091; rs750155; rs9282861

(continued)

Table 16.3 (continued)

Herbal medicine	Herb-drug interactions	Toxicological mechanism	Clinically validated genes and variants affecting toxicity and drug-herb, herb-herb interactions
Glycyrrhiza glabra (Licorice)	Digoxin Ethinylestradiol Prednisolone Spironolactone Loratidine Quinidine Procainamide	Hypokalemia Hypertension Edema Fluid retention Arrhythmias	**ABCB1** – rs1045642; rs1128503; rs2032582 **KCNH2** – rs12720441; rs1805123; rs36210421; rs3807375; rs3815459 **KCNJ11** – rs5219 **SCN5A** – rs1805124; rs6791924; rs7626962 **ACE** – rs1799752 **ADRB1** – rs1801252; rs1801253 **ADRB2** – rs1042713; rs1042714; rs1800888 **CYP1A2** – rs12720461; rs2069514; rs762551 **CYP2A6** – rs1801272; rs28399433; rs28399444; rs28399454; rs28399468; rs5031016; rs8192726 **CYP2C19** – rs12248560; rs4244285; rs4986893 **CYP2C8** – rs10509681; rs1058930; rs11572080; rs11572103; rs17110453; rs7909236 **CYP2C9** – rs1057910; rs1799853 **CYP2D6** – rs1065852; rs16947; rs28371706; rs28371725; rs35742686; rs3892097; rs5030655; rs5030656; rs59421388; rs61736512 **CYP3A4** – rs12721627; rs12721634; rs2740574; rs4986907; rs4986908; rs4986909; rs4986910; rs4986913; rs4987161 **CYP3A5** – rs10264272; rs76293380; rs776746 **CYP4F2** – rs2108622 **PTGIS** – chr20:48184659; rs5629

Harpagophytum procumbens (devil's claw)	Digoxin Warfarin Beta blockers	Arrhythmias Increase bleeding	**ABCB1** – rs1045642; rs1128503; rs2032582 **CYP1A2** – rs12720461; rs2069514; rs762551 **CYP2C19** – rs12248560; rs4244285; rs4986893 **CYP2C8** – rs10509681; rs1058930; rs11572080; rs11572103; rs17110453; rs7909236 **CYP2C9** – rs1057910; rs1799853 **CYP2D6** – rs1065852; rs16947; rs28371706; rs28371725; rs35742686; rs3892097; rs5030655; rs5030656; rs59421388; rs61736512 **CYP3A4** – rs12721627; rs12721634; rs2740574; rs4986907; rs4986908; rs4986909; rs4986910; rs4986913; rs4987161 **CYP4F2** – rs2108622 **P2RY1** – rs1065776; rs701265 **P2RY12** – rs2046934 **KCNH2** – rs12720441; rs1805123; rs36210421; rs3807375; rs3815459 **KCNJ11** – rs5219 **SCN5A** – rs1805124; rs6791924; rs7626962
Helichrysum italicum	All drugs that are metabolized by cytochrome P450 3A4	Allergic effect and inhibition CYP3A4	**ABCB1** – rs1045642; rs1128503; rs2032582 **CYP3A4** – rs12721627; rs12721634; rs2740574; rs4986907; rs4986908; rs4986909; rs4986910; rs4986913; rs4987161 **CYP2C9** – rs1057910; rs1799853 **CYP2C8** – rs10509681; rs1058930; rs11572080; rs11572103; rs17110453; rs7909236

(continued)

Table 16.3 (continued)

Herbal medicine	Herb-drug interactions	Toxicological mechanism	Clinically validated genes and variants affecting toxicity and drug-herb, herb-herb interactions
Hidrocotyle asiatica (centella)	Hypnotics Phenothiazine type antipsychotics	Sedative effect	**ABCB1** – rs1045642; rs1128503; rs2032582 **CYP1A2** – rs12720461; rs2069514; rs762551 **CYP2C19** – rs12248560; rs4244285; rs4986893 **CYP2C9** – rs1057910; rs1799853 **CYP2D6** – rs1065852; rs16947; rs28371706; rs28371725; rs35742686; rs3892097; rs5030655; rs5030656; rs59421388; rs61736512 **SLCO1B1** – rs2306283; rs4149015; rs4149056 **CYP3A4** – rs12721627; rs12721634; rs2740574; rs4986907; rs4986908; rs4986909; rs4986910; rs4986913; rs4987161 **CYP3A5** – rs10264272; rs76293380; rs776746 **NAT2** – rs1041983; rs1208; rs1495741; rs1799929; rs1799930; rs1799931; rs1801279; rs1801280; rs4271002; rs4646244
Humulus lupulus (Hops)	Hypnotics Phenothiazine type antipsychotics Carbamazepine Sertraline	Sedative effect	**ABCB1** – rs1045642; rs1128503; rs2032582 **CYP1A2** – rs12720461; rs2069514; rs762551 **CYP2C19** – rs12248560; rs4244285; rs4986893 **CYP2D6** – rs1065852; rs16947; rs28371706; rs28371725; rs35742686; rs3892097; rs5030655; rs5030656; rs59421388; rs61736512 **SLCO1B1** – rs2306283; rs4149015; rs4149056 **CYP3A4** – rs12721627; rs12721634; rs2740574; rs4986907; rs4986908; rs4986909; rs4986910; rs4986913; rs4987161 **CYP3A5** – rs10264272; rs76293380; rs776746 **NAT2** – rs1041983; rs1208; rs1495741; rs1799929; rs1799930; rs1799931; rs1801279; rs1801280; rs4271002; rs4646244

Huperzia serrata (Qian Ceng Ta)	Tacrine Donepezil	Inhibition anticholinesterase	**ABCB1** – rs1045642; rs1128503; rs2032582 **CYP1A2** – rs12720461; rs2069514; rs762551 **CYP2D6** – rs1065852; rs16947; rs28371706; rs28371725; rs35742686; rs3892097; rs5030655; rs5030656; rs59421388; rs61736512 **CYP3A4** – rs12721627; rs12721634; rs2740574; rs4986907; rs4986908; rs4986909; rs4986910; rs4986913; rs4987161 **CYP3A5** – rs10264272; rs76293380; rs776746 **ACHE** – rs2571598
Hydrastis Canadensis (goldenseal)	Aspirin Clopidogrel Dipyridamole Fexofenidine Heparin Ticlopidine	Procoagulant effects	**ABCB1** – rs1045642; rs1128503; rs2032582 **KCNH2** – rs12720441; rs1805123; rs36210421; rs3807375; rs3815459 **KCNJ11** – rs5219 **SCN5A** – rs1805124; rs6791924; rs7626962 **CYP1A2** – rs12720461; rs2069514; rs762551 **CYP2C19** – rs12248560; rs4244285; rs4986893 **CYP2C8** – rs10509681; rs1058930; rs11572080; rs11572103; rs17110453; rs7909236 **CYP2C9** – rs1057910; rs1799853 **CYP3A4** – rs12721627; rs12721634; rs2740574; rs4986907; rs4986908; rs4986909; rs4986910; rs4986913; rs4987161 **CYP4F2** – rs2108622 **P2RY1** – rs1065776; rs701265 **P2RY12** – rs2046934 **F5** – rs6025

(continued)

Table 16.3 (continued)

Herbal medicine	Herb-drug interactions	Toxicological mechanism	Clinically validated genes and variants affecting toxicity and drug-herb, herb-herb interactions
Hypericum perforatum (St. John's wort)	Amytryptiline, anesthetics, Amprenavir, anticonvulsants Benzodiazepines, buspirone Contraceptives, cyclosporine Digoxin, indinavir, irinotecan Lamivudine, loperamide, warfarin Fenoxfenadine, methadone Midazolam, nifedipine, quazepam 5HT reuptake inhib, verapamil Phenprocoumon, simvastatin Tacrolimus, theophylline	Inductor CYP3A4 enzyme	**ABCB1** – rs1045642; rs1128503; rs2032582 **CYP3A4** – rs12721627; rs12721634; rs2740574; rs4986907; rs4986908; rs4986909; rs4986910; rs4986913; rs4987161 **CYP1A2** – rs12720461; rs2069514; rs762551 **CYP2D6** – rs1065852; rs16947; rs28371706; rs28371725; rs35742686; rs3892097; rs5030655; rs5030656; rs59421388; rs61736512 **CYP2C9** – rs1057910; rs1799853 **HMGCR** – rs17238540; rs17244841; rs3846662 **SLCO1B1** – rs2306283; rs4149015; rs4149056
Hypoxis hemerocallidea (African potato)	All drugs which are metabolized by cytochrome P450 3A4	Inhibition CYP3A4 enzyme	**ABCB1** – rs1045642; rs1128503; rs2032582 **CYP3A4** – rs12721627; rs12721634; rs2740574; rs4986907; rs4986908; rs4986909; rs4986910; rs4986913; rs4987161

Lavandula spp. (lavender)	Barbiturates Benzodiazepines Chloral hydrate	Sedative effects	**ABCB1** – rs1045642; rs1128503; rs2032582 **CYP1A2** – rs12720461; rs2069514; rs762551 **CYP2C19** – rs12248560; rs4244285; rs4986893 **CYP2D6** – rs1065852; rs16947; rs28371706; rs28371725; rs35742686; rs3892097; rs5030655; rs5030656; rs59421388; rs61736512 **SLCO1B1** – rs2306283; rs4149015; rs4149056 **CYP3A4** – rs12721627; rs12721634; rs2740574; rs4986907; rs4986908; rs4986909; rs4986910; rs4986913; rs4987161 **CYP3A5** – rs10264272; rs76293380; rs776746 **NAT2** – rs1041983; rs1208; rs1495741; rs1799929; rs1799930; rs1799931; rs1801279; rs1801280; rs4271002; rs4646244
Linum usitatissimum (flax)	Warfarin Digoxin Tamoxifen	Estrogen effects	**ABCB1** – rs1045642; rs1128503; rs2032582 **CYP3A4** – rs12721627; rs12721634; rs2740574; rs4986907; rs4986908; rs4986909; rs4986910; rs4986913; rs4987161 **CYP1A2** – rs12720461; rs2069514; rs762551 **COMT** – rs4680 **UGT1A1** – rs4148323; rs8175347 **SULT1A1** – rs1801030; rs3760091; rs750155; rs9282861 **KCNH2** – rs12720441; rs1805123; rs36210421; rs3807375; rs3815459 **KCNJ11** – rs5219 **SCN5A** – rs1805124; rs6791924; rs7626962
Lupinus	Antidiabetic drugs	Hypoglycemic effects	**ABCB1** – rs1045642; rs1128503; rs2032582 **CYP1A2** – rs12720461; rs2069514; rs762551 **SLC22A1** – rs12208357; rs34059508; rs34130495; rs72552763 **SLCO1B1** – rs2306283; rs4149015; rs4149056 **CYP2C8** – rs10509681; rs1058930; rs11572080; rs11572103; rs17110453; rs7909236 **CYP2C9** – rs1057910; rs1799853 **KCNJ11** – rs5219

(continued)

Table 16.3 (continued)

Herbal medicine	Herb-drug interactions	Toxicological mechanism	Clinically validated genes and variants affecting toxicity and drug-herb, herb-herb interactions
Lycium barbarum (Chinese wolfberry)	Warfarin	Increase bleeding	**ABCB1** – rs1045642; rs1128503; rs2032582 **CYP1A2** – rs12720461; rs2069514; rs762551 **CYP2C19** – rs12248560; rs4244285; rs4986893 **CYP2C8** – rs10509681; rs1058930; rs11572080; rs11572103; rs17110453; rs7909236 **CYP2C9** – rs1057910; rs1799853 **CYP3A4** – rs12721627; rs12721634; rs2740574; rs4986907; rs4986908; rs4986909; rs4986910; rs4986913; rs4987161 **CYP4F2** – rs2108622 **VKORC1** – rs7294; rs9923231; rs9934438
Marsdenia condurango (Condurango)	Carbamazepine Paroxetine Ritonavir Sertraline	Increase bleeding	**ABCB1** – rs1045642; rs1128503; rs2032582 **CYP1A2** – rs12720461; rs2069514; rs762551 **CYP2B6** – rs2279343; rs28399499; rs3211371; rs3745274 **CYP2C19** – rs12248560; rs4244285; rs4986893 **CYP2C8** – rs10509681; rs1058930; rs11572080; rs11572103; rs17110453; rs7909236 **CYP2C9** – rs1057910; rs1799853 **CYP2D6** – rs1065852; rs16947; rs28371706; rs28371725; rs35742686; rs3892097; rs5030655; rs5030656; rs59421388; rs61736512 **CYP2E1** – CYP2E1 *1B; CYP2E1 *5B; CYP2E1 *5A; CYP2E1 *6; CYP2E1 *1A **CYP3A4** – rs12721627; rs12721634; rs2740574; rs4986907; rs4986908; rs4986909; rs4986910; rs4986913; rs4987161 **CYP3A5** – rs10264272; rs76293380; rs776746 **CYP4F2** – rs2108622 **VKORC1** – rs7294; rs9923231; rs9934438

Matricaria recutita (chamomile)	Aspirin Clopidogrel Dipyridamole Heparin Ticlopidine Warfarin	Coumarin compounds may increase bleeding	**ABCB1** – rs1045642; rs1128503; rs2032582 **CYP1A2** – rs12720461; rs2069514; rs762551 **CYP2B6** – rs2279343; rs28399499; rs3211371; rs3745274 **CYP2C19** – rs12248560; rs4244285; rs4986893 **CYP2C8** – rs10509681; rs1058930; rs11572080; rs11572103; rs17110453; rs7909236 **CYP2C9** – rs1057910; rs1799853 **CYP3A4** – rs12721627; rs12721634; rs2740574; rs4986907; rs4986908; rs4986909; rs4986910; rs4986913; rs4987161 **CYP4F2** – rs2108622 **VKORC1** – rs7294; rs9923231; rs9934438 **P2RY1** – rs1065776; rs701265 **P2RY12** – rs2046934
Mentha pulegium spp	Oxytocin	Kidney disease Pregnancy and breast-feeding	**ABCB1** – rs1045642; rs1128503; rs2032582 **CYP3A4** – rs12721627; rs12721634; rs2740574; rs4986907; rs4986908; rs4986909; rs4986910; rs4986913; rs4987161 **CYP3A5** – rs10264272; rs76293380; rs776746 **CYP1A2** – rs12720461; rs2069514; rs762551 **CYP2E1** – CYP2E1 *1B; CYP2E1 *5B; CYP2E1 *5A; CYP2E1 *6; CYP2E1 *1[A] **CYP2C19** – rs12248560; rs4244285; rs4986893
Momordica charantia (bitter melon)	Chlorpropamide	Hyperglycemia	**SLC22A1** – rs12208357; rs34059508; rs34130495; rs72552763 **SLCO1B1** – rs2306283; rs4149015; rs4149056 **CYP2C8** – rs10509681; rs1058930; rs11572080; rs11572103; rs17110453; rs7909236 **CYP2C9** – rs1057910; rs1799853 **KCNJ11** – rs5219

(continued)

Table 16.3 (continued)

Herbal medicine	Herb-drug interactions	Toxicological mechanism	Clinically validated genes and variants affecting toxicity and drug-herb, herb-herb interactions
Nepeta cataria (Catnip)	Barbiturates Benzodiazepines	Sedative effect	**ABCB1** – rs1045642; rs1128503; rs2032582 **CYP1A2** – rs12720461; rs2069514; rs762551 **CYP2C19** – rs12248560; rs4244285; rs4986893 **CYP2D6** – rs1065852; rs16947; rs28371706; rs28371725; rs35742686; rs3892097; rs5030655; rs5030656; rs59421388; rs61736512 **SLCO1B1** – rs2306283; rs4149015; rs4149056 **CYP3A4** – rs12721627; rs12721634; rs2740574; rs4986907; rs4986908; rs4986909; rs4986910; rs4986913; rs4987161 **CYP3A5** – rs10264272; rs76293380; rs776746 **NAT2** – rs1041983; rs1208; rs1495741; rs1799929; rs1799930; rs1799931; rs1801279; rs1801280; rs4271002; rs4646244
Nerium oleander (Oleander)	Digoxin	Inhib. cardiomuscular Na,K-adenosine phosphatase (ATPase)	**ABCB1** – rs1045642; rs1128503; rs2032582 **KCNH2** – rs12720441; rs1805123; rs36210421; rs3807375; rs3815459 **KCNJ11** – rs5219 **SCN5A** – rs1805124; rs6791924; rs7626962

Oenothera biennis (evening primrose)	Anticonvulsants (e.g.,, barbiturates, phenytoin) Fluphenazine	Reduction of the seizures' thresholds	**CYP1A2** – rs12720461; rs2069514; rs762551 **CYP2A6** – rs1801272; rs28399433; rs28399444; rs28399454; rs28399468; rs5031016; rs8192726 **CYP2B6** – rs2279343; rs28399499; rs3211371; rs3745274 **CYP2C19** – rs12248560; rs4244285; rs4986893 **CYP2C8** – rs10509681; rs1058930; rs11572080; rs11572103; rs17110453; rs7909236 **CYP2C9** – rs1057910; rs1799853 **CYP2D6** – rs1065852; rs16947; rs28371706; rs28371725; rs35742686; rs3892097; rs5030655; rs5030656; rs59421388; rs61736512 **CYP2E1** – CYP2E1 *1B; CYP2E1 *5B; CYP2E1 *5A; CYP2E1 *6; CYP2E1 *1A **CYP3A4** – rs12721627; rs12721634; rs2740574; rs4986907; rs4986908; rs4986909; rs4986910; rs4986913; rs4987161 **CYP3A5** – rs10264272; rs76293380; rs776746 **NAT2** – rs1041983; rs1208; rs1495741; rs1799929; rs1799930; rs1799931; rs1801279; rs1801280; rs4271002; rs4646244 **NQO1** – rs1800566 **NR1I2** – rs12721608; rs3814055 **PTGIS** – chr20:48184659; rs5629 **UGT1A1** – rs4148323; rs8175347
Paeonia officinalis	Hypnotic	Sedative effect	**ABCB1** – rs1045642; rs1128503; rs2032582 **CYP1A2** – rs12720461; rs2069514; rs762551 **CYP2C19** – rs12248560; rs4244285; rs4986893 **CYP2D6** – rs1065852; rs16947; rs28371706; rs28371725; rs35742686; rs3892097; rs5030655; rs5030656; rs59421388; rs61736512 **SLCO1B1** – rs2306283; rs4149015; rs4149056 **CYP3A4** – rs12721627; rs12721634; rs2740574; rs4986907; rs4986908; rs4986909; rs4986910; rs4986913; rs4987161 **CYP3A5** – rs10264272; rs76293380; rs776746 **NAT2** – rs1041983; rs1208; rs1495741; rs1799929; rs1799930; rs1799931; rs1801279; rs1801280; rs4271002; rs4646244

(continued)

Table 16.3 (continued)

Herbal medicine	Herb-drug interactions	Toxicological mechanism	Clinically validated genes and variants affecting toxicity and drug-herb, herb-herb interactions
Panax ginseng (ginseng)	Bumetanide Ethacrynic acid Furosemide Isocarboxazid Nifedipine Estrogens Corticosteroids Phenelzine Torasemide Tranylcypromine Warfarin Antidiabetic agents	Insomnia Headache Tremors Manic symptoms Increase bleeding Hypoglycemia Fluid retention	**ABCB1** – rs1045642; rs1128503; rs2032582 **CYP1A2** – rs12720461; rs2069514; rs762551 **CYP2C19** – rs12248560; rs4244285; rs4986893 **CYP2C8** – rs10509681; rs1058930; rs11572080; rs11572103; rs17110453; rs7909236 **CYP2C9** – rs1057910; rs1799853 **CYP2D6** – rs1065852; rs16947; rs28371706; rs28371725; rs35742686; rs3892097; rs5030655; rs5030656; rs59421388; rs61736512 **CYP3A4** – rs12721627; rs12721634; rs2740574; rs4986907; rs4986908; rs4986909; rs4986910; rs4986913; rs4987161 **CYP4F2** – rs2108622 **VKORC1** – rs7294; rs9923231; rs9934438 **SLC22A1** – rs12208357; rs34059508; rs34130495; rs72552763 **SLCO1B1** – rs2306283; rs4149015; rs4149056 **KCNJ11** – rs5219 **DRD2** – rs1799732; rs1800497; rs1801028; rs6277 **COMT** – rs4680 **UGT1A1** – rs4148323; rs8175347 **SULT1A1** – rs1801030; rs3760091; rs750155; rs9282861
Passiflora incarnata (passion flower)	Hypnotics	Sedative effects	**ABCB1** – rs1045642; rs1128503; rs2032582 **CYP1A2** – rs12720461; rs2069514; rs762551 **CYP2C19** – rs12248560; rs4244285; rs4986893 **CYP2D6** – rs1065852; rs16947; rs28371706; rs28371725; rs35742686; rs3892097; rs5030655; rs5030656; rs59421388; rs61736512 **SLCO1B1** – rs2306283; rs4149015; rs4149056 **CYP3A4** – rs12721627; rs12721634; rs2740574; rs4986907; rs4986908; rs4986909; rs4986910; rs4986913; rs4987161 **CYP3A5** – rs10264272; rs76293380; rs776746 **NAT2** – rs1041983; rs1208; rs1495741; rs1799929; rs1799930; rs1799931; rs1801279; rs1801280; rs4271002; rs4646244

Paullinia cupana (Guarana)	Theophylline	Palpitations Tachycardia	**ABCB1** – rs1045642; rs1128503; rs2032582 **KCNH2** – rs12720441; rs1805123; rs36210421; rs3807375; rs3815459 **KCNJ11** – rs5219 **SCN5A** – rs1805124; rs6791924; rs7626962 **ADRB1** – rs1801252; rs1801253 **ADRB2** – rs1042713; rs1042714; rs1800888 **CYP1A2** – rs12720461; rs2069514; rs762551 **CYP2C19** – rs12248560; rs4244285; rs4986893 **CYP2C9** – rs1057910; rs1799853 **CYP2D6** – rs1065852; rs16947; rs28371706; rs28371725; rs35742686; rs3892097; rs5030655; rs5030656; rs59421388; rs61736512 **CYP2E1** – CYP2E1 *1B; CYP2E1 *5B; CYP2E1 *5A; CYP2E1 *6; CYP2E1 *1A **CYP3A4** – rs12721627; rs12721634; rs2740574; rs4986907; rs4986908; rs4986909; rs4986910; rs4986913; rs4987161

(continued)

Table 16.3 (continued)

Herbal medicine	Herb-drug interactions	Toxicological mechanism	Clinically validated genes and variants affecting toxicity and drug-herb, herb-herb interactions
Pausinystalia johimbe (Yohimbe)	Tricyclic antidepressants Tetracyclines Venlafaxine	Hypertension Increase stimulation of sympathetic nervous system	**ABCB1** – rs1045642; rs1128503; rs2032582 **KCNH2** – rs12720441; rs1805123; rs36210421; rs3807375; rs3815459 **KCNJ11** – rs5219 **SCN5A** – rs1805124; rs6791924; rs7626962 **ACE** – rs1799752 **ADRB1** – rs1801252; rs1801253 **ADRB2** – rs1042713; rs1042714; rs1800888 **CYP1A2** – rs12720461; rs2069514; rs762551 **CYP2A6** – rs1801272; rs28399433; rs28399444; rs28399454; rs28399468; rs5031016; rs8192726 **CYP2C19** – rs12248560; rs4244285; rs4986893 **CYP2C8** – rs10509681; rs1058930; rs11572080; rs11572103; rs17110453; rs7909236 **CYP2C9** – rs1057910; rs1799853 **CYP2D6** – rs1065852; rs16947; rs28371706; rs28371725; rs35742686; rs3892097; rs5030655; rs5030656; rs59421388; rs61736512 **CYP3A4** – rs12721627; rs12721634; rs2740574; rs4986907; rs4986908; rs4986909; rs4986910; rs4986913; rs4987161 **CYP3A5** – rs10264272; rs76293380; rs776746 **CYP4F2** – rs2108622 **PTGIS** – chr20:48184659; rs5629

Persea Americana (avocado)	Warfarin	Reduction xenobiotics absorption and increase xenobiotics metabolism	**ABCB1** – rs1045642; rs1128503; rs2032582 **CYP1A2** – rs12720461; rs2069514; rs762551 **CYP2C19** – rs12248560; rs4244285; rs4986893 **CYP2C8** – rs10509681; rs1058930<, rs11572080<<, rs11572103<<<, rs17110453<, rs7909236 **CYP2C9** – rs1057910; rs1799853 **CYP3A4** – rs12721627; rs12721634; rs2740574; rs4986907; rs4986908; rs4986909; rs4986910; rs4986913; rs4987161 **CYP4F2** – rs2108622 **VKORC1** – rs7294; rs9923231; rs9934438
Peumus boldus (Boldo)	Warfarin	Increase bleeding	**ABCB1** – rs1045642; rs1128503; rs2032582 **CYP1A2** – rs12720461; rs2069514; rs762551 **CYP2C19** – rs12248560; rs4244285; rs4986893 **CYP2C8** – rs10509681; rs1058930; rs11572080; rs11572103; rs17110453; rs7909236 **CYP2C9** – rs1057910; rs1799853 **CYP3A4** – rs12721627; rs12721634; rs2740574; rs4986907; rs4986908; rs4986909; rs4986910; rs4986913; rs4987161 **CYP4F2** – rs2108622 **VKORC1** – rs7294; rs9923231; rs9934438
Pimpinella anisum (Anise)	Warfarin Aspirin	Increase bleeding	**ABCB1** – rs1045642; rs1128503; rs2032582 **CYP1A2** – rs12720461; rs2069514; rs762551 **CYP2C19** – rs12248560; rs4244285; rs4986893 **CYP2C8** – rs10509681; rs1058930; rs11572080; rs11572103; rs17110453; rs7909236 **CYP2C9** – rs1057910; rs1799853 **CYP3A4** – rs12721627; rs12721634; rs2740574; rs4986907; rs4986908; rs4986909; rs4986910; rs4986913; rs4987161 **CYP4F2** – rs2108622 **VKORC1** – rs7294; rs9923231; rs9934438 **P2RY1** – rs1065776; rs701265 **P2RY12** – rs2046934

(continued)

Table 16.3 (continued)

Herbal medicine	Herb-drug interactions	Toxicological mechanism	Clinically validated genes and variants affecting toxicity and drug-herb, herb-herb interactions
Piper betle (betel nut)	Beta blockers, digoxin Flupentixol, fluphenazine Prednisolone, procyclidine Salbutamol	Bronchoconstriction Bradycardia	**ADRB1** – rs1801252; rs1801253 **ADRB2** – rs1042713; rs1042714; rs1800888 **ABCB1** – rs1045642; rs1128503; rs2032582 **KCNH2** – rs12720441; rs1805123; rs36210421; rs3807375; rs3815459 **KCNJ11** – rs5219 **SCN5A** – rs1805124; rs6791924; rs7626962
Piper methysticum (kava Kava)	Acetaminophen, benzodiazepines Barbiturates	Sedation effects Increase GABA effects	**ABCB1** – rs1045642; rs1128503; rs2032582 **CYP1A2** – rs12720461; rs2069514; rs762551 **CYP2C19** – rs12248560; rs4244285; rs4986893 **CYP2D6** – rs1065852; rs16947; rs28371706; rs28371725; rs35742686; rs3892097; rs5030655; rs5030656; rs59421388; rs61736512 **SLCO1B1** – rs2306283; rs4149015; rs4149056 **CYP3A4** – rs12721627; rs12721634; rs2740574; rs4986907; rs4986908; rs4986909; rs4986910; rs4986913; rs4987161 **CYP3A5** – rs10264272; rs76293380; rs776746 **NAT2** – rs1041983; rs1208; rs1495741; rs1799929; rs1799930; rs1799931; rs1801279; rs1801280; rs4271002; rs4646244

Piper nigrum (black pepper)	Phenytoin Propranolol Rifampicin Theophylline	Inhibition of P-gp transport proteins Inhibition of CYP3A4, CYP2C9, CYP1A1, CYP1A2	**ABCB1** – rs1045642; rs1128503; rs2032582 **CYP1A2** – rs12720461; rs2069514; rs762551 **CYP2A6** – rs1801272; rs28399433; rs28399444; rs28399454; rs28399468; rs5031016; rs8192726 **CYP2B6** – rs2279343; rs28399499; rs3211371; rs3745274 **CYP2C19** – rs12248560; rs4244285; rs4986893 **CYP2C8** – rs10509681; rs1058930; rs11572080; rs11572103; rs17110453; rs7909236 **CYP2C9** – rs1057910; rs1799853 **CYP2D6** – rs1065852; rs16947; rs28371706; rs28371725; rs35742686; rs3892097; rs5030655; rs5030656; rs59421388; rs61736512 **CYP2E1** – CYP2E1 *1B; CYP2E1 *5B; CYP2E1 *5A; CYP2E1 *6; CYP2E1 *1A **CYP3A4** – rs12721627; rs12721634; rs2740574; rs4986907; rs4986908; rs4986909; rs4986910; rs4986913; rs4987161 **CYP3A5** – rs10264272; rs76293380; rs776746
Plantago ovata (Psyllium)	Lithium	Reduction absorption	**ABCB1** – rs1045642; rs1128503; rs2032582 **SLC19A1** – rs1051266; rs1051296; rs1051298; rs1131596; rs12659 **SLC22A1** – rs12208357; rs34059508; rs34130495; rs72552763 **SLCO1B1** – rs2306283; rs4149015; rs4149056
Primula officinalis (Cowslip)	Antihypertensives	Hypertension	**ABCB1** – rs1045642; rs1128503; rs2032582 **ACE** – rs1799752 **ADRB1** – rs1801252; rs1801253 **ADRB2** – rs1042713; rs1042714; rs1800888 **CYP3A4** – rs12721627; rs12721634; rs2740574; rs4986907; rs4986908; rs4986909; rs4986910; rs4986913; rs4987161 **CYP3A5** – rs10264272; rs76293380; rs776746 **CYP4F2** – rs2108622 **PTGIS** – chr20:48184659; rs5629

(continued)

Table 16.3 (continued)

Herbal medicine	Herb-drug interactions	Toxicological mechanism	Clinically validated genes and variants affecting toxicity and drug-herb, herb-herb interactions
Psyllium spp. (Ispaghula)	Many drugs	Diarrhea Reduction absorption	**ABCB1** – rs1045642; rs1128503; rs2032582 **SLC19A1** – rs1051266; rs1051296; rs1051298; rs1131596; rs12659 **SLC22A1** – rs12208357; rs34059508; rs34130495; rs72552763 **SLCO1B1** – rs2306283; rs4149015; rs4149056
Pueraria lobata (Kudzu)	Verapamil Triptans Methotrexate	Hypotension	**ABCB1** – rs1045642; rs1128503; rs2032582 **KCNH2** – rs12720441; rs1805123; rs36210421; rs3807375; rs3815459 **KCNJ11** – rs5219 **SCN5A** – rs1805124; rs6791924; rs7626962 **ACE** – rs1799752 **ADRB1** – rs1801252; rs1801253 **ADRB2** – rs1042713; rs1042714; rs1800888 **CYP1A2** – rs12720461; rs2069514; rs762551 **CYP2A6** – rs1801272; rs28399433; rs28399444; rs28399454; rs28399468; rs5031016; rs8192726 **CYP2C19** – rs12248560; rs4244285; rs4986893 **CYP2C8** – rs10509681; rs1058930; rs11572080; rs11572103; rs17110453; rs7909236 **CYP2C9** – rs1057910; rs1799853 **CYP2D6** – rs1065852; rs16947; rs28371706; rs28371725; rs35742686; rs3892097; rs5030655; rs5030656; rs59421388; rs61736512 **CYP3A4** – rs12721627; rs12721634; rs2740574; rs4986907; rs4986908; rs4986909; rs4986910; rs4986913; rs4987161 **CYP3A5** – rs10264272; rs76293380; rs776746 **CYP4F2** – rs2108622 **PTGIS** – chr20:48184659; rs5629
Punicas granatum (pomegranate)	Tolbutamide Carbamazepine	Antioxidant effects	**ABCB1** – rs1045642; rs1128503; rs2032582 **CYP3A4** – rs12721627; rs12721634; rs2740574; rs4986907; rs4986908; rs4986909; rs4986910; rs4986913; rs4987161 **CYP3A5** – rs10264272; rs76293380; rs776746 **CYP2C9** – rs1057910; rs1799853 **CYP1A2** – rs12720461; rs2069514; rs762551

Rhamnus frangula and R. purshiana (cascara)	Multiple drugs	Diarrhea Decrease absorption	**ABCB1** – rs1045642; rs1128503; rs2032582 **SLC19A1** – rs1051266; rs1051296; rs1051298; rs1131596; rs12659 **SLC22A1** – rs12208357; rs34059508; rs34130495; rs72552763 **SLCO1B1** – rs2306283; rs4149015; rs4149056
Rheum officinale (Rhubard)	Cardiac glycosides Diuretics	Hypokalemia Reduction absorption	**ABCB1** – rs1045642; rs1128503; rs2032582 **SLC19A1** – rs1051266; rs1051296; rs1051298; rs1131596; rs12659 **SLC22A1** – rs12208357; rs34059508; rs34130495; rs72552763 **SLCO1B1** – rs2306283; rs4149015; rs4149056 **KCNH2** – rs12720441; rs1805123; rs36210421; rs3807375; rs3815459 **KCNJ11** – rs5219
Rhodiola rosea (golden root)	Digoxin	Inhibition P glycoprotein and CYP3A4	**ABCB1** – rs1045642; rs1128503; rs2032582 **CYP3A4** – rs12721627; rs12721634; rs2740574; rs4986907; rs4986908; rs4986909; rs4986910; rs4986913; rs4987161 **KCNH2** – rs12720441; rs1805123; rs36210421; rs3807375; rs3815459 **KCNJ11** – rs5219 **SCN5A** – rs1805124; rs6791924; rs7626962
Rosmarinus officinale (rosemary)	Antidiabetic agents	Hyperglycemia	**ABCB1** – rs1045642; rs1128503; rs2032582 **CYP1A2** – rs12720461; rs2069514; rs762551 **CYP2C19** – rs12248560; rs4244285; rs4986893 **CYP2C8** – rs10509681; rs1058930; rs11572080; rs11572103; rs17110453; rs7909236 **CYP2C9** – rs1057910; rs1799853 **CYP2D6** – rs1065852; rs16947; rs28371706; rs28371725; rs35742686; rs3892097; rs5030655; rs5030656; rs59421388; rs61736512 **CYP3A4** – rs12721627; rs12721634; rs2740574; rs4986907; rs4986908; rs4986909; rs4986910; rs4986913; rs4987161 **SLC22A1** – rs12208357; rs34059508; rs34130495; rs72552763 **SLCO1B1** – rs2306283; rs4149015; rs4149056 **KCNJ11** – rs5219

(continued)

Table 16.3 (continued)

Herbal medicine	Herb-drug interactions	Toxicological mechanism	Clinically validated genes and variants affecting toxicity and drug-herb, herb-herb interactions
Salix spp. (Willon)	Warfarin NSAIDS Phenytoin	Increase bleeding	**ABCB1** – rs1045642; rs1128503; rs2032582 **CYP1A2** – rs12720461; rs2069514; rs762551 **CYP2C19** – rs12248560; rs4244285; rs4986893 **CYP2C8** – rs10509681; rs1058930; rs11572080; rs11572103; rs17110453; rs7909236 **CYP2C9** – rs1057910; rs1799853 **CYP3A4** – rs12721627; rs12721634; rs2740574; rs4986907; rs4986908; rs4986909; rs4986910; rs4986913; rs4987161 **CYP4F2** – rs2108622 **VKORC1** – rs7294; rs9923231; rs9934438
Salvia miltiorrhiza (Danshen)	Warfarin	Increase bleeding	**ABCB1** – rs1045642; rs1128503; rs2032582 **CYP1A2** – rs12720461; rs2069514; rs762551 **CYP2C19** – rs12248560; rs4244285; rs4986893 **CYP2D6** – rs1065852; rs16947; rs28371706; rs28371725; rs35742686; rs3892097; rs5030655; rs5030656; rs59421388; rs61736512 **CYP2C8** – rs10509681; rs1058930; rs11572080; rs11572103; rs17110453; rs7909236 **CYP2C9** – rs1057910; rs1799853 **CYP3A4** – rs12721627; rs12721634; rs2740574; rs4986907; rs4986908; rs4986909; rs4986910; rs4986913; rs4987161 **CYP3A5** – rs10264272; rs76293380; rs776746 **CYP4F2** – rs2108622 **VKORC1** – rs7294; rs9923231; rs9934438
Sambucus nigra (elder)	Diuretics	Increase diuresis	**ACE** – rs1799752 **CYP3A4** – rs12721627; rs12721634; rs2740574; rs4986907; rs4986908; rs4986909; rs4986910; rs4986913; rs4987161 **UGT1A1** – rs4148323; rs8175347

Scilla spp. (squill)	Digoxin, quinidine, methylxanthines Phosphodiesterase inhib sympathomimetics	Inhib. cardiomuscular Na,K-adenosine phosphatase (ATPase) Arrhythmias	**ABCB1** – rs1045642; rs1128503; rs2032582 **KCNH2** – rs12720441; rs1805123; rs36210421; rs3807375; rs3815459 **KCNJ11** – rs5219 **SCN5A** – rs1805124; rs6791924; rs7626962
Scopolia carniolica (scopolia)	Tricyclic antidepressants Amantadine	Sedative effects	**ABCB1** – rs1045642; rs1128503; rs2032582 **CYP1A2** – rs12720461; rs2069514; rs762551 **CYP2C19** – rs12248560; rs4244285; rs4986893 **CYP2D6** – rs1065852; rs16947; rs28371706; rs28371725; rs35742686; rs3892097; rs5030655; rs5030656; rs59421388; rs61736512 **SLCO1B1** – rs2306283; rs4149015; rs4149056 **CYP3A4** – rs12721627; rs12721634; rs2740574; rs4986907; rs4986908; rs4986909; rs4986910; rs4986913; rs4987161 **CYP3A5** – rs10264272; rs76293380; rs776746 **NAT2** – rs1041983; rs1208; rs1495741; rs1799929; rs1799930; rs1799931; rs1801279; rs1801280; rs4271002; rs4646244
Scutellaria baicalensis (Huang gin)	Irinotecan	Abdominal pain Hepatotoxicity	**ABCB1** – rs1045642; rs1128503; rs2032582 **CYP3A4** – rs12721627; rs12721634; rs2740574; rs4986907; rs4986908; rs4986909; rs4986910; rs4986913; rs4987161 **SLCO1B1** – rs2306283; rs4149015; rs4149056 **UGT1A1** – rs4148323; rs8175347
Senna spp. (Senna)	Multiple drugs	Reduction absorption Diarrhea	**ABCB1** – rs1045642; rs1128503; rs2032582 **SLC19A1** – rs1051266; rs1051296; rs1051298; rs1131596; rs12659 **SLC22A1** – rs12208357; rs34059508; rs34130495; rs72552763 **SLCO1B1** – rs2306283; rs4149015; rs4149056

(continued)

Table 16.3 (continued)

Herbal medicine	Herb-drug interactions	Toxicological mechanism	Clinically validated genes and variants affecting toxicity and drug-herb, herb-herb interactions
Serenoa serrulata (Saw palmetto)	Finasteride Flutamide Oral contraceptives Disulfiram Warfarin Ibuprofen Naproxen Metronidazole	Increase bleeding Inhibition of 5 alpha reductase	**ABCB1** – rs1045642; rs1128503; rs2032582 **CYP1A2** – rs12720461; rs2069514; rs762551 **CYP2C19** – rs12248560; rs4244285; rs4986893 **CYP2C8** – rs10509681; rs1058930; rs11572080; rs11572103; rs17110453; rs7909236 **CYP2C9** – rs1057910; rs1799853 **CYP3A4** – rs12721627; rs12721634; rs2740574; rs4986907; rs4986908; rs4986909; rs4986910; rs4986913; rs4987161 **CYP4F2** – rs2108622 **VKORC1** – rs7294; rs9923231; rs9934438
Schisandra chinensis (wǔ wèi zi)	Tacrolimus	Cough and dysentery Inhibition P glycoprotein	**ABCB1** – rs1045642; rs1128503; rs2032582 **CYP1A2** – rs12720461; rs2069514; rs762551 **CYP2C19** – rs12248560; rs4244285; rs4986893 **CYP2C8** – rs10509681; rs1058930; rs11572080; rs11572103; rs17110453; rs7909236 **CYP2C9** – rs1057910; rs1799853 **CYP3A4** – rs12721627; rs12721634; rs2740574; rs4986907; rs4986908; rs4986909; rs4986910; rs4986913; rs4987161 **CYP2D6** – rs1065852; rs16947; rs28371706; rs28371725; rs35742686; rs3892097; rs5030655; rs5030656; rs59421388; rs61736512 **CYP3A5** – rs10264272; rs76293380; rs776746
Silybum spp. (milk thistle)	Amiodarone Indinavir	Inhibition cytochrome P450 3A4	**ABCB1** – rs1045642; rs1128503; rs2032582 **CYP3A4** – rs12721627; rs12721634; rs2740574; rs4986907; rs4986908; rs4986909; rs4986910; rs4986913; rs4987161 **CYP3A5** – rs10264272; rs76293380; rs776746 **CYP2C9** – rs1057910; rs1799853 **CYP2C8** – rs10509681; rs1058930; rs11572080; rs11572103; rs17110453; rs7909236

Stephania tetrandra (Han Fang ji)	Calcium channel blocking agents Antidiabetic drugs	Hyperglycemia Kidney toxicity	**ABCB1** – rs1045642; rs1128503; rs2032582 **CYP1A2** – rs12720461; rs2069514; rs762551 **CYP3A4** – rs12721627; rs12721634; rs2740574; rs4986907; rs4986908; rs4986909; rs4986910; rs4986913; rs4987161 **CYP2D6** – rs1065852; rs16947; rs28371706; rs28371725; rs35742686; rs3892097; rs5030655; rs5030656; rs59421388; rs61736512 **ADRB1** – rs1801252; rs1801253 **ADRB2** – rs1042713; rs1042714; rs1800888
Tanacetum parthenium (feverfew)	Warfarin	Coumarin compounds may increase bleeding	**ABCB1** – rs1045642; rs1128503; rs2032582 **CYP1A2** – rs12720461; rs2069514; rs762551 **CYP2C19** – rs12248560; rs4244285; rs4986893 **CYP2C8** – rs10509681; rs1058930; rs11572080; rs11572103; rs17110453; rs7909236 **CYP2C9** – rs1057910; rs1799853 **CYP3A4** – rs12721627; rs12721634; rs2740574; rs4986907; rs4986908; rs4986909; rs4986910; rs4986913; rs4987161 **CYP4F2** – rs2108622 **VKORC1** – rs7294; rs9923231; rs9934438
Taraxacum officinalis (dandelion)	Diuretics Antidiabetic drugs Antihytpertensives	Reduction xenobiotics absorption	**ABCB1** – rs1045642; rs1128503; rs2032582 **SLC22A1** – rs12208357; rs34059508; rs34130495; rs72552763 **SLCO1B1** – rs2306283; rs4149015; rs4149056
Taxus chinensis (yew)	Paclitaxel	Arrhythmogenic effect	**ABCB1** – rs1045642; rs1128503; rs2032582 **CYP3A4** – rs12721627; rs12721634; rs2740574; rs4986907; rs4986908; rs4986909; rs4986910; rs4986913; rs4987161 **KCNH2** – rs12720441; rs1805123; rs36210421; rs3807375; rs3815459 **KCNJ11** – rs5219 **SCN5A** – rs1805124; rs6791924; rs7626962

(continued)

Table 16.3 (continued)

Herbal medicine	Herb-drug interactions	Toxicological mechanism	Clinically validated genes and variants affecting toxicity and drug-herb, herb-herb interactions
Tetradium ruticarpum (Euodia)	Caffeine Benzodiazepines	Excitatory effect	**ABCB1** – rs1045642; rs1128503; rs2032582 **CYP1A2** – rs12720461; rs2069514; rs762551 **CYP3A4** – rs12721627; rs12721634; rs2740574; rs4986907; rs4986908; rs4986909; rs4986910; rs4986913; rs4987161 **CYP2D6** – rs1065852; rs16947; rs28371706; rs28371725; rs35742686; rs3892097; rs5030655; rs5030656; rs59421388; rs61736512
Teucrium spp (germanders)	Anti-cancer drugs	Hepatotoxicity	**ABCB1** – rs1045642; rs1128503; rs2032582 **CYP3A4** – rs12721627; rs12721634; rs2740574; rs4986907; rs4986908; rs4986909; rs4986910; rs4986913; rs4987161
Trigonella foenum-graecum (fenugreek)	Warfarin	Coumarin compounds may increase bleeding	**ABCB1** – rs1045642; rs1128503; rs2032582 **CYP1A2** – rs12720461; rs2069514; rs762551 **CYP2C19** – rs12248560; rs4244285; rs4986893 **CYP2C8** – rs10509681; rs1058930; rs11572080; rs11572103; rs17110453; rs7909236 **CYP2C9** – rs1057910; rs1799853 **CYP3A4** – rs12721627; rs12721634; rs2740574; rs4986907; rs4986908; rs4986909; rs4986910; rs4986913; rs4987161 **CYP4F2** – rs2108622 **VKORC1** – rs7294; rs9923231; rs9934438
Tripterygium wilfordii (thunder duke vine)	Immunosuppressive drugs	Osteoporosis Weakened immune system	**ABCB1** – rs1045642; rs1128503; rs2032582 **CYP3A4** – rs12721627; rs12721634; rs2740574; rs4986907; rs4986908; rs4986909; rs4986910; rs4986913; rs4987161 **CYP2C19** – rs12248560; rs4244285; rs4986893

Uncaria tomentosa (cat's claw)	Ritonavir Benzodiazepines Cyclosporine	CYP3A4 inhibition	**ABCB1** – rs1045642; rs1128503; rs2032582 **CYP3A4** – rs12721627; rs12721634; rs2740574; rs4986907; rs4986908; rs4986909; rs4986910; rs4986913; rs4987161 **CYP2D6** – rs1065852; rs16947; rs28371706; rs28371725; rs35742686; rs3892097; rs5030655; rs5030656; rs59421388; rs61736512 **CYP2C19** – rs12248560; rs4244285; rs4986893 **CYP3A5** – rs10264272; rs76293380; rs776746 **DRD2** – rs1799732; rs1800497; rs1801028; rs6277
Vaccinum spp. (bilberry leaf)	Warfarin Aspirin	Increases bleeding	**ABCB1** – rs1045642; rs1128503; rs2032582 **CYP1A2** – rs12720461; rs2069514; rs762551 **CYP2B6** – rs2279343; rs28399499; rs3211371; rs3745274 **CYP2C19** – rs12248560; rs4244285; rs4986893 **CYP2C8** – rs10509681; rs1058930; rs11572080; rs11572103; rs17110453; rs7909236 **CYP2C9** – rs1057910; rs1799853 **CYP3A4** – rs12721627; rs12721634; rs2740574; rs4986907; rs4986908; rs4986909; rs4986910; rs4986913; rs4987161 **CYP4F2** – rs2108622 **VKORC1** – rs7294; rs9923231; rs9934438 **P2RY1** – rs1065776; rs701265 **P2RY12** – rs2046934

(continued)

Table 16.3 (continued)

Herbal medicine	Herb-drug interactions	Toxicological mechanism	Clinically validated genes and variants affecting toxicity and drug-herb, herb-herb interactions
Valeriana officinalis (valerian)	Benzodiazepines	Sedative effects	**ABCB1** – rs1045642; rs1128503; rs2032582 **CYP1A2** – rs12720461; rs2069514; rs762551 **CYP2C19** – rs12248560; rs4244285; rs4986893 **CYP2D6** – rs1065852; rs16947; rs28371706; rs28371725; rs35742686; rs3892097; rs5030655; rs5030656; rs59421388; rs61736512 **SLCO1B1** – rs2306283; rs4149015; rs4149056 **CYP3A4** – rs12721627; rs12721634; rs2740574; rs4986907; rs4986908; rs4986909; rs4986910; rs4986913; rs4987161 **CYP3A5** – rs10264272; rs76293380; rs776746 **NAT2** – rs1041983; rs1208; rs1495741; rs1799929; rs1799930; rs1799931; rs1801279; rs1801280; rs4271002; rs4646244
Vitex agnus-castus (chaste tree)	Metoclopramide Bromocriptine	Nausea and vomiting	**DRD2** – rs1799732; rs1800497; rs1801028; rs6277
Vitis vinifera (grape)	Warfarin Aspirin	Increase bleeding	**ABCB1** – rs1045642; rs1128503; rs2032582 **CYP1A2** – rs12720461; rs2069514; rs762551 **CYP2B6** – rs2279343; rs28399499; rs3211371; rs3745274 **CYP2C19** – rs12248560; rs4244285; rs4986893 **CYP2C8** – rs10509681; rs1058930; rs11572080; rs11572103; rs17110453; rs7909236 **CYP2C9** – rs1057910; rs1799853 **CYP3A4** – rs12721627; rs12721634; rs2740574; rs4986907; rs4986908; rs4986909; rs4986910; rs4986913; rs4987161 **CYP4F2** – rs2108622 **VKORC1** – rs7294; rs9923231; rs9934438 **P2RY1** – rs1065776; rs701265 **P2RY12** – rs2046934

Withania somnifera (ashwagandha)	Barbiturates Benzodiazepines	Sedative effect	**ABCB1** – rs1045642; rs1128503; rs2032582 **CYP1A2** – rs12720461; rs2069514; rs762551 **CYP2C19** – rs12248560; rs4244285; rs4986893 **CYP2D6** – rs1065852; rs16947; rs28371706; rs28371725; rs35742686; rs3892097; rs5030655; rs5030656; rs59421388; rs61736512 **SLCO1B1** – rs2306283; rs4149015; rs4149056 **CYP3A4** – rs12721627; rs12721634; rs2740574; rs4986907; rs4986908; rs4986909; rs4986910; rs4986913; rs4987161 **CYP3A5** – rs10264272; rs76293380; rs776746 **NAT2** – rs1041983; rs1208; rs1495741; rs1799929; rs1799930; rs1799931; rs1801279; rs1801280; rs4271002; rs4646244
Zingiber officinale (ginger)	Cardiac glycosides Phenprocoumon Saquinavir Warfarin	Inhibition of the membranous Na,K-adenosine triphosphatase (ATPase) of cardiomuscular tissue Inhibition platelet aggregation	**ABCB1** – rs1045642; rs1128503; rs2032582 **KCNH2** – rs12720441; rs1805123; rs36210421; rs3807375; rs3815459 **KCNJ11** – rs5219 **SCN5A** – rs1805124; rs6791924; rs7626962 **CYP1A2** – rs12720461; rs2069514; rs762551 **CYP2C19** – rs12248560; rs4244285; rs4986893 **CYP2C8** – rs10509681; rs1058930; rs11572080; rs11572103; rs17110453; rs7909236 **CYP2C9** – rs1057910; rs1799853 **CYP3A4** – rs12721627; rs12721634; rs2740574; rs4986907; rs4986908; rs4986909; rs4986910; rs4986913; rs4987161 **CYP4F2** – rs2108622 **P2RY1** – rs1065776; rs701265 **P2RY12** – rs2046934

(continued)

The CYP3A enzyme group constitutes the largest number of CYP enzymes and is highly expressed, not only in the liver but also in the small intestine. Pharmacokinetic interferences can occur through the inhibition or induction of CYP3A enzymes by herbs or natural substances such as grapefruit juice (*Citrus paradise*) and St. John's wort (*Hypericum perforatum*) (Ainslie et al. 2014; Markert et al. 2014). Organic compounds like furanocoumarins, found in grapefruit juice, are responsible for the inhibitory action on CYP3A, resulting in reduced pre-systemic metabolism of many drugs. At the same time, grapefruit juice also inhibits intestinal P-glycoprotein (Pgp), a member of the ATP-binding cassette family encoded by the ABCB1 gene (also called MDR1) (Ainslie et al. 2014; Cock 2015). Pgp is an excretory transporter and is also found in the enterocyte membrane, where it transports lipophilic molecules back into the intestinal lumen. Certain drugs that possess lipophilic properties are either metabolized by CYP3A4 or moved to the intestine by the Pgp transporter. Both the Pgp and CYP3A4 may act synergistically as a barrier to many orally administered drugs. Therefore, their inhibition can markedly increase the bioavailability of a drug, thus affecting the final effect. St. John's wort is a medicinal plant widely used in clinical practice, especially for depression, and is known to interact with many drugs by increasing intestinal and hepatic CYP3A4 activity through activation of the nuclear pregnane X receptor (PXR). Plasma concentrations of CYP3A4 substrates are thus reduced, especially in some PXR genotypes and, conversely, subjects harboring the ABCB1 haplotype comprising 1236C>T, 2677G>T/A, and 3435C>T polymorphisms have been shown to have an attenuated inductive response to St. John's wort. The St. John's wort constituent hyperforin is probably responsible for CYP3A4 induction, while hypericin is a P-glycoprotein-inducing compound and accounts for the other drug interference of *Hypericum perforatum* (Markert et al. 2014; Cock 2015).

Tianqi Jiangtang is an herbal medicine widely used for diabetes treatment in China; it consists of ten Chinese herbal medicines: *Radix Astragali, Radix Trichosanthis, Fructus Ligustri Lucidi, Caulis D*endrobii, *Radix Ginseng, Cortex Lycii Radicis* bone, *Rhizoma Coptidis*, Asiatic Cornelian cherry fruit, Ecliptae Herba, and Chinese gall. One isolated component of Tianqi Jiangtang has been identified as having a major effect, berberine hydrochloride from *Rhizoma Coptidis*, which has been successfully used in antidiabetic treatments. The polymorphism, rs1142345 (A>G) SNP, in the thiopurine S-methyltransferase gene, has been shown to be associated with the hypoglycemic effect of the drug. and as such it is considered a companion diagnostic, predictive of the efficacy of treatment (Li et al. 2013).

Genotyping enzymes and transporters involved in pharmacokinetics and the mapping of molecular targets involved in pharmacodynamics, allows the prediction of responses to treatment and detection of potential adverse reactions to drugs. The consultation of DNA and pathways databases allows access to all the information necessary to perform targeted genomic investigations. The primary function of human DNA databases includes the establishment of the reference genome (e.g., NCBI RefSeq), the profiling of human genetic variation (e.g.,, dbSNP), and the association of genotype with phenotype (e.g., EGA). Useful tools for the clinical use of pharmacogenomics can also be found in PharmGKB, a publicly available

Internet research tool developed by Stanford University (Guo et al. 2014; Solomon et al. 2013).

The search of genetic clues, which today can track pathological information even from circulating cells and platelets from "liquid biopsies" (Best et al. 2015), is thus important during the general characterization of the patient, searching for SNPs as suggested by the family health history, and afterwards, before prescribing a drug or a herbal preparation. The genetic testing must then be integrated with the other medical information so that the data can be consolidated within the entire dynamic biological makeup of each individual, as suggested (see the following). Together with the environmental and lifestyle factors that interface with this makeup, a complex, personal phenotype can be generated, and the best therapeutic or preventive actions made, thus fully realizing the shift from reactive medicine to proactive, pre-emptive, and preventive healthcare. An important phase towards this goal is the patient examination, which will increasingly focus on integrating information from multiple sources, not only genomics and other 'omics technologies, but also environmental and lifestyle data.

Patient Interview and Medication Reconciliation

During physical examination, an important element for building an integrative personal profile is the "classic" patient interview designed to obtain comprehensive information about personal health-related behaviors and lifestyle information, such as dietary habits, caffeine intake, use of tobacco, and alcohol and exercise habits as well as personal history, i.e., educational level, composition of the family of origin, and current household and personal interests. An assessment of the activities of daily living to determine baseline function is also part of the interview, especially in disabled or older patients.

Once a therapeutic decision is made that might include the administration of a phytocomplex, attention to the possibility of adverse events has to be made. Medication reconciliation, a formal process of identifying the complete and accurate list of medications that the patient takes, is of pivotal pharmaco-toxicological importance. Other drug-related information is also collected, such as drug allergies (Best Medication History, BMH). The list is then used to evaluate whether the drugs or herbal preparations to be prescribed or administered are compatible with the other medications being taken by the patient (reconciliative phase), to reduce the risk of toxicological interactions. BMH is a vital component of the interview, and is the area where the physician must dedicate the most time. Medications include prescription and over-the-counter drugs as well as herbal products. The medication history provides insight into the patient's current and past medications, adverse drug reactions or allergies, adherence to therapies, and the patient's own understanding about his or her medications (Ramjaun et al. 2015).

Herbal products often represent an underestimated risk, especially in the patient's view, so medication reconciliation must include herbal preparations for the effective prevention of adverse events. Special precautions need to be taken in patients in particular circumstances, such as in pregnant women, when the maternal bloodstream is shared with the fetal bloodstream (Cock 2015). Patients with existing medical conditions should also be very carefully advised when taking phytocomplexes. An important example is that of diabetics; more than 400 plants have been identified as being capable of lowering blood glucose levels, such as the commonly used rehmannia (*Rehmannia glutinosa*) and American ginseng (*Panax quinquefolius*) (Zhou et al. 2015; Mucalo et al. 2013). Bitter melon (*Momordica charantia*) and *Gymnema sylvestre*, used in traditional Asian medicines as remedies for diabetes, or aloe species including *Aloe vera* and fenugreek (*Trigonella foenum-graecum*), used in the Middle East as anti-diabetes drugs, can interfere with hypoglycemic drug therapies (Cock 2015; Yang et al. 2015; Tiwari et al. 2014; Alinejad-Mofrad et al. 2015; Bahmani et al. 2015). Many herbal preparations have cardiopulmonary effects that may be dangerous for people with heart disease. Some medications directly increase the heart rate; these include Belladonna (*Atropa belladonna*), a plant used in traditional medicine systems (Cock 2015). Licorice (*Glycyrrhiza glabra*) treatment may induce the production of mineralocorticoids, resulting in the retention of sodium and water, and this may result in increased blood pressure, hypertension, and pulmonary edema (Cock 2015; Schröder et al. 2015).

Today, electronic platforms are available, containing real-time updated databases about drug-drug and herb-drug interactions that can be linked to a patient's electronic record, so that prescription incompatibilities can be flagged and signalled to the physician and known predictable risks can be virtually eliminated (Lowry et al. 2003).

Personalized P4 Precision Medicine at Work. What's Available Beyond Genomics?

Whether a phytocomplex has been prescribed as a cure or as a preventative measure, regular monitoring ,which is also based on the analysis of medically relevant variants and health risk assessment, is required. In clinical toxicology, the current diagnostic techniques, although quite efficient in correlating with diseases or serious poisoning, are incapable of identifying the dysfunctions at preliminary stages. Diagnosis usually depends on one or a few highly correlated markers, many of which reach detectable levels only at the advanced stages. Today the advancements of real applications of diagnostic-omics in medical practice are still mostly limited to predictive genomics and pharmacogenomics which, although essential for diagnosis, risk prediction and pharmaco-toxicological predisposition, cannot be used for monitoring wellness, disease, or pharmacological treatment. There are few exceptions, but systems medicine platforms are still used mostly at an experimental level; although clinical, this makes large proteomic or metabolomic molecular networks not commonly exploitable in the diagnostic routine. The consequence is that single biomarkers are still today the main tool to monitor a patient's state or a pharmacological treatment, especially for outpatients and patients in the ambulatory care environment. Still, besides bringing up some important examples of applied diagnostic-omics in specific fields that are also outside the pharmaco-toxicology of herbal medicines, it is possible to take a quick look at what is happening in the systems medicine field to have an idea of how far we might be from widespread application.

The analysis of the transcripts in a cell (e.g.,, mRNA, non-coding RNA, and small RNAs), called the "transcriptome," is the first analytical step downstream from DNA. Gene expression profiling, or transcriptomics, can be used in clinical practice to delineate disease classification, improve diagnostic accuracy and provide new biological insights, in particular, in the diagnosis of various types of cancer. High-throughput technologies, such as microarrays and sequencing platforms, allow the measurement of thousands of transcripts simultaneously, to look for different pattern changes across subsets that help characterize a particular physiological state or clinical phenotype. Today it is possible to use the diagnostic and screening transcriptomic biomarkers that have been developed in subjects at high risk of cancer. A number of transcriptomic biomarkers for the early detection of cancer (e.g.,, lung cancer) have leveraged the so-called "field cancerization" or "field effect paradigm," in which abnormalities in gene expression in the normal tissue are shared with those found in the cancer (Montani et al. 2015). Transcriptome profiling can be performed for specific cancers, including breast cancer, gastrointestinal tumors, prostate cancer, and lung cancer. MicroRNAs are non-coding RNAs that play an important role in regulating gene expression; since miRNAs are relatively more stable than mRNAs, any miRNA profiles of cancer risk or diagnosis promise to be more accurate and useful in the clinical practice (Sturla et al. 2014; Mias and Snyder 2013; Dopazo 2014). Another promising opportunity for early diagnosis is today represented by RNA from "tumor-educated platelets," where circulating platelets contain genetic material transferred from or induced by tumoral cells (Best et al. 2015), a model that can be extended to other pathological conditions.

Together with the analysis of drug- or herb-related metabolites in bio-fluids, monitoring global changes in the proteome, metabonome, and lipidome, and correlating them to health and disease, is steadily spreading. Highly correlated network markers for many physiological conditions have been obtained. Today, a mass-spectrometric analysis of bio-fluids, such as serum or plasma, could accurately identify, and in most cases quantify, the majority of its constituent molecules to high precision. One can apply various statistical tools to extract significantly deviating sets of molecules. An interesting approach to classifying a range of molecular patterns across various sample types would be to follow standard clustering techniques, identifying and grouping the majority of covariating molecules (Sturla et al. 2014). When a particular cluster correlates well with a pathophysiological state, this can then represent a marker for that particular condition. With a little more progress, one can catalogue all possible deviations for various stages of a given state and develop an algorithm that can provide a support vector machine platform and predict the state of new data sets generated from blind samples (Robertson 2005; Clayton et al. 2006; Kell and Goodacre 2014; Mias and Snyder 2013).

An example of a personalized medicine clinical application of metabonomics is that of lipidomics. The lipidome is the complete lipid profile of a biological system, and is an example of a specialized subset of the metabonome. Lipidomics is a systematic approach to characterizing and quantifying lipids in biological samples using analytical methods based on NMR and MS. It has been recognized that alterations of lipid homeostasis contribute to several pathophysiological conditions (atherosclerosis, hepatic steatosis, obesity, etc.) (Sturla et al. 2014; Robertson 2005; Clayton et al. 2006). Lipids are not only an important energy store but are also essential constituents of all cellular membranes and exert a number of signalling functions. Indeed, profound changes in the cellular and tissue lipid composition occur in response to exposure to active substances and environmental factors. Lipidomics is thus a powerful method for monitoring the overall lipid composition of biological matrixes, and it has been shown to have good potential to identify and detect candidate biomarker signatures indicative of toxicity (Sturla et al. 2014; Wenk 2010). The lipid extracted from various biological matrices can then be analyzed by multiple MS platforms, either by detecting the lipids by shotgun lipidomics, or after separation by liquid chromatography, to detect and quantify lipids of lower abundance. Lipidomic profiling enables the analysis of data sets to derive mechanistic information, and it is one of the validated metabolomic profiling methods used to investigate a patient's lipid alterations for pharmacological and toxicological purposes (Sturla et al. 2014). The lipidomic profile of erythrocyte membranes has been introduced into the diagnostic clinical routine where, for example, it can be used to monitor the pharmaco-toxicological effects of herbal medicines like *Borago officinalis* and other extracts from microalgae, sesame, linseed, etc., on the fatty acids pathway (omega-6 and omega-3, in particular), by monitoring the fatty acid concentrations and the relative network of enzymatic activities involved in the pathway. Looking at the lipidomic asset of the erythrocyte membrane, or of serum, allows the detection of pattern alterations that are correlated with pathophysiological events, and this information can be used not only for monitoring the effectiveness, but also for the early identification of toxic effects on the fatty acids pathway (Wenk 2010; Mackay et al. 2015).

'Omics profiling can also include mapping of the personal microbiome, the analysis of the complete set of microbes in a given area of the individual (found mainly on the skin or in the gut, stool, conjunctiva, saliva, and mucosa), possibly using a combined 'omics approach to look at genetic makeup and metabolic components. With the emergence of next-generation sequencing technology, the role of the gut microbiome in human health and disease is becoming clearer, and the dysbiosis of gut microbiota have been associated with numerous diseases (obesity, type 2 diabetes, inflammatory bowel disease, irritable bowel syndrome, cardiovascular diseases, non-alcoholic fatty liver diseases, atopic dermatitis, psychiatric disorders, etc.) (Integrative HMP Research Network Consortium 2014). It has also been suggested that the human gut microbiota (http://www.human-microbiome.org) plays an active role in immunity and in maintaining an important cross-communication with the central nervous system (Min and Rhee 2015; Burokas et al. 2015). The dynamic monitoring of microbiome-related changes can help identify the specific microbiota involved in disease responses and help elucidate microbiome-host interactions.

Notably, a variety of drugs have the potential to modulate the microbiome, and the microbiome has been shown to be capable of modulating drug degradation and adsorption. The establishment of characteristic and validated signatures of the intestinal microbiome will allow the development of new prophylactic, therapeutic, and preventive strategies for wellness and for targeted modifications of the patient's intestinal microbiome (Qi and Kelley 2014). Most metagenomic tools required to address these important questions are already available, standard operating procedures are under development, and insights into the human microbiome are rapidly evolving (Turnbaugh et al. 2007; Ji and Nielsen 2015; D'Argenio and Salvatore 2015).

Looking towards the full application of personalized, precise, and preventive medicine, data integration from all the 'omics finds the right framework in systems biology. Various diagnostic-omics profiling performed longitudinally in the same subject can trace the temporal molecular patterns associated with healthy and diseased states. Among the analytical integrated models proposed, Mias and Snyder have suggested that all this information be used to define an integrative Personal 'Omics Profiling (iPOP), integrating the diagnostic-omic components in a longitudinal approach with three essential steps, as illustrated in Fig. 16.1:

1. Risk estimation: The personal and common genomic variants determined in an individual genome can be linked to disease, with pharmacogenomic evaluation to determine possible drug response. This may be done in combination with a complete medical and family history and whole-genome sequencing, in conjunction with classical clinical risk factor profiling.
2. Dynamic profiling of multiple 'omics: Starting with a "healthy" baseline, by monitoring changes in the molecular components over multiple points in time, changes in pathophysiological states might be assessed, and the dynamic early onset of disease or toxicity profiled, and possibly even prevented.
3. Data integration and biological impact assessment: The multiple 'omics data can first be analyzed individually to characterize their temporal response profile using standard statistical time-series analysis. The different temporal responses can then be analyzed in their biological pathways by associating the various components in networks and making the corresponding disease associations. The elaborations of this amount of data take place through the wide range of open-source software solutions that are continuously being developed and made available to store and manage data, along with the information to describe accurately the associated clinical conditions.

The analysis of complex data sets is greatly facilitated by integrated workflows that automate a sequence of computational tasks. iPOP thus heavily relies on computational approaches for managing, analyzing, and interpreting the many data generated by the large-scale diagnostic-omics. This evaluation is done using a network analysis that permits the representation of the relevant mechanisms leading to adverse outcomes upon exposure of substances and agents. The quantification of the perturbations of these biological networks upon exposure, and the assessment of their overall biological impact, are finally obtained (Mias and Snyder 2013). The

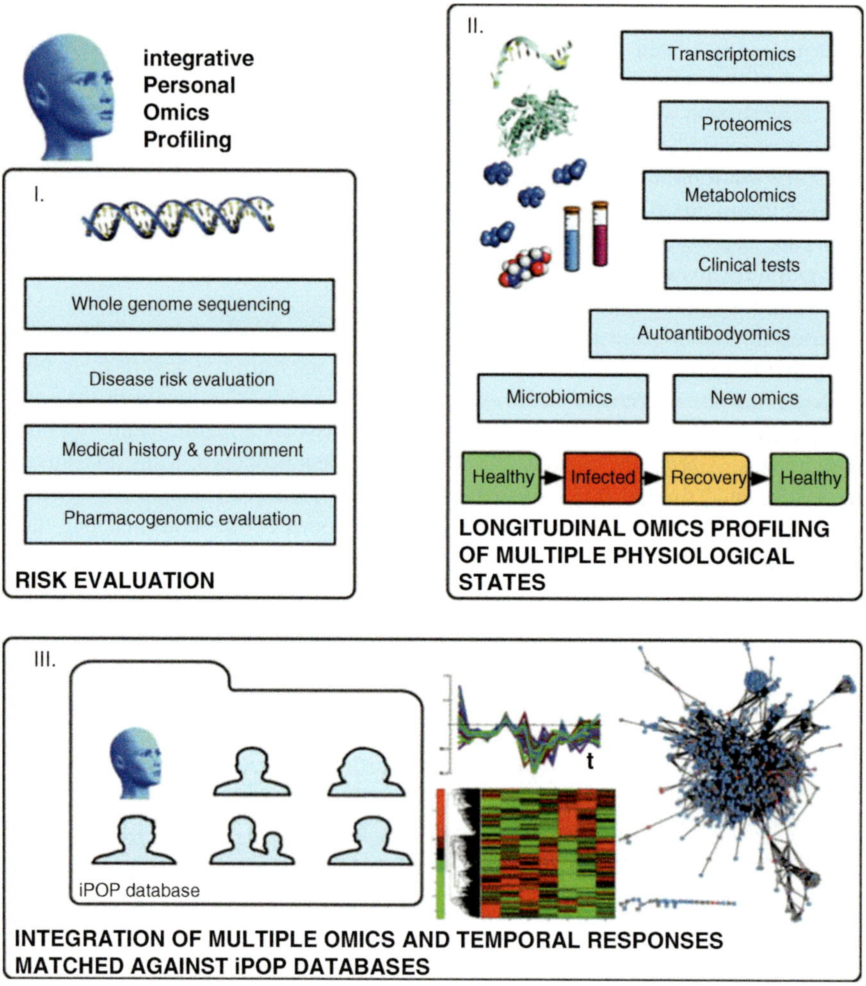

Fig. 16.1 iPOP for personalized medicine. The framework mentioned in the text employs multi-omics analyses that may be implemented for individuals. In step (1), risk estimation for disease is carried out using a whole genome sequencing to perform variant analysis coupled with medical history, environmental considerations and pharmacogenomics evaluations. In step (2), dynamic profiling of multiple omics using an array of technologies follows multiple omics longitudinally in a subject as they progress through their different physiological states, including healthy, diseased, and recovery states. Thus, thousands of molecular components are collected over time for (3) data integration and biological impact assessment, using temporal patterns to obtain matched 'omics information, correlate, and classify responses, compare with pathway databases, and visualize components. The future iPOP implementations may be gathered into a curated database of iPOP-disease associations that may help in categorizing an 'omics dynamic response to a catalogued physiological state and disease onset, with potential diagnostic capabilities (figure is reproduced with the kind permission of Mias and Snyder (2013). 1:33, where a more detailed explanation can be found)

iPOP analysis allows the development of an individual predictive *in silico* model for each patient that can then be used for risk assessment and to monitor effectiveness and toxicity of drugs and phytocomplexes (Fig. 16.1).

Today there is an increasing number of Web-based systems biology platforms that physicians can use to evaluate 'omics data for the prediction of effectiveness, or for monitoring the toxicity of phytocomplexes. They provide a view of all of the biological systems modulated by phytocomplexes, while the pathway profiling tools allow the prediction of optimal treatments for patients. The use of clinical databases will in the future even allow the identification of potential links between phytocomplexes and human disease or toxicological effects, once the databases are fed with significant numbers of clinical association data from EHR networks systems (Sturla et al. 2014; Zhou et al. 2015; Castaneda et al. 2015; Ritchie et al. 2015). Network analysis has been used for the development of the "pathways of toxicity" (PoT), defined as "a molecular definition of cellular processes shown to mediate adverse outcomes of toxicants," aiming at the completion of the "human toxome" mentioned in the preceding paragraphs (Bouhifd et al. 2014). After collecting a critical amount of quali-quantitative diagnostic-omics data on molecular responses and clinical conditions, a wealth of interrelated information will be ready to be used to develop predictive *in silico* models for at-risk assessment and toxicity prediction.

Conclusion

The future of toxicological applications in the field of herbal drugs will be characterized by the fast growing use of diagnostic-omics in the clinic. If genomics can help the identification of risk factors for a particular individual, proteomics and metabonomics will assist the physician to monitor risk factors, improve the characterization of a given pathological condition and to closely monitor therapeutic interventions. "'Omics" techniques are particularly appropriate for analyzing the biological effects, both pharmacological and toxicological, of herbal drugs, which exert multiple concomitant actions on several molecular targets. 'Omics techniques can assess simultaneous molecular effects, building up a large number of biological data that can be managed with systems biology driven bio-informatics, allowing a global view of the biological network under examination, and ultimately potentiating the actionability of diagnostic data. All this in a context of personalized medicine, characterized by a more precise use of therapeutic interventions, and the possibility to focus closely on the molecular prediction of toxicological risks. Thus the advances of systems medicine and network pharmacotoxicology is providing new strategies and tools for a guided and assisted use of phytocomplexes. Despite the increasingly information available from systems medicine, research is not yet fully exploited in the clinic, although the introduction of diagnosticomics in routine medical care is becoming more relevant and meaningful every day, and there are several meaningful examples and proposals for a global effective use of this approach, like in the case of the integrative Personal Omics Profiling (iPOP) clinical model.

Acknowledgements We would like to thank Roberta Sato at the Library of the Department of Pharmaceutical and Pharmacological Sciences of the University of Padova for her technical assistance with database search and bibliography, and Mariagnese Barbera for text revision.

Conflict of Interest Alessandro Buriani and Stefano Fortinguerra are in charge of the Personalized Medicine Service of the Gruppo Data Medica Padova.

References

Abdel-Rahman A, Anyangwe N, Carlacci L et al (2011) The safety and regulation of natural products used as foods and food ingredients. Toxicol Sci 123:333–348. doi:10.1093/toxsci/kfr198

Agarwal A, Ressler D, Snyder G (2015) The current and future state of companion diagnostics. Pharmgenomics Pers Med 8:99–110. doi:10.2147/PGPM.S49493

Ahn AC, Tewari M, Poon CS et al (2006) The clinical applications of a systems approach. PLoS Med 3:956–960. doi:10.1039/C0AN00746C

Ainslie GR, Wolf KK, Li Y et al (2014) Assessment of a candidate marker constituent predictive of a dietary substance-drug interaction: case study with grapefruit juice and CYP3A4 drug substrates. J Pharmacol Exp Ther 51:576–584. doi:10.1124/jpet.114.216838

Alinejad-Mofrad S, Foadoddini M, Saadatjoo SA et al (2015) Improvement of glucose and lipid profile status with *Aloe vera* in pre-diabetic subjects: a randomized controlled-trial. J Diabetes Metab Disord 14:22. doi:10.1186/s40200-015-0137-2

Allyse M, Minear MA, Berson E et al (2015) Non-invasive prenatal testing: a review of international implementation and challenges. Int J Womens Health 7:113–126. doi:10.2147/IJWH.S67124

Alyass A, Turcotte M, Meyre D (2015) From big data analysis to personalized medicine for all: challenges and opportunities. BMC Med Genomics 8:33. doi:10.1186/s12920-015-0108y

Ashley EA (2015) The precision medicine initiative: a new national effort. JAMA 313:2119–2120. doi:10.1001/jama.2015.3595

Ashley EA, Butte AJ, Wheeler MT et al (2010) Clinical evaluation incorporating a personal genome. Lancet 375:1525–1535. doi:10.1016/S0140-6736(10)60452-7

Auffray C, Charron D, Hood L (2010) Predictive, preventive, personalized and participatory medicine: back to the future. Genome Med 2:57. doi:10.1186/gm178

Bahmani M, Shirzad H, Mirhosseini M et al (2015) A review on ethnobotanical and therapeutic uses of Fenugreek (*Trigonella foenum-graceum* L). J Evid Based Complement Altern Med. doi:10.1177/2156587215583405

Ball MP, Thakuria JV, Zaranek AW et al (2012) A public resource facilitating clinical use of genomes. Proc Natl Acad Sci USA 109:11920–11927. doi:10.1073/pnas.1201904109

Ball MP, Bobe JR, Chou MF et al (2014) Harvard Personal Genome Project: lessons from participatory public research. Genome Med 6:10. doi:10.1186/gm527

Banerjeea S, Debnathb P, Kumar Debnathc P (2015) Ayurnutrigenomics: Ayurveda-inspired personalized nutrition from inception to evidence. J Tradit Complement Med 5:228–233. doi:10.1016/j.jtcme.2014.12.009

Bauer C, Stec K, Glintschert A et al (2015) BioMiner: paving the way for personalized medicine. Cancer Informat 14:55–63. doi:10.4137/CIN.S20910

Best MG, Sol N, Kooi I et al (2015) RNA-Seq of tumor-educated platelets enables blood-based pan-cancer, multiclass, and molecular pathway cancer diagnostics. Cancer Cell 28:1–11. doi:10.1016/j.ccell.2015.09.018

Boissel JP, Auffray C, Noble D et al (2015) Bridging systems medicine and patient needs. CPT Pharmacometrics Syst Pharmacol 4:135–145. doi:10.1002/psp4.26

Bouhifd M, Hogberg HT, Kleensang A et al (2014) Mapping the human toxome by systems toxicology. Basic Clin Pharmacol Toxicol 115:24–31. doi:10.1111/bcpt.12198

Bragazzi NL (2013) From P0 to P6 medicine, a model of highly participatory, narrative, interactive, and "augmented" medicine: some considerations on Salvatore Iaconesi's clinical story. Patient Prefer Adherence 7:353–359. doi:10.2147/PPA.S38578

Buriani A, Fortinguerra S (2015) The systems biology oriented, holistic vision of personalized medicine and the emerging concept of proactive herbal medicine. J Int Med 1(2):007–009

Buriani A, Garcia-Bermejo ML, Bosisio E et al (2012) Omic techniques in systems biology approaches to traditional Chinese medicine research: present and future. J Ethnopharmacol 140:535–544. doi:10.1016/j.jep.2012.01.055

Burokas A, Moloney RD, Dinan TG et al (2015) Microbiota regulation of the Mammalian gut-brain axis. Adv Appl Microbiol 91:1–62. doi:10.1016/bs.aambs.2015.02.001

Calabrese B, Cannataro M (2015) Bioinformatics and microarray data analysis on the cloud. In: Methods Mol Biol. Springer Ed. Walker JM. pp 1–15. doi:10.1007/7651_2015_236

Castaneda C, Nalley K, Mannion C et al (2015) Clinical decision support systems for improving diagnostic accuracy and achieving precision medicine. J Clin Bioinforma 26:5–14. doi:10.1186/s13336-015-0019-3

Chadwick LH (2012) The NIH roadmap epigenomics program data resource. Epigenomics 4:317–324. doi:10.2217/epi.12.18

Chatterjee R, Gupta B, Bose S (2015) Oral screening for pre-cancerous lesions among areca-nut chewing population from rural India. Oral Health Prev Dent. doi:10.3290/j.ohpd.a34052

Chen R, Mias G, Li-Pook-Than J et al (2012) Personal omics profiling reveals dynamic molecular and medical phenotypes. Cell 148:1293–1307. doi:10.1016/j.cell.2012.02.009

Chen YC, Douville C, Wang C et al (2014) A probabilistic model to predict clinical phenotypic traits from genome sequencing. PLoS Comput Biol 10:e1003825. doi:10.1371/journal.pcbi.1003825

Cheng R, Leung RK, Chen Y et al (2015) Virtual pharmacist: a platform for pharmacogenomics. PLoS One 10:e0141105. doi:10.1371/journal.pone.0141105

Chiappelli F, Prolo P, Cajulis OS (2005) Evidence-based research in complementary and alternative medicine I: history. Evid Based Complement Altern Med 2:453–458. doi:10.1093/ecam/neh106

Clayton TA, Lindon JC, Cloarec O et al (2006) Pharmaco-metabonomic phenotyping and personalized drug treatment. Nature 440:1073–1077

Cock IE (2015) The safe usage of herbal medicines: counter-indications, cross-reactivity and toxicity. Pharmacogn Commun 5:2–38. doi:10.5530/pc.2015.1.2

Collins FS, Varmus H (2015) A new initiative on precision medicine. N Engl J Med 372(9):793–795. doi:10.1056/NEJMp1500523

Cutica I, Vie GM, Pravettoni G (2014) Personalised medicine: the cognitive side of patients. Eur J Intern Med 25:685–688. doi:10.1016/j.ejim.2014.07.002

D'Argenio V, Salvatore F (2015) The role of the gut microbiome in the healthy adult status. Clin Chim Acta 451(Pt A):97–102. doi:10.1016/j.cca.2015.01.003

Dopazo J (2014) Genomics and transcriptomics in drug discovery. Drug Discov Today 19:126–132. doi:10.1016/j.drudis.2013.06.003

Dunn WB, Lin W, Broadhurst D et al (2015) Molecular phenotyping of a UK population: defining the human serum metabolome. Metabolomics 11:9–26. doi:10.1007/s11306-014-0707-1

Efferth T, Koch E (2011) Complex interactions between phytochemicals. The multi-target therapeutic concept of phytotherapy. Curr Drug Targets 12:122–123

ESF Forward Look (2012) Personalised medicine for the European citizen. Towards more precise medicine for the diagnosis, treatment and prevention of disease (iPM). European Science Foundation, Strasbourg, http://www.esf.org/fileadmin/Public_documents/Publications/Personalised_Medicine.pdf

Evans WE, McLeod HL (2003) Pharmacogenomics – drug disposition, drug targets, and side effects. N Engl J Med 348:538–549. doi:10.1056/NEJMra020526

Fasinu PS, Bouic PJ, Rosenkranz B (2012) An overview of the evidence and mechanisms of herb-drug interactions. Front Pharmacol 3:69. doi:10.3389/fphar.2012.00069

Flores M, Glusman G, Brogaard K et al (2013) P4 medicine: how systems medicine will transform the healthcare sector and society. Pers Med 10:565–576

Fortinguerra S, Varano G, Buriani A (2015) Antiplatelet pharmacogenomic assessment in a case of coronary artery bypass. J Int Med 1(2):009–012

Fox JL (2015) Obama catapults patient-empowered precision medicine. Nat Biotechnol 33:325. doi:10.1038/nbt0415-325

Fox CS, Hall JL, Arnett DK et al (2015) Future translational applications from the contemporary genomics era: a scientific statement from the American Heart Association. Circulation 131:1715–1736. doi:10.1161/CIR.0000000000000211

Fryburg DA, Song DH, Laifenfeld D et al (2014) Systems diagnostics: anticipating the next generation of diagnostic tests based on mechanistic insight into disease. Drug Discov Today 19:108–112. doi:10.1016/j.drudis.2013.07.006

Gardeux V, Achour I, Li J et al (2014) 'N-of-1-pathways' unveils personal deregulated mechanisms from a single pair of RNA-Seq samples: towards precision medicine. J Am Med Inform Assoc 21:1015–1025. doi:10.1136/amiajnl-2013-002519

Ghosh S, Matsuoka Y, Asai Y et al (2011) Software for systems biology: from tools to integrated platforms. Nat Rev Genet 12:821–832. doi:10.1038/nrg3096

Gibbs DL, Gralinski L, Baric RS et al (2014) Multi-omic network signatures of disease. Front Genet 4:309. doi:10.3389/fgene.2013.00309

Gibson JB, Jiang R, Yu F (2013) The 1000 Genomes Project: paving the way for personalized genomic medicine. Pers Med 10:321–324

Gierman HJ, Fortney K, Roach JC et al (2014) Whole-genome sequencing of the world's oldest people. PLoS One 9:e112430. doi:10.1371/journal.pone.0112430

Gonzalez-Garay ML (2014) The road from next-generation sequencing to personalized medicine. Pers Med 11:523–544

Gorini A, Pravettoni G (2011) P5 medicine: a plus for a personalized approach to oncology. Nat Rev Clin Oncol 8:444. doi:10.1038/nrclinonc.2010.227-c1

Greshake B, Bayer PE, Rausch H et al (2014) openSNP – a crowdsourced web resource for personal genomics. PLoS One 9:e89204. doi:10.1371/journal.pone.0089204

Gudbjartsson DF, Helgason H, Gudjonsson SA et al (2015) Large-scale whole-genome sequencing of the Icelandic population. Nat Genet 47:435–444. doi:10.1038/ng.3247

Guo L, Du Y, Chang S et al (2014) rSNPBase: a database for curated regulatory SNPs. Nucleic Acids Res 42(Database issue):D1033–D1039. doi:10.1093/nar/gkt1167

Gupta PD (2015) Pharmacogenetics, pharmacogenomics and ayurgenomics for personalized medicine: a paradigm shift. Indian J Pharm Sci 77:135–141

Harvey AL, Edrada-Ebel R, Quinn RJ (2015) The re-emergence of natural products for drug discovery in the genomics era. Nat Rev Drug Discov 14:111–129. doi:10.1038/nrd4510

Henry VJ, Bandrowski AE, Pepin AS et al. (2014) OMICtools: an informative directory for multiomic data analysis. Database 2014. pii: bau069. doi:10.1093/database/bau069

Hernansaiz-Ballesteros RD, Salavert F, Sebastián-León P et al (2015) Assessing the impact of mutations found in next generation sequencing data over human signaling pathways. Nucleic Acids Res 43(W1):W270–W275. doi:10.1093/nar/gkv349

Hood L (2008) A personal journey of discovery: developing technology and changing biology. Annu Rev Anal Chem (Palo Alto Calif) 1:1–43. doi:10.1146/annurev.anchem.1.031207.113113

Hood L, Flores M (2012) A personal view on systems medicine and the emergence of proactive P4 medicine: predictive, preventive, personalized and participatory. N Biotechnol 29:613–624. doi:10.1016/j.nbt.2012.03.004

Hood L, Lovejoy JC, Price ND (2015) Integrating big data and actionable health coaching to optimize wellness. BMC Med 13:4. doi:10.1186/s12916-014-0238-7

Hu M, Fan L, Zhou HH, Tomlinson B (2012) Theranostics meets traditional Chinese medicine: rational prediction of drug-herb interactions. Expert Rev Mol Diagn 12:815–830. doi:10.1586/erm.12.126

Huang H, Yang P, Xue J et al (2014) Evaluating the individualized treatment of traditional Chinese medicine: a pilot study of N-of-1 trials. Evid Based Complement Altern Med 2014, 148730. doi:10.1155/2014/148730

Hun Lee J, Shu L, Fuentes F et al (2013) Cancer chemoprevention by traditional Chinese herbal medicine and dietary phytochemicals: targeting nrf2-mediated oxidative stress/anti-inflammatory responses, epigenetics, and cancer stem cells. J Tradit Complement Med 3:69–79. doi:10.4103/2225-4110.107700

Integrative HMP (iHMP) Research Network Consortium (2014) The Integrative Human Microbiome Project: dynamic analysis of microbiome-host omics profiles during periods of human health and disease. Cell Host Microbe 16:276–289. doi:10.1016/j.chom.2014.08.014

Jafari S, Abdollahi M, Saeidnia S (2014) Personalized medicine: a confluence of traditional and contemporary medicine. Altern Ther Health Med 20:31–40

Jameson JL, Longo DL (2015) Precision medicine – personalized, problematic, and promising. NEJM 372:2229–2234. doi:10.1056/NEJMsb1503104

Ji B, Nielsen J (2015) New insight into the gut microbiome through metagenomics. Adv Genom Genet 5:77–91. doi:10.2147/AGG.S57215

Jiang M, Lu C, Zhang C et al (2012) Syndrome differentiation in modern research of traditional Chinese medicine. J Ethnopharmacol 140:634–642. doi:10.1016/j.jep.2012.01.033

Kell DB, Goodacre R (2014) Metabolomics and systems pharmacology: why and how to model the human metabolic network for drug discovery. Drug Discov Today 19:171–182. doi:10.1016/j.drudis.2013.07.014

Khoury MJ, Gwinn ML, Glasgow RE et al (2012) A population approach to precision medicine. Am J Prev Med 42:639–645. doi:10.1016/j.amepre.2012.02.012

Landis-Piwowar KR, Iyer NR (2014) Cancer chemoprevention: current state of the art. Cancer Growth Metastasis 7:19–25. doi:10.4137/CGM.S11288

Li A, Meyre D (2014) Jumping on the train of personalized medicine: a primer for non-geneticist clinicians: part 3. Clinical applications in the personalized medicine area. Curr Psychiatry Rev 10:101–117

Li Y, Wu YL (2003) An over four millennium story behind qinghaosu (artemisinin) – a fantastic antimalarial drug from a traditional Chinese herb. Curr Med Chem 10:2197–2230. doi:10.2174/0929867033456710

Li X, Lian FM, Guo D et al (2013) The rs1142345 in TPMT affects the therapeutic effect of traditional hypoglycemic herbs in prediabetes. Evid Based Complement Altern Med 2013, 327629. doi:10.1155/2013/327629

Lin LL, Wang YH, Lai CY et al (2012) Systems biology of meridians, acupoints, and Chinese herbs in disease. Evid Based Complement Altern Med 2012, 372670. doi:10.1155/2012/372670

Liu J, Sun K, Zheng C et al (2015) Pathway as a pharmacological target for herbal medicines: an investigation from reduning injection. PLoS One 10:e0123109. doi:10.1371/journal.pone.0123109

Lowry CM, Kostka-Rokosz MD, McCloskey WW (2003) Evaluation of personal digital assistant drug information databases for the managed care pharmacist. J Manag Care Pharm 9:441–448

Lv W, Piao JH, Jiang JG (2012) Typical toxic components in traditional Chinese medicine. Expert Opin Drug Saf 11:985–1002. doi:10.1517/14740338.2012.726610

Mackay DS, Gebauer SK, Eck PK et al (2015) Lathosterol-to-cholesterol ratio in serum predicts cholesterol-lowering response to plant sterol consumption in a dual-center, randomized, single-blind placebo-controlled trial. Am J Clin Nutr 101:432–439. doi:10.3945/ajcn.114.095356

Mahapatra S, Kamath R, Shetty BK et al (2015) Risk of oral cancer associated with gutka and other tobacco products: a hospital-based case-control study. J Cancer Res Ther 11:199–203. doi:10.4103/09731482.143332

Margolis R, Derr L, Dunn M et al (2014) The National Institutes of Health's Big Data to Knowledge (BD2K) initiative: capitalizing on biomedical big data. J Am Med Inform Assoc 21:957–958. doi:10.1136/amiajnl-2014-002974

Markert C, Ngui P, Hellwig R et al (2014) Influence of St. John's wort on the steady-state pharmacokinetics and metabolism of bosentan. Int J Clin Pharmacol Ther 52:328–336. doi:10.5414/CP202048

Mason CE, Porter SG, Smith TM (2014) Characterizing multi-omic data in systems biology. Adv Exp Med Biol 799:15–38. doi:10.1007/978-1-4614-8778-4_2
Meisel SF, Carere DA, Wardle J et al (2015) Explaining, not just predicting, drives interest in personal genomics. Genome Med 7:74. doi:10.1186/s13073-015-0188-5
Mias GI, Snyder M (2013) Personal genomes, quantitative dynamic omics and personalized medicine. Quant Biol 1:71–90. doi:10.1007/s40484-013-0005-3
Min YW, Rhee PL (2015) The role of microbiota on the gut immunology. Clin Ther 37:968–975. doi:10.1016/j.clinthera.2015.03.009
Montani F, Marzi MJ, Dezi F et al (2015) miR-test: a blood test for lung cancer early detection. J Natl Cancer Inst 107:djv063. doi:10.1093/jnci/djv063
Mucalo I, Jovanovski E, Rahelić D et al (2013) Effect of American ginseng (Panax quinquefolius L.) on arterial stiffness in subjects with type-2 diabetes and concomitant hypertension. J Ethnopharmacol 150:148–153. doi:10.1016/j.jep.2013.08.015
Nicholson JK (2006) Global systems biology, personalized medicine and molecular epidemiology. Mol Syst Biol 2:52. doi:10.1038/msb4100095
Normile D (2015) DRUG DISCOVERY. Nobel for antimalarial drug highlights East-West divide. Science 350:265. doi:10.1126/science.350.6258.265
Panahiazar M, Taslimitehrani V, Jadhav A (2014) Empowering personalized medicine with big data and semantic web technology: promises, challenges, and use cases. Proc IEEE Int Conf Big Data 2014:790–795. doi:10.1109/BigData.2014.7004307
Personalized Medicine Coalition (2014) The case for personalized medicine, 4th edn. Personalized Medicine Coalition, Washington, DC, http://www.personalizedmedicinecoalition.org/Userfiles/PMC-Corporate/file/pmc_case_for_personalized_medicine.pdf
Qi Z, Kelley E (2014) The WHO traditional medicine strategy 2014-2023: a perspective. Science 346:S5–S6
Ramjaun A, Sudarshan M, Patakfalvi L et al (2015) Educating medical trainees on medication reconciliation: a systematic review. BMC Med Educ 15:33. doi:10.1186/s12909-015-0306-5
Rasool M, Malik A, Naseer MI et al (2015) The role of epigenetics in personalized medicine: challenges and opportunities. BMC Med Genomics 8:S5. doi:10.1186/1755-8794-8-S1-S5
Ritchie MD, de Andrade M, Kuivaniemi H (2015) The foundation of precision medicine: integration of electronic health records with genomics through basic, clinical, and translational research. Front Genet 6:104. doi:10.3389/fgene.2015.00104
Roberts JS, Ostergren J (2013) Direct-to-consumer genetic testing and personal genomics services: a review of recent empirical studies. Curr Genet Med Rep 1:182–200
Robertson DG (2005) Metabonomics in toxicology: a review. Toxicol Sci 85:809–822. doi:10.1093/toxsci/kfi102
Schröder T, Hubold C, Muck P et al (2015) A hypertensive emergency with acute visual impairment due to excessive liquorice consumption. Neth J Med 73:82–85
Schroën Y, van Wietmarschen HA, Wang M et al (2015) East is East and West is West, and never the twain shall meet? Science 346(6216 Suppl):S10–S12
Scott SA (2011) Personalizing medicine with clinical pharmacogenetics. Genet Med 13:987–995. doi:10.1097/GIM.0b013e318238b38c
Shankar S, Kumar D, Srivastava RK (2013) Epigenetic modifications by dietary phytochemicals: implications for personalized nutrition. Pharmacol Ther 138:1–17. doi:10.1016/j.pharmthera.2012.11.002
Shukla SK, Murali NS, Brilliant MH (2015) Personalized medicine going precise: from genomics to microbiomics. Trends Mol Med 21:461–462. doi:10.1016/j.molmed.2015.06.002
Siva N (2015) UK gears up to decode 100,000 genomes from NHS patients. Lancet 385(9963):103–104. doi:10.1016/S0140-6736(14)62453-3
Skipper M, Eccleston A, Gray N et al (2015) Presenting the epigenome roadmap. Nature 518:313. doi:10.1038/518313a
Smith TD, Vihinen M on Behalf of the Human Variome Project (2015) Standard development at the human variome project. Database 2015:1–5. doi:10.1093/database/bav024

Solomon BD, Nguyen AD, Bear KA et al (2013) Clinical genomic database. Proc Natl Acad Sci USA 110:9851–9855. doi:10.1073/pnas.1302575110
Sookoian S, Pirola CJ (2015) Liver enzymes, metabolomics and genome-wide association studies: from systems biology to the personalized medicine. World J Gastroenterol 21:711–725. doi:10.3748/wjg.v21.i3.711
Sturla SJ, Boobis AR, FitzGerald RE et al (2014) Systems toxicology: from basic research to risk assessment. Chem Res Toxicol 27:314–329. doi:10.1021/tx400410s
Su XZ, Miller LH (2015) The discovery of artemisinin and the Nobel Prize in physiology or medicine. Sci China Life Sci. doi:10.1007/s11427-015-4948-7
Sudlow C, Gallacher J, Allen N et al (2015) UK biobank: an open access resource for identifying the causes of a wide range of complex diseases of middle and old age. PLoS Med 12:e1001779. doi:10.1371/journal.pmed.1001779
Swen JJ, Nijenhuis M, de Boer A et al (2011) Pharmacogenetics: from bench to byte – an update of guidelines. Clin Pharmacol Ther 89:662–673. doi:10.1038/clpt.2011.34
Tada T (2004) Toward the philosophy of CAM: super-system and epimedical sciences. Evid Based Complement Altern Med 1:5–8. doi:10.1093/ecam/neh012
Tammen SA, Friso S, Choi SW (2013) Epigenetics: the link between nature and nurture. Mol Aspects Med 34:753–764. doi:10.1016/j.mam.2012.07.018
Tiwari P, Mishra BN, Sangwan NS (2014) Phytochemical and pharmacological properties of Gymnema sylvestre: an important medicinal plant. Biomed Res Int 2014, 830285. doi:10.1155/2014/830285
Tu Y (2011) The discovery of artemisinin (qinghaosu) and gifts from Chinese medicine. Nat Med 17:1217–1220. doi:10.1038/nm.2471
Turnbaugh PJ, Ley RE, Hamady M et al (2007) The human microbiome project. Nature 449:804–810
Uhlén M, Fagerberg L, Hallström BM et al (2015) Tissue-based map of the human proteome. Science 347:1260419. doi:10.1126/science.1260419Proteomics
Van Galen E (2014) Traditional herbal medicines worldwide, from reappraisal to assessment in Europe. J Ethnopharmacol 158(Pt B):498–502. doi:10.1016/j.jep.2014.07.013
Wang J, Xiong X (2012) Current situation and perspectives of clinical study in integrative medicine in China. Evid Based Complement Altern Med 2012, 268542. doi:10.1155/2012/268542
Wang Y, Xu A (2015) Zheng: a systems biology approach to diagnosis and treatments. Science 346:S13–S15
Wang S, Penchala S, Prabhu S et al (2010) Molecular basis of traditional Chinese medicine in cancer chemoprevention. Curr Drug Discov Technol 7:67–75
Wang L, McLeod HL, Weinshilboum RM (2011) Genomics and drug response. N Engl J Med 364:1144–1153. doi:10.1056/NEJMra1010600
Wang X, Zhang A, Sun H (2012) Future perspectives of Chinese medical formulae: chinmedomics as an effector. OMICS 16:414–421. doi:10.1089/omi.2011.0138
Wang C, Sen A, Plegue M et al (2015a) Impact of family history assessment on communication with family members and health care providers: a report from the Family Healthware™ Impact Trial (FHITr). Prev Med. doi:10.1016/j.ypmed.2015.04.007, pii:S0091-7435(15)00113-9
Wang RS, Maron BA, Loscalzo J (2015b) Systems medicine: evolution of systems biology from bench to bedside. Wiley Interdiscip Rev Syst Biol Med. doi:10.1002/wsbm.1297
Waters MD, Fostel JM (2004) Toxicogenomics and systems toxicology: aims and prospects. Nat Rev Genet 5:936–948. doi:10.1038/nrg1493
Wei WQ, Denny JC (2015) Extracting research-quality phenotypes from electronic health records to support precision medicine. Genome Med 7:41. doi:10.1186/s13073-015-0166-y
Wenk MR (2010) Lipidomics: new tools and applications. Cell 143:888–895. doi:10.1016/j.cell.2010.11.033
Weston AD, Hood LJ (2004) Systems biology, proteomics, and the future of health care: toward predictive, preventative, and personalized medicine. Proteome Res 3:179–196

Whirl-Carrillo M, McDonagh EM, Hebert JM et al (2012) Pharmacogenomics knowledge for personalized medicine. Clin Pharmacol Ther 92:414–417. doi:10.1038/clpt.2012.96

Williamson EM, Chan K, Xu Q (2015) Evaluating the safety of herbal medicines: integrated toxicological approaches. Science 347:S47–S49

Witt CM, Liu J, Robinson N (2015) Combining 'omics and comparative effectiveness research: evidence-based clinical research decision-making for Chinese medicine. Science 346(6216 Suppl):S10–S12

Wu X, Patterson S, Hawk E (2011) Chemoprevention – history and general principles. Best Pract Res Clin Gastroenterol 25:445–459. doi:10.1016/j.bpg.2011.10.012

Wu IC, Lin CC, Hsiung CA (2015) Emerging roles of frailty and inflammaging in risk assessment of age-related chronic diseases in older adults: the intersection between aging biology and personalized medicine. Biomedicine (Taipei) 5:1. doi:10.7603/s40681-015-0001-1

Xu H, Chen KJ (2010) Making evidence-based decisions in the clinical practice of integrative medicine. Chin J Integr Med 16:483–485. doi:10.1007/s11655-010-0560-7

Yang CC (2014) Patient centered healthcare informatics. IEEE Intell Informa Bull 15:1–5

Yang B, Li X, Zhang C et al (2015) Design, synthesis and biological evaluation of novel peptide MC2 analogues from Momordica charantia as potential anti-diabetic agents. Org Biomol Chem 13:4551–4561. doi:10.1039/c5ob00333d

Yip V, Hawcutt DB, Pirmohamed M (2015) Pharmacogenetic markers of drug efficacy and toxicity. Clin Pharmacol Ther 98:61–70. doi:10.1002/cpt.135

Zhou J, Xu G, Yan J et al (2015) Rehmannia glutinosa (Gaertn.) DC. polysaccharide ameliorates hyperglycemia, hyperlipemia and vascular inflammation in streptozotocin-induced diabetic mice. J Ethnopharmacol 164:229–238. doi:10.1016/j.jep.2015.02.026

Index

O. Pelkonen et al. (eds.), *Toxicology of Herbal Products*,
DOI 10.1007/978-3-319-43806-1

Printed by Printforce, the Netherlands